Roderick

Asmundson

5/29/03

Handbook of Fermented Functional Foods

FUNCTIONAL FOODS AND NUTRACEUTICALS SERIES

Series Editor
G. Mazza, Ph.D.
Senior Research Scientist and Head
Food Research Program
Pacific Agri-Food Research Centre
Agriculture and Agri-Food Canada
Summerland, British Columbia

Functional Foods: Biochemical and Processing Aspects
Volume 1
Edited by G. Mazza, Ph.D.

Herbs, Botanicals, and Teas
Edited by G. Mazza, Ph.D. and B.D. Oomah, Ph.D.

Functional Foods: Biochemical and Processing Aspects
Volume 2
Edited by John Shi, Ph.D., G. Mazza, Ph.D., and Marc Le Maguer, Ph.D.

Methods of Analysis for Functional Foods and Nutraceuticals
Edited by W. Jeffrey Hurst, Ph.D.

Handbook of Fermented Functional Foods

EDITED BY
Edward R. Farnworth, Ph.D.
Food Research and Development Centre
Agriculture and Agri-Food Canada

CRC PRESS

Boca Raton London New York Washington, D.C.

Library of Congress Cataloging-in-Publication Data

Handbook of fermented functional foods / edited by Edward R. Farnworth.
p. cm.
Includes bibliographical references and index.
ISBN 0-8493-1372-4 (alk. paper)
1. Functional foods—Handbooks, manuals, etc. 2. Fermented foods—Health aspects—Handbooks, manuals, etc. I. Farnworth, Edward R.

QP144.F85 H357 2003
613.2′8—dc21 2002191169

Direct all inquiries to CRC Press LLC, 2000 N.W. Corporate Blvd., Boca Raton, Florida 33431.

Visit the CRC Press Web site at www.crcpress.com

International Standard Book Number 0-8493-1372-4
Library of Congress Card Number 2002191169
Printed in the United States of America 1 2 3 4 5 6 7 8 9 0
Printed on acid-free paper

Series Preface

The Functional Foods and Nutraceuticals Book Series, launched in 1998, was developed to provide a timely and comprehensive treatment of the emerging science and technology of functional foods and nutraceuticals that are shown to play a role in preventing or delaying the onset of diseases, especially chronic diseases. The first four titles in the series: *Functional Foods: Biochemical and Processing Aspects, Volumes 1 and 2*; *Herbs, Botanicals, and Teas*; and *Methods of Analysis for Functional Foods and Nutraceuticals* have received widespread acceptance by food, nutrition and health professionals.

Functional Foods: Biochemical and Processing Aspects, Volume 1, the first in the series, is a bestseller, and is devoted to functional food products from oats, wheat, rice, flaxseed, mustard, fruits, vegetables, fish, and dairy products.

In *Volume 2*, the focus is on the latest developments in the chemistry, biochemistry, pharmacology, epidemiology, and engineering of tocopherols and tocotrienols from oil and cereal grain, isoflavones from soybeans and soy foods, flavonoids from berries and grapes, lycopene from tomatoes, limonene from citrus, phenolic diterpenes from rosemary and sage, organosulfur constitutes from garlic, phytochemicals from Echinacea, pectin from fruit, and omega-3 fatty acids and docosahexaenoic acid from flaxseed and fish products. *Volume 2* also covers solid–liquid extraction technologies for manufacturing nutraceuticals and dietary supplements.

Herbs, Botanicals, and Teas provides the latest scientific and technical information on the chemical, pharmacological, epidemiological, and clinical aspects of garlic, ginseng, Echinacea, ginger, fenugreek, St. John's wort, Ginkgo biloba, kava kava, goldenseal, saw palmetto, valerian, evening primrose, liquorice, bilberries and blueberries, and green and black teas. The book also contains chapters on international regulations and quality assurance and control for the herbal and tea industry.

Methods of Analysis for Functional Foods and Nutraceuticals presents advanced methods of analysis for carotenoids, phytoestrogens, chlorophylls, anthocyanins, amino acids, fatty acids, flavonoids, water soluble vitamins, and carbohydrates.

The current volume, *Handbook of Fermented Functional Foods*, edited by Edward R. Farnworth, provides a comprehensive, state-of-the-art treatment of the scientific and technological information on the production of fermented foods, the microorganisms involved, the changes in composition that occur during fermentation and, most importantly, the effect of these foods and their active ingredients on human health. Many of the foods covered in the book — yoghurt, cheese, pickles, sauerkraut, and fermented meat — are well known to western consumers, and their production methods and health effects are reasonably well documented. Several chapters in the book deal with foods such as gaio, kefir, natto, miso, kimchi, kocho, ogi, and togwa, which are much better known in Eastern Europe, Asia, and Africa. These foods have long traditions and are only now being tested scientifically. The book provides a good history of fermented foods, which shows that these foods have been an important part of the human diet for many centuries. Modern research, especially on the immune system, is revealing how these foods and their active ingredients impact human health. Also discussed is how fermented foods are likely to become an even more important part of our diet both on Earth and as we explore beyond Earth.

Dr. Farnworth has assembled a group of outstanding international contributors in the forefront of fermented food science and technology. It is hoped that their efforts will be valuable to food, nutrition, and health practitioners, and to students and researchers in industry, government, and university laboratories.

G. Mazza, Ph.D., FCIFST
Series Editor

Preface

Fermented foods are consumed in every country throughout the world. In some cases, these foods have played important roles in the diet for centuries. Fermented foods have often been produced as a means of preserving perishable primary products such as milk and meat, when other options such as refrigeration or pasteurization were unavailable or too expensive. In many cases, the science to explain what was happening to the food as it fermented was unknown and most likely not considered important in years gone by. However, the fact that similar foods can be found around the world — with only the name changing — shows that fermentation is a process that has universal appeal and application.

It has only been recently, with advances in food microbiology, that the organisms responsible have been isolated and identified and their effects on foods during fermentation understood. Food manufacturers are now able to produce large quantities of uniform products year round. From a food processing point of view, fermentation is in itself an important technology, as it can not only preserve foods but also changes the characteristics of the food to make it more desirable to the consumer. Food products are now consumed that use in their fabrication, or contain when eaten, bacteria, yeasts, and molds.

The microorganisms that carry out the fermentation process can affect the nutritional quality of food in several ways. During the fermentation process, nutrients that are already constituents of the food may be made more available. In some cases, the microorganisms carrying out the fermentation can synthesize new, nutritionally important constituents. The microorganisms themselves may contain cellular enzymes or other biologically active components that are consumed when the fermented food is eaten and serve a useful purpose to the host. Foods that contain microorganisms and are good for health are termed probiotics; probiotics are a subset of the larger category called functional foods — conventional foods that have physiological benefits and/or reduce the risk of chronic disease beyond basic nutritional functions. The realization that the microorganisms we eat in some foods can impact on our nutrition, health, and disease resistance is an important part of the current interest in functional foods.

As the chapters of this book show, fermented foods are an important part of our diet and, because of their composition, can have major impacts on many of the nutrition/disease/health problems that are common worldwide. In addition, as we start to embark on voyages that go beyond the protective boundaries of Earth, fermented foods may also become an important part of the diet of space travelers.

The history of fermented foods is a long one; they are an important part of our diet today. Advances in food technology, microbiology, and nutrition will give us even more fermented foods to eat and more reasons to eat them in the future.

Editor

Edward (Ted) Farnworth is a senior research scientist with Agriculture and Agri-food Canada. He currently heads a team at the Food Research and Development Centre at St. Hyacinthe, Quebec, that is carrying out research on the microbiology, chemical composition, and health properties of a wide range of fermented foods and beverages. He has a B.Sc. (Brock University) and M.Sc. (McMaster University) in chemistry and a Ph.D. (University of Guelph) in nutrition. Over his more than 24 years of research, he has published papers on a wide variety of subjects including mycotoxins in feeds and food, nutritional value of canola oil, sow nutrition, fetal pig nutrition, fat metabolism in swine, prebiotics for animals, food flavors, and probiotics. He is an adjunct professor at three universities and lectures in both French and English on prebiotics, probiotics, and human nutrition. He is a past president of the Canadian Society for Nutritional Sciences and currently is editor-in-chief of the online magazine *Medicinal Food News* (www.medicinalfoodnews.com).

Contributors

Marcelo Chiara Bertolami, M.D., Ph.D.
Medical Section
Instituto Dante Pazzanese de Cardiologia
São Paulo, Brazil

Wilhelm Bockelmann, Ph.D.
Institute for Microbiology
Federal Dairy Research Centre
Kiel, Germany

Herbert J. Buckenhüskes, Ph.D.
Gewürzmüller GmbH
Stuttgart, Germany

Mairead Coakley, M.Sc.
TEAGASC
Dairy Products Research Centre
Moorepark
Fermoy, Ireland

J. Kevin Collins, Ph.D.
Microbiology Department
University College Cork
Cork, Ireland

Colette Desmond, B.Sc.
TEAGASC
Dairy Products Research Centre
Moorepark
Fermoy, Ireland

Edward R. Farnworth, Ph.D.
Food Research and Development Centre
Agriculture and Agri-Food Canada
St. Hyacinthe, Quebec, Canada

Gerald Fitzgerald, Ph.D.
Microbiology Department
University College Cork
Cork, Ireland

Michael G. Gänzle, Ph.D.
Technische Universität München
Lehrstuhl für Technische Mikrobiologie
Freising-Weihenstephan, Germany

Dirk Haller, Ph.D.
Department of Medicine
Center for Digestive Disease and Nutrition
University of North Carolina at Chapel Hill
Chapel Hill, North Carolina

Walter P. Hammes, Ph.D.
Institute of Food Technology
Hohenheim University
Stuttgart, Germany

Knut J. Heller, Ph.D.
Institut für Mikrobiologie
Bundesanstalt für Milchforschung
Kiel, Germany

Wilhelm Holzapfel, Ph.D.
Institut für Hygiene und Toxikologie/BFE
Karlsruhe, Germany

Tomohiro Hosoi, Ph.D.
Tokyo Metropolitan Food Technology Research Centre
Tokyo, Japan

Young-Kyung Lee Kim, Ph.D.
Research Institute of Home Ecology
Seoul National University
Seoul, Korea

Kan Kiuchi, Ph.D.
Department of Food Science and Nutrition
Kyoritsu Women's University
Tokyo, Japan

Hoonjeong Kwon, Ph.D.
Food Toxicology Division
Department of Food and Nutrition
Seoul National University
Seoul, Korea

Jean Guy LeBlanc, Ph.D.
Department of Chemistry and Biochemistry
Université de Moncton
Moncton, New Brunswick, Canada

Isabelle Mainville
Food Research and Development Centre
Agriculture and Agri-Food Canada
St. Hyacinthe, Quebec, Canada

Luc Martin
Department of Chemistry and Biochemistry
Université de Moncton
Moncton, New Brunswick, Canada

Chantal Matar, Ph.D.
Department of Chemistry and Biochemistry
Université de Moncton
Moncton, New Brunswick, Canada

Takeshi Matsuzaki, Ph.D.
Yakult Central Institute for Microbiological Research
Tokyo, Japan

Yukiko Minamiyama, Ph.D.
First Department of Pathology
Okayama University Medical School
Okayama, Japan

Göran Molin, Ph.D.
Laboratory of Food Hygiene
Division of Food Technology
Lund University
Lund, Sweden

Baboo M. Nair, Ph.D.
Department of Applied Nutrition and Food Chemistry
Lund University
Lund, Sweden

Shigeru Okada, Ph.D.
Department of Pathological Research
Graduate School of Medicine and Dentistry
Okayama University
Okayama, Japan

Gabriela Perdigón, Ph.D.
Facultad de Bioquimica, Quimica y Farmacia
Argentina Centro de Referencias por lactobacillos
Universidad Nacional de Tucuman
Argentina

Jashbhai B. Prajapati, Ph.D.
Dairy Microbiology Department
SMC College of Dairy Science
Gujarat Agricultural University
Anand, India

R. Paul Ross, Ph.D.
TEAGASC
Dairy Products Research Centre
Moore Park
Fermoy, Ireland

Ulrich Schillinger, Ph.D.
Institut für Hygiene und Toxicologie, BFE
Karlsruhe, Germany

Juergen Schrezenmeir, Ph.D.
Institute for Physiology and Biochemistry of Nutrition
Federal Dairy Research Centre
Kiel, Germany

Catherine Stanton, Ph.D.
TEAGASC
Dairy Products Research Centre
Moore Park
Fermoy, Ireland

Judy Van de Water, Ph.D.
Division of Rheumatology/Allergy
University of California, Davis
Davis, California

Michael deVrese, Ph.D.
Institute for Physiology and Biochemistry of Nutrition
Federal Dairy Research Centre
Kiel, Germany

Table of Contents

1 The History of Fermented Foods

Jashbhai B. Prajapati and Baboo M. Nair

CONTENTS

0-8493-1372-4/03/$0.00+$1.50

1.1 INTRODUCTION

The origin of fermented foods is lost in antiquity. It may have been a mere accident when people first experienced the taste of fermented food. After drying, fermentation is the world's oldest food preservation method. Fermentation became popular with the dawn of civilization because it not only preserved food but also gave it a variety of tastes, forms, and other sensory sensations. Slowly, people have realized the nutritional and therapeutic value of fermented foods and drinks, and this has made fermented foods even more popular.

It seems that the art of fermentation originated in the Indian subcontinent, in the settlements that predate the great Indus Valley civilization. During the Harappan spread or pre-Vedic times, there are indications of a highly developed system of agriculture and animal husbandry. Artifacts from Egypt and the Middle East also suggest that fermentation was known from ancient times in that region of the world. It is believed that the knowledge written in the four Vedas (sacred Hindu writings) came from the experiences, wisdom, and foresight of sages that had been preserved by verbal tradition. As there are no written proofs, controversies exist among historians in predicting the probable date of the Vedas. Based on astronomy, Lokmanya Tilak estimated it as the period between 6000 to 4000 VP (VP stands for the Hindu calendar of Vikram, 2002 A.D.); using other methods of calculation, it is approximately 2500 VP.[1]

The cow was referred to 700 times in Rigveda (the oldest and most important of the sacred books of the Hindus) alone. It is a symbol of endless bounty in numerous contexts, and the importance of milk, curds, buttermilk, and country butter was emphasized. For the most part, the cultures of the South Asian countries have left relatively few artifacts, and this has led to an overemphasis on the cultural advances of other regions, such as the Near East, Central America, and even sub-Saharan Africa in comparison to South Asia.[2] The skills of food preservation existed among the native people of many areas, and the knowledge was propagated orally. During the Middle Ages, the varieties of fermented foods and drinks developed depended upon the availability of raw materials, environmental conditions, and the taste preferences of the local people. Knowledge of some of the products that originated in ancient times has developed, and these products are now being manufactured on a large commercial scale. However, many products are still poorly understood, and their technology needs to be refined and commercialized in order for their health benefits to be realized by society. The important milestones in the history of fermented foods are presented in Table 1.1.

As a process, fermentation consists of transformation of simple raw materials into a range of value-added products by utilizing the phenomenon of growth of microorganisms and/or their activities on various substrates.[3] This means that knowledge of microorganisms is essential to understand the process. Such knowledge has only existed since 1680, when Antony Van Leeuwenhoek first demonstrated the use of a microscope and described the existence of microorganisms. Louis Pasteur, in the middle of the nineteenth century, contributed significantly to the understanding of the phenomenon of fermentation; he established the role of microbes in fermentation and also proved that there are many different kinds of fermentations.

TABLE 1.1
Milestones in the History of Fermented Foods

Milestone	Development/Location
ca. 10,000 B.C. to Middle Ages	Evolution of fermentation from salvaging the surplus, probably by pre-Aryans
ca. 7000 B.C.	Cheese- and breadmaking practiced
ca. 6000 B.C.	Winemaking in the Near East
ca. 5000 B.C.	Nutritional and health value of fermented milk and beverages described
ca. 3500 B.C.	Breadmaking in Egypt
ca. 1500 B.C.	Preparation of meat sausages by ancient Babylonians
2000 B.C.–1200 A.D.	Different types of fermented milks from different regions
ca. 300 B.C.	Preservation of vegetables by fermentation by the Chinese
500–1000 A.D.	Development of cereal–legume based fermented foods
1881	Published literature on koji and sake brewing
1907	Publication of book *Prolongation of Life* by Eli Metchnikoff describing therapeutic benefits of fermented milks
1900–1930	Application of microbiology to fermentation; use of defined cultures
1970–present	Development of products containing probiotic cultures or friendly intestinal bacteria

Source: Data compiled from Joshi, V.K. and Pandey, A., *Biotechnology: Food Fermentations*, Vol. I, Educational Publishers and Distributors, New Delhi, 1999, pp. 1–24; Pederson, C.S., *Microbiology of Food Fermentations*, AVI, Westport, CT, 1971, pp. 1–274; IDF, Fermented milks, *IDF Bull.*, No. 179, 16–32, 1984; Metchnikoff, E., *The Prolongation of Life,* G.P. Putnam's Sons, New York, 1908; Steinkraus, K.H., *Handbook of Indigenous Fermented Foods,* Marcel Dekker, Inc. New York, 1983; Padmaja, G. and George, M., in *Biotechnology: Food Fermentations*, Vol. II, Joshi V.K. and Pandey A., Eds., Educational Publishers and Distributors, New Delhi, 1999, pp. 523–582.

Since the time of Pasteur, there have been manifold increases in the knowledge of the microbiology, biochemistry, technology, and food engineering aspects of food fermentations. At present, we have a number of fermented foods and drinks including fermented milks, cereals, fruits, vegetables, fish, meat, and many mixed products, which have emerged from very early times.

1.2 FERMENTED MILKS

Rock drawings discovered in the Libyan Desert, believed to have been made about 9000 B.C., depict cow worship and cows being milked.[4] Some of the oldest records suggest development of dairying in ancient India, Mesopotamia, and Egypt. It is apparent from writings, drawings, and friezes dating back to 6000 B.C. from the Sumerians of Mesopotamia that dairying was highly developed. A sculptured relief, which dates back to 2900 to 2460 B.C., found at Teil Ubaid in the Middle East, in the territory of Ancient Babylonia, shows development of a system for processing milk. It could be deduced from all these pieces of evidence that the souring of milk was used to produce butter, and probably milk was also consumed in a soured form.[5]

A great many of today's fermented milk products were originally developed by nomadic Asian cattle breeders.

Nearly every civilization has developed fermented milk of some type. The terms dahi, buttermilk, yogurt, leben, and acidophilus milk are familiar to many people, but those who first produced these foods did not know that they were fermented by bacteria. Fermented milk products originating from different countries are listed in Table 1.2.

1.2.1 Dahi

Dahi (Sanskrit: Dadhi) is a popular Indian fermented milk product, which is quite analogous to plain yogurt in appearance and consistency. It is popular with consumers due to its distinctive flavor and because it is believed to have good nutritional and therapeutic value. It is utilized in various forms in many Indian culinary preparations. The use of dahi has been prevalent since Vedic times, and it is mentioned in ancient scriptures such as Vedas, Upanishads, and various hymns.[6] During Lord Krishna's time (ca. 3000 B.C.), dahi, buttermilk, and country butter were highly regarded. Dahi is also traditionally used as an article in rituals and an ingredient of panchamrut (five nectars). Ayurveda, the traditional scientific system of Indian medicine, in its treatises *Charaka Samhita* and *Sushruta Samhita,* discusses various properties of cow and buffalo milk dahi and emphasizes its therapeutic characteristics.[7,8] Ayurveda also describes the properties of various types of chhash (stirred diluted dahi) and their role in control of intestinal disorders.[9] Dahi, which came into use as a means of preserving milk nutrients, was probably used by Aryans in their daily diet as it reduced putrefactive changes and provided an acidic, refreshing taste. Dahi is consumed with rice in South India and with wheat preparations in the north; it is also used as a beverage or dessert. It is also prepared from the milk of the yak and the zomo in the Himalayas.[10] Dahi is still made by local halwais, shops, and restaurants and in homes by traditional methods. Some dairies have started its commercial manufacture in India.

Chakka is a concentrated product obtained after draining the whey from dahi. When it is blended with sugar and other condiments, it becomes shrikhand, referred to as shikhrini in old Sanskrit literature. This has been a very popular dessert in western India for several hundred years.

1.2.2 Kefir

Kefir is a refreshing drink that originated on the northern slopes of the Caucasus mountains. The product is made using kefir grains which, according to legend, were given to orthodox people by Mohammad. Mohammad strictly forbade the secret of kefir preparation to be given outside the faith; otherwise the grains would lose their magic strength. This may be the reason why the method of kefir preparation was kept a secret for such a long time.[11] Traditional kefir was made in skin bags. The milk was poured in daily, and a natural fermentation took place. It was customary to hang the bag near the door, and everyone who came in or out had to push/kick the bag in order to mix the liquid. The finished product has a high acidity and varying

TABLE 1.2
Origin of Some Important Fermented Milk Products

Product	Country of Origin	Period	Characteristics and Use
Dahi	India	6000–4000 B.C.	Coagulated sour milk eaten as a food item; an intermediate product for making country butter and ghee (clarified butter)
Chhash (buttermilk)	India	6000–4000 B.C.	Diluted dahi or the buttermilk left after churning of dahi into butter; used as beverage after/with meal
Laban zeer/khad	Egypt	5000–3000 B.C.	Sour milk; traditionally coagulated in earthenware vessels
Leben	Iraq	ca. 3000 B.C.	Traditional fermented milk containing yogurt bacteria; whey partially drained by hanging the curd
Zabady	Egypt and Sudan	2000 B.C.	Natural type yogurt; firm consistency and cooked flavor
Cultured cream	Mesopotamia	1300 B.C.	Naturally soured cream
Shrikhand	India	400 B.C.	Concentrated sour milk, sweetened and spiced; semisolid mass eaten with meals as sweet dish
Kishk	Egypt and Arab world	—	Dry fermented product made from laban zeer and parboiled wheat; small round irregular pieces, yellowish brown in color with hard texture; highly nutritious with high amino acid and vitamin content
Kumys, kumiss	Central Asia (Mongol, Russia)	400 B.C. (probably known around 2000 B.C.)	Traditionally, mares' milk fermented by lactobacilli and yeast; sparkling beverage containing lactic acid, alcohol, and carbon dioxide
Mast	Iran	—	Natural type yogurt; firm consistency and cooked flavor
Villi	Finland	—	High viscosity fermented milk with lactic acid bacteria and mold
Taette	Norway	—	Viscous fermented milk also known as cellarmilk
Langfil, tattemjolk	Sweden	—	Milk fermented with slime-producing culture of lactococci
Ymer	Denmark	—	Protein-fortified milk fermented by Leuconostocs and lactococci; whey is separated
Skyr	Iceland	870 A.D.	Made from ewes' milk by addition of rennet and starter; today concentrated by membrane technology.

TABLE 1.2 ***(Continued)***
Origin of Some Important Fermented Milk Products

Product	Country of Origin	Period	Characteristics and Use
Prostokvasha	Countries of the former Soviet Union	—	Fermented milk made since ancient times by fermenting raw milk with mesophilic lactic bacteria
Kefir	Caucasian China	—	Milk fermented with kefir grains; foamy effervescent product with acid and alcoholic taste
Yogurt (kisle mliako)	Bulgaria	—	Cows' or ewes' milk fermented by *Streptococcus thermophilus* and *Lactobacillus bulgaricus*
Yogurt	Turkey	800 A.D.	Custard-like sour fermented milk
Bulgarian milk	Bulgaria	500 A.D.	Very sour milk fermented by *Lb. bulgaricus* alone or with *Str. thermophilus*
Trahana	Greece	—	Traditional Balkan fermented milk; fermented ewes' milk mixed with wheat flour and then dried
Churpi	Nepal	—	Fermented milk is churned and the buttermilk remaining is heated to form a solid curd; may be further dried
Airan	Central Asia, Bulgaria	1253–1255 A.D.	Cows' milk soured by *Lb. bulgaricus*, used as a refreshing beverage
Yakult	Japan	1935 A.D.	Highly heat-treated milk fermented by *Lb. casei* strain Shirota; beverage and health supplement

Source: Data compiled from Pederson, C.S., *Microbiology of Food Fermentations*, AVI, Westport, CT, 1971, pp. 1–274; IDF, *IDF Bull.*, No. 179, 16–32, 1984; Yegna Narayan Aiyar, A.K., *Indian Dairyman,* 5, 77–83, 1953; Koroleva, N.S., *IDF Bull.,* No. 227, 96–100, 1988; Rasic, J.L.J. and Kurmann, J.A., *Yoghurt — Scientific Grounds, Technology, Manufacture and Preparations,* Technical Dairy Pub. House, Copenhagen, 1978, pp. 11–15.

amounts of alcohol and carbon dioxide. Kefir can be produced by a type of continuous process where the kefir is taken out and fresh milk is added.[11,12] Commercial production of kefir now occurs in many countries, particularly in Eastern Europe.

(See Chapter 4 for more details on the production and health properties of kefir.)

1.2.3 Kumys

Kumys (kumiss) prepared from mares' milk is an ancient drink widely consumed throughout Eastern Europe and the Asiatic regions. The name kumys was derived from a tribe called Kumanes, who lived along the river Kumane in the Asiatic steppes. Scythian tribes that roamed in southeast Russia and Middle Asia used to drink mares' milk in the form of kumys some 25 centuries ago.[11] It was consumed as food as well as an alcoholic drink. Marco Polo mentioned kumys as being a pleasant milk drink.

1.2.4 Yogurt (Yoghurt)

As is the case with many other fermented milk products, no precise records are available regarding the origin of yogurt. It is believed that the ancient Turkish people in Asia, where they lived as nomads, first made yogurt. The first Turkish name for this product appeared in the eighth century as "yoghurut," and the name was subsequently changed in the eleventh century to its present spelling. One legend tells that an angel brought down a pot that contained the first yogurt, while another source claims that the ancient Turks, who were Buddhists, used to offer yogurt to the angels and stars who protected them.[13] According to Chomakow[14] and others, yogurt originates from the Balkans. The inhabitants of Thrace used to make soured milks called "prokish" from sheep's milk, which later became yogurt. In the Bible, it is recorded that when the patriarch Abraham entertained three angels, he put before them soured and sweet milk (Genesis VIII, 8). The ancient Greeks and Romans were also acquainted with preparations of soured milks. The biography of Roman Emperor Elagabalus (204–222 A.D.) mentions two recipes for soured milk.

Ancient physicians of the Near and Middle East prescribed yogurt or related soured milks for curing disorders of the stomach, intestines, and liver and for stimulation of the appetite.[13] Records also exist of the use of soured milks, particularly yogurt, for preservation of meat against spoilage during the summer.[15] Earlier writers of the Middle East mentioned the use of soured milks as cosmetics for Persian women. However, systematic studies on the therapeutic properties of fermented milks started after the publication of the book *Prolongation of Life* by Metchnikoff.[16] In this book, Metchnikoff attributed the long life of the Bulgarian people to the consumption of large quantities of Bulgarian milk containing *Lactobacillus bulgaricus*. Later, it was found that *Lb. bulgaricus* cannot be implanted in the intestines. In the search for another milk-souring organism, Moro in 1900 described *Lactobacillus acidophilus,* which was isolated from the feces of infants and is a normal inhabitant of human intestines. This organism could be implanted in the intestinal tract and hence was selected as a more suitable candidate for making fermented milk with a higher therapeutic value.

One of the first industrial productions of yogurt in Europe was undertaken by Danone in 1922 at Madrid, Spain.[13] After World War II and particularly since 1950, the technology of yogurt and understanding of its properties have advanced rapidly. The yogurt made in the United States for many years was a soft-curd product quite different from the custard-like yogurt prepared in the Middle East.[4] The organisms involved in this first commercial yogurt were *Lb. bulgaricus* and *Streptococcus thermophilus*. Fermentation was carried out at a lower temperature than those prevalent in the Middle East. This product resembled the soft-curd product commonly used in northern areas of Europe. The method of preparation varied considerably, but the basic process, using high acid producing lactic acid bacteria, was the same.

New criteria have been introduced for culture selection for yogurt production. Supplementing yogurt flora with *Lb. acidophilus* and *Bifidobacterium bifidus* for the purpose of increasing the product's health-promoting value resulted in new cultured milks called Aco-yoghurt, Acidophilus–Bifidus Yoghurt, Biogarde, and Bioghurt.

(See Chapters 5 and 7 for more details on the production and health properties of yogurt.)

1.2.5 Cheese

According to an ancient legend, cheese was accidentally made by an Arabian merchant when he put his supply of milk into a pouch made of a sheep's stomach when he set out on a long day's journey across the desert. The rennet in the lining of the pouch combined with the heat of the sun caused the milk to separate into curd and whey. This story seems to have occurred approximately 7000 years B.C. in the so-called Fertile Crescent situated between the rivers Euphrates and Tigris in Iraq. The earlier records in Vedic hymns in India (6000 to 4000 B.C.), Egyptian records (4000 B.C.), and Babylonian records (2000 B.C.) clearly show references to milk, butter, and cheese. However, it is believed that with the advance of civilization, the art of cheesemaking spread via the Mediterranean basin to the rest of the world.[17] There is reference to cheese in biblical times (Job X.10 [ca. 1520 B.C.] and Samuel 1.XVII.18 and 2.XVII.29 [ca. 1017 B.C.]), but written history is scarce until the periods of the Greek and Roman empires, when various authors left written evidence.[18] Greek records go back to about 1550 B.C. and Roman records to 750 B.C., indicating that milk and cheese were important components of the diets of these peoples. By the beginning of the Christian era, milk and cheese were used as food throughout Europe.[4]

Milking operations and the curdling of milk are depicted in an early Sumerian frieze from El-Ubaid. A food material found in the tomb of Hories Aha (ca. 3000 B.C.) has been proven to be cheese.[18] A scene on the walls of a Ramesid tomb (100 B.C.) depicts goats being led to pasture and also skin bags suspended from poles. Such bags were traditionally used to ferment milk by nomadic tribes. During fermentation, drainage of whey through cloth or perforated bowls allowed the collection of curds, which when salted became cheese. Bowls with perforated bases, presumably used for draining whey, have been found in several locations in Europe and Asia. Baskets from reeds and other stems have also been found. Such baskets are used today in India for making both Surati Paneer and Dacca curds.

Impressions of baskets found at Windmill Hill in Dorset, England (ca. 1800 B.C.) indicate that cheese was made in England well before the arrival of the Romans. Cheese was included in the offering of ancient Greeks to the gods at Mount Olympus, and cheesemaking was clearly a well-established craft at the time of Homer's writing. Homer, ca. 1184 B.C., referred to cheese made in caves by the "Cyclops" Polyphemus from the milk of sheep and goats.[19] Such cheese may have been the ancestor of the feta cheese made widely in Greece today.[8] Later, Herodotus, 484 to 408 B.C., referred to the "Scythian" cheese made from mares' milk, while Aristotle (384 to 322 B.C.) noted that "Phrygian" cheese was made from the milk of mares and asses. The trade of cheese between countries became important during the rule of the Roman emperor Diocletian (284 to 305 A.D.).

By the fourteenth century, cheesemaking was a considerable industry in Switzerland, but export was forbidden. At this time, a cheese market was operating in

Gouda, Holland. It is reported that the first cooperative cheese factory was started at Voralberg in the Balkans in about 1380.[18] By 1500, it is recorded that the expansion of cheesemaking in England, France, Germany, and Holland resulted in Italy losing its dominant position as a cheesemaker.[17]

Cheddar cheese originated from the village Cheddar in Somerset, England and was popular during the reign of Queen Elizabeth I (1558 to 1603), although it has been said to have been known from three centuries earlier. The period 1860 to 1880 saw the introduction of a factory system throughout the cheesemaking world. In 1851, the first cheese factory was established in Oneida County, New York, and it proved so successful that within a few years several other factories were established.[20] In 1870, the first British cheese factory was opened in Derbyshire.[21]

Some cheeses were developed only later to become extinct. Changes in agricultural practice, environment, food habits, sociological conditions, etc. have been given as reasons for the disappearance of some cheese varieties, but consumer reaction is probably the major reason. The craft was traditionally handed down, usually from mother to daughter, by word of mouth, or by practical teaching, but the art of cheesemaking was kept alive by monasteries in Europe. Occasionally, some new varieties of cheese were produced, but these were not as successful as those recorded by monks. The monasteries, through the interchange of monks, spread the knowledge of methods of cheesemaking. Six countries — Italy, France, England, Holland, Switzerland, and Germany — are considered as the "big six" in the history of cheesemaking. Most of the cheese varieties consumed today come from these countries.

Sir Joseph Lister first isolated the milk bacterium now known as *Lactococcus lactis* in 1878. After his discovery, attempts were made to use selected cultures of lactic acid bacteria for making cheese, butter, and fermented milks. The first instance of the use of a selected culture to make fermented milk was reported in 1890 by the Danish scientist Storch,[22] who used selected strains for souring cream for butter making. Prior to the use of cultures, cheese was made by:

1. Natural souring with adjustment of temperature
2. Addition to milk of sour whey or buttermilk
3. Adding homemade starter

About 1870, Hansen in Denmark put a commercial rennet preparation on the market, and in the beginning of 1900, he put commercial cultures for cheesemaking on the market. This gave a boost to the manufacture of cheese on a wider scale.[17]

There are about 2000 names assigned to cheeses based on the area of the cheese's origin, country, source of milk, method of production, moisture content, cultures used, inventor, method of ripening, etc. Of these, about 800 varieties have been well established. They can be classified in 18 distinct types. Descriptions of more than 400 varieties have been provided by the U.S. Department of Agriculture (USDA).[23] Table 1.3 shows some of the cheese varieties, their countries of origin, and estimates of when they were first made.

(See Chapter 8 for more details on the production and health properties of cheese.)

TABLE 1.3
History of Some Important Cheese Varieties

Cheese	Period of Origin	Country of Origin	Characteristics
Karish cheese	ca. 3200 B.C.	Egypt	Soft cheese made from sour milk; consumed fresh or after pickling
Mish cheese	ca. 3200 B.C.	Egypt	Made by pickling karish cheese in pickling medium in earthenware for more than one year; yellowish brown color, sharp flavor and high salt content
Feta (old cyclops)	ca. 1184 B.C.	Greece	White, pickled soft cheese
Domiati	332 B.C.	Egypt	Soft, white, pickled cheese
Emmenthal	ca. 58 B.C.	Switzerland	Hard, pressed-curd cheese with eyes
Gorgonzola	879 A.D.	Italy	Blue-green veined
Roquefort	1070 A.D.	France	Blue-veined, semisoft to hard
Marolles	1174 A.D.	France	Soft cows' milk cheese cured for 3–5 months
Grana (parmesan)	1200 A.D.	Italy	Granular body and texture, sharp flavor, small eyes
Cheddar	1500 A.D.	England	Hard cheese ripened for 3–6 months
Gouda	1697 A.D.	Holland	Sweet curd, semisoft to hard
Stilton	1785 A.D.	England	Hard, mild, blue-veined
Camembert	1791 A.D.	France	Soft, surface mold ripened
Limburger	1800 A.D.	Belgium	Semisoft surface bacterial ripened with strong aroma and flavor

Source: Data compiled from Davis, J.G., *Cheese Vol. I. Basic Technology,* Churchill Livingstone, Edinburgh, 1964, pp. 1–16; Scott, R., *Cheese Making Practice*, 2nd ed., Elsevier, London, 1986, pp. 1–11; Galloway, J.H. and Crawford, R.J.M., in *Microbiology of Fermented Foods* Vol. 1, Wood, B.J.B., Ed., Elsevier, London, 1985, pp. 111–166; USDA, Cheese Varieties and Descriptions, USDA, Washington, D.C., 1978, pp. 1–140.

1.3 CEREAL- AND LEGUME-BASED FERMENTED FOODS

Cereals and legumes are important contributors of carbohydrates and proteins to the diet, especially for the vegetarian population of the world. Ancient peoples in Asia used techniques of hydrolyzing starch and proteins in these products to improve the digestibility and organoleptic properties of their food. Asians have been pioneers in the development of fermented plant proteins to produce meat-like flavors; Indonesians developed fermentation methods to introduce a meat-like texture to vegetable products; Egyptians developed wheat bread leavened with yeasts; and Indians discovered methods for souring and leavening cereal–legume batters.[24]

The traditional methods for fermenting cereals and legumes are simple and inexpensive. However, these methods are changing rapidly through modern microbial technology. Soybeans, black grams, mung beans, and Bengal gram are the principal legumes, and rice is the main cereal used in the preparation of a variety of fermented foods in different parts of the world. Wheat is mainly used for making breads. Some of these fermented cereal/legume based foods, which originated from different regions of the world, are listed in Table 1.4.

1.3.1 Bread

Bread is one of the most widespread and ancient cereal products fermented by yeasts. The art of modern breadmaking came from the Egyptians about 3500 years ago.[25] The Egyptians were probably the first to observe fermentation and leavening when bread dough was allowed to stand for hours. Bread was the principal food of the Egyptians, and it was also given out in lieu of wages. The Romans were probably the first to commercialize breadmaking by using yeasts separated from wine. It is estimated that 250 bakeries existed in Rome around 100 B.C.[4] The ancient Greeks prepared unleavened bread that contained barley flour. With the development of leavened bread, the use of barley declined because it does not produce the light airy bread typical of wheat bread.

The early Europeans made a flat sour rye bread using sour rye starter cultures as early as 800 B.C.[26] Sour rye bread has survived the centuries and is still very popular in many parts of Europe as well as in North America. One of the most unusual starter cultures is "mother sponge," which is used to make San Francisco sourdough French bread. This culture contains yeasts and bacteria in a ratio of 1:100. The origin of this natural culture is not known, but it has been used continuously for over 140 years.[25]

1.3.2 Idli

While the ancient Egyptians developed wheat breads, the people of India discovered methods of leavening cereal and legume batters with bacterial and yeast fermentations. The people of the Middle East discovered that sour milks combined with wheat resulted in dried soup ingredients with superior nutritional value and excellent keeping quality. Cereal- and legume-based mixed fermented products have complementary nutritional value. Idli (Figure 1.1a) and dosa (Figure 1.1b), staple foods of Southern India prepared from rice and legumes, have each had a long history, though not every detail can be clearly traced. Idli is frequently mentioned; in 1025 A.D., the poet Chavundaraya described it unequivocally as urad dal (black gram) soaked in buttermilk, ground to a fine paste, mixed with the clear water of curds, cumin, coriander, pepper, and asafoetida and then shaped.[27] The Manasollasa of about 1130 A.D., written in Sanskrit, describes idli as made from fine urad flour, fashioned into small balls, fried in ghee, and then spiced with pepper powder, cumin powder, and asafoetida.[28]

1.3.3 Dosa

Dosa is a pancake made from the batter of rice and black gram; it is first noted in the Tamil (India) Sangam literature about the sixth century A.D.[29] The use of pulses to make the well-known dhokla (steam-cooked fermented spicy cake) of today was first mentioned in 1066 A.D. (Figure 1.1c). In northern India, dried spicy hollow balls

TABLE 1.4
Some Cereal- and Legume-Based Fermented Foods

Product	Country/Region	Substrate	Microorganism(s) Involved	Nature of Product	Product Use
Ang-kak	China	Red rice	*Monascus purpureus*	Powder	Dry red powder used as colorant
Bagni	Caucasus	Millet	—	Liquid	Drink
Banku	Ghana	Maize and cassava	Lactic acid bacteria, yeasts	Solid	Used as a staple food
Bhallae	India	Black gram	Lactic acid bacteria, yeasts	Deep-fried patties	Snack after soaking in water
Bhatura	India	White wheat flour	Lactic acid bacteria, yeasts	Deep-fried bread	Breakfast food
Bongkrek	Central Java	Coconut press cake	*Rhizopus oligosporus*	Solid	Roasted or fried in oil, meat substitute
Burukutu	Savannah region of Nigeria	Sorghum and cassava	Lactic acid bacteria, *Candida* sp., *Saccharomyces cerevisiae*	Liquid	Liquid creamy drink
Chee-fan	China	Soybean whey curd	*Mucor* spp., *Aspergillus glaucus*	Solid	Eaten fresh, like cheese
Chickwangue	Congo	Cassava roots	Bacteria	Paste	Staple food
Darassum	Mongolia	Millet	—	Liquid	Drink
Dawadawa	West Africa, Nigeria	African locust bean	Spore forming bacteria, lactic acid bacteria, yeast	Solid	Eaten fresh or in stews
Dhokla	India	Bengal gram and wheat	Lactic acid bacteria, yeasts	Solid/spongy	Spongy condiment
Dosai/dosa	India	Black gram and rice	*Leuconostocs, Lactobacillus fermentum, Saccharomyces*	Solid	Spongy fried breakfast food
Fermented rice	India	Rice	Lactic acid bacteria	Semisolid	Breakfast food
Fufu	Africa	Cassava roots	*Lactobacillus* sp., *Leuconostoc* sp., *S. cerevisiae*	Paste	Eaten with soup, sauce or stews
Gari	West Africa	Cassava roots	*Corynebacterium, Geotrichum candidum, Lactobacillus plantarum, Leuconostocs, Alcaligenes* sp., *Candida* sp.	Granular powder	Granular wet paste eaten as a staple with stews

Hama natto	Japan	Soybeans, wheat flour	*Aspergillus oryzae*, *Streptococcus* sp., *Pediococcus* sp.	Soft	Raisin-like flavoring agent for meat or fish or eaten as a snack
Hopper (appa)	Sri Lanka	Rice or wheat flour and coconut water	Baker's yeast, acid-producing bacteria	Semisolid	Breakfast food
Idli	India	Black gram and rice	*Leuconostocs, Saccharomyces* sp.	Solid	Spongy, steam-cooked breakfast food
Injera	Ethiopia	Wheat, barley, teff, maize	*Candida guiliermondii*	Solid/spongy	Bread substitute
Jalebies	India, Nepal, Pakistan	Wheat flour	Yeasts, lactobacilli	Solid	Syrup-filled confectionery
Kanji	India	Rice and carrots	*Hansenula anomala*	Liquid	Sour liquid added to vegetables
Kecap	Indonesia and nearby regions	Soybeans, wheat	*A. oryzae, Lactobacillus* sp., *Hansenula* sp., *Saccharomyces* sp.	Liquid	Condiment and seasoning agent
Kenima	Nepal, northeast India	Soybeans	—	Solid	Snack food
Kenkey	Ghana	Maize	*Corynebacteria, Saccharomyces,* molds	Solid	Steamed, eaten with vegetables
Ketjap	Indonesia	Black soybeans	*A. oryzae*	Syrup	Seasoning agent
Khaman	India	Bengal gram	*Leuconostocs,* lactobacilli, yeasts	Solid	Cake-like breakfast food
Kisra	Sudan	Sorghum flour	Yeasts, lactobacilli, *Acetobacter*	Spongy bread	Staple food
Kulcha	North India and Pakistan	White wheat flour	Lactic acid bacteria, yeasts	Flat bread	Staple food
Lafun	West Africa and Nigeria	Cassava roots	*Leuconostocs, Corynebacteria, Candida* sp.	Paste	Staple food
Lao-chao	China, Indonesia	Rice	*Rhizopus oryzae, R. chinensis, Chlamydomucor oryzae, Saccharomyces* sp.	Solid	Soft, glutinous, eaten with vegetables
Mahewu	South Africa	Maize	Lactic acid bacteria	Liquid	Drink
Meitauza	China, Taiwan	Soybean cake	*Actinomucor elegans*	Solid	Fried in oil or cooked with vegetables
Meju	Korea	Soybeans	*A. oryzae, Rhizopus* sp.	Paste	Seasoning agent
Merissa	Sudan	Sorghum	*Saccharomyces* sp.	Liquid	Drink

TABLE 1.4 *(Continued)*
Some Cereal- and Legume-Based Fermented Foods

Product	Country/Region	Substrate	Microorganism(s) Involved	Nature of Product	Product Use
Minchin	China	Wheat gluten	*Paecilomyces* sp., *Aspergillus* sp., *Cladosporium* sp., *Fusarium* sp.	Solid	Condiment
Miso	Japan, China	Rice and soybeans	*Aspergillus* sp., *Torulopsis etchellsii, Lactobacillus* sp., *Saccharomyces rouxii*	Paste	Paste, soup base
Nan	India, Pakistan, Afghanistan, Iran	Unbleached wheat flour	Yeasts	Solid	Snack food
Natto	Japan	Soybeans	*Bacillus natto*	Solid	Cake used as a meat substitute
Ogi	Nigeria, West Africa	Maize	Lactic acid bacteria, *Cephalosporium* sp., *Fusarium* sp., *Aspergillus* sp., *Penicillium* sp., *S. cerevisiae*	Paste	Staple breakfast food
Ontjom	Indonesia	Peanut cake	*Neurospora intermedia, R. oligosporous*	Solid	Roasted or fried food used as meat substitute
Peujeum	Java	Cassava roots	Yeasts, molds	Solid	Acidic product with alcoholic flavor, eaten as such or after baking
Poi	Hawaii	Taro corms	*Lactobacillus* sp., *Candida vini, Geotrichum candidum*	Semisolid	Taken with fish or meat
Pozol	Mexico	White maize	Molds, yeasts	Solid	Beverage or porridge
Puda/pudla	India	Bengal gram, mung, wheat	Lactic acid bacteria, yeasts	Solid	Pancake snack food
Puto	Philippines	Rice	*Leuconostocs, Streptococcus faecalis, S. cerevisiae*	Solid	Snack food
Shamsy bread	Egypt	Wheat flour	Yeasts	Spongy bread	Staple food

Soy sauce	Japan, China, Philippines and other Oriental countries	Soybeans and wheat	*A. oryzae, A. sojae, Lactobacillus* sp., *Saccharomyces rouxii*	Liquid	Seasoning agent for meat, fish and cereals
Sufu	China, Taiwan	Soybean whey curd	*Aspergillus elegans, Mucor hiemalis, Mucor silvaticus, M. subtilissimus*	Solid	Soybean cake, condiment
Tao-si	Philippines	Wheat flour, soybeans	—	Semisolid	Seasoning agent
Taotjo	East Indies	Roasted wheat meal or glutinous rice, soybeans	—	Semisolid	Condiment
Tape	Indonesia and nearby regions	Cassava roots or rice	*S. cerevisiae, H. anomala, R. oryzae, Mucor* sp., *Endomycopsis fibuliger*	Solid/paste	Soft, solid eaten as a staple
Tempeh	Indonesia and nearby regions	Soybeans	*Rhizopus* sp.	Solid	Fried in oil; roasted; as a meat substitute
Uji	Kenya, Uganda and Tanzania	Maize, sorghum or millet flour	Lactobacilli, pediococci, *Leuconostocs*	Semisolid	Breakfast and lunch food
Vadai	India	Black gram	*Leuconostocs, H. anomala, Saccharomyces*	Deep-fried patties	Snack food
Waries	India	Black gram flour	*Candida* sp., *Saccharomyces* sp.	Solid	Spongy, spicy condiment

Source: Modified from Soni, S.K. and Sandhu, D.K., in *Biotechnology: Food Fermentations*, Vol. II, Joshi, V.K. and Pandey, A., Eds., Educational Publishers and Distributors, New Delhi, 1999, pp. 895–950; Padmaja, G. and George, M., in *Biotechnology: Food Fermentations*, Vol. II, Joshi, V.K. and Pandey, A., Eds., Educational Publishers and Distributors, New Delhi, 1999, pp. 523–582.

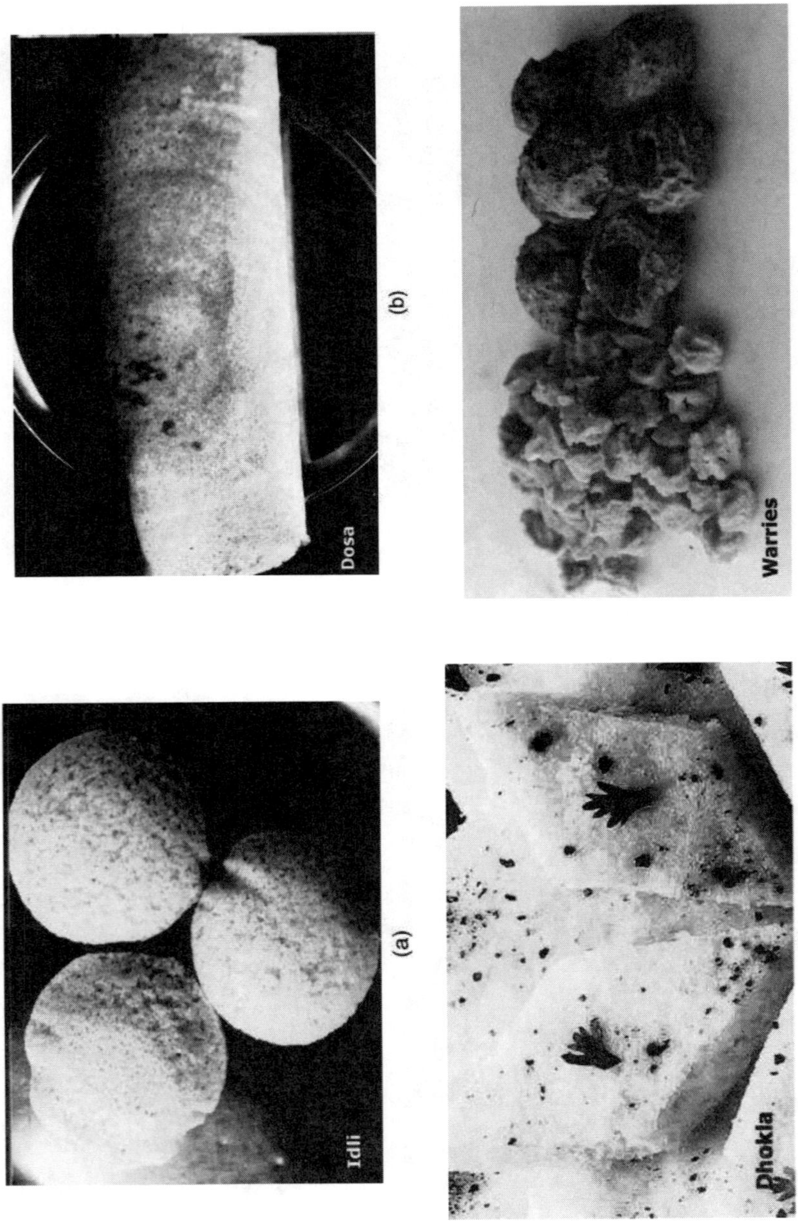

FIGURE 1.1 Some fermented foods of Asia.

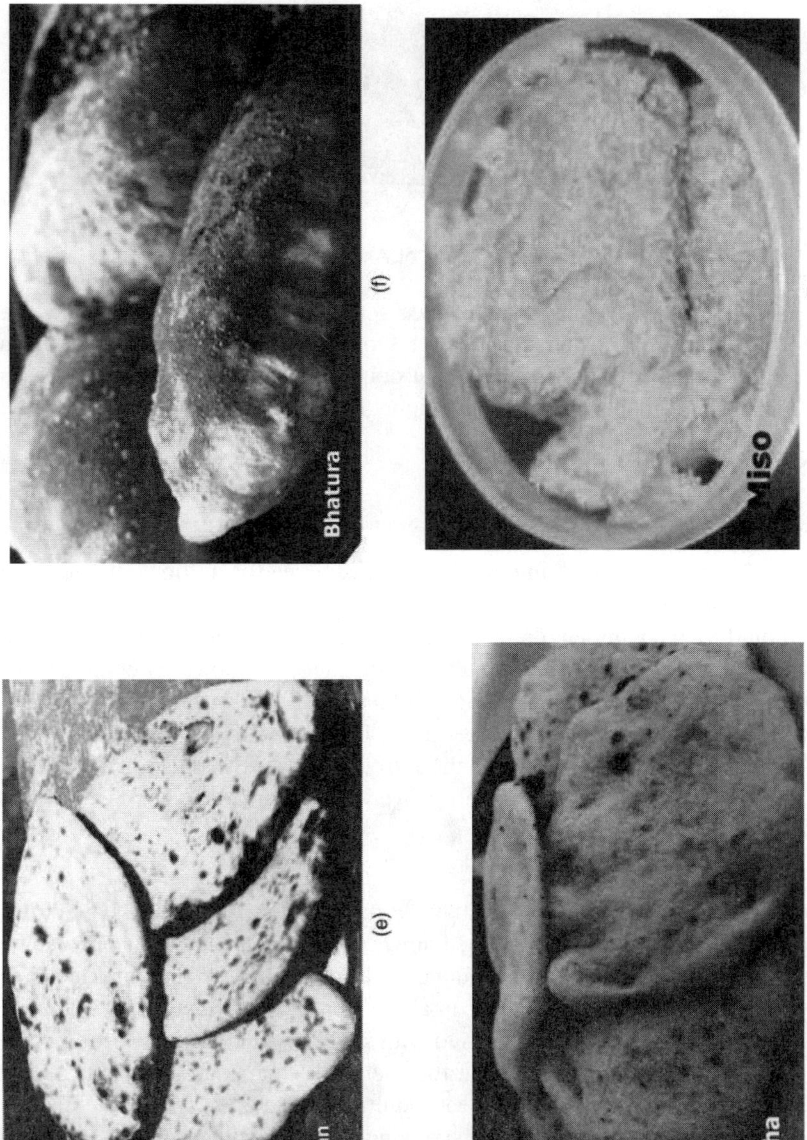

FIGURE 1.1 *Continued.*

(i)

FIGURE 1.1 *Continued.* Some fermented foods of Asia.

called warries (Figure 1.1d) have been made for more than 100 years. They are prepared from black gram paste fermented for 3 to 10 days. Nan, bhatura, and kulcha are staple foods made from fermented wheat dough and shaped as flattened breads in India, Pakistan, Afghanistan, and Iran (Figures 1.1e to g).

1.3.4 Soy Foods

1.3.4.1 Soy Sauce

The soybean is one of the most important protein sources for millions of people in the Orient; one of the most popular products for centuries has been soy sauce. Soy sauce is a liquid food condiment prepared from fermented rice or wheat and soybean with the help of molds, bacteria, and yeasts. Soy sauce is known as ch'au yau or pak yau in China, shoyu in Japan, ketjap in Malaysia, kecap in Indonesia, kenjang in Korea, toyo in the Philippines, and see-ieu in Thailand.[30] It is said that soy sauce became popular in Japan as a result of the introduction of Buddhism from China. The Chinese have been using soy sauce for over 3000 years.[31]

1.3.4.2 Miso

Miso (see Figure 1.1h) is a fermented soybean paste that is believed to have originated in China in 600 A.D. or earlier. It is known as chiang in China, miso in Japan, jang or deoenjang in Korea, tauco in Indonesia, taochieo in Thailand, and tao-si in the Philippines. Most of these products contain rice or barley fermented by *Aspergillus oryzae*, which makes koji. This is mixed with soybeans and fermented for several months. Miso has been popular in Japan for over 1000 years and is used as a base for soups and as a sauce served with meat, poultry, seafood, and vegetable dishes.

(See Chapter 11 for more details on the production and health properties of miso.)

1.3.4.3 Tempeh

One of the most important products of Indonesia is tempeh, which is made from soybeans fermented by the mold *Rhizopus*. It is particularly important in Java and Bali. It is also produced in Malaysian villages, Singapore, Canada, Holland, the

West Indies and the United States.[31] Tempeh is a mold-fermented cake. It has been consumed as a meat analogue for several centuries. Prinsen Geerlings[32] was the first to identify the tempeh mold. Later, Boorsma[33] analyzed tempeh and soybeans to determine the changes that were occurring to the substrates during fermentation.

1.3.4.4 Natto

Natto in Japan and thua-nao in Thailand are ancient whole soybean fermented products. This fermented food is also known as tu-si in China, tao-si in the Philippines, and tao-tjo in East India (see Figure 1.1i). Foods of this type are consumed with boiled rice or used as a seasoning agent with cooked meat, seafood, and vegetables. Natto products are dark in color and have a pungent but pleasant aroma. They are inexpensive and highly nutritious foods, which serve as a substitute for fermented fish and meat. Hama natto is reported to have come to Japan by way of Korea approximately 350 years ago at the time of the Japanese invasion.[34] The word natto means contributed beans. It is believed that the ancestors of the owners of the Yamaya Brewery and the Saito Mido plant of Hamanatsn inherited the process of natto making from Buddhist monks.[31] (See Chapter 9 for more details on the production and health properties of natto.)

1.3.4.5 Sufu

The manufacture of fermented soybean curd, known as sufu, began during the era of the Han Dynasty in China. In the *Pen Ts'as* or Chinese *Materia Medica* of 1596, it was implied that soybean curd was invented by Lin An (179 to 122 B.C.), king of Wainan. Literally, sufu means "molded milk" and tosufu means "molded bean milk." In the west, sufu has been referred to as Chinese cheese. Because of the numerous dialects used in China and the difficulties of phonetic rendering from Chinese to English, the synonyms tosufu, fu-su, fu-ru, foo-yue, etc. are found. It is called chao in Vietnam, tahuri in the Philippines, takaoan in Indonesia, and tao-hu-yi in Thailand and Taiwan. The major molds used in sufu fermentation are *Actinomucor, Rhizopus,* and *Mucor.*[35]

1.4 FERMENTED PLANT ROOT PRODUCTS

Several tuber crops in Africa and other countries are traditionally fermented to produce nutritious and safe foods. Cassava (*Manihot esculenta* ssp. *esculenta*) is the most abundant and important staple in tropical regions of Africa, Latin America, and Asia, where it is a common food for more than half a billion people.[36] Cassava is a perishable crop and contains toxigenic cyanogenic glucosides, linamarin and lotaustralin. Many traditional fermented foods are made from this crop, as they have the dual advantage of preserving it for a longer time and reducing the cyanogen content. Other tuber crops, such as potato, sweet potato, yams, and taro, have a longer shelf life, and fermented food items are not usually prepared from them.

1.4.1 Gari

Gari is one of the most popular fermented cassava products known traditionally in many West African countries. The indigenous technology of gari making was introduced into West Africa over a century ago by immigrant freed slaves from Brazil, who were used to "farinha de Mandioca," an analogue of gari in South America.[37] The traditional preparation of gari is mainly done by village women by fermenting peeled and grated cassava pulp. The prepared pulp is put into cloth bags, the bags are tied, heavy stones are placed on the bag to press out cassava juice, and the remaining solids are allowed to ferment for 3 to 5 days. The fermented mash is sifted and then roasted. The final product is a dry, farinaceous, cream-colored powder. Candi and kpokpogari are similar traditional products made from cassava.[38]

1.4.2 Fufu

Fufu, produced in West Africa, is a solid cake, ball-shaped, or granular product of fermented cassava. Lafun is a similar powdery product made in Nigeria, while chickwangue is popular in Zaire, and penjeum is a traditional product of Java.[38] In south India and Sri Lanka, the mother liquor prepared from toddy (coconut wine) and curd are used to ferment cassava and make sour cassava flour.[39]

1.5 FERMENTED FRUITS AND VEGETABLES

It seems that the development of fermented fruit and vegetable products took place from the time ancient people started collecting and storing food. Fresh fruits and vegetables are difficult to store. Fruits are naturally rich in juices and sugars and are slightly acidic. These components induce growth of yeasts and are naturally used for making alcoholic beverages. In the case of vegetables, people first added salt or seawater that resulted in extended shelf life. Before history was recorded, it was known that salt preserved foods and enhanced their organoleptic qualities.

Vegetable fermentation may have started in China; this can be deduced from references to the mixing of vegetables, including cabbage, radishes, turnips, cucumber, and beets, which were given as rations to coolies during the construction of the great Chinese wall in the third century B.C.[4] Vegetables in the Orient are often fermented in salt brines. The pickling of cucumbers probably originated in Southeast Asia.

1.5.1 Sauerkraut

Cabbage was a common vegetable in both Greek and Roman gardens. Artifacts from ancient Egypt depict the use of cabbage as an offering to the gods. Greek doctors used cabbage as a general cure for illness.[4] Sauerkraut is prepared by fermenting shredded cabbage in salt solution. Sauerkraut is a German term meaning "sour cabbage"; this food has become popular in the United States and various European countries. At first, the cabbage leaves were dressed with sour wine or vinegar. Later, the cabbage was broken or cut into pieces, packed into containers, and covered with sour juice from grapes or other fruits, sour wine, or vinegar. When the first acid liquids were replaced by salt and spontaneous fermentation resulted is not precisely known. Vaughn[40] speculated that the method used today was developed between 1550 and 1750 A.D.

Pederson[4] suggested that cabbage became the only ingredient used in preparation of sauerkraut in view of the health benefits ascribed to it by the Greeks and Romans and the plentiful supply of that vegetable in several areas of Europe. Related vegetables were included, probably cauliflower at about 1600 A.D., broccoli at about 1700 A.D., and Brussels sprouts at an earlier date.[4] Today, sauerkraut making is an important industry that makes use of the latest knowledge in microbiology and fermentation technology. *Leuconostoc mesenteroides* is the principal organism involved in sauerkraut fermentation, as it grows in vegetables more rapidly over a wide range of temperature and salt concentrations than any other lactic acid bacterium.[41]

(See Chapter 14 for more details on the production and health properties of sauerkraut.)

1.5.2 Kimchi

Kimchi is the general name given to a group of acid fermented vegetable foods that have a long tradition in Korea. More specific names are used for these pickled vegetables depending on the raw material, processing methods, season of the year, and locality.[42] Kimchi is a popular side dish served at every meal along with cooked rice and other dishes and is made primarily from cabbage or radishes.

(See Chapter 12 for more details on the production and health properties of kimchi.)

1.5.3 Pickled Vegetables

Pickled vegetables, made in households or small factories, have been popular in Egypt for centuries.[43] The vegetables pickled in Egypt include carrots, cucumbers, turnips, cauliflower, green and black olives, onions, and peppers. Pickled vegetables are used as appetizers and served with practically every meal. Homemade pickles made from fruits and vegetables are called jeruk in Malaysia; they have been popular since very early times. Cucumber is one of the oldest vegetables cultivated continuously by people. It is thought to have originated in India more than 3000 years ago.[45] It is utilized both as a fresh vegetable and a pickled product.

1.5.4 Olives

Olives are one of the oldest fruit crops in the Mediterranean area. The exact date when olive fermentation started is not known. However, the more recent history of the table olive industry in California has been well documented.[44] Between 1870 and 1900, many varieties of olives were imported from the Mediterranean area. Olives were used for oil production in the Californian missions as early as 1780. The olive literature of California contains directions for pickling of ripe and green olives that were used in the home for many years; olive pickling became commercialized by 1900. The processes of pickling were standardized later on. Cruess[45] reports five processes for pickling olives in Mediterranean countries: Spanish green olive, French brine, dry salt, water, and Italian dried. In the first two processes, lye is used to destroy the bitter glucoside found in olives. Four types of fermented olives — California ripe, brined Greek type, Siciliano-type green, and Spanish-type green — are reported by Vaughn.[46] The extent of lye treatment, salting, and the period of fermentation vary for different types of olives, but the main organisms causing fermentation in all these varieties are *Lactobacillus plantarum, Lactobacillus casei,* and *Leuconostoc mesenteroides.*[46]

1.6 FERMENTED FISH AND FISH PRODUCTS

Fermented fish products such as fish sauces, fish paste, or salted fish have been consumed since ancient times. Because of poor roads and other methods of transport, the provision of fresh fish to potential inland consumers was impossible, and this encouraged fermentation as a preservation technique. In Southeast Asian countries such as Thailand, Kampuchea, Malaysia, the Philippines, and Indonesia, the use of fermentation as a preservation method for fish has been of great value since earliest times. In the countries of northern Europe, fermented fish products are used mainly as condiments, whereas in Southeast Asia, various fish products are regarded as staples.[47]

The earliest reported fermented fish sauce is garum, which is known to have been popular in the Roman era.[48] It is made from the viscera and blood of mackerel. Other fish sauces, for example, botargue and ootarides, were produced in Italy and Greece in the nineteenth century. Another sauce reported to be produced in Greece was aimeteon, which was made from tunny viscera and blood.[47] Nuoc-mam is a fish sauce prepared from small fish in Southeast Asia. The fish are fermented in earthenware containers in a high concentration of salt for several months. The clear amber liquid that rises is separated and consumed. Shoittsuru is the fermented fish sauce of Japan, sometimes referred to as fish soy; its origin may predate that of soy sauce.[4] Burong dalag is a blend of rice and the fish dalag prepared by fermentation in the Philippines.[4]

At present, a number of fermented fish sauces exist in the world. However, fish pastes are more popular than fish sauces and are consumed as condiments. In general, the fermentation time for pastes is shorter than for sauces. In southern India and Sri Lanka, pickled or Colombo curd fish have been known for many years. In this food, fish and salt in a 3:1 ratio are mixed in concrete tanks, dried tamarind fruit is added, and the product is pickled.[47]

1.7 FERMENTED MEAT PRODUCTS

The Sushrut Samhita, an old Indian treatise written in about the third or fourth century A.D. based on knowledge prevailing many hundreds of years earlier, describes seven types of meat preparations. One of them is sour meat prepared using ghee (clarified butter), curd, rice gruel soured by fermentation, acid fruits, and pungent and aromatic ingredients. There are also indications of fermented meat products in the ancient Roman literature. These products originated in the Mediterranean region, where Romans added salt, sugar, and spices to ground meat and ripened it for varying periods of time to get a palatable product with a long shelf life. Ripening probably found favor due to the moderate temperature and frequent rainfall in these regions.[4]

Salting and drying of unground meat was the traditional method of meat preservation in Germany and other European countries. In Germany, the manufacture of fermented sausages commenced only some 150 years ago, and most of the sausages are smoked, while in Mediterranean countries, France, Hungary, and the Balkan countries, air-dried spicy sausages are predominant.[49]

The recorded history of sausage manufacture begins in the ninth century B.C,. as established by statements in Homer's *Odyssey.* However, it has been stated that sausage was prepared and consumed by ancient Babylonians as far back as 1500 B.C. and by the people of ancient China.[4] Grecian literature after Homer's time makes frequent mention of sausage or oryae. Fermented dry sausages probably had their origin in Italy about 250 years ago.

The term "salami" might have originated from the city Salamis, located on the east coast of Cyprus, that was destroyed in 449 B.C. It is also believed that sausage making was practiced in many areas of Europe during the Middle Ages.

Over time, some unique varieties of meat products developed in other regions of the world. All these fermentations were dependent on natural flora present in the raw materials, which decrease the pH and increase shelf life of the meat. After the successful use of starters in the cheese industry, the same cultures were tried for the fermentation of meat around 1940, but they did not proliferate in meat mixtures probably because of their lack of tolerance to salt and nitrite.[50]

As with many other fermented foods, intensive research into the microbiology and the chemistry of sausage ripening was triggered when traditional empirical methods of manufacture no longer met the requirements of large-scale, consistent-quality, low-cost industrial production. It is therefore not surprising that such research commenced in the United States in the 1930s, whereas in Europe, the first systematic studies on the microbiology and chemistry of sausage ripening were published in the 1950s.[50]

(See Chapter 10 for more details on the production and health properties of fermented meats.)

ACKNOWLEDGMENT

We are thankful to Manisha Ashar for her assistance in compilation of literature.

REFERENCES

1. Upadhyay B., *Vedic Sahitya aur Sanskruti,* 3rd ed., Sharda Mandir, Varanasi, India, 1967, pp. 436–450.
2. Stanton, W.R., Food fermentation in the tropics, in *Microbiology of Fermented Foods,* Vol. 2, Wood, B.J.B., Ed., Elsevier, London, 1985, pp. 193–211.
3. Joshi, V.K. and Pandey, A., *Biotechnology: Food Fermentations*, Vol. I, Educational Publishers and Distributors, New Delhi, 1999, pp. 1–24.
4. Pederson, C.S., *Microbiology of Food Fermentations*, AVI, Westport, CT, 1971, pp. 1–274.
5. IDF, Fermented milks, *IDF Bull.*, No. 179, 16–32, 1984.
6. Yegna Narayan Aiyar, A.K., Dairying in ancient India, *Indian Dairyman,* 5, 77–83, 1953.

7. Bhimshegranta, K.K., *Sushruta Samhita,* Calcutta, 1907. Cited in Ganguli, 1984 (reference 9).
8. Kaviratna, J.S., *The Kalpa-Sthana of Charak-samhita,* Calcutta, 1912. Cited in Ganguli, 1984 (reference 9).
9. Ganguli, N.C., Present status and potentials of cultured dairy products in India, *Indian Dairyman,* 36, 233–252, 1984.
10. Tokita, F., Hosono, A., Takahashi, F., Ishida, T., and Otani, H., Animal products in Nepal. II. Milk products and their manufacturing methods, *Jpn. J. Dairy Sci.,* 29, 179–184, 1980.
11. Koroleva, N.S., Technology of kefir and kumys, *IDF Bull.,* No. 227, 96–100, 1988.
12. IDF, Cultured dairy products in human nutrition, *IDF Bull.,* No. 225, 1–32, 1991.
13. Rasic, J.L.J. and Kurmann, J.A., *Yoghurt — Scientific Grounds, Technology, Manufacture and Preparations,* Technical Dairy Pub. House, Copenhagen, 1978, pp. 11–15.
14. Chomakow, H., The Dairy Industry in the People's Republic of Bulgaria, booklet, Agric. Acad., Bulgaria, 1973.
15. Kern, F., *Veterin,* 3, reprint, Vjesnik, 1912. Cited in Rasic and Kurmann, 1978 (reference 13).
16. Metchnikoff, E., *The Prolongation of Life,* G.P. Putnam's Sons, New York, 1908.
17. Davis, J.G., *Cheese, Vol. I. Basic Technology,* Churchill Livingstone, Edinburgh, 1964, pp. 1–16.
18. Scott, R., *Cheese Making Practice*, 2nd ed., Elsevier, London, 1986, pp. 1–11.
19. Homer, *The Odyssey,* (trans. AT Murray), GP Putnam's Sons, New York, 1919. Cited in Scott, 1986 (reference 18).
20. Fussell, G.E., *Dairy Eng.* 72(2), 53, 1955. Cited in Galloway and Crawford, 1985 (reference 21).
21. Galloway, J.H. and Crawford, R.J.M., Cheese fermentations, in *Microbiology of Fermented Foods* Vol. 1, Wood, B.J.B., Ed., Elsevier, London, 1985, pp. 111–166.
22. Storch, V., *Nogle Undersogelser over Flodens, Syrning,* Fra Forsoglab, 1890. Cited in Galloway and Crawford, 1985 (reference 21).
23. USDA, Cheese Varieties and Descriptions, USDA, Washington, D.C., 1978, pp. 1–140.
24. Soni, S.K. and Sandhu, D.K., Fermented cereal products, in *Biotechnology: Food Fermentations,* Vol. II, Joshi, V.K. and Pandey, A., Eds., Educational Publishers and Distributors, New Delhi, 1999, pp. 895–950.
25. Sugihara, T.F., Microbiology of bread making, in *Microbiology of Fermented Foods,* Vol. 1, Wood, B.J.B., Ed., Elsevier, London, 1985, pp. 249–262.
26. Pyler, E.J., *Baking Science and Technology,* Siebel Publishing Co., Chicago, 1952.
27. Iyengar, S.H., *Lokopakara of Chavundarava,* Oriental Manuscripts Library, Madras, India, 1950, pp. 120–134.
28. Shrigondekar, K.L., Manasollasa of King Someswara, *Gaekwad's Oriental Ser.,* 84, 21–23, 1939.
29. Srinivasa, P.T.I., *Pre-Aryan Tamil Culture,* University of Madras, Madras, India, 1930, pp. 57–70.
30. Wang, H.L. and Hesseltine, C.W., Sufu and lao-chao, *J. Agric. Food Chem.,* 18, 572–575, 1970.
31. Steinkraus, K.H., *Handbook of Indigenous Fermented Foods,* Marcel Dekker, Inc., New York, 1983.
32. Prinsen Geerlings H.C., *Enige Chinese voedingsmiddelenuit Soyabean bereid. Pharm.,* 32, 33, 1985. Cited in Soni and Sandhu, 1999 (reference 24).

33. Boorsma, P.A., Scheikundig onderzoek van Ned. Indie inheemse voedingsmiddelen, *Geneesk Tijdchr V. Ned-Indie,* 40, 247, 1990. Cited in Soni and Sandhu, 1999 (reference 24).
34. Hayashi, U., Japanese natto, in *Handbook of Indigenous Fermented Foods,* Steinkraus, K.H., Ed., Marcel Dekker, Inc., New York, 1983, p. 530.
35. Su, Y.C., Sufu production in Taiwan, in *Handbook of Indigenous Fermented Foods,* Steinkraus, K.H., Ed., Marcel Dekker, Inc,. New York, 1983, pp. 553–571.
36. IFDA/FAO, The World Cassava Economy: Facts, Trends and Outlook, IFDA/FAO, Rome, 2000, pp. 1–58.
37. Odunfa, S.A., African fermented foods, in *Microbiology of Fermented Foods,* Vol. 2, Wood, B.J.B., Ed., Elsevier, London, 1985, pp. 155–192.
38. Onabolu, A., Cassava Processing, Consumption and Dietary Cyanide Exposure, Ph.D. thesis, Karolinska Insitute, Stockholm, 2001, pp. 7–19.
39. Padmaja, G. and George, M., Oriental fermented foods, in *Biotechnology: Food Fermentations,* Vol. II, Joshi, V.K. and Pandey, A., Eds., Educational Publishers and Distributors, New Delhi, 1999, pp. 523–582.
40. Vaughn, R.H., Lactic acid fermentation of cabbage, cucumbers, olives and other produce, in *Industrial Microbiology,* 4th Ed., Reid. G., Ed., AVI, Westport, CT, 1982.
41. Pederson, C.S. and Albury, M.N., Factors affecting the bacterial flora in fermenting vegetables, *Food Res.,* 18, 290–300, 1953.
42. Mheen, T-I., Lee, K-H., Chang, C-K., Lee, S-R., Park, K-I., and Kwon, T-W., Known Korean kimchi, in *Handbook of Indigenous Fermented Foods,* Steinkraus, K.H., Ed., Marcel Dekker, Inc., New York, 1983, p. 114.
43. Morcos, S.R., Egyptian fermented foods, in *Handbook of Indigenous Fermented Foods,* Steinkraus, K.H., Ed., Marcel Dekker, Inc., New York, 1983, p. 267.
44. Vaughn, R.H., The microbiology of vegetable fermentation, in *Microbiology of Fermented Foods,* Vol. 1, Wood, B.J.B., Ed., Elsevier, London, 1985, pp. 49–110.
45. Cruess, V.W., Olive pickling in Mediterranean countries, *California State Agr. Exptl. Sta. Ctr.* 278, 1–33, 1924.
46. Vaughn, R.H., Lactic acid fermentation of cucumbers, sauerkraut and olives, in *Industrial Fermentations,* Underkofler, L.A. and Hickey, R.J., Eds., Chemical Publ. Co., New York, 1954.
47. Beddows, C.G., Fermented fish and fish products, in *Microbiology of Fermented Foods,* Vol. 2, Wood, B.J.B., Ed., Elsevier, London, 1985, pp. 1–40.
48. Badham, C.D., *Prose Halieutics or Ancient and Modern Fish Tattle,* Parker and Sons, London, 1854, p. 552.
49. Lucke, F., Fermented sausages, in *Microbiology of Fermented Foods,* Vol. 2, Wood, B.J.B., Ed., Elsevier, London, 1985, pp. 41–84.
50. Jensen, L.B. and Paddock, L., Sausage treatment with lactobacilli, US Patent 2,225,783, 1940.

2 Challenges Facing Development of Probiotic-Containing Functional Foods

Catherine Stanton, Colette Desmond, Mairead Coakley, J. Kevin Collins, Gerald Fitzgerald, and R. Paul Ross

CONTENTS

2.1 INTRODUCTION

There is ever-increasing evidence to support potential clinical applications of probiotics in the prevention and treatment of diseases of the gastrointestinal, respiratory, and urogenital tracts. Research is showing potential for probiotics in human trials in the prevention of diarrhea caused by certain pathogenic bacteria and viruses[1] and in the management of inflammatory bowel diseases, such as Crohn's disease.[2] As a result of such research, there is increasing interest in promoting the consumption of foods containing beneficial bacteria, in particular milk products.[3] In the marketplace, too, there is increasing demand for such products, with probiotic dairy foods such as yogurt and milk being the most sought-after functional food products in Europe.[4]

The beneficial effects of foods containing probiotic bacteria on human health have been demonstrated in a number of studies. Probiotics are described as "live microorganisms, which when consumed in adequate amounts as part of food, confer a health benefit on the host."[3] A major challenge associated with the application of probiotic cultures in the development of functional foods is the retention of viability during processing. In order to exert their health benefits within the consumer's body, probiotics must be able to grow and/or proliferate in the human intestine and therefore they should possess the capability to not only survive passage through the gastrointestinal tract (GIT), which involves exposure to hydrochloric acid in the stomach and bile in the small intestine, but also the harsh conditions often encountered during food processing. Given that probiotics are generally of intestinal origin, many such strains are unsuitable for growth in dairy-based media and are inactivated upon exposure to high heat, acid, or oxygen during dairy and food processing. The challenges of maintaining viability and activity of probiotic cultures in foods to the end of shelf life are two important criteria that must be fulfilled in order to provide effective probiotic food products for general consumption.

High levels of viable microorganisms are recommended for efficacy of probiotic foods.[5] Technological challenges associated with the introduction and maintenance of high numbers of probiotic organisms in foods include the form of the probiotic inoculant, ability of the probiotic culture to retain viability in the environment of the food matrix, and maintenance of probiotic characteristics in the food product through to the time of consumption. In addition, the probiotic culture must sustain some level of viability during gastric transit. Fermented dairy foods, including milk and yogurt, are among the most accepted food carriers for delivery of viable probiotic cultures to the human GIT. Spray dried and freeze-dried cultures are useful means of introducing the probiotic culture into these food systems. However, the use of such approaches in preparing cultures may impair viability and probiotic functionality. Given the extent of cell injury that may occur during processes such as spray drying, which may impair probiotic characteristics and further technological performance, it is vital to evaluate the suitability of such dried cultures in further functional food developments. Few or no data are currently available in the literature addressing the technological characteristics of stored cultures, nor indeed is much information available on the retention of probiotic characteristics following functional food and dairy processing. This chapter will review developments in probiotic foods, with particular emphasis on the introduction of probiotic lactobacilli and bifidobacteria into foods for human consumption and approaches that have been tested for the enhancement of probiotic viability in food systems to the end of shelf life.

2.2 HISTORY OF CONSUMPTION OF FERMENTED DAIRY FOODS

The consumption of fermented milks containing bacterial cultures has long been associated with beneficial health effects, and probiotic cultures have had a long association with dairy food products. The Roman historian Plinio advocated the use of fermented milks for treating gastrointestinal infection as early as 76 A.D.[6] In 1907, the Russian scientist Elie Metchnikoff suggested that the consumption of foods such as yogurt, kefir, and sour milk containing lactic acid bacteria was associated with good health and longevity. Metchnikoff, in his book *The Prolongation of Life,* reported that Bulgarian peasants who consumed large quantities of Bulgarian sour milk lived longer.[7] This milk contained the microorganism "*Bulgarican bacillus*" which was later renamed *Lactobacillus bulgaricus.* Metchnikoff reasoned that these bacteria eliminated putrefactive bacteria from the GIT.[8] The works of Metchnikoff are regarded as the birth of probiotics.[9] Also around the beginning of the twentieth century, Tissier, in parallel with Metchnikoff, proposed that bifidobacteria might be effective in preventing infections in infants, since they were the predominant component of the intestinal microflora of breast-fed infants.[10]

In Japan, bifidobacteria research began in the 1950s. In 1971, the Morinaga Milk Industry Company developed the first bifidus product, which was a fermented milk containing *Bifidobacterium longum* and *Streptococcus thermophilus.*[10,11] Throughout the 1970s, technology was developed that was capable of delivering products containing viable bifidobacteria on a commercial basis, and probiotic consumption changed from being purely for therapeutic benefit to reasons related more to general health improvement.[11] The Morinaga Milk Industry Company launched a bifidus milk in Japan in 1977 and a bifidus yogurt in 1979, while Yakult launched a fluid yogurt called MilMil™ in Japan in 1978 containing *Bifidobacterium breve, Bifidobacterium bifidum,* and *Lactobacillus acidophilus.*[10] Today, probiotic products are available in a variety of forms, including conventional dairy foods (such as yogurt, fluid milk, and cottage cheese), fermented milks, food supplements, and dietary supplements (capsules, tablets),[12] and the range of products continues to expand. In parallel, the market for such foods continues to develop, with most activity in developed countries, in particular in Europe, Japan, and the United States.[4]

2.3 DEFINITION OF PROBIOTICS

The term "probiotic," which comes from the Greek meaning "for life," was first used to describe substances produced by one microorganism that stimulate the growth of another microorganism.[13,14] Fuller defined a probiotic as "a live microbial feed supplement which beneficially affects the host animal by improving its intestinal microbial balance."[15] Fuller's definition has since been broadened to state that "a probiotic is a mono- or mixed culture of live microorganisms which, when applied to animal or man, affect the host beneficially by improving the properties of the indigenous microflora."[16] This definition stresses the importance of live microorganisms that improve the health status of either man or animal and that occur in the mouth, GIT, or upper respiratory or urogenital tracts.[16] Salminen et al.[17] proposed that probiotics be defined as microbial cell preparations or components of microbial cells that have a beneficial effect on the health and well-being of the host. This definition emphasizes that probiotics can be either nonviable cells or parts of cells, as probiotics in these forms, as well as certain fermentation end products and enzymes, have been shown to have health benefits.[17] In 2001, a joint Food and Agriculture Organization of the United Nations/World Health Organization (FAO/WHO) expert consultation on health and nutritional properties of powder milk with live lactic acid bacteria redefined probiotics as "live microorganisms which, when administered in adequate amounts (as part of food), confer a health benefit on the host," again highlighting the importance of viability in this most recent definition. This group recognized that probiotics should be capable of exerting health benefits on the host through growth and/or activity in the body.[3]

2.4 INTESTINAL MICROFLORA AND ECOLOGY OF THE GUT

The microflora of the human GIT is comprised of an extremely diverse microbial community encompassing both facultative anaerobic and obligate anaerobic microorganisms[14] and is one of the preferred sources of potential probiotic microorganisms destined for human use. While there are estimated to be more than 400 different species of bacteria in the gut, only 30 to 40 species represent 99% of the microflora found in any one human subject.[18] Throughout the GIT, the numbers and types of bacteria present exhibit significant variation right through from stomach to colon.[8] Initially, bacterial cells arrive in the stomach washed with saliva from the oral cavity, and only the most acid-resistant microorganisms survive.[14,19] Fewer than 10^3 cells per gram of mainly *Streptococcus, Staphylococcus,* and *Lactobacillus* species are found in the stomach.[19] Gram-negative and anaerobic bacteria predominate in the distal ileum, reaching levels of 10^5 and 10^7 cells/ml of contents.[20] In the colon, bacterial numbers reach levels of 10^{11} to 10^{12} cells/g of gut contents, with approximately 10^{14} bacterial cells present in the human colon, which is ten times greater than the total number of cells in the human body.[19]

Obligate anaerobic species of bacteria predominate in the large intestine and include both Gram-positive and Gram-negative bacteria.[21] *Bacteroides fragilis* is the most numerically predominant culturable bacterium found in the colon. The other predominant intestinal bacteria include bifidobacteria, clostridia, peptococci, streptococci, eubacteria, lactobacilli, peptostreptococci, ruminicocci, enterococci, coliforms, methanogens, dissimilatory sulfate-reducing bacteria, and acetogens.[8]

The influence of intestinal bacteria on human health can be considered harmful, beneficial, or neutral. *Bifidobacterium* and *Lactobacillus* species are beneficial microorganisms and can contribute to digestion, immune stimulation, and inhibition of pathogens. *Bacteroides, Escherichia, Clostridium,* and *Proteus* species are examples of potentially harmful bacteria found in the GIT. These bacteria are capable of producing harmful substances, including amines, indole, hydrogen sulfide, and phenols, from food components.[10] Harmful bacteria in the intestine have been linked to a number of clinical disorders such as cancer, inflammatory disease, and ulcerative colitis and also to an increase in the host's susceptibility to infection by enteropathogens such as *Salmonella, Campylobacter, Escherichia coli,* and *Listeria.*[8] For example, *Clostridium difficile* has been implicated as the primary cause of pseudomembraneous colitis. Some gut microflora undoubtedly protect the host animal from disease;[9] for example, germ-free guinea pigs can be killed by only 10 cells of *Salmonella enteritidis,* while it requires up to 10^9 cells to kill an animal with a normal gut microflora.[22] In addition, it is important to stress that the correct "balance" of bacteria must be maintained in order to allow the intestine to operate optimally.[8]

Lactobacillus and *Bifidobacterium* species are the most commonly used probiotics in foods for human consumption; for this reason, this chapter will deal only with members of these two genera. In contrast, the genera *Lactobacillus, Bifidobacterium, Bacillus, Streptococcus, Pediococcus, Enterococcus, Saccharomyces, Aspergillus,* and *Torulopsis* have all been tested as animal probiotics.[23] Lactobacilli are typically used as human probiotics because they are easy to cultivate in bulk and have a long history of safe use in fermented foods, even though they are not predominant members of the human gastrointestinal microflora.[24] Two well-documented probiotic strains, namely, *Lactobacillus acidophilus* LC1 and *Lactobacillus* GG, which have been shown to have natural immunity-enhancing properties, have heightened the awareness of lactobacilli as potential probiotic cultures.[25] Bifidobacteria have also been used in probiotic products because evidence has indicated that the human GIT is colonized by large numbers of this microorganism, and numerous studies have indicated that ingestion of bifidobacteria can enhance health.[23,25]

2.5 DESIRABLE PROBIOTIC CHARACTERISTICS

Ideally, a microorganism should meet a number of predefined criteria in order to be considered as probiotic (see Table 2.1).[26] Strains of human origin are most suitable because some health-promoting benefits may be species specific, and microorganisms may perform optimally in the species from which they were isolated. However, it is the specificity of the action, not the source of the microorganism, that is recognized as being most important when selecting probiotic strains for particular applications.[3] Probiotic microorganisms must be proven to be safe and efficacious in humans, following specific protocols for their isolation. Adherent probiotic strains are desirable because they have a greater chance of becoming established in the GIT, thus enhancing their probiotic effect.[27] All probiotic strains should have generally recognized as safe (GRAS) status, be nonpathogenic, and cause no adverse health effects to the recipient.[3,26]

Probiotic microorganisms should also be technologically suitable for incorporation into food products, such that they retain both viability and efficacy in that food product (on a commercial scale) through to and following consumption. Pro-

TABLE 2.1
Desirable Characteristics of Probiotic Microorganisms

1. Human origin
2. Generally recognized as safe (GRAS) status
3. Possession of antibiogram profiles (sensitivity of strain to antibiotics)
4. Production of antagonistic antibacterial profiles
5. Desirable metabolic activity
6. Technological suitability
7. Nonpathogenic
8. Noninflammatory
9. Survival in association with the adult mucosal immune system
10. Immunostimulatory for the mucosal immune system with appropriate cytokine stimulation
11. Antimutagenic and anticarcinogenic properties
12. Potential vehicle for the delivery of recombinant proteins and peptides in a site-specific fashion to the human gastrointestinal tract (GIT)

Source: From Collins, J.K., Thornton, G., and O'Sullivan, G.O., *Int. Dairy J.,* 8, 487–490, 1998. With permission.

biotics should be capable of surviving industrial applications (e.g., common dairy processing or pharmaceutical manufacturing protocols) and be able to grow/survive at high levels in the product to the end of shelf life.[28,29] Above all, probiotic food products must demonstrate efficacy in controlled validated clinical trials to prove that the probiotic characteristics were not altered or lost following subjection to the technological processes involved in probiotic food manufacture.

2.6 ISOLATION AND ENUMERATION OF PROBIOTIC BACTERIA

The isolation and enumeration of probiotic strains from the mammalian intestine or the carrier food product can pose a significant challenge in itself. These microorganisms are very fastidious and can be oxygen-sensitive and difficult to grow in laboratory media. Since the efficient isolation and enumeration of probiotics is a critical step in the development of functional foods, advances in the formulation of media for bifidobacteria and lactobacilli will first be considered.

A means of distinguishing bifidobacteria from other microflora is a prerequisite to their enumeration from food products. Extensive literature exists on culture media suitable for the isolation, detection, and enumeration of bifidobacteria.[30,31] However, to date no medium has been developed that is fully selective for bifidobacteria while simultaneously having no negative effect on their growth.[25] *Bergey's Mannual of Determinative Bacteriology* reports that no one selective medium is suitable for all species of bifidobacteria.[32] Isolation or enumeration of bifidobacteria is complicated by the fact that the species of interest often occurs in environments containing mixed populations of bacteria, such as the GIT or fermented dairy products.

Bifidobacteria media often contain special growth factors, substances that lower the redox potential (cysteine, cysteine hydrochloride, cystine, ascorbic acid, sodium sulfite, liver extract), and antimicrobial substances that inhibit the growth of other bacteria (e.g., lactobacilli and other lactic acid bacteria [LAB], propionibacteria, and *Actinomyces* sp.).[30,33] Tryptone phytone yeast extract (TPY) medium is traditionally recommended for the cultivation of bifidobacteria, as it is an elective medium for all known bifidobacteria species.[32] Several selective media have been developed for the enumeration of bifidobacteria.[34–42] Enumerating bifidobacteria alone necessitates the use of a rich medium such as glucose–blood–liver (BL) agar, de Man, Rogosa, and Sharpe (MRS) agar, or reinforced clostridial medium (RCM).[30] For example, a selective medium for the enumeration of bifidobacteria from fermented dairy products was described that also inhibits strains of lactobacilli and streptococci.[43] The basis of this selective medium is BL agar, supplemented with oxgall (0.2 mg/ml) and gentamycin (30 mg/ml), with recovery of bifidobacteria on this selective medium being approximately 90% of that obtained on BL agar. An inexpensive whey-based medium supplemented with *N*-acetylglucosamine and yeast extract, in the presence of sodium thioglycolate, has been developed for *B. bifidum* ATCC 11863, promoting good growth of this strain at 37°C.[44]

In products containing bifidobacteria and other LAB, neomycin–nalidixic acid–lithium chloride–paromomycin agar (NNLP agar) may be the medium of choice, while modified Rogosa selective agar may also be considered suitable.[30] The NNLP medium contains the antibiotics neomycin sulfate, paromomycin sulfate, and nalidixic acid and the reducing agent lithium chloride.[34] Bifidobacteria grow well in this medium, which offers good discrimination against dairy-derived microorganisms,[41] and this medium has been used recently in the enumeration of bifidobacteria from probiotic Cheddar cheese.[45]

The media used for isolating and enumerating lactobacilli depend on the type of sample, the specificity required, and to some extent, the characteristics of the particular *Lactobacillus* culture. For most lactobacilli, various requirements for essential nutrients are met when the medium contains fermentable carbohydrates, peptone, meat extract, and yeast extract. Supplementation with tomato juice, manganese, acetate, and oleic acid esters, especially Tween 80, is stimulatory or essential for most species. A widely used selection medium that includes these compounds is MRS, as mentioned previously.[46] Lactobacilli adapted to particular substrates may require special growth factors, for example, MRS supplemented with maltose, raffinose, or melibiose in place of dextrose is used for the enumeration of *Lb. acidophilus* in yogurt.[47] The isolation of *Lb. acidophilus* from yogurt is typical of the type of mixed culture population that exists in fermented foods from which lactobacilli are isolated. Starter yogurt bacteria generally cannot utilize and grow on a wide range of carbohydrates, unlike *Lb. acidophilus.* A simple differential medium that can be used in this case is one containing a single sugar suitable for growth of *Lb. acidophilus* but not the other yogurt bacteria.[48] This approach has also been used to enumerate lactobacilli in Swiss[49] and mozzarella[50] cheese varieties. Other media that have been described for the enumeration of *Lb. acidophilus* include lactobacillus selection agar (LBS),[51] cellobiose esculin agar,[52] bile medium,[53] and an agar medium based on X-glu.[54]

2.7 CULTIVATION OF PROBIOTICS FOR INCORPORATION INTO FUNCTIONAL FOODS

The use of probiotics in fermentation has numerous advantages. The fermentation acts to retain and optimize microbial viability and productivity, while simultaneously preserving the probiotic properties.[55] Lactic acid fermentation in food is routinely associated with milk and dairy products, particularly when considering species of probiotic lactobacilli such as *Lb. paracasei, Lb. acidophilus,* and *Lb. rhamnosus.* Although the most important characteristics of probiotic bacteria are their positive effects on host health, evaluation of technological traits such as growth and survival in milk-based media and during product manufacture and shelf life can be important considerations for selection of strains for food applications. Since the ability to culture the probiotic of interest to high cell density in a suitable medium for food applications is an essential prerequisite to its incorporation into foods, the cultivation of these strains will be dealt with in detail in this chapter. Some probiotic strains have demonstrated the ability to grow in the food product after manufacture, for example human-derived *Lb. paracasei* in Cheddar cheese,[56] thus avoiding the necessity for their incorporation at high cell numbers during the manufacturing process.

2.7.1 Culturing of Bifidobacteria

Bifidobacteria are generally known to be nutritionally fastidious microorganisms, requiring certain amino acids and vitamins for their growth.[57] The growth of bifidobacteria in culture media is often related to the presence of various growth factors.[58] Bifidobacteria are difficult to propagate as they are not acid tolerant and cannot grow in a medium with a high oxidative potential.[59,60] Reducing agents such as cysteine, cysteine hydrochloride, and ascorbic acid are often added to growth media for bifidobacteria.[59] Synthetic media such as MRS and TPY broth are too complex and expensive for generating large quantities of bifidobacteria cultures for commercial applications.[61] The addition of bifidobacteria, grown in synthetic media, to dairy products such as yogurt and ice cream can contribute to off flavor unless they are extensively washed and may be in breach of regulations for adding bacterial cultures to dairy products.[61,62] Thus, the potentially most useful media for the delivery of probiotic bifidobacteria to the human GIT are milk and yogurt.[63] A milk-based medium is the most suitable for producing an inoculum for a quality dairy product while maintaining the texture of the product.[61] However, bifidobacteria are often difficult to propagate in bovine milk because of its deficiency in necessary growth factors.[57,61,62,64] Bifidobacteria cells have been shown to change their morphology from typical branched, bifurcated Y-form cells in MRS to a variable morphology in milk, which has been attributed to a lack of certain nutrients or anaerobic conditions.[61] While milk contains many of the essential nutrients necessary for the growth of bifidobacteria, these may not be at optimal concentrations and are often in a form inaccessible to the microorganism.[57] For example, there is a lack of available nitrogen in milk for bifidobacteria because of the low concentrations of free amino acids and peptides, despite its relatively high content of (inaccessible to the microorganism) casein.[57,58]

A variety of approaches have been used to improve the growth of bifidobacteria in milk-based media, including the addition of growth factors.[58] These growth factors, referred to as "bifidus factors," are thought to be present in the intestines of breast-fed infants and are considered to be responsible for the predominance of bifidobacteria in the intestine of the young infant.[58] A variety of complex oligosaccharides (termed prebiotics), many of which are found naturally in human milk, are also thought to be responsible for promoting the growth of bifidobacteria.[65] Studies have indicated that consumption of these complex carbohydrates can result in significant increases in bifidobacteria numbers in the human gut microflora and in feces.[66–68] Oligosaccharides, such as fructooligosaccharides and galactooligosaccharides, are used as bifidogenic oligosaccharides in the food and infant food industries. The term synbiotic is used to describe the combination of a probiotic and a prebiotic and is defined as "a mixture of a probiotic and a prebiotic that beneficially affects the host by improving the survival and the implantation of live microbial dietary supplements in the GIT, by selectively stimulating the growth and/or by activating the metabolism of one or a number of health-promoting bacteria."[69,70] Foods such as chicory, garlic, onion, Jerusalem artichoke, and asparagus all contain fructooligosaccharides, and these have been associated with improvements in the human gut including increased lactobacilli, bifidobacteria, and short-chain fatty acid levels as well as decreased clostridia, fusobacteria, bacteroides, and pH levels.[71] Prebiotics have also been suggested to have a beneficial effect on reducing blood lipids, although clinical evidence so far has been limited.[72]

In addition to oligosaccharides and carbohydrates, other biologically complex materials such as bovine casein digest, yeast extract, amino sugars, and peptides have been examined in efforts to improve the growth of bifidobacteria.[58,63,64] Poch and Bezkorovainy demonstrated that yeast extract and bovine casein digest promoted the growth of bifidobacteria in TPY medium.[63] Klaver et al. reported that 15 out of 17 *Bifidobacterium* strains tested grew poorly in milk; this was attributed to lack of proteolytic activity in these strains; addition of casitone (casein hydrolysate) or a mixture of amino acids to the milk-based medium resulted in good growth of most of the strains.[58] Another approach has involved the co-culture of proteolytic species such as lactobacilli with bifidobacteria, which has resulted in the enhanced availability of sufficient nitrogenous compounds leading to improved growth of bifidobacteria in milk.[58] For example, *B. lactis* demonstrated enhanced growth when co-cultured with *Lb. acidophilus* in milk.[57] Bifidobacteria species have also shown variable growth in media, as studies by Desjardins et al. have indicated that bifidobacteria of infant origin grow much better in milk than those of adult origin,[64] while Poch and Bezkorovainy indicated that bifidobacteria species of infant origin grow better than those of adult origin in the presence of growth factors.[63]

2.7.2 Culturing of Lactobacilli

Lactobacilli are extremely fastidious, adapted to complex organic substrates. For energy, they require carbohydrate and carbon sources; they also require amino acids, nucleotides, and vitamins.[73] The composition of milk satisfies some of the growth requirements of lactobacilli, containing more than 87% water, ~4.7% lactose, ~3.8%

fat, ~3.3% protein, ~0.2% citrate, and ~0.6% minerals.[74] Typically, lactobacilli can be cultivated quite successfully in milk and will reach maximum numbers after 24 h incubation at 37°C.[75,76] Lactobacilli grown in milk can usually reach levels of up to 10^8 or 10^9 colony forming units (CFU)/ml, by which time they will also have entered stationary phase.[75]

During the fermentation of milk with lactobacilli, the pH is typically in the range of 3.9 to 4.4. Thus in some cases, it is the acidity of the final fermentate that can prove inhibitory to lactobacilli when cultured in milk-based media; however, this does not affect the survival of aciduric or acid-tolerant species of lactobacilli.[73] At this point, fermentation is usually stopped by cooling and/or neutralization, which will prevent acid injury[75] during subsequent processing or storage.

In cases where the probiotic does not grow well in milk, the level of inoculum can be increased to overcome the shortfall in numbers, or the milk can be fortified with various additives to promote growth of the culture. Additives such as tomato juice,[77,78] casein peptone,[77] whey protein,[79] sucrose,[80] papaya pulp,[81] manganese and magnesium ions,[82] simple fermentable sugars,[83] and a combination of casitone and fructose[84] have all been used to promote the growth of lactobacilli in milk. In a study by Saxena et al., supplementation of milk with a combination of casitone and fructose enhanced viable numbers of *Lb. acidophilus* by 1.5- to 2.0-fold during 21 days of storage. These supplements effectively reduced the generation time for all *Lb. acidophilus* strains tested and enhanced acid production and sugar utilization, when compared with the growth and metabolism of the control culture.[84] In a similar study by Rana et al., maximum growth of *Lb. acidophilus* was obtained in skim milk supplemented with 0.5% yeast extract and 1.0% glucose, among various whey and skim milk preparations. Continuous neutralization of the medium during fermentation to the initial pH (6.5) at periodic intervals (8 h) resulted in a further increase in cell numbers.[85]

Not all strains of lactobacilli perform equivalently in milk during growth and storage, indicating that performance of strains should be evaluated individually prior to commercial use. For example, in a study by Sanders et al., the performance of six commercially available lactobacilli in fluid milk was compared during storage at 4°C and 10°C for 21 days and during frozen storage at –20°C for 6 weeks. The cultures tested remained stable in fluid milk, with less than tenfold decline in numbers; however, during frozen storage, performance of the cultures varied, with no significant change in viability for four of the six strains, whereas the numbers of the two of the strains were reduced by more than tenfold following 6 weeks of frozen storage.[86]

If probiotic lactobacilli are to be added as adjunct cultures to fermented dairy products such as yogurt and cheese, it must be taken into consideration that living bacteria will interact with their environment. The chemical makeup of the dairy product is therefore an essential element when considering the metabolic activities of the probiotic.[74] This means that essential variables for the propagation of live microorganisms in milk and milk products are the type and quantities of available carbohydrates and the degree of hydrolysis of milk proteins and lipids.[87,88] On the other hand, the proteolytic, lipolytic, and saccharolytic properties of probiotics in milk-based products would be potentially important for further degradation of proteins, lipids, and complex carbohydrates, leading to changes in the taste and/or flavor of the dairy product. From a commercial point of view, it is essential that the flavor and texture of the probiotic fermented product remain appealing to the consumer.

2.8 PHYSIOLOGICAL FACTORS AFFECTING GROWTH AND SURVIVAL OF PROBIOTICS IN FUNCTIONAL FOODS

A number of physiological traits have been identified that make the incorporation of probiotic lactobacilli and bifidobacteria into dairy foods difficult; methods are being sought to overcome such constraints. This section examines some of the most important characteristics of probiotic microorganisms, such as their acid, bile, and oxygen stress tolerance.

Essential determinants in the choice of a suitable probiotic *Lactobacillus* strain for commercial use are ability to survive transit through the small intestine and bile tolerance.[89] The terminal ileum and colon have proven to be the sites of colonization for intestinal lactobacilli; however, few data are available on the resistance of potential probiotic lactobacilli to small intestinal secretions.[89] Lactobacilli of intestinal origin appear to be more bile resistant than those of fermented food origin.[90] An estimate of the numbers of *Lb. acidophilus* cells in a fermented milk capable of surviving intestinal transit was reported by Marteau et al. to be 1.3 to 1.5% of an oral inoculum.[91] Interestingly, lactobacilli have shown strain variation in their resistance to bile salts, a trait that is considered important for selection of probiotic lactobacilli. Many lactobacilli are able to deconjugate bile acids[92,93] using the enzyme bile salt hydrolase, although the significance of this activity *in vivo* is not completely understood.

Acid tolerance is important to survive passage through the GIT and also for probiotic survival in fermented foods.[27] Lactobacilli are mainly acid tolerant or aciduric, particularly when isolated from the harsh environment of the GIT where the gastric pH frequently falls below 2.0.[94] Even at pH $\geq$ 2.0, *Lb. acidophilus* is able to maintain cytoplasmic pH at values near neutrality.[95] Upregulation of genes involved in stress protection, such as F1F0-ATPase (an important element in the response and tolerance to low pH in *Lb. acidophilus*)[96] can produce dramatic changes in culture performance. For example, acid adaptation of *Lb. acidophilus* has been successfully used to enhance survival of the culture in normally lethal acid conditions and in yogurt.[97] Acid-tolerant variants may be selected through long-term subculturing, particularly if the culture medium is acidified during fermentation, which would make selection of less acid-tolerant variants unlikely.[55]

Because the intestinal tract is considered to be the natural environment of many probiotic bacteria, the oxygen content and redox potential of their growth medium must be considered.[74] Oxygen can easily dissolve in milk, thus viability in fermented dairy foods is influenced by oxygen content in the product in addition to oxygen permeation through the package. Dave and Shah[98] showed that survival of *Lb. acidophilus* in yogurt was directly affected by the dissolved oxygen content, which was found to be higher in yogurts made in plastic containers than in glass. Thus, it may be important to store the products in glass containers or to increase the thickness of the packaging materials.[97] Interestingly, two antioxidative strains of lactobacilli, tentatively identified as *Lb. fermentum,* were isolated from intestinal microflora of a healthy child.[99] Survival time of these strains in the presence of reactive oxygen species was significantly increased compared with a nonoxidative strain. Such resistance to oxidative stress may enhance the survival of these potential probiotic lactobacilli in both the intestinal microbial ecosystem and under exogenous oxidative stress conditions. Also, understanding the molecular mechanisms governing the resistance of these lactobacilli to oxidative stress could potentially lead to the development of other aerotolerant probiotic lactobacilli.

Similar to lactobacilli, bifidobacteria also show considerable strain variation in their resistance to acid and bile stress.[100] Selection of probiotic bifidobacteria is sometimes limited by their intrinsic inability to survive harsh conditions in the gut and the acid conditions of fermented foods. For this reason, strains such as *B. longum, B. pseudolongum* and *B. animalis* are frequently used in fermented foods because of their aciduric nature and resistance to bile salts.[100–102] In fact, in a study by Clark and Martin, it was reported that *B. longum* could survive bile concentrations as high as 4.0%.[103]

As with lactobacilli, environmental adaptation of bifidobacteria may prove to be an important survival mechanism during exposure to a variety of stresses. The ability of *Bifidobacterium* spp. to adapt to acid stress and the general resistance of acid-adapted cells to other environmental stresses (including bile salts, H_2O_2, and cold storage) was investigated by Park et al.[104] Acid adaptation of *B. breve* ATCC 15700 (pH 5.2 for 2 h) was found to enhance survival of the culture 100-fold at pH 2.0 for 60 min, compared with the survival of an unadapted control culture. Furthermore, acid-adapted cells were also better able to survive exposure to normally lethal conditions of H_2O_2 (1000 ppm for 60 min) and cold storage (4°C for 7 days). Similarly, bile salt adaptation of *B. adolescentis* NCC251 generated homologous protection against normally lethal concentrations of bile salts, in addition to stimulating cross-protection against freeze–thawing cycles and lethal heat stress.[105] A possible relationship between cell surface hydrophobicity (CSH) and this tolerance to environmental stresses in *Bifidobacterium* spp. has been suggested.[106] CSH was determined using BATH (bacterial adherence to hydrocarbons assay) for seven bifidobacteria strains, and it was found that strains with the highest CSH demonstrated significantly more resistance to stresses such as bile salt, H_2O_2, heat, cold storage, and acid.

Because bifidobacteria are anaerobic, oxygen toxicity is also an important consideration. Ahn et al. examined the effect of oxygen stress on *B. longum* and found that in the presence of oxygen, the lag phase became extended and cell growth was limited.[107] Morphology during oxygen stress was also altered, with the *Bifidobacterium* cells becoming longer in size; the formation of nodes on the surface of the cells due to incomplete cell division was also observed. Cellular fatty acid profiles changed, such that the carbon chain was shortened and dimethyl acetals originating from plasmalogen were reduced, while cyclopropane fatty acids were increased. This group also identified a 35.5 kDa protein, Osp, which was upregulated in an oxygen-tolerant *Bifidobacterium* strain, and it was considered that the protein may have a role in defense against oxygen stress.

2.9 CHALLENGES ASSOCIATED WITH THE DEVELOPMENT OF DRIED PROBIOTIC CULTURES

Food products or supplements containing viable probiotic strains are frequently supplied on the market in a lyophilized form.[108] Probiotic cultures are typically supplied in the dried form for traditional food uses and as starter cultures or are applied as such in different food products. Dried preparations of live probiotic bacteria have the advantages of long-term preservation and convenience in handling, storage, marketing, and consumption.[75] Of the numerous drying technologies available, convective drying and freeze-drying are the methods most frequently used.[109] Most probiotic lactobacilli do not survive well, however, during the temperature and osmotic extremes that they are exposed to during the drying process.[110–114] This section focuses on the freeze-drying and spray drying of probiotic bacteria, and on the modifications used to enhance viability of probiotic cultures during lyophilization.

2.9.1 Freeze-Drying

Freeze-drying is a process involving the removal of water from a product by sublimation and desorbtion. It consists of three phases; freezing (where the mobile water of the product is frozen), primary drying (where heat is applied to the product to cause the frozen mobile water to sublime), and secondary drying (where the temperature is increased to desorb bound water such as water of crystallization until the residual water content falls to the range required for optimum product stability). Unfortunately, when this process is applied to the preservation of bacterial cultures such as lactobacilli[115] and bifidobacteria,[116] much of their activity is typically lost after a few weeks of storage at room temperature. Loss of viability during freeze-drying is associated with stress that is induced by temperature changes, phase changes, and drying, the combination of which tend to damage cell membranes and proteins. To overcome inactivation during drying and poor stability during storage, cryoprotectants such as glycerol and cysteine[117] or sucrose[118] are added during freeze-drying of lactobacilli. Other disadvantages associated with this process include the time it takes, the expense, and the high transport and storage costs associated with frozen concentrates.[113] Alternatively, convective drying (i.e., spray drying) offers some of the advantages of freeze-drying but is much more economical, especially when applied on a commercial scale.

2.9.2 Spray Drying

Spray drying involves the transformation of feed from fluid into a dried particulate form by spraying the feed into a hot drying medium. It is one of the predominant processing tools used in the dairy industry, as it allows economical production of large amounts of dairy ingredients; it has been estimated that the cost of spray drying is six times lower per kilogram of water removed than that of freeze-drying.[5] The process of spray drying also has the advantage that it is easily scaled up and uses equipment readily available in the food industry.[119] During spray drying, the feed solution is atomized and is introduced in the drying chamber along with hot air. This mixture of hot air and atomized feed moves towards the air exhaust of the drying chamber; the time taken for this to happen is called the residence time. During this residence time, the feed droplets lose moisture to hot air, and the resultant dry powder falls onto a conical portion of the drying chamber, slides down through a rotary valve located at the bottom of the chamber, and is collected in a collection bag/bottle.

The successful spray drying of lactobacilli and bifidobacteria has previously been reported for a number of different strains: *Lb. paracasei*,[113,120] *Lb. curvatus* and *Lb.* sp. 8Z,[121] *Lb. acidophilus*,[75] *Lb. bulgaricus*,[110,122] *Lb. helveticus*,[115] and *B. ruminantium*.[123] Of the lactobacilli, the probiotic *Lb. paracasei* was successfully spray dried, with up to 49% survival obtained following drying at an outlet temperature of 80 to 85°C.[113] Espina and Packard[124] manufactured a spray dried powder containing up to 10^9 CFU/g of *Lb. acidophilus,* which was subsequently used as an inoculum for milk solids non-fat (MSNF), while Prajapati et al.[75,125] successfully produced a blended spray dried probiotic *Lb. acidophilus* preparation that contained up to 10^7 CFU/g *Lb. acidophilus* LB1HH3.

The main reason why spray drying of live cultures has not been developed commercially for many strains is that the organisms do not survive the treatment well. Problems such as low survival rates during drying, low stability under storage, and difficulties in rehydrating the product are common.[121] During spray drying of live cells, low survival rates may arise due to dehydration and high temperature leading to injury or death of bacterial cells.[126–129] Previous reports have shown that the destruction of bacteria during heat stress and spray drying cannot only be ascribed to a thermal effect, but also to a nonthermal drying effect caused by the loss of bound water at the cell surface.[130] Teixeira et al. and Daemen and van der Stege found that probiotic lactobacilli showed increased sensitivity to lysozyme and NaCl, indicators of cell wall and cell membrane damage following spray drying.[110,130] Other possible sites in the cell where damage may occur as a result of heat stress and spray drying include DNA,[131] where the nature of the damage remains unknown, and ribosomes,[132] which are damaged possibly due to the loss of the stabilizing effect of Mg^{2+}, which can escape from the heat-compromised cell membrane.

To overcome inactivation during drying and poor stability during storage, a number of approaches have been examined to date. Prajapati et al. successfully used a concentrate of banana, tomato juice, and sugar as a protective agent for the probiotic *Lb. acidophilus* during spray drying,[125] while Johnson and Etzel showed that the addition of dextrin and pH adjustment of a lactic acid bacteria culture prior to spray drying did not offer protection during spray drying.[115] Control of the resistance of lactobacilli to temperature stress may have potential practical benefits in industrial fermentation processes in which bacteria with enhanced thermotolerance are required. Heat-inducible thermotolerance allows bacteria, after a nonlethal heat shock, to tolerate a second heat stress higher in intensity.[133] Teixeira et al.[134] and Gouesbet and Boyaval[135] reported that heat adaptation increased the thermotolerance of lactobacilli. Similarly, a heat-adapted probiotic, *Lb. paracasei* NFBC 338, exhibited greater thermotolerance (survival at a lethal temperature of 60°C) compared with controls in MRS and RSM and during spray drying (Figure 2.1a), where viability of the adapted culture was enhanced 18-fold.[120]

Bacterial survival of a specific stress can also be improved by pretreatment of cells to a sublethal heterologous condition, such as moderate temperature, low pH,

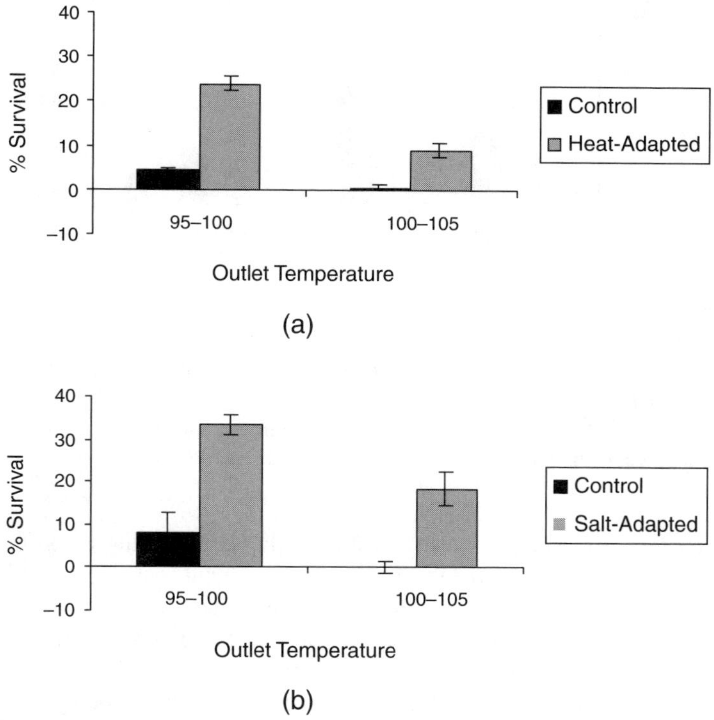

FIGURE 2.1 (a) Percent survival (CFU/g) of control and heat-adapted (52°C × 15 min) *Lb. paracasei* NFBC 338 during spray drying at outlet temperatures of 95 to 100 and 100 to 105°C. (b) Percent survival (CFU/g) of control and salt-adapted (0.3 *M* NaCl × 30 min) *Lb. paracasei* NFBC 338 during spray drying at outlet temperatures of 95 to 100 and 100 to 105°C. (From Desmond, C., et al., *Int. Dairy J.,* 11, 801, 2001. With permission.)

or moderate osmolarity. The positive effect of suboptimal growth temperature (25°C) on resistance of *Lb. acidophilus* to environmental stresses such as freezing, heating, osmotic stress, and exposure to ethanol, peroxide, or acid was reported by Lorca and de Valdez.[136] The cloning of cold shock genes in *Lb. plantarum* has also been reported, with the objective of protecting the culture against low-temperature stress.[137] In addition, Gouesbet and Boyaval[135] and Desmond et al.[120] demonstrated the acquisition of a cross-stress tolerance (to heat) in lactobacilli by exposure to mild osmotic stress. The effect of salt adaptation (0.3 *M* NaCl for 30 min) on the heat resistance of *Lb. paracasei* NFBC 338 during spray drying (Figure 2.1b) was also investigated, and it was found that there was a 16-fold higher viability of the salt-adapted culture compared to the viability of an untreated control culture dried under the same conditions.[120]

Encapsulation, as a protection mechanism, allows the active core ingredient, or substrate, to be separated from its environment by a protective film or coating. This separation occurs until the release of the functional ingredient is desired. In the case of probiotics, this would be in the jejunum and ileum in the human body.[138] For the incorporation of probiotics into food products, microencapsulation offers protection

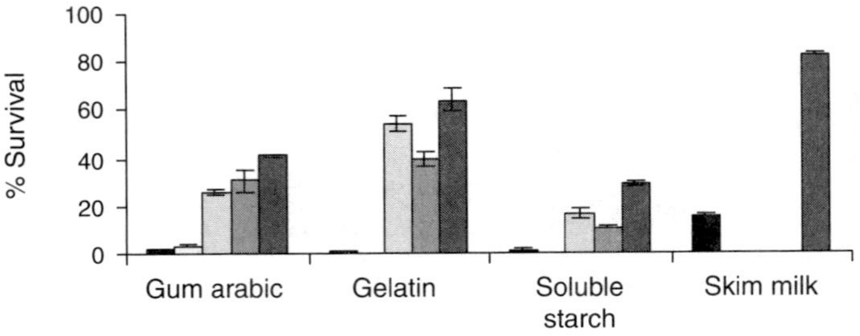

FIGURE 2.2 Survival of bifidobacteria after spray drying with various carriers, where the outlet temperature during drying was 50°C. The five strains tested were *B. infantis* CCRC 14633 (■), *B. infantis* CCRC 14661 (□), *B. longum* ATCC 15708 (◫), *B. longum* CCRC 14634 (⊟), and *B. longum* B6 (▨). The survival of *B. infantis* CCRC 14661, *B. longum* ATCC 15708, and *B. longum* CCRC 14634 was not determined using skim milk as the carrier material. (Adapted from Wen-Chian, L., Hung-Chi, H., and Cheng-Chun, C, *Int. J. Food Microbiol.,* 74, 79–86, 2002. With permission.)

to fine particles such as those produced during the spray drying of probiotic concentrates. In a study by Selmer-Olsen et al., it was found that encapsulating lactobacilli in calcium alginate beads improved their heat tolerance.[112] Subsequently, this technology was used to prolong the viability during storage of a spray dried *Bifidobacterium* strain, *B. ruminatium.*[123] In this study the spray-coating process for the production of starch-encapsulated bifidobacteria was optimized; however, the strain was not protected against adverse environmental conditions such as acid shock or storage in two dry food preparations. In a recent study by Lian et al.,[139] four strains of bifidobacteria were successfully spray dried and encapsulated in gum acacia, gelatin, and soluble starch. Survival of these probiotic bacteria varied with strains and was highly dependent on the carriers used (Figure 2.2). Thus, for the efficient microencapsulation of probiotic strains, there is a need to examine the application of other readily available encapsulating materials. An appreciation of these systems of preserving probiotic viability either alone or in combination should allow probiotics to be incorporated into food systems more readily, and most importantly, provide a greater assurance to manufacturers and consumers regarding the viability and quantity of probiotics in the final products.

2.10 PROBIOTIC PRODUCT DEVELOPMENT

As stated previously, probiotic strains destined for incorporation into human foods should have demonstrable benefits to human health and be GRAS. Furthermore, from a food processing perspective, it is desirable that such strains be suitable for large-scale industrial production by being capable of withstanding the processing conditions described above, such as freeze-drying or spray drying, as well as being capable of survival in food products throughout shelf life.

Bifidobacteria are extremely fastidious and are generally sensitive to oxygen and therefore pose particular challenges for incorporation into food products. Process modifications that limit oxidative stress and the use of growth factors to ensure cultivation to high numbers for inoculation purposes are therefore desirable when working with these organisms. From a cultivation point of view, a *Bifidobacterium* strain should ideally be acid and oxygen tolerant and should rapidly grow and acidify milk so as to reduce the incubation time necessary, thus limiting the possibility of contamination.[57,64]

Incorporation of lactobacilli into the food chain can also be difficult. An important technological reason for the use of dairy products as carriers of lactobacilli is that many of these products have already been optimized to some extent for survival of live fermentation microorganisms. Thus, the existing technologies can be readily adapted to allow the incorporation of probiotic lactobacilli.[74] There are, however, problems associated with the incorporation of lactobacilli into milk-based products. Some of the most constraining drawbacks include the poor temperature, salt, oxygen, and bile tolerance of some species. Solutions to some of these problems include selection of acid- and bile-resistant strains, use of oxygen impermeable containers, microencapsulation, stress adaptation, and incorporation of nutrients such as peptides and complex carbohydrates.[97]

A number of probiotic dairy food products are available commercially, and the range of such products continues to expand (Table 2.2). In the sections that follow, physiological constraints associated with probiotic cultures and the technological challenges associated with the development of some fermented dairy products incorporating these probiotic strains, in particular probiotic lactobacilli and bifidobacteria, are described.

2.10.1 Yogurt and Fermented Milk Drinks

Yogurt has long been recognized as a product with many desirable attributes for consumers, making it an obvious choice as a carrier of probiotic strains. In recent years, the popularity of "bio-yogurts," which contain *Lb. acidophilus* and species of *Bifidobacterium* (these are referred to as AB-cultures), in addition to the traditional yogurt organisms, *S. thermophilus* and *Lb. bulgaricus,*[140] has increased significantly. For the manufacture of these probiotic yogurts, it is particularly attractive that conventional yogurt processing procedures can be applied, with the probiotic bacteria added prior to fermentation, simultaneously with the traditional yogurt cultures, or after fermentation to the cooled product before packaging.[141] The methods used to manufacture stirred yogurt and drink yogurt, in particular, are well suited to the addition of probiotics after fermentation.[74]

Live cultures of probiotic lactobacilli and bifidobacteria have been reported to remain viable in yogurt during refrigerated storage at levels of $\geq 10^6$ CFU/g.[141,142] However, problems with the stability of probiotic bacteria in yogurt and fermented milk products have been reported.[143] Probiotic viability is dependent on such varied factors as the production method and the strain, the associative yogurt microorganisms, culture conditions, chemical composition of the fermentation medium, final acidity, milk solids content, availability of nutrients, the presence of growth promoters and inhibitors, concentration of sugars, dissolved oxygen (especially for *Bifidobacterium* species), and storage temperature of the fermented dairy product.[97,98,140,144] Among these, the main factors responsible for loss of viability of lactobacilli appear

TABLE 2.2
Examples of Food Carriers for Probiotic Lactobacilli and Bifidobacteria

Species	Strain	Carrier	Ref.
Lb. acidophilus	2409	Yogurt	142
Lb. acidophilus	2401	Yogurt	147
Lb. acidophilus	LAI	Yogurt	155
Lb. casei	GG	Yogurt	194
B. bifidum	BBI	Yogurt	155
B. lactis	Laftitrade mark B94	Yogurt	150
B. longum	B6 and ATCC15708	Yogurt	152
B. infantis	1912	Yogurt	147
Lb. acidophilus	Ki	Ki cheese	163
Lb. acidophilus	A1 and A2	Fresco soft cheese	155
Lb. acidophilus	La-5	Tallaga cheese	171
Lb. acidophilus	La-5	Ras cheese	196
Lb. casei	C1 and C2	Fresco soft cheese	155
Lb. helveticus	I	Cheddar cheese	197
Lb. paracasei	NFBC 338	Cheddar cheese	56
Lb. paracasei	M3	Bulgarian yellow cheese	195
B. bifidum	ATCC 15696	Cheddar cheese	162
B. bifidum	bo	Ki cheese	163
B. bifidum	B3 and B4	Fresco soft cheese	155
B. bifidum	Bb02	Canestrato Pugilese cheese	170
B. bifidum	Bb-12	Ras	180
B. lactis	Bb-12	Tallaga	171
B. longum	B1 and B2	Fresco soft cheese	155
Lb. acidophilus	La-5	Ice cream	174
Lb. rhamnosus GG	ATCC53103	Ice cream	174
B. bifidum	10LF	Ice cream	173
B. bifidum	Bb-12	Ice cream	174
Lb. plantarum	299V	Oatmeal gruel	189

to be the decrease in the pH of the medium and accumulation of organic acids as a result of growth and fermentation.[145,146] In a recent study, microencapsulation of *Lb. acidophilus* in an alginate–starch mix was found to protect the culture from acid stress, and enhanced viability of the culture in yogurt by 0.5 log over an 8-week period.[147] In yogurts containing lactobacilli added as supplements, the production of hydrogen peroxide by starter microorganisms may also adversely affect viability.[47,97] Also, the effect of refrigeration on the viability of lactobacilli in fermented milk and yogurt was examined by Nighswonger et al.,[144] who found that three out of five strains showed no significant loss of viability during storage. Similarly, in a number of commercially available probiotic yogurts, viable numbers of lactobacilli varied greatly during refrigerated storage; the majority, however, contained viable counts greater than 10^5/g even at the end of shelf life (about 2 or 3 weeks in most cases).[148]

Maintaining the viability of bifidobacteria in yogurt-type products has been a major challenge to dairy processors because of the inability of bifidobacteria to grow in milk.[58] However, synergistic growth-promoting effects between *Lb. acidophilus* and *B. bifidum* are known to occur, providing the necessary growth stimulants for bifidobacteria.[149] A number of growth-promoting substances are also known to improve the growth of bifidobacteria in yogurt. Dave and Shah[140] successfully improved the viability of bifidobacteria (to a variable extent) using supplements such as cysteine, acid hydrolysates, and tryptone. The term synbiotics is used when referring to the use of prebiotics and probiotics in combination.[69] Such prebiotics have potential in bio-yogurts to increase bifidobacteria levels, not only in the colon, but also during shelf life of the product.[150] Viability of the probiotic bacteria has been shown to be enhanced by rupturing of the yogurt bacteria, causing intracellular beta-galactosidase to be released from the yogurt cultures and reducing their viability.[142] In this study, the β-galactosidase that was released hydrolyzed the lactose in milk to galactose and glucose, which could then be utilized by probiotic lactobacilli and bifidobacteria in the yogurt.

Since *Bifidobacterium* is strictly anaerobic, oxygen toxicity presents another important and critical problem. During yogurt production, oxygen dissolves in milk in addition to permeating through packages during storage. To alleviate this problem, it has been suggested that *S. thermophilus* and *Bifidobacterium* be simultaneously inoculated during fermentation, as *S. thermophilus* acts as an oxygen scavenger in yogurt, which results in the depletion of dissolved oxygen in the product.[10] The survival of *B. longum* in milk can also be improved by the addition of baker's yeast.[151] However, in a similar study by Dave and Shah, ascorbic acid did not prove as effective when used as an oxygen scavenger in bio-yogurt.[98]

The potential use of microencapsulation of bifidobacteria to protect against high acidity and preserve viability in yogurt has also been investigated. Microencapsulation in κ-carrageenan was successfully employed to preserve the viability of bifidobacteria in set yogurt for 30 days during refrigerated storage.[152] However, in this case, sensory evaluation indicated that the product was adversely affected by encapsulation, which underlines the need to maintain or improve the sensory characteristics of a product when reformulating processing conditions.

The production of probiotic yogurts generates the need to selectively enumerate both *Bifidobacterium* spp. and probiotic lactobacilli in the initial product after manufacture and also in the product during refrigerated storage. Therefore, in order to assess probiotic viability, simple and reliable methods for routine enumeration of specific probiotic cultures are needed.[97,141] Media used for the differential enumeration of probiotic cultures must take into consideration the type of food, the species or strains to isolate and enumerate, and the nature of the competing genera.[153] Therefore, individual selective media cannot be applied in all situations and should be evaluated for the specific strain of interest in a given situation.[153]

Culture media for the enumeration of probiotic bacteria in yogurt can essentially be divided into four main groups:

1. General media that will give an overall total colony count, without distinguishing different genera or species, e.g., MRS medium[46]
2. Media for the enumeration of yogurt culture organisms, e.g., Lee's agar, RCM adjusted to pH 5.5, and M17 agar[97]
3. Media used to selectively grow each probiotic genus, e.g., bile medium and media based on X-glu for the enumeration of *Lb. acidophilus,* LC agar[154] or B-MRS agar[155] for the selective enumeration of *Lb. casei,* and NNLP agar[156] or LP-MRS agar for isolating *Bifidobacterium* sp. (In a study by Nebra et al., the recovery of oxygen-stressed bifidobacteria was increased by the addition of mixtures of reducing agents [including L-cysteine, sodium pyruvate, and sodium thioglycolate] and preincubation for 4 h at 37°C.[157])
4. Media that allow the enumeration of all four bacterial types found in yogurts with visually distinguishable colonies on the same plate, e.g., tryptose-proteose-peptone yeast extract prussian blue (TPPYPB) agar.[34]

Currently, a number of commercially available probiotic products containing bifidobacteria are on the market, both in Europe and the United States, including Bifidus fermented milk containing *B. longum,* manufactured by Arla (Sweden), and the dietary supplement Primadophilus Junior, containing *B. adolescentis* and *B. infantis,* manufactured by Nature's Way, Springville, Utah.[158,159] Yakult is a fermented milk drink that was originally developed in 1935 by a Japanese company (Honsha Co., Ltd.), and contains *Lb. casei Shirota* (LcS). (See Chapter 6 for more details about LcS.) Cultura is a cultured whole milk product from MD Foods, and is made using *Lb. acidophilus* and *B. bifidum.*[160] It is a product that has been well accepted in Sweden and Denmark, especially by younger consumers. A new probiotic product launched by Glanbia in Ireland, "Everybody," is the first product to contain a combination of *Lb.* GG and 15 vitamins and minerals.[161]

2.10.2 Probiotic Cheese

Many cheese varieties require ripening for periods greater than 6 months and consequently, the manufacture of probiotic cheese necessitates either that the strains grow to high numbers during ripening or that they are incorporated during manufacture at high levels and survive the cheese ripening period. A number of cheese varieties have been investigated as carriers of probiotic microorganisms, including Cheddar,[45,56,162–164] white brined,[165] goats',[166] Crescenza,[167] cottage,[168] Kariesh,[169] Canestrato Pugliese,[170] fresco,[155] Tallaga,[171] and fresh cheeses.[172] (See Chapter 8 for more details about probiotic cheese.)

2.10.3 Frozen Dairy Products

Bifidobacteria have been incorporated into a wide variety of products, including sour cream, ice cream, buttermilk, yogurt, powdered milks, cheese, frozen desserts, mayonnaise, and pharmaceutical preparations. A Biogarde® ice cream was marketed in Germany in the mid-1980s, which was reported to contain 10^8 CFU/g of *Lb. acido-*

philus, 10^7 CFU/g of *B. bifidum,* and also *S. thermophilus.*[65] A *B. bifidum* strain was incorporated into ice cream initially at 2.5×10^8 CFU/ml, and cell numbers of 1×10^7 CFU/ml were maintained during 17 weeks of storage at –20°C.[173] *B. lactis* Bb-12 was also incorporated into ice cream and survived at levels greater than 10^6 CFU/g throughout 52 weeks of frozen storage without significantly affecting flavor.[174]

Dairy products such as ice cream, frozen yogurt, frozen desserts, and sour milk have also been proposed to be effective carriers for lactobacilli.[154,173,175,176] Ice cream and frozen yogurt in particular are considered suitable carriers of viable bacteria because these products can be stored longer than other dairy foods.[177] In addition, frozen storage has been shown to have little effect on the survival of lactobacilli.[178] *Lb. acidophilus* has been found to survive at between 10^6 and 10^8 CFU/g or ml in ice cream,[173,174,179] frozen yogurt,[175,178] and other frozen dairy desserts.[180] As with many stress conditions, the ability of lactobacilli to withstand frozen storage is strain dependent.[179] In an attempt to examine the effects of short-term frozen storage on lactobacilli, Holocomb et al. reported that frozen storage at –5°C for 6 h had no adverse effect on the bile salt sensitivity of the organism.[175] The ability of lactobacilli to withstand more long-term freezing has also been reported,[173,174,179,180] with probiotic viability maintained at ~10^6 CFU/g or ml during up to 52 weeks frozen storage at –20°C.

The numbers of starter and probiotic bacteria in these types of frozen fermented dairy desserts or yogurt preparations are reduced dramatically by acid, freezing injury, high temperatures, oxygen toxicity, or moisture content. Alterations to cell membrane permeability, and intracellular dehydration caused by ice crystal formation that may rupture cells, are also likely causes of microbial inactivation during freezing.[181] Frozen dairy products such as ice cream contain various natural substances with cryoprotective properties; these include casein, sucrose, and fat.[174] However, in some cases, additional protection is required to improve the survival of cells during freezing. The effect of pH[173,182] on probiotic survival and the addition of cryoprotectants such as sucrose[182] and glycerol[174] have also been examined in attempts to maintain viability during frozen storage. Microencapsulation is a process in which the cells are retained within the encapsulating membrane to reduce cell injury or cell loss and may have applications in these products. The successful encapsulation of live bacteria in gelatin or vegetable gums has been reported,[97] including lactobacilli in calcium alginate gum.[183] In a study by Ravula and Shah, the survival of sodium alginate–encapsulated *Lb. acidophilus* in fermented frozen dairy desserts was enhanced by two logs compared with the control (free) cells.[182]

2.10.4 Nondairy Products

Active cultures may be used to add function to foods that are not milk based such as mayonnaise,[184] fruit drinks,[185] cereals,[186] and meat.[187]

Mayonnaise was manufactured containing *B. bifidum* and *B. infantis* as encapsulated cells that survived to 1×10^5 CFU/g and 1×10^4 CFU/g, respectively, for 12 weeks.[184] While these bifidobacteria survived only at low levels, their addition to mayonnaise was responsible for reducing the total bacterial count and inhibiting the growth of yeasts and molds for up to 10 days; it also improved sensory properties.

Lb. plantarum is typically associated with fermented foods of plant origin. Johansson et al. used *Lb. plantarum* in a drink that was made with rose hips and oats and subsequently demonstrated positive health benefits of this probiotic drink during human feeding trials.[188] In addition, a probiotic lactic acid fermented oatmeal gruel that is mixed in a fruit drink was successfully launched in Sweden in 1994; it contained 5×10^{10} *Lb. plantarum* 299V/L.[189] This functional food, ProViva, is marketed as being the world's first nondairy probiotic drink; it contains fruit blended with oatmeal and fermented with the patented *Lb. plantarum* 299V bacterium.[190] Primavita is another probiotic fruit drink, based on cereal and fruit, that has been available in Germany since July 1997 in strawberry, bilberry, and rose-hip flavors.[191] (See Chapter 13 for more details about *Lb. plantarum.*)

The potential of producing a probiotic fermented gruel based on a cereal and legume blend combined with a fresh tomato pulp has been investigated.[192] The mixture was fermented with *Lb. acidophilus,* and feeding of this fermentate to mice with *E. coli*–induced diarrhea demonstrated clinical benefits. Feeding arrested the diarrhea, increased lactobacilli counts, and reduced counts of *E. coli* as well as moisture, ash, and protein contents in the murine feces. In another study, the use of three probiotic strains, *Lb. casei, Lb. acidophilus,* and *B. lactis,* in the production of fermented dry sausages was investigated.[187] The sausages were produced with a traditional starter culture in combination with one of the three probiotic strains, and it was found that *Lb. casei* and *B. lactis* were suitable for use in the sausages, producing acceptable color, flavor, texture, and aroma. (See Chapter 10 for more details about fermented meats.)

A number of pharmaceutical preparations containing bifidobacteria either alone or in combination with other probiotic strains are available worldwide.[193] Such products include Bifidogène®(*Bifidobacterium* sp.), Synerlac® (*B. bifidum, Lb. acidophilus,* and *Lb. delbrueckii* ssp. *bulgaricus*), and Lyobifidus® (*B. bifidum*), produced in France and Eugalan®, Euga-Lein®, and Lactopriv®, all containing *Bifidobacterium* sp., produced in Germany.[193]

2.11 CONCLUSIONS

Undoubtedly, the most important characteristics of probiotic strains relate to their clinically proven health-promoting effects in humans. In addition, however, it is very desirable that the strain chosen lend itself to food processing through efficient large-scale cultivation, concentration, incorporation, and survival in food products, for it to be successful as a candidate functional food component. This chapter has outlined some of the problems that are encountered during the development of efficacious functional foods containing viable probiotic strains, using specific examples. These objectives can be particularly challenging when it comes to large-scale production of bifidobacteria, which are strictly anaerobic microorganisms with complex nutritional requirements. In many cases, our aspirations for the development of products containing such strains will only be realized by a better understanding of the physiology of these microorganisms, which should lead to strategies whereby the growth and stability of these cultures can be improved in situations suitable for functional food development. In this respect, the advances that are currently being achieved in the genomic analysis of probiotic cultures promise to yield detailed insight into the physiological performance of health-promoting strains in a variety of environments. Such improved understanding may offer novel solutions to many of the challenges that currently hamper the commercial-scale development of functional foods containing particular probiotic strains.

ACKNOWLEDGMENTS

This work was funded by the Irish government under the National Development Plan 2000–2006, by the European Research and Development Fund, and by EU Project QLK1-CT-2000–30042. C. Desmond is in receipt of a Teagasc Walsh Fellowship.

REFERENCES

1. Andersson, H., Asp, N.G., Bruce, A., Roos, S., Wadstrom, T., and Wold, A.E., Health effects of probiotics and prebiotics: a literature review on human studies, *Scand. J. Nutr.,* 45, 58–75, 2001.
2. Shanahan, F., Probiotics and inflammatory bowel disease: is there a scientific rationale? *Inflamm. Bowel Dis.,* 6, 107–115, 2000.
3. FAO/WHO, Evaluation of Health and Nutritional Properties of Powder Milk with Live Lactic Acid Bacteria, Report from FAO/WHO Expert Consultation, Cordoba, Argentina, Oct. 1–4, 2001.
4. Stanton, C., Gardiner, G., Meehan, H., Collins, K., Fitzgerald, G., Lynch, P.B., and Ross, R.P., Market potential for probiotics, *Am. J. Clin. Nutr.* 73, 476S-483S, 2001.
5. Knorr, D., Technology aspects related to microorganisms in functional foods, *Trends Food Sci. Technol.,* 9, 295–306, 1998.
6. Fox, P.F., McSweeney, P.L.H., Wallace, J., and Law, J., Cheese, an overview, in *Cheese: Chemistry, Physics and Microbiology,* Vol. 1, 2nd ed., Fox, P.F., Ed., Chapman and Hall, London, 1993, chap. 1.
7. Metchnikoff, E., Lactic acid as inhibiting intestinal putrefaction, in *The Prolongation of Life: Optimistic Studies*, W. Heinemann, London, 1907, pp. 161–201.
8. Fooks, L.J., Fuller, R., and Gibson, G.R., Prebiotics, probiotics and human gut microbiology, *Int. Dairy J.,* 9, 53–61, 1999.
9. Fuller, R., History and development of probiotics, in *Probiotics—The Scientific Basis*, Fuller, R., Ed., Chapman and Hall, New York, 1992, chap. 1.
10. Ishibashi, N. and Shimamura, S., Bifidobacteria: research and development in Japan, *Food Technol.,* 46, 126–135, 1993.
11. Hughes, D.B. and Hoover, D.G., Bifidobacteria: their potential for use in American dairy products, *Food Technol.,* 45, 74–83, 1991.
12. Sanders, M.E. and Huis in't Veld, J., Bringing a probiotic-containing functional food to the market: microbiological, product, regulatory and labelling issues, *Anton. Leeuw. Int. J. G.,* 76, 293–315, 1999.
13. Ouwehand, A.C., Kirjavainen P.V., Shortt, C., and Salminen, S., Probiotics: mechanisms and established effects, *Int. Dairy J.,* 9, 43–52, 1999.
14. Naidu, A.S., Biblack, W.R., and Clemens, R.A., Probiotic spectra of lactic acid bacteria (LAB), *Crit. Rev. Food Sci. Nutr.,* 39, 13–16, 1999.
15. Fuller, R., Probiotics in man and animals, *J. Appl. Bacteriol.,* 66, 365–378, 1989.

16. Havenaar, R. and Huis in't Veld, J H.J., Probiotics: a general view, in *The Lactic Acid Bacteria: The Lactic Acid Bacteria in Health and Disease*, Vol. 1, Wood, B.J.B., Ed., Chapman and Hall, New York, 1992, pp. 209–248.
17. Salminen, S., Ouwenhand, A., Benno, Y., and Lee, Y.K., Probiotics: how should they be defined? *Trends Food Sci. Tech.,* 10, 107–110, 1999.
18. Kimura, K., McCartney, A.L., McConnell., M.A., and Tannock, G.W., Analysis of fecal populations of bifidobacteria and lactobacilli and investigation of the immunological responses of their human hosts to the predominant strains, *Appl. Environ. Microbiol.,* 63, 3394–3398, 1997.
19. Parodi, P.W., The role of intestinal bacteria in the causation and prevention of cancer: modulation by diet and probiotics, *Aust. J. Dairy Technol.,* 54, 103–121, 1999.
20. Mitsuoka, T., Intestinal flora and ageing, *Nutr. Rev.,* 50, 438–446, 1992.
21. McCartney, A.L., Wenzhi, W., and Tannock, G.W., Molecular analysis of the composition of the bifidobacterial and *Lactobacillus* microflora of humans, *Appl. Environ. Microbiol.,* 62, 4608–4613, 1997.
22. Collins, F.M. and Carter, P.B., Growth of salmonellae in orally infected germ free mice, *Infect. Immun.,* 21, 41–47, 1978.
23. Tannock, G.W., Probiotic properties of lactic-acid bacteria: plenty of scope for fundamental R & D, *Trends Biotechnol.,* 15, 270–274, 1997.
24. Berg, R.D., Probiotics, prebiotics or "conbiotics," *Trends Microbiol.,* 6, 89–92, 1998.
25. O'Sullivan, D.J. and Kullen, M.J., Tracking of probiotic bifidobacteria in the intestine, *Int. Dairy J.,* 8, 513–518, 1998.
26. Collins, J.K, Thornton, G., and O'Sullivan, G.O., Selection of probiotic strains for human application, *Int. Dairy J.,* 8, 487–490, 1998.
27. Lee, Y.K. and Salminen, S., The coming of age of probiotics, *Trends Food Sci. Tech.,* 6, 241–245, 1995.
28. Rogelj, I., Lactic acid bacteria as probiotics, *Mljekarstvo,* 44, 277–279, 1994.
29. Stanton, C., Gardiner, G., Lynch, P.B., Collins, J.K., Fitzgerald, G., and Ross, R.P., Probiotic cheese, *Int. Dairy J.,* 8, 491–496, 1998.
30. Rasic, J.L., Culture media for detection and enumeration of bifidobacteria in fermented milk products, *Bull. IDF,* 252, 24–27, 1990.
31. Charteris, W.P., Kelly, P.M., and Morelli, L.C., Selective detection, enumeration and identification of potentially probiotic *Lactobacillus* and *Bifidobacterium* species in mixed bacterial populations, *Int. J. Food Microbiol.,* 35, 1–27, 1997.
32. Scardovi, V., Genus *Bifidobacterium*: Orla-Jensen, in *Bergey's Manual of Determinative Bacteriology,* Vol. 2, Sneath, P.H.A., Ed., The Williams & Wilkins Co., Baltimore, 1986, pp. 1418–1450.
33. Rada, V., Sirotek, K., and Petr, J., Evaluation of selective media for bifidobacteria in poultry and rabbit caecal samples, *Zentralbl. Veterinarmed.,* 46, 369–373, 1999.
34. Teraguchi, S., Uehara, M., Ogasa, K., and Mitsuoka, T., Enumeration of bifidobacteria in dairy products, *Jpn. J. Bacteriol.,* 33, 753–761, 1978.
35. Resnick, I.G. and Levin, M.A., Quantitative procedure for enumeration of bifidobacteria, *Appl. Environ. Microbiol.,* 42, 427–432, 1981.
36. Munoa, F.J. and Pares R., Selective medium for isolation and enumeration of *Bifidobacterium* spp., *Appl. Environ. Microbiol.,* 54, 1715–1718, 1988.
37. Wijsman, M.R., Hereijgers, J.L.P.M., and de Groote, J.M.F.H., Selective enumeration of bifidobacteria in fermented dairy products, *Neth. Milk Dairy J.,* 43, 395–405, 1989.
38. Roy, D. and Ward, P., Evaluation of rapid method for differentiation of *Bifidobacterium* species, *J. Appl. Bacteriol.,* 69, 739–749, 1990.

39. Sozzi, T., Brigidi, P., Mignot, O., and Matteuzzi, D., Use of dicloxacillin for the isolation and counting of bifidobacteria from dairy products, *Lait,* 70, 357–361, 1990.
40. Samona, A. and Robinson, R.K., Enumeration of bifidobacteria in dairy products, *Jpn. Soc. Dairy Tech.,* 44(3), 64–66, 1991.
41. Lapierre, L., Undeland, P., and Cox, L.J., Lithium chloride–sodium propionate agar for the enumeration of bifidobacteria in fermented dairy products, *J. Dairy Sci.,* 75, 1192–1196, 1992.
42. Onggo, I. and Fleet, G.H., Media for the isolation and enumeration of lactic acid bacteria from yoghurts, *Aust. J. Dairy Technol.,* 48(2), 89–92, 1993.
43. Lim, K.S., Huh, C.S., Baek, Y.J., and Kim, H.U., A selective enumeration medium for bifidobacteria in fermented dairy products, *J. Dairy Sci.,* 78, 2108–2112, 1995.
44. Mahalakshmi, R. and Murthy, V.V.P.S., Growth of *Bifidobacterium bifidum* in whey-based media, *J. Ind. Microbiol. Biotechnol.,* 25(4), 177–179, 2000.
45. McBrearty, S., Ross, R.P., Fitzgerald, G.F., Collins, J.K., and Wallace, J.M., Cheese quality, *Int. Dairy J.,* 11, 599–610, 2001.
46. de Man, J.C., Rogosa, M., and Sharpe, M.E., A medium for the cultivation of lactobacilli, *J. Appl. Bacteriol.,* 23, 130–135, 1960.
47. Hull, R.R. and Roberts, A.V., Differential enumeration of *Lactobacillus acidophilus* in yoghurt, *Aust. J. Dairy Sci.,* 39, 160–163, 1984.
48. Lee, S.Y., An agar medium for the differential enumeration of yoghurt starter bacteria, *J. Milk Food Technol.,* 37, 272–276, 1974.
49. Gilles, J., Turner, K.W., and Martley, F.G., Swiss-type cheese. 1. Manufacturing and sampling procedures, *N. Z. J. Dairy Sci.,* 18, 109–115, 1983.
50. Hull, R.R., Roberts, A.V., and Mayes, J.J., The association of *Lactobacillus casei* with a soft-body defect in commercial mozzarella cheese, *Aust. J. Dairy Technol.,* 38, 78–80, 1983.
51. Speck, M.L., Enumeration of viable *Lactobacillus acidophilus* organisms in dairy products, *J. Food Prot.*, 41, 135–137, 1978.
52. Hunger, W., Aesculin–cellobiose agar for the isolation and counting of *Lactobacillus acidophilus, Milchwissenschaft,* 41, 283–286, 1986.
53. Collins, E.B., Enumeration of *Lactobacillus acidophilus* with the agar plate count, *J. Food Prot.,* 41, 439–441, 1998.
54. Kneifel, W. and Pacher, B., An X-Glu based agar medium for the selective enumeration of *L. acidophilus* in yogurt related milk products, *Int. Dairy J.,* 3, 277–291, 1993.
55. Tuomola, E., Crittenden, R., Isolauri, E., and Salminen, S., Quality assurance for probiotic bacteria, *Am. J. Clin. Nutr.,* 73, 393S–398S, 2001.
56. Gardiner, G., Ross, R.P., Collins, J.K.G., Fitzgerald, G.F., and Stanton, C., Development of a probiotic Cheddar cheese containing human-derived *Lactobacillus paracasei* strains, *Appl. Environ. Microbiol.,* 64, 2192–2199, 1998.
57. Gomes, A.M.P. and Malcata, F.X., Use of small ruminants' milk supplemented with available nitrogen as growth media for *Bifidobacterium lactis* and *Lactobacillus acidophilus, J. Appl. Microbiol.,* 85, 839–849, 1998.
58. Klaver, F.A.M., Kingma, F., and Weerkamp, A.H., Growth and survival of bifidobacteria in milk, *Neth. Milk Dairy J.,* 47, 151–164, 1993.
59. Calicchia, M.L., Wang, C.I.E., Nomura, T., Yotsuzuka, F., and Osato, D.W., Selective enumeration of *Bifidobacterium bifidum, Enterococcus faecium* and streptomycin-resistant *Lactobacillus acidophilus* from a mixed probiotic product, *J. Food Prot.,* 56, 954–957, 1993.
60. Shah, N.P., Bifidobacteria: characteristics and potential for application in fermented milk products, *Milchwissenschaft,* 52, 16–21, 1997.

61. Ventling, B.L. and Mistry, V.V., Growth characteristics of bifidobacteria in ultrafiltered milk, *J. Dairy Sci.,* 76, 962–971, 1993.
62. Abu-Taraboush, H.M., Al-Dagal, M.M., and Al-Royli, M.A., Growth, viability and proteolytic activity of bifidobacteria in whole camel milk, *J. Dairy Sci.,* 81, 354–361, 1998.
63. Poch, M. and Bezkorovainy, A., Growth enhancing supplements for various species of the genus *Bifidobacterium, J. Dairy Sci.,* 71, 3214–3221, 1988.
64. Desjardins, M.L., Roy, D., and Goulet, J., Uncoupling of growth and acid production in *Bifidobacterium, J. Dairy Sci.,* 73, 1478–1484, 1990.
65. Tamime, A.Y., Marshall, V.M., and Robinson, R.K., Microbiological and technological aspects of milks fermented by bifidobacteria, *J. Dairy Res.,* 62, 151–187, 1995.
66. Gibson, G.R. and Wang, X., Regulatory effects of bifidobacteria on the growth of other colonic bacteria, *J. Appl. Bacteriol.,* 77, 412–420, 1994.
67. Bouhnik, Y., Flourie B., Andrieux, C., Bisetti, N., Briet, F., and Rambaud, J.C., Effects of *Bifidobacterium* sp. fermented milk ingested with or without inulin on colonic bifidobacteria and enzymatic activities in healthy humans, *Eur. J. Clin. Nutr.,* 50, 269–273, 1996.
68. Rivero-Urgell, M. and Santamaria-Orleans, A., Oligosaccharides: application in infant food, *Early Hum. Dev.,* 65, S43–S47, 2001.
69. Roberfroid, M.B., Prebiotics and synbiotics: concepts and nutritional properties, *Br. J. Nutr.,* 80, S197–S202, 1998.
70. Roberfroid, M. and Slavin, J., Nondigestible oligosaccharides, *Crit. Rev. Food Sci. Nutr.,* 40, 461–480, 2000.
71. Ziemer, C. and Gibson, G.R., An overview of probiotics, prebiotics and synbiotics in the functional food concept: perspectives and future strategies, *Int. Dairy J.,* 8, 473–479, 1998.
72. Taylor, G.R.J. and Williams, C.M., Effects of probiotics and prebiotics on blood lipids, *Br. J. Nutr.,* 80, S225–S230, 1998.
73. Kandler, O. and Weiss, N., Regular, non-sporing Gram-positive rods, in *Bergey's Manual of Determinative Bacteriology,* Sneath, P.H.A., Ed., Vol. 2, The Williams and Wilkins Co., Baltimore, 1989, pp. 1208–1311.
74. Heller, K.J., Probiotic bacteria in fermented foods: product characteristics and starter organisms, *Am. J. Clin. Nutr.,* 73, 374S–379S, 2001.
75. Prajapati, J.B., Shah, R.K., and Dave, J.M., Survival of *Lactobacillus acidophilus* in blended spray dried acidophilus preparations, *Aust. J. Dairy Tech.,* 42, 17–21, 1987.
76. Gonzalez, S.N., Apella, M.C., Romero, N.C., Nader de Marcias, M.E., and Guillermo, O., Inhibition of enteropathogens by *Lactobacillus* strains used in fermented milk, *J. Food Prot.,* 56, 773–776, 1993.
77. Miller, B. and Puhan, Z., Possibilities of shortened fermentation of acidophilus milk, *Schweiz. Milchw. Forschung.,* 9, 49–55, 1981.
78. Babu, V., Mital, B.K., and Garg, S.K., Effect of tomato juice addition on the growth and activity of *Lactobacillus acidophilus, Int. J. Food Microbiol.,* 17, 67–70, 1992.
79. Marshall, W.M., Cole, W.M., and Vega, J.R., A yoghurt-like product made by fermenting ultrafiltered milk containing elevated whey proteins with *Lactobacillus acidophilus, J. Dairy Res.,* 49, 665–670, 1982.
80. Agrawal, V., Usha, M.S., and Mital, B.K., Preparation and evaluation of acidophilus milk, *Asian J. Dairy Res.,* 5, 33–37, 1986.
81. Kumar, M., Mital, B.K., and Garg, S.K., Effect of papaya pulp addition on the growth of *Lactobacillus acidophilus, J. Food Safety,* 10, 63–73, 1989.

82. Ahemed, B., Mital, B.K., and Garg, S.K., Effect of magnesium and manganese ions on the growth of *Lactobacillus acidophilus, J. Food Sci. Technol.*, 27, 228–233, 1990.
83. Srinivas, D., Mital, B.K., and Garg, S.K., Utilization of sugars by *Lactobacillus acidophilus* strains, *Int. J. Food Microbiol.,* 10, 51–57, 1990.
84. Saxena, S.N., Mital, B.K., and Garg, S.K., Effect of casitone and fructose on the growth of *Lactobacillus acidophilus* and its survival during storage, *Int. J. Food Microbiol.,* 21, 271–276, 1994.
85. Rana, R., Ghandi, D.N., and Rana, R., Effect of basal medium and pH on the growth of *Lactobacillus acidophilus, Indian J. Dairy Sci.,* 53, 338–342, 2000.
86. Sanders, M.E., Walker, D.C., Walker, K.M., Aoyama, K., and Klaenhammer, T.R., Performance of commercial cultures in fluid milk applications, *J. Dairy Sci.,* 79, 943–955, 1996.
87. Fox, P.F., McSweeney, P.L.H., Wallace, J., and Law, J., Biochemistry of cheese ripening, in *Cheese: Chemistry, Physics and Microbiology,* Vol. 1, Fox, P.F., Ed., Chapman and Hall, London, 1993, pp. 389–438.
88. Fox, P.F., Wallace, J.M., Morgan, S., Lynch, C.M., Niland, E., and Tobin, J., Acceleration of cheese ripening, *Antonie Van Leeuwenhoek,* 70, 175–297, 1996.
89. Charteris, W.P., Kelly, P.M., Morelli, L., and Collins, J.K., Development and application of an *in vitro* methodology to determine the transit tolerance of potentially probiotic *Lactobacillus* and *Bifidobacterium* species in the upper human gastrointestinal tract, *J. Appl. Microbiol.,* 84, 759–768, 1998.
90. Haller, D., Colbus, H., Ganzle, M.G., Scherenbacher, P., Bode, C., and Hammes, W.P., Metabolic and functional properties of lactic acid bacteria in the gastro-intestinal ecosystem: a comparative *in vitro* study between bacteria of intestinal and fermented food origin, *Syst. Appl. Micriobiol.,* 24, 218–226, 2001.
91. Marteau, P., Pochart, P., Bouhnik, Y., and Rambaud, J.C., The fate and effects of transiting, nonpathogenic microorganisms in the human intestine, *World Rev. Nutr. Diet.,* 74, 1–21, 1993.
92. Christiaens, H., Leer, R.J., Pouwels, P.H., and Verstraete, W., Cloning and expression of a conjugated bile acid hydrolase gene from *Lactobacillus plantarum* by using a direct plate assay, *Appl. Environ. Microbiol.,* 58, 3792–3798, 1992.
93. de Smet, I., Van Hoorde, L., Vande Woestyne, M., Christiaens, H., and Verstraete, W., Significance of bile salt hydrolytic activities of lactobacilli, *J. Appl. Bacteriol.,* 79, 292–301, 1995.
94. McLauchlan, G., Fullarton, G.M., Crean, G.P., and McColl, K.E., Comparison of gastric body and antral pH: a 24 hour ambulatory study in healthy volunteers, *Gut,* 30, 573–578, 1998.
95. Kashket, E.R., Bioenergetics of lactic acid bacteria: cytoplasmic pH and osmotolerance, *FEMS Microbiol. Rev.,* 46, 233–244, 1987.
96. Kullen, M.J. and Klaenhammer, T.R., Identification of the pH inducible, proton-translocating F_1F_0-ATPase (atpBEFHAGDC) operon of *Lactobacillus acidophilus* by differential display: gene structure, cloning and characterization, *Mol. Microbiol.,* 33, 1152–1161, 1999.
97. Shah, N.P., Probiotic bacteria: selective enumeration and survival in dairy foods, *J. Dairy Sci.,* 83, 894–907, 2000.
98. Dave, R.I. and Shah, N.P., Viability of yoghurt and probiotic bacteria in yoghurts made from commercial starter cultures, *Int. Dairy J.,* 7, 31–41, 1997.
99. Kulisaar, T., Zilmer, M., Mikelsaar, M., Vihalemm, T., Annuk, H., Kairane, C., and Kilk, A., Two antioxidative strains as promising probiotics, *Int. J. Food Microbiol.,* 72, 215–224, 2002.

100. Lankaputhra W.E.V. and Shah, N.P., A simple method for the enumeration of *Lactobacillus acidophilus* in yoghurt supplemented with *L. acidophilus* and *Bifidobacterium* spp., *Milschwissenschaft,* 51, 446–451, 1996.
101. Iwana, H., Masuda, S., Fujisawa, T., Suzuki, H., and Mitsuoka, T., Isolation and identification of *Bifidobacterium* spp. in commercial yogurt sold in Europe, *Bifidobact. Microflor.,* 12, 39, 1993.
102. Clark, P.A., Cotton, L.N., and Martin, J.H., Selection of bifidobacteria for use as dietary adjuncts in cultured dairy foods: II. Tolerance to simulated pH of human stomachs, *Cult. Dairy Prod. J.,* 28, 11–14, 1993.
103. Clark, P.A. and Martin, J.H., Selection of bifidobacteria for use as dietary adjuncts in cultured dairy foods: III. Tolerance to simulated bile concentrations of human small intestines, *Cult. Dairy Prod. J.,* 29, 18–21, 1994.
104. Park, H-K., So, J-S., and Heo, T-R., Acid adaptation promotes survival of *Bifidobacterium breve* against environmental stress, *Foods and Biotechnol.,* 4, 226–230, 1995.
105. Schmidt, G. and Zink, R., Basic features of the stress response in three species of bifidobacteria: *B. longum, B. adolescentis,* and *B. breve, Int. J. Food Microbiol.,* 55, 41–45, 2000.
106. Nam, K.L., Ji, C.Y., Hee, K.P., Tae, R H., and Jae, S.S., The relationship between cell surface hydrophobicity (CSH) and stress tolerance in *Bifidobacterium* spp., *Food Sci. Biotechnol.,* 7, 66–70, 1998.
107. Ahn J-B., Hwang H-J., and Park J-H., Physiological responses of oxygen tolerant anaerobic *Bifidobacterium longum* under oxygen, *J. Microbiol. Biotechnol.,* 11, 443–451, 2001.
108. Holzapfel, W.H., Haberer, P., Geisen, R., Bjorkroth, J., and Schillinger, U., Taxonomy and important features of probiotic microorganisms in food and nutrition, *Am. J. Clin. Nutr.,* 73, 365S–373S, 2001.
109. Lievense, L.C. and Van't Reit, K., Convective drying of bacteria. The drying process, *Adv. Biochem./Eng. Biotechnol.,* 50, 46–52, 1993.
110. Teixeira, P., Castro, H., and Kirby, R., Spray drying as a method for preparing concentrated cultures of *Lactobacillus bulgaricus, J. Appl. Bacteriol.,* 78, 456–462, 1995.
111. Teixeira, P.C., Castro, M.H., Malcata, F.X., and Kirby, R.M., Survival of *Lactobacillus delbrueckii* ssp. *bulgaricus* following spray drying, *J. Dairy Sci.,* 78, 1025–1031, 1995.
112. Selmer-Olsen, E., Sorhaug, T., Birkeland, S.E., and Pehrson, R., Survival of *Lactobacillus helveticus* entrapped in Ca-alginate in relation to water content, storage and rehydration, *J. Ind. Microbiol. Biotechnol.,* 23(2), 79–85, 1999.
113. Gardiner, G., O'Sullivan, E., Kelly, J., Auty, M.A.E., Fitzgerald, G.F., and Collins, J.K., Comparative survival of human-derived probiotic *Lactobacillus paracasei* and *L. salivarius* strains during heat treatment and spray drying, *Appl. Environ. Microbiol.,* 66, 2605–2612, 2000.
114. Silva, J., Carvahlo, A.S., Teixeira, P., and Gibbs, P.A., Bacteriocin production by spray dried lactic acid bacteria, *Lett. Appl. Microbiol.,* 34, 77–81, 2002.
115. Johnson, J.A.C. and Etzel, M.R., Properties of *Lactobacillus helveticus* CNRZ-32 attenuated by spray drying, freeze-drying, or freezing, *J. Dairy Sci.,* 78, 761–768, 1995.
116. Gagne, J., Roy, D., and Gauthier, S.F., Production of frozen and freeze-dried concentrates of *Bifidobacterium infantis* by membrane filtration, *Milchwissenschaft,* 48, 501–505, 1993.

117. de Valdez, F.G., de Savoy, G.G., de Pesce, R.H.A., and Oliver, G., Comparative study of the efficiency of some additives in protecting lactic acid bacteria against freeze drying, *Cryobiology,* 20, 560–566, 1983.
118. Hyun, C.K. and Shin, H.K., Utilization of bovine blood plasma obtained from a slaughterhouse for economic production of probiotics, *J. Ferment. Bioeng.,* 86, 34–37, 1998.
119. Gibbs, B.F., Kermasha, S., Alli, I., and Mulligan, C.N., Encapsulation in the food industry: a review, *Int. J. Food Sci. Nutr.,* 50, 213–224, 1999.
120. Desmond, C., Stanton, C., Fitzgerald, G.F., Collins, K., and Ross, R.P., Environmental adaptation of probiotic lactobacilli towards improved performance during spray drying, *Int. Dairy J.,* 11, 801–808, 2001.
121. Mauriello, G., Aponte, M., Andolfi, R., Moschetti, G., and Villani, F., Spray drying of bacteriocin-producing lactic acid bacteria, *J. Food Prot.,* 62, 773–777, 1999.
122. Kim, S.S. and Bhowmik, S.R., Survival of lactic acid bacteria during spray drying of yoghurt, *J. Food Sci.,* 55, 1008–1010, 1990.
123. O'Riordan, K.O., Andrews, D., Buckle, K., and Conway, P., Evaluation of microencapsulation of a *Bifidobacterium* strain with starch as an approach to prolonging viability during storage, *J. Appl. Microbiol.,* 91, 1059–1066, 2001.
124. Espina, F. and Packard, V.S., Survival of *Lactobacillus acidophilus* in a spray drying process, *J. Food Prot.,* 42, 149–152, 1979.
125. Prajapati, J.B., Shah, R.K., and Dave, J.M., Nutritional and therapeutic benefits of a blended–spray dried acidophilus preparation, *Cult. Dairy Prod. J.,* 21, 16–21, 1986.
126. Lievense, L.C., Verbeck, M.A.M., Noomen, A., and van't Reit, K., Mechanism of dehydration inactivation of *Lactobacillus plantarum, Appl. Microbiol. Biotechnol.,* 41, 90–94, 1994.
127. Linders, L.J.M., de Jong, G.I.W., Meerdink, G., and van't Riet, K., The effect of disaccharide addition on the dehydration inactivation of *Lactobacillus plantarum* during drying and the importance of water activity, in *Proc. 9th International Drying Symposium,* Keey, R.B., Ed., Gold Coast, Australia, 1994, pp. 945–952.
128. Linders, L.J.M., Meerdink, G., and van't Reit, K., The influence of temperature and drying rate on the dehydration inactivation of *Lactobacillus plantarum, Food Bioprod. Proc.,* 74C, 110–114, 1996.
129. Linders, L.J.M., Meerdink, G., and van't Reit, K., Effect of growth parameters on the residual activity of *Lactobacillus plantarum* after drying, *J. Appl. Microbiol.,* 82, 683–688, 1997.
130. Daemen, A.L.H. and van der Stege, H.J., 1. The destruction of enzymes and bacteria during the spray drying of milk and whey. 2. The effect of the drying conditions, *Neth. Milk Dairy J.,* 36, 211–242, 1982.
131. Teixeira, P., Castro, H., Mohásci-Farkas, C., and Kirby, R., Identification of sites of injury in *Lactobacillus bulgaricus* during heat stress, *J. Appl. Microbiol.,* 83, 219–226, 1997.
132. Abee, T. and Wouters, J.A., Microbial stress response in minimal processing, *Int. J. Food Microbiol.,* 50, 65–91, 1999.
133. Boutibonnes, P., Tranchard, C., Hartke, A., Thammavongs, B., and Auffray, Y., Is thermotolerance correlated to heat shock protein synthesis in *Lactococcus lactis* subsp. *lactis*? *Int J. Food Microbiol.,* 16, 227–236, 1992.
134. Teixeira, P., Castro, H., and Kirby, R., Inducible thermotolerance in *Lactobacillus bulgaricus, Lett. Appl. Microbiol.,* 18, 218–221, 1994.

135. Gouesbet, G., Jan, G., and Boyaval, P., Two-dimensional electrophoresis study of *Lactobacillus delbrueckii* sbsp. *bulgaricus* thermotolerance, *Appl. Environ. Microbiol.,* 68, 1055–1063, 2002.
136. Lorca, G.L. and de Valdez, G.F., The effect of suboptimal growth temperature and growth phase on resistance of *Lactobacillus acidophilus* to environmental stress, *Cryobiology,* 39, 144–147, 1999.
137. Mayo, B., Derzelle, S., Fernandez, M., Leonard, C., Ferain, T., Hols, P., Suarez, J.E., and Delious, J., Cloning and characterization of cspL and cspP, two cold-inducible genes from *Lactobacillus plantarum, J. Bacteriol.,* 179, 3039–3042, 1997.
138. Siuta-Cruce, P. and Goulet, J., Improving probiotic survival rates: microencapsulation preserves the potency of probiotic microorganisms in food systems, *Food Technol.,* 55(10), 36–42, 2001.
139. Lian, W.C., Hsiao, H.C., and Chou, C.C., Survival of bifidobacteria after spray drying, *Int. J. Food Microbiol.,* 74, 79–86, 2002.
140. Dave, R.I. and Shah, N.P., Ingredient supplementation effects on viability of probiotic bacteria in yogurt, *J. Dairy Sci.,* 81, 2804–2816, 1998.
141. Lourens-Hattingh, A. and Viljoen, B.C., Yogurt as probiotic carrier food, *Int. Dairy J.,* 11, 1–17, 2001.
142. Shah, N.P. and Lankaputhra, E.V., Improving *Lactobacillus acidophilus* and *Bifidobacterium* spp. in yogurt, *Int. Dairy J.,* 7, 349–356, 1997.
143. Shah, N.P., Lankapurthra, E.V., Britz M.L., and Kyle, W.S.A., Survival of *Lactobacillus acidophilus* and *Bifidobacterium bifidum* in commercial yogurt during refrigerated storage, *Int. Dairy J.,* 5, 515–521, 1995.
144. Nighswonger, B.D., Brashears, M.M., and Gilliland, S.E., Viability of *Lactobacillus acidophilus* and *Lactobacillus casei* in fermented milk products during refrigerated storage, *J. Dairy Sci.,* 79, 212–219, 1996.
145. Hood, S.K. and Zottola, M.L., Effect of low pH on the ability of *Lactobacillus acidophilus* to survive and adhere to human intestinal cells, *J. Food Sci.,* 53, 1514–1516, 1988.
146. Shah, N.P. and Jelen, P., Survival of lactic acid bacteria and their lactases under acidic conditions, *J. Food Sci.,* 55, 506–509, 1990.
147. Sultana, K., Godward, G., Reynolds, N., Arumugaswamy, R., Peiris, P., and Kailasapathy, K., Encapsulation of probiotic bacteria with alginate–starch and evaluation of survival in simulated gastrointestinal conditions and in yoghurt, *Int. J. Food Microbiol.,* 62, 47–55, 2000.
148. Schillinger, U., Isolation and identification of lactobacilli from novel-type probiotic and mild yoghurts and their stability during refrigerated storage, *Int. J. Food Microbiol.,* 47, 79–87, 1999.
149. Hansen, R., Bifidobacteria have come to stay, *N. Eur. Dairy J.,* 51(3), 79–83, 1985.
150. Crittenden, R., Laitila, A., Forssell, P., Matto, J., Saarela, M., Mattila Sandholm, T., and Myllarinen, P., Adhesion of bifidobacteria to granular starch and its implications in probiotic technologies, *Appl. Environ. Microbiol.,* 67, 3469–3475, 2001.
151. Kailasapathy, K. and Rybka, S., *L. acidophilus* and *Bifidobacterium* spp.: their therapeutic potential and survival in yogurt, *Aust. J. Dairy Tech.,* 52, 28–35, 1997.
152. Adhikari, K., Mustapha, A., Grun, I.U., and Fernando, L., Viability of microencapsulated bifidobacteria in set yogurt during refrigerated storage, *J. Dairy Sci.,* 83, 1946–1951, 2000.
153. IDF, Fermented and non-fermented dairy products. Detection and enumeration of *Lactobacillus acidophilus, Culture media, Bull. IDF.,* 306, 23–25, 1995.

154. Ravula, R.R. and Shah, N.P., Viability of probiotic bacteria in fermented frozen dairy desserts, *Food Austral.*, 50(3), 136–139, 1998.
155. Vinderola, C.G., Prosello, W., Ghiberto, D., and Reinheimer, J.A., Viability of probiotic (*Bifidobacterium, Lactobacillus acidophilus* and *Lactobacillus casei*) and non-probiotic microflora in Argentenian fresco cheese, *J. Dairy Sci.*, 83, 1905–1911, 2000.
156. Laroia, S. and Martin, J.H., Methods for enumerating and propagating *Bifidobacteria, Cult. Dairy Prod. J.*, 5, 32–33, 1991.
157. Nebra, Y., Jofre, J., and Blanch, A.R., The effect of reducing agents on the recovery of *Bifidobacterium* cells, *J. Microbiol. Meth.*, 49(3), 247–254, 2002.
158. Sanders, M.E., Development of consumer probiotics for the US market, *Br. J. Nutr.*, 80, S213–S218, 1998.
159. Young, J., European market developments in prebiotic- and probiotic-containing foodstuffs, *Br. J. Nutr.*, 80, S231–S233, 1998.
160. Moller, M., Digestive cultured products doing well, *Nord. Mejeriinform.*, 20(4), 115, 1993.
161. www.ukfoodserviceinfo.com
162. Dinakar, P. and Mistry V.V., Growth and viability of *Bifidobacterium bifidum* in cheddar cheese, *J. Dairy Sci.*, 77(10), 2854–2864, 1994.
163. Gomes, A.M.P., Malcata, F.X., Klaver, F.A.M., and Grande, H.J., Incorporation and survival of *Bifidobacterium* sp. strain Bo and *Lactobacillus acidophilus* strain Ki in a cheese product, *Neth. Milk Dairy J.*, 49, 71–95, 1995.
164. Daigle, A., Belanger, G., and Vuillemard, J.C., Production of probiotic cheese (Cheddar-like cheese) using enriched cream fermented by *Bifidobacterium infantis, J. Dairy Sci.*, 82, 1081–1091, 1999.
165. Ghoddusi, H.B. and Robinson, R.K., The test of time, *Dairy Ind. Int.*, 61(7), 25–26, 28, 1996.
166. Gomes, A.M.P., Silva, M.L.P.C., and Malcata, F.X., Caprine cheese with probiotic strains: the effects of ripening temperature and relative humidity on proteolysis and lipolysis, *Z. fur Lebens. Untersuchung und Forschung A.*, 207, 386–394, 1998.
167. Gobbetti, M., Corsetti, A., Smacchi, E., Zocchetti, Z., and de Angelis, M., Production of Crescenza cheese by incorporation of bifidobacteria, *J. Dairy Sci.*, 81, 37–47, 1998.
168. Blanchette, L., Belanger, G., and Gauthier, S.F., Production of cottage cheese using dressing fermented by bifidobacteria, *J. Dairy Sci.*, 79, 8–15, 1996.
169. Murad, H.A., Zeinab, Sadek, I., and Fathy, F.A., Production of bifidus Kariesh cheese, *Deutsche Lebens. Rundschau*, 94, 409–412, 1998.
170. Corbo, M.R., Albenzio, M., De-Angelis, M., Sevi, A., and Gobbetti, M., Microbiological and chemical properties of Canestro Pugliese hard cheese supplemented with bifidobacteria, *J. Dairy Sci.*, 84, 551–561, 2001.
171. El Zayatt, A.I. and Osman, M.M., The use of probiotics in Tallaga cheese, *Egypt. J. Dairy Sci.*, 29, 99–106, 2001.
172. Roy, D., Mainville, I., and Mondou, F., Selective enumeration and survival of bifidobacteria in fresh cheese, *Int. Dairy J.*, 7, 785–793, 1997.
173. Hekmat, S. and McMahon, D.J., Survival of *Lactobacillus acidophilus* and *Bifidobacterium bifidum* in ice cream for use as a probiotic food, *J. Dairy Sci.*, 75, 1415–1422, 1992.
174. Hagen, M. and Narvhus, J.A., Production of ice cream containing probiotic bacteria, *Milchwissenschaft*, 54, 265–268, 1999.
175. Holocomb, J.E., Frank, J.F., and McGregor, J.U., Viability of *Lactobacillus acidophilus* and *Bifidobacterium bifidum* in soft-serve frozen yogurt, *Cult. Dairy Prod. J.*, 26(3), 4–5, 1991.

176. Gutknecht, K.W., Microbes confer healthful benefits on ice cream and yogurt, *Utah Sci.,* 52(4), 134–135, 1992.
177. Hong S.H. and Marshall, R.T., Natural exopolysaccharides enhance survival of lactic acid bacteria in frozen dairy desserts, *J. Dairy Sci.,* 84, 1367–1374, 2001.
178. Davidson, R.H., Duncan, S.E., Hackney, C.R., Eigel, W.N., and Boling, J.W., Probiotic culture survival and implications in fermented frozen yogurt characteristics, *J. Dairy Sci.,* 83, 666–673, 1999.
179. Duthie, A.H., Duthie, A.E., Nilson, K.M., and Atherton. H.V., An ideal vehicle for *Lactobacillus acidophilus, Dairy Field,* 164(11), 139–140, 1981.
180. Laroia, S. and Martin, J.H., Effect of pH on survival of *Bifidobacterium bifidum* and *Lactobacillus acidophilus* in frozen fermented dairy desserts, *Cult. Dairy Prod. J.,* 26(4), 13–21, 1991.
181. Moss, C.W. and Speck, M.L., Injury and death of *Streptococcus lactis* due to freezing and frozen storage, *Appl. Microbiol.,* 11, 326–329, 1963.
182. Ravula, R.R. and Shah, N.P., Survival of microencapsulated *L. acidophilus* and *Bifidobacterium* spp. in fermented frozen dairy desserts, *J. Dairy Sci.,* 82, 666–668, 1999.
183. Sheu, T.Y. and Marshall, R.T., Microentrapment of *Lactobacilli* in calcium alginate gels, *J. Food Sci.,* 54, 557–567, 1993.
184. Khalil, A.H. and Mansour, E.H., Alginate encapsulated bifidobacteria survival in mayonnaise, *J. Food Sci.,* 63, 702–705, 1998.
185. Katz, F., Active cultures add function to yoghurt and other foods, *Food Technol.,* 55(3), 46–50, 2001.
186. Cavadani, C., Ballevre, O., and Gaier, W., Cereal product containing probiotics, *EPO,* 0, 862–863, 1997.
187. Crates, E., Sausage science (Probiotic sausage), *Functional Food,* 1(2), 26–27,29, 1998.
188. Johansson, M.L., Nobaek, S., Berggren, A., Nyman, M., Bjorck, I., Ahrne, S., Jeppsson, B., and Molin, G., Survival of *Lactobacillus plantarum* DSM 9843 (299V), and effect on the short chain fatty acid content of faeces after ingestion of a rose hip drink with fermented oats, *Int. J. Food Microbiol.,* 42(1–2), 29–38, 1998.
189. Molin, G., Probiotics in foods not containing milk or milk constituents, with special reference to *Lactobacillus plantarum* 299v., *Am. J. Clin. Nutr.,* 73, 380S-385S, 2001.
190. Anonymous, Probing the probiotics, *Liq. Food, Int.,* 3(3), 31, 1999.
191. Anonymous, Probiotics with a difference. Primavita, a fruit drink based on cereal and fruit, comes to Germany, *Milch Marketing,* 14, 32, 1997.
192. Rani, B. and Khetarpaul, N., Probiotic fermented food mixtures: possible applications in clinical anti-diarrhoea usage, *Nutr. Health,* 12, 97–105, 1998.
193. Tamime, A.Y. and Marshall, V.M., Microbiology and technology of fermented milks, in *Microbiology and Biochemistry of Fermented Milk,* Law, B., Ed., Chapman and Hall, New York, 1997, pp. 57–152.
194. Elmadfa, I., Heinzle, C., Majchrzak, D., and Foissy, H., Influence of a probiotic yogurt on the status of vitamins B_1, B_2 and B_6 in the healthy adult human, *Ann. Nutr. Metab.,* 45, 13–18, 2001.
195. Atanassova, M.R., Chipeva, V., Ivanova, I., and Haertle, T., Determination of the growth stages of *Lactobacillus paracasei* subsp. *paracasei* M3 from Bulgarian yellow cheese by electroconductivity, *J. Microbiol. Meth.,* 46, 227–233, 2001.
196. Osman, M.M. and Abbas, F.M., Fate of *Lactobacillus acidophilus* La-5 and *Bifidobacterium lactis* Bb-12 in probiotic Ras cheese, 8th Egyptian Conference for Dairy Science and Technology, 2001, pp. 653–664.
197. Madkor, S.A., Tong, P.S., and El Soda, M., Ripening of cheddar cheese with added attenuated adjunct cultures of lactobacilli, *J. Dairy Sci.,* 83, 1684–1691, 2000.

3 The Properties of *Enterococcus faecium* and the Fermented Milk Product Gaio®

Marcelo Chiara Bertolami and
Edward R. Farnworth

CONTENTS

0-8493-1372-4/03/$0.00+$1.50

3.1 INTRODUCTION

Many fermented foods are produced throughout the world. Fermentation is a process that transforms the starting material into a product that may have enhanced nutritional and/or organoleptic characteristics. With the advent of probiotics, many researchers have analyzed the microflora in traditional fermented foods in attempts to find foods that contain bacteria that may be beneficial to health, metabolism, and disease resistance. In a few cases, an opposite approach has been taken. Based on studies testing individual bacteria in animals and humans, new products have been developed that include these bacteria, thereby creating new probiotic foods. (See Chapter 6 on *Lactobacillus casei* strain Shirota [LcS] for such an example.)

Early studies on *Enterococcus faecium* and its effects against diarrhea and more importantly on cholesterol metabolism showed that *E. faecium* might be an ideal candidate to include in a fermented milk probiotic product. Gaio® (which contains both *E. faecium* and *Streptococcus thermophilus*) was developed and is now distributed in at least two European countries. This chapter reviews studies where *E. faecium* and Gaio were tested for their effects on serum cholesterol, diarrhea, and mutagens.

3.2 SERUM CHOLESTEROL AND CARDIOVASCULAR DISEASE

3.2.1 Background

Coronary artery disease is one of the most frequent causes of morbidity and mortality all over the world.[1] In the United States, it accounts for fully one-half of the nearly one million deaths each year from cardiovascular disease and is the leading cause of death for both genders.[2] Each year, about 1.5 million Americans suffer acute myocardial infarction, and almost all myocardial infarctions are due to atherosclerosis of the coronary arteries.[3] It is known that individuals with some conditions, designated as risk factors, have a higher chance of prematurely developing this disease. Among the risk factors, hypercholesterolemia is one of the most important. A continuous and graded positive relation was demonstrated between serum total cholesterol level and coronary artery disease mortality in the more than 350,000 men screened for the Multiple Risk Factor Intervention Trial (MRFIT).[4] However, the numbers of people with very high serum cholesterol levels are not expressive, so they do not account for a large number of cases of symptomatic coronary disease. The vast majority of these cases are individuals presenting cholesterol considered to be in the "normal" (average) range or with a slight increase.[5,6] According to the National Cholesterol Education Program — Adult Treatment Panel III,[7] the optimal levels in a human blood lipid profile are: total cholesterol below 200 mg/dl, low-density-lipoprotein (LDL) cholesterol below 100 mg/dl, high-density-lipoprotein (HDL) cholesterol above 40 mg/dl, and triglycerides below 150 mg/dl. Such patients in general also have other risk factors, including smoking, hypertension, diabetes mellitus, sedentary life, weight excess, and psychosocial stress, because of the modern lifestyle typical of Western industrialized countries.

It was demonstrated that healthy people presenting serum cholesterol levels within the "normal" range have a reduction in the risk of future cardiovascular events when their cholesterol levels are decreased.[8] One meta-analysis recently performed showed that for a 10% cholesterol reduction, the mortality risk due to cardiovascular disease is decreased by 15%, and total mortality risk is decreased by 11%.[9] Consequently, any dietary intervention that could help to decrease serum cholesterol levels, particularly in people who do not have highly elevated levels, probably will be helpful in the prevention of coronary heart disease.

3.2.2 Cardiovascular Disease and Probiotics

Modulation of the microbial community (approximately 10^{14} bacterial cells/g) in the gut by probiotics and prebiotic foodstuffs has been considered as an important opportunity to positively influence human health.[10–12] Prebiotics are defined as a nondigestible food ingredient that beneficially affects the host by selectively stimulating the growth and/or the activity of one or a limited number of bacteria in the colon and thus improves health.[11] Probiotics are defined as live microbial food ingredients that are beneficial to health.[13]

The question of whether lactic acid bacteria (LAB) can be beneficial to health was raised a long time ago. Metchnikoff[14] was the first to show an interest in LAB and their possible beneficial effects on health. He claimed that lactobacilli from fermented milk were able to prolong life through a reduced formation of toxins in the gut. Mann and Spoerry[15] described a hypocholesterolemic effect of fermented milk in the natives of the African Masai tribe. These authors proposed that dietary habits were the explanation for the low incidence of ischemic heart disease among these people, since they traditionally had a high intake of milk and fermented milk. However, this means that they also were consuming large quantities of saturated fat and cholesterol. After these seminal observations, many studies addressed this topic, as has been reviewed.[16–19] Until now, no clear answer has been available, and even more difficult is the identification of a possible biochemical mechanism that can explain how this improvement of health is really achieved. One possibility is that milk and fermented milk products have an effect on the metabolism of serum lipids and lipoproteins, resulting in a better serum lipid profile. Not all studies have come to the same conclusion, as discussed by Eichholzer and Stähelin[20] in an extensive review of this subject and more recently by the Mission Scientifique de Syndifrais[21] and by St-Onge et al.[22] A controversy exists because milk and milk products have been shown to be atherogenic due to their concentration of saturated fat and cholesterol,[18] but they also possibly play a role in the prevention of coronary heart disease. There appear to be scientific reports to support both effects. Other benefits of milk products such as yogurt have been reported, including beneficial intestinal flora changes, better immunological response, production of antibiotic substances, improved calcium absorption, prevention of osteoporosis, cataract prevention, and prolongation of life,[16,17,19] that would justify increased consumption of milk products.

A group at Kiev University (unpublished results) demonstrated a significant hypocholesterolemic effect utilizing a milk product fermented with bacteria isolated from Abkhasia in the Caucasus. This region has a reputation for the longevity of its people, and it is known that fermented milk is a major part of the traditional diet of this population, suggesting a link between these two observations.

The studies analyzed in this chapter utilized a new yogurt-like product, similar to the one used in the study at Kiev University, but produced in Denmark. The product Causido® (Gaio) was constituted as a fermented (probiotic) dairy product, containing a bacterial culture (*E. faecium, S. thermophilus*). Gaio was consumed for different periods of time.

3.3 *ENTEROCOCCUS FAECIUM*

3.3.1 Characteristics and Occurrence

Enterococci are included in the broad category of LAB.[23] The identificaton of the members of the genus *Enterococcus* using traditional classification tests is difficult because there are no phenotypic characteristics that can be used to distinguish them from other Gram-positive, catalase-negative, coccus-shaped bacteria.[24,25] Enterococci are generally considered to be hardy because they can survive a wide range of temperatures, pH levels, saline solutions, and environments, such as are found in the human gastrointestinal tract.[26] An important distinguishing characteristic (from a human health perspective) of various enterococci is their ability to resist many antibiotics.[27]

Enterococci occur on plants and in the feces of animals and man. *Enterococcus faecalis* and *Enterococcus faecium* are the two most common species found in human gastrointestinal tract. *E. faecalis* counts can reach 10^5 to 10^7 colony forming units (CFU)/g, while *E. faecium* levels of 10^4 to 10^5 CFU/g are found.[27,28] Levels of these two species can vary among individuals; diet and other factors are believed to alter the proportions of *Enterococcus* species. *E. faecalis* has been isolated in feces of neonates, but *E. faecium* has not.[28,29]

3.3.2 Antibiotic Resistance

Many authors have presented criteria that should be used when choosing potential probiotic bacteria.[30–32] Above all, it is agreed that any microorganism that is intentionally added to a food should be generally recognized as safe (GRAS). For example, the microorganism should not be pathogenic, should not produce toxins or metabolites that could adversely affect the health and metabolism of the host, and should not negatively impact on bacterial populations already resident in the host. A large number of fermented food products that contain LAB are now eaten because they have a long history of safe use. Within the LAB, enterococci and streptococci have been identified as causes for concern.[30] With the introduction of a fermented milk product on the market that contains *E. faecium,* the question of antibiotic resistance has been raised.[33]

Plasmids are genetic elements independent of the cell chromosome. Over time, some plasmids have developed genes that make them resistant to selected antibiotics. Plasmids are easily transferred between cells, and this allows the efficient spread between bacteria of resistance to an antibiotic. Antibiotics originally were substances produced by fungal microorganisms to eliminate competition from bacteria for survival reasons. However, the number of antibiotics now available and their widespread use in prescription drugs and for general disinfection purposes has raised concerns about the potential for the development of antibiotic resistance. The introduction of probiotic foods into the diet raises the possibility of the ingested probiotic becoming antibiotic resistant from related intestinal bacteria that already have acquired resistance, or more seriously, the reverse — a probiotic product containing antibiotic-resistant bacteria might pass the characteristic on to resident bacteria.

Most strains of enterococci are resistant to tetracycline, erythromycin, clindamycin, chloramphenicol, and sulfonamides.[27] This, together with the fact that the incidence of infections attributed to enterococci appears to be increasing, and along with the difficulty in treating such infections, places these organisms as important human pathogens of concern. The hospital environment is one where enterococcal contamination has received much attention. However, since vancomycin-resistant enterococci (VRE), in particular, have been identified in a wide variety of farm animals and birds, it is not clear whether food is a major vector in the transfer of VRE.[34]

With respect to Gaio, Lund et al.[35] were able to show that when Gaio was taken together with vancomycin, the total number of enterococci decreased at day 10 of the feeding trial, but by 3 weeks after the cessation of the experiment, enterococci numbers were 100 times those counted at the beginning of the experiment. Subjects who had received only the Gaio had no such increase in enterococci numbers, indicating that the antibiotic treatment may have given the ingested probiotic bacteria in the Gaio an advantage due to reduced colonization resistance. No major overgrowth of VRE occurred in any subject, including those receiving the vancomycin treatment. However, some subjects (20%) were transient carriers of VRE. Resistant strains isolated in subjects were not associated with the consumption of Gaio, since pulse-field gel electrophoresis analysis showed that the strains of *E. faecium* found in Gaio and VRE fecal samples were different. Lund et al. were able to state that "no resistance against vancomycin emerged in intestinal enterococci, and *E. faecium* from the Gaio product was not found to acquire vancomycin resistance during the study period."

Adams[36] pointed out that there have been no reports of infection contracted as a result of consumption of foods containing enterococci. However, the Lactic Acid Bacteria Industrial Platform recommended that enterococci should not be used in foods unless there was a demonstrable (possibly health) benefit.[37]

3.4 PRODUCTION OF GAIO

Gaio was first produced by the Danish dairy corporation MD Foods A/S, established in Aarhus in Denmark. Recently, MD Foods merged with Arla Foods; currently, the product is only produced by Arla Foods and is consumed in Denmark and Sweden.

The production of Gaio uses a fermentation of milk at a temperature of 37°C. The level of starter inoculated is approximately 5×10^{12} CFU per 1000 l of milk. The fermentation time is approximately 9 h, to a final pH of 4.5. The final product is very viscous and has a mild, slightly acid taste. The product is sold in plastic containers of 500 g as "natural" and with different fruit flavors. The product is distributed and sold refrigerated.

The original Ukrainian bacterial culture (Causido) is used to produce Gaio. This culture contains one human species of *E. faecium* and two strains of *S. thermophilus.* The CFUs of the fresh product are 10^5 to 10^9/ml for *E. faecium* and 5 to 20 × 10^8/ml for *S. thermophilus.* One hundred grams of the product has an energy content of 240 kJ and contains 4.9 g of protein, 5.4 g of carbohydrate, and 1.5 g of fat (66% as milk fat and 33% as soybean fat). The cholesterol content is about 5 mg for every 100 g of the product. Vitamins E and C are added in accordance with the original Ukrainian recipe, giving final concentrations of 0.5 mg and 10 mg, respectively, per 100 g.

3.5 CHOLESTEROL STUDIES

3.5.1 EARLY WORK

The observation that the Masai people had low serum cholesterol levels in spite of their high dietary cholesterol intakes[15] initiated interest in the beneficial effects of fermented milk on cholesterol metabolism, because the Masai consumed large quantities of fermented cows' milk. Several studies were then carried out in animal models (rat, rabbit) and humans to investigate the properties of milk fermented with a variety of bacteria on cholesterol metabolism,[38–40] with mixed results.

Zacconi et al.[41] used axenic mice to show that when mice were fed a hypercholesterolemic diet for 60 days, animals reared in sterile boxes had higher serum cholesterol levels than those dosed with various bacteria. Although the quantity of bacteria given to each animal was not indicated, it was evident that the largest reductions in serum cholesterol levels occurred in mice given *E. faecium* (females: –16.9%; males: –7.8%) and *Lactobacillus acidophilus* (females: –11.4%; males: –5.3%). Zacconi et al.[41] found that the reductions were greater in female mice and that the animals receiving the *E. faecium* were healthier than the axenic mice.

3.5.2 STUDIES WITH *ENTEROCOCCUS FAECIUM*

Mikeš et al.[42] carried out a human feeding trial in which subjects were given lyophilized *E. faecium* M-74 (5 × 10^9/day) in capsules for 6 weeks. The mean number of *E. faecium* in fecal samples generally rose and plateaued during the period of dosing and then fell slowly during the following 5 weeks. However, large individual differences in the numbers of *E. faecium* recovered in fecal samples were noted. Fecal ß-D-glucuronidase activity was measured during the experiment, and it was found that activity decreased during the dosing period and remained low 5 weeks after the dosing with *E. faecium* was stopped. The serum LDL and total serum cholesterol rose during the initial weeks of the study; both parameters dropped significantly (compared to values obtained before the study) 14 days after consumption of the bacteria was discontinued. Conversely, serum HDL values rose after the bacteria treatment stopped. There were no changes in other blood parameters.

To better understand how the bacteria might be altering cholesterol metabolism, the metabolic activation of neutrophils was measured, since it is known that neutrophils are capable of producing reactive oxygen intermediates that can oxidize lipoproteins and thereby contributing to atherosclerosis. Mikeš et al. found that during the time when LDL and total cholesterol were lowered, stimulated neutrophils from subjects had increased ability to reduce 3-(4-iodophenyl)-2-(4-nitrophenyl)-5-phenyltetrazolium chloride (INT) and to produce superoxide. This finding was consistent with an earlier observation that *E. faecium* was able to stimulate human peripheral neutrophils resulting in the production of reactive oxygen intermediates *in vitro*.[43] The observed reciprocal correlation between cholesterol levels and neutrophil INT reductase activity and superoxide production raised the possibility that *E. faecium* might be affecting some oxidative process, which in turn reduced cholesterol levels.

3.5.3 Studies with *Streptococcus thermophilus*

Streptococcus thermophilus and *Lactobacillus bulgaricus* are the two bacteria added to milk to produce yogurt. The effect, if any, of *S. thermophilus* on serum cholesterol is not clear because feeding trials with animals and humans have tested *S. thermophilus* in combination with other LAB. For example, Akalin et al.[44] reported a study where they fed mice a chow diet and yogurt containing either *S. thermophilus* and *Lactobacillus delbrueckii* ssp. *bulgaricus* or *S. thermophilus* and *Lb. acidophilus*. The fresh yogurt had 10^7 lactobacilli/ml. The number of *S. thermophilus* in the yogurt was not measured. The animals eating the yogurts had lower serum cholesterol levels and LDL cholesterol (significantly lower in the case of *S. thermophilus* and *Lb. acidophilus*) than control mice that ate chow. Serum triglycerides did not appear to be affected. These authors concluded that the effects on cholesterol metabolism were attributable to the *Lb. acidophilus* in the yogurt, and that the *S. thermophilus* had no effect.

Kawase et al.[45] showed that *Lactobacillus casei* TM0409, *S. thermophilus* TMC1543, and whey protein concentrate had a synergistic effect on lowering serum cholesterol in rats. They then used the two bacteria to produce a fermented milk that contained 6.1×10^8 *Lb. casei*/ml and 2.6×10^7 *S. thermophilus*/ml and fed it to male human volunteers for 8 weeks (200 ml, twice a day). HDL cholesterol was significantly increased after 4 weeks, and the increase continued until week 8. Total serum cholesterol was lower, but not statistically lower, in the group receiving the fermented milk compared to the group receiving a placebo. No mechanism was proposed to explain the hypocholestrolemic effect of the fermented milk.

3.5.4 Human Trials with Gaio

The first study to utilize a milk product fermented with bacteria isolated from Abkhasia was carried out by Sarkisov at Kiev University in Ukraine. However, the results were not published. These authors showed a cholesterol-lowering effect of approximately 39 mg/dl, an increment in HDL-cholesterol, and a reduction in triglycerides after 4 weeks of dietary supplementation with the new product in a very heterogeneous group of males and females.[46]

Four studies in which the effect of Gaio consumption on blood lipid profiles was evaluated can be found in the literature. The first published trial to test the effects of Gaio (a product practically identical with the Ukrainian one, though produced in Denmark) on LDL cholesterol levels was performed by Agerbaek et al.[47] They studied 58 male volunteers of Danish descent, all 44 years old, with normal fasting cholesterol values. All were selected from a cohort examined in 1989 and again in 1990 in a study of the prevalence of risk factors for coronary heart disease at the University Hospital of Aarhus. They were selected on the basis of having had normal fasting values of serum cholesterol (5.0 to 6.5 mmol/l) and triglycerides (<5.0 mmol/l) at both examinations; no history of cardiovascular, cerebrovascular, or metabolic disease; normal weight (body mass index [BMI] < 27.5) with a stable weight; alcohol consumption <315 g/week; and normal blood pressure (<150/95 mm Hg). During the intervention period, the subjects maintained their habitual diets supplemented with 200 ml/d of either Gaio or a similar placebo product (chemically fermented). Fasting blood samples were drawn initially and after 3 and 6 weeks and analyzed for plasma values of total cholesterol, HDL cholesterol, and triglycerides. LDL cholesterol was estimated by the Friedewald formula (LDL cholesterol = total cholesterol – HDL cholesterol – VLDL cholesterol [triglycerides/5]).[48] After 6 weeks, the consumption of Gaio produced an average reduction of 6% in total serum cholesterol levels, completely ascribed to a decrease in LDL cholesterol level of 10%, since HDL cholesterol and triglycerides were unchanged. The authors stated that although the findings were promising, the results could not be extrapolated to other human subjects who were not included in their study. They suggested further studies to determine the potential effects of the new fermented milk product on lipoproteins among women, the elderly, and subjects with manifest hypercholesterolemia. They also stated that the potential beneficial effect for middle-aged men would be true only if the cholesterol-lowering effect persisted for longer periods than the 6 weeks investigated in their study. This is based in the findings of prevention studies using drugs. Those studies showed that improvement in vascular risk only begins to appear after 1 to 2 years of drug use and cholesterol reduction.[9]

Following this line of reasoning, Richelsen et al.[49] studied the consumption of Gaio for a longer period of time in a randomized, double-blind, placebo-controlled trial that included 87 nonobese and normocholesterolemic females and males, aged 50 to 70 years. The volunteers were recruited through an announcement in the local newspaper. Before inclusion in the experiment, blood samples were drawn to eliminate subjects with liver disease, diabetes, kidney disease, anemia, and hypercholesterolemia (total cholesterol > 8.0 mmol/l). Inclusion criteria were: healthy men and women age 50 to 70 years (the women were all postmenopausal); body mass index < 32 kg/m^2; no medication influencing plasma lipids. Participants were instructed not to change their ordinary diet, alcohol intake, level of physical exercise, and tobacco consumption during the study period. They consumed 200 ml of either the fermented milk product (Gaio) or a placebo (chemically fermented). The study showed a rapid reduction in LDL cholesterol level by about 8% after 1 month, but after 6 months, although the effect remained, the reduction in LDL cholesterol was similar to the reduction observed in the placebo group. The authors reasoned that after 1 month of the placebo use there was a gradual fall in total and LDL cholesterol in both genders, making the interpretation of the results less clear cut. Thus, after 6 months the levels of total cholesterol and LDL cholesterol were significantly lower than initial values in both groups, but the reduction and the absolute values were not different in the two groups (treatment vs. placebo). They speculated that the placebo product without the bacterial culture but chemically fermented with an organic acid (gluconic acid-Δ-lactone), containing 1.5% fat, itself could have cholesterol-lowering effects or alternatively that the seasonal variation in subjects' lipid levels[50] could explain the results. The authors also suggested that some people participating in the trial could be "responders" while others could be "nonresponders," but the basis for this biological (genetic) phenomenon is still unknown. Another possibility suggested by the authors was that the absence of effect could be due to lack of statistical power to detect a 4 to 5% reduction in blood cholesterol due to the small number of subjects included in the study.

Bertolami et al.[51] tested the effect of Gaio in a group of patients with primary hypercholesterolemia (11 men and 21 women, 36 to 65 years old) who had not shown a significant improvement in LDL cholesterol level after dietetic modifications alone (phase I of the American Heart Association — National Cholesterol Education Program (NCEP) Adult Treatment Panel II). A prospective, randomized, double-blind, 8-week crossover design, controlled by placebo (a chemically fermented milk) was used. After initial clinical and laboratory analysis, the patients began to consume 200 g daily of Gaio or the placebo in a randomized and double-blind manner. Seventeen patients started the trial using the active product and 15 began with placebo. After 8 weeks, blood was collected again for lipid profile evaluation and the crossover was made (those consuming Gaio changed to placebo and vice versa). After an additional 8-week period, blood was collected for the last lipid profile determination. The results showed that Gaio was able to significantly reduce total cholesterol by 5.3% and LDL cholesterol by 6.15% in these hypercholesterolemic subjects compared with the placebo product. Bertolami et al. also suggested the possibility of different patient responses ("responders" and "nonresponders") to the use of Gaio and the placebo product (Tables 3.1 and 3.2).[51]

Recently, Agerholm-Larsen[52] studied the effects of Gaio in 70 overweight and obese subjects in a randomized, double-blind, placebo- and compliance-controlled, parallel protocol. The study was designed as a five-armed parallel study in which Gaio was compared with two other fermented milk products, a chemically acidified milk product with an organic acid (delta-acid-lactone) instead of a living bacterial culture, and an inert placebo pill. One comparison milk product was fermented with two strains of *S. thermophilus* (1×10^8 CFU/ml) and two strains of *Lb. acidophilus* ($\sim 2 \times 10^7$ CFU/ml), and the other was fermented with two strains of *S. thermophilus* ($\sim 8 \times 10^8$ CFU/ml) and one strain of *Lactobacillus rhamnosus* ($\sim 2 \times 10^8$ CFU/ml). The protocol involved, besides lipid level evaluation, the determination of fibrinogen and C reactive protein because these two acute-phase proteins are also implicated as risk factors for coronary artery disease in healthy men.[53,54] The authors planned to offer an increased amount of fermented milk product (450 ml daily) to obtain a greater effect on LDL cholesterol. Participants were healthy, weight stable, overweight or obese ($25.0 < BMI < 37.5$ kg/m^2), 20 men and 50 women, 18 to 55 years old. Participants were instructed not to change their habitual diets, level of physical exercise, tobacco and alcohol habits, or body weight during the study period. Compliance was tested at home every second week (weeks 2, 4, 6, and 8) by analyzing

TABLE 3.1
Effects of the Fermented Milk Product Gaio on Serum Lipid and Lipoprotein Levels in an 8-Week Trial in Hypercholesterolemic Subjects

	Diet Only	Diet + Placebo	Diet + Gaio
Total cholesterol (mg/dl)	248.47 (±26.75)	249.09 (±28.45)	235.75 (±35.03)[a,b]
Triglycerides (mg/dl)	119.16 (±49.28)	118.38 (±39.47)	116.66 (±38.79)
HDL cholesterol (mg/dl)	52.38 (±14.00)	56.91 (±15.96)[c]	54.41 (±14.97)
LDL cholesterol (mg/dl)	172.22 (±21.17)	168.59 (±24.18)	158.00 (±31.04)[d,e]
Weight (kg)	66.55 (±10.57)	66.05 (±10.37)[f]	65.97 (±10.60)[g]

Note: All values are medians ± standard deviation.

[a] Comparison between diet only and active product; $P = 0.012$.
[b] Comparison between placebo and active product; $P = 0.004$.
[c] Comparison between diet only and placebo; $P = 0.001$.
[d] Comparison between diet only and active product; $P = 0.002$.
[e] Comparison between placebo and active product; $P = 0.012$.
[f] Comparison between diet only and placebo; $P = 0.026$.
[g] Comparison between diet only and active product; $P = 0.014$.

Source: Bertolami, M.C., Faludi, A.A., and Batlouni, M., *Eur. J. Clin. Nutr.*, 53, 97–101, 1999. With permission.

TABLE 3.2
Percent Changes in Total Cholesterol and LDL Cholesterol in an 8-Week Trial of Gaio in Hypercholesterolemic Subjects

	Phase Comparison	Mean (SD)	Maximum Decrease	Maximum Increase
Total cholesterol	D vs. P	+0.39 (6.82)	−11.38	+13.45
	D vs. Y	−5.04 (10.44)	−24.26	+23.10
	P vs. Y	−5.30 (9.38)	−23.40	+18.77
LDL cholesterol	D vs. P	−1.90 (9.39)	−19.87	+16.02
	D vs. Y	−8.29 (13.29)	−37.69	+23.78
	P vs. Y	−6.15 (12.73)	−30.00	+18.65

Note: D = diet only; P = diet plus placebo; Y = diet plus yogurt; SD = standard deviation.

Source: Bertolami, M.C., Faludi, A.A., and Batlouni, M., *Eur. J. Clin. Nutr.*, 53, 97–101, 1999. With permission.

sample product bags labeled with ^{13}C-acetate by a nondispersive infrared spectroscopy method.

After 4 weeks of consuming 450 ml of Gaio a day, there was no decrease in LDL cholesterol levels. After 8 weeks, these levels decreased significantly by 8.4%, and fibrinogen increased significantly compared to the placebo group, while C reactive protein did not change. The authors expected to see differences in lipids after 4 weeks as demonstrated in previous comparable studies but found none. Although they had no obvious explanation for the lack of reduction in LDL cholesterol levels after 4 weeks, they speculated that a possible mechanism behind this finding could perhaps be the small number of subjects in each group. According to the authors, it is likely that there is some between-subject variability in intestinal colonization of the active bacteria that could influence the results. As fibrinogen is an acute-phase protein, the authors speculated that the increased fibrinogen concentration found in the Gaio group could be attributed to immunostimulation. They also suggested that it was not possible to exclude the idea that a transient colonic inflammation caused by the bacterial strains in Gaio was the reason for the increase in fibrinogen in subjects consuming this fermented milk product. However, the lack of any increase in C reactive protein does not support this possibility.

3.6 META-ANALYSIS OF CHOLESTEROL EXPERIMENTS

In a recently published meta-analysis, Agerholm-Larsen et al.[55] questioned the conflicting results pertaining to this product's efficacy in reducing plasma cholesterol. They analyzed six studies conducted with Gaio involving 425 subjects of both genders and different initial cholesterol levels, concluding that five studies showed a small beneficial short-term effect of 6 to 10% on serum LDL cholesterol (Sarkisov et al.,[46] Agerbaek et al.,[47] Richelsen et al.,[49] Bertolami et al.,[51] Agerholm-Larsen et al.[52]). However, the long-term effect was inconclusive (Richelsen et al.[49]), and one study failed to demonstrate any effect at all (Sessions et al.[56]). The authors pointed out that a meta-analysis suffers from unpublished data material because negative studies often remain unpublished, leading to publication bias. However, they affirmed that they had no knowledge of other unpublished material on the Causido culture. As a conclusion, they suggested that the meta-analysis based on the five controlled study interventions showed that the intake of the fermented milk product (Gaio) produced a statistically significant and clinically important reduction in plasma cholesterol. They found a reduction of 5% in LDL cholesterol, which is considered large enough to have a beneficial effect on risk factors for coronary heart disease.[9] However, they emphasized that long-term studies on Causido are required to document whether a sustained effect on blood lipids occurs.

A summary of the results of the studies evaluating the effects of consumption of Gaio on plasma lipid profiles is shown in Table 3.3. Studies are listed by author, type of population enrolled (normo- or hypercholesterolemic), and the percent changes of total cholesterol and LDL cholesterol comparing active treatment vs. placebo (only the study from Sarkisov was open and did not involve a placebo control).

TABLE 3.3
Human Studies of the Effect of Gaio on Total and LDL Cholesterol

Study Group	% Decrease in Total Cholesterol	% Decrease in LDL Cholesterol	Ref.
Hypercholesterolemic	15.87	16.85	Sarkisov (unpublished results, 1991)
Normocholesterolemic	6.08	9.77	47
Normocholesterolemic	3.51 (not significant)	5.96 (not significant)	49
Hypercholesterolemic	0.17 (not significant)	3.6 (not significant)	56
Hypercholesterolemic	5.3	6.15	51
Normocholesterolemic	5.14	6.64	52

3.7 MECHANISM OF ACTION

The reason for the observed hypocholesterolemic effect in subjects consuming Gaio is not fully understood at this time. It has been proposed that the cholesterol-lowering effect is related to the bacterial culture in the product.[47] A potential explanation given by Agerholm-Larsen et al.[52] is that an association exists between the gut microflora and cholesterol absorption in the small intestine. The intestinal bacteria can bind bile acids to cholesterol, resulting in the excretion of bile acid–cholesterol complexes in the feces. Decreased bile acid recycling through the enterohepatic circulation would result in cholesterol uptake from the circulation into the liver for *de novo* synthesis of bile acids. Another possible explanation was provided by St-Onge et al.[22] High numbers of bacteria in products such as yogurt when consumed will ensure passage of sufficient numbers of bacteria into the intestine to exert effects on metabolism. Because the bacteria contained in a fermented milk product are consumed with macronutrients that alter the stomach's pH (buffering effect), bacterial survival is increased, and the bacteria pass into the small intestine and then into the large intestine. Once in the large intestine, the bacteria ferment indigestible carbohydrates and produce short-chain fatty acids. The relative proportions of this production are likely to alter cholesterol synthesis. Gaio contains *S. thermophilus* and *E. faecium. In vitro* studies have shown that *S. thermophilus* is acid sensitive and cannot survive the passage through to the small intestine. However, *E. faecium* is known to have good bile tolerance. Consequently, Agerholm-Larsen et al. suggest that *E. faecium* is the bacterial strain with the cholesterol-lowering effect.[52]

3.8 OTHER PROPERTIES OF *ENTEROCOCCUS FAECIUM*

3.8.1 Treatment of Diarrhea

One of the most obvious applications of probiotics is to restore the intestinal microflora population during diarrhea. A variety of LAB, including *Lb. rhamnosus* GG,

Lb. reuteri, Streptococcus boulardii, and *E. faecium,* have been tested for their ability to reduce the severity and/or duration of various diarrheas.[57] A group (78 people) of sufferers of acute diarrhea were found to have a lower frequency of diarrhea after 7 days of treatment with *E. faecium,* compared to those receiving a placebo.[58] Results were not as positive in a 3-day study involving patients suffering from diarrhea due to infection with *Vibrio cholerae* (114 subjects) and enterotoxigenic *Escherichia coli* (41 subjects). In this study, there was no difference in the duration of the diarrhea between the placebo group and those receiving the *E. faecium.*[59]

3.8.2 Antimutagenicity

Milk fermented with *E. faecium* has been investigated for its antimutagenic properties. Belicová et al.[60] used an ether extract of milk that had been fermented with *E. faecium* to carry out a variety of tests. They reported that their milk extract showed a dose-dependent inhibition of mutagenesis induced by chemical and physical mutagens. Using both *Salmonella typhimurium* TA97 and TA100, they showed that the fermented milk extract (4 μl/plate) could reduce UV irradiation damage by 72 and 55% (for TA97 and TA100, respectively). A 10 μl dose of extract produced about a 30% reduction in the mutagenicity of nitrovine; the same dose reduced 5-nitro-2-furylacrylic acid mutagenicity by up to 25% and *N*-methyl-*N'*-nitro-*N*-nitrosoguanidine (MNNG) mutagencity by 28%.

Earlier, Ebringer et al.[43] showed that viable *E. faecium* M-74 showed significant antimutagenic effects on nitrovin and 2-aminofluorene mutagenicity (Ames test), but heat treated nonviable cells did not. They also reported that *E. faecium* cells themselves showed no immunostimulatory activity, but mixtures of *E. faecium* and phagocytes significantly stimulated mean INT-reductase activities. Ebringer et al. concluded that *E. faecium* contained factors that could reduce the effect of the mutagens they tested and that heat stable proteins might be responsible.

3.8.3 Cheesemaking

Enterococcus faecium has been shown to be a good starter adjunct in the production of cheddar cheese. Gardiner et al.[61] showed that 9 months after adding 2×10^7 CFU/ml (0.1%) *E. faecium* PR88 to a commercial lactococcal starter, 3×10^8 CFU/g were viable. Cheese with the added bacteria was found to have increased proteolysis and higher levels of some odor-active compounds. The cheddar cheese containing the *E. faecium* was judged (by a commercial grader) to be ripening faster and had a "better flavor" than the control cheese. This study confirmed previous reports about the positive effects of enterococci on ripening and flavor development in cheddar cheese.[62–64]

3.9 CONCLUSIONS

After analyzing the results of published studies using Gaio as a possible option to obtain reduction in total and LDL cholesterol levels, it can be concluded that more studies are needed to ascertain:

- Whether the effects of prolonged consumption of Gaio on blood lipid profiles are the same as those observed during shorter periods of consumption
- Whether the final beneficial effect will lead to an improvement in coronary heart disease prevention with fewer events and fewer deaths due to this worldwide health problem of modern societies

REFERENCES

1. Tunstall-Pedoe, H., Kuulasmaa, H., Amouyel, P., Arveiler, D., Rajakangas, A.M., and Pajak, A., WHO-MONICA Project: Myocardial infarction and coronary deaths in the World Health Organization Monica Project: registration procedures, event rates, and case-fatality rates in 36 populations from 21 countries in four continents, *Circulation,* 90, 583–612, 1994.
2. American Heart Association, *Heart and Stroke Facts: 1995 Statistical Supplement,* Dallas, American Heart Association, 1994.
3. Farmer, J.A. and Gotto, A.M. Jr., Dyslipidemia and other risk factors for coronary artery disease, in *Heart Disease: A Textbook of Cardiovascular Medicine*, 5th ed., Brawnwald, W.B., Ed., Saunders Company, Philadelphia, 1997, pp. 1126–1160.
4. Stamler, J., Wentworth, D., and Neaton, J.D. for the MRFIT Research Group, Is the relationship between serum cholesterol and risk of premature death from coronary heart disease continuous and graded? Findings in 356,222 primary screenees of the Multiple Risk Factor Intervention Trial (MRFIT), *JAMA,* 256, 2823–2828, 1986.
5. Kannel, W.B., Range of serum cholesterol values in the population developing coronary artery disease, *Am. J. Cardiol.,* 76, 69C–77C, 1995.
6. Martin, M.J., Browner, W.S., Wentworth, D., Hulley, S.B., and Kuller, L.H., Serum cholesterol, blood pressure and mortality — implications from a cohort of 361,662 men, *Lancet,* 2, 933–936, 1986.
7. Executive Summary of the Third Report of the National Cholesterol Education Program (NCEP) Expert Panel on Detection, Evaluation, and Treatment of High Blood Cholesterol in Adults (Adults Treatment Panel III), *JAMA,* 285, 2486–2497, 2001.
8. Downs, J.R., Clearfield, M., Weis, S., Whitney, E., Shapiro, D.R., Beere, P.A., Langendorfer, A., Stein, E.A., Kruyer, W., and Gotto, A.M. Jr. for the AFCAPS/TexCAPS Research Group, Primary prevention of acute coronary events with lovastatin in men and women with average cholesterol levels. Results of AFCAPS/TexCAPS, *JAMA,* 279, 1615–1622, 1998.
9. Gould, A.L., Rossouw, J.E., Santanello, N.C., Heyse, J.F., and Furberg, C.D., Cholesterol reduction yields clinical benefit: impact of statin trials, *Circulation,* 97, 946–952, 1998.
10. Simmering, R. and Blaut, M., Pro- and prebiotics: the tasty guardian angels? *Appl. Microbiol. Biotechnol.,* 56, 19–28, 2001.
11. Delzenne, N.M. and Williams, C.M., Prebiotics and lipid metabolism, *Curr. Opin. Lipidol.,* 13, 61–67, 2002.
12. Walker, W.A. and Duffy, L.C., Diet and bacterial colonization: role of probiotics and prebiotics, *J. Nutr. Biochem.,* 9, 668–675, 1998.
13. Salminen, S., Bouley, C., Boutron–Ruault, M.C., Cummings, J.H., Franck, A., Gibson, G.R., Isolauri, E., Moreau, M.C., Roberfroid, M., and Rowland, I., Functional food science and gastrointestinal physiology and function, *Br. J. Nutr.,* 80, S147–S171, 1998.

14. Metchnikoff, E., *The Prolongation of Life,* 1st ed., G.P. Putnam's Sons, New York, 1908.
15. Mann, G.V. and Spoerry, A., Studies of a surfactant and cholesterolemia in the Masai, *Am. J. Clin. Nutr.,* 27, 464–469, 1974.
16. Gilliland, S.E., Acidophilus milk products: a review of potential benefits to consumers, *J. Dairy Sci.,* 72, 2483–2494, 1989.
17. Hitchins, A.D. and McDonough, F.E., Prophylactic and therapeutic aspects of fermented milk, *Am. J. Clin. Nutr.,* 49, 675–684, 1989.
18. Berner, J.A., Roundtable discussion on milkfat, dairy foods, and coronary heart disease risk, *J. Nutr.,* 123, 1175–1184, 1993.
19. Sanders, M.E., Summary of conclusions from a consensus panel of experts on health attributes of lactic cultures: significance to fluid milk products containing cultures, *J. Dairy Sci.,* 76, 1819–1828, 1993.
20. Eicholzer, M. and Stähelin, H., Is there a hypocholesterolemic factor in milk and milk products? *Int. J. Vit. Nutr. Res.,* 63, 159–167, 1993.
21. Mission Scientifique de Syndifrais, Yaourts, laits fermentés, *Lait,* 77, 321–358, 1997.
22. St-Onge, M.P., Farnworth, E.R., and Jones, P.J.H., Consumption of fermented and nonfermented dairy products: effects on cholesterol concentrations and metabolism, *Am. J. Nutr.,* 71, 674–681, 2000.
23. Frobisher, M., Ed., *Fundamentals of Microbiology,* 9th ed., W.B. Saunders Co., Philadelphia, 1974.
24. Ruoff, K.L., Recent taxonomic changes in the genus *Enterococcus, Eur. J. Clin. Microbiol. Infect. Dis.,* 9, 75–79, 1990.
25. Devriese, L.A., Pot, B., and Collins, M.D., Phenotypic identification of the genus *Enterococcus* and differentiation of phylogenetically distinct enterococcal species and species groups, *J. Appl. Bacteriol.,* 75, 399–408, 1993.
26. Hardie, J.M. and Whiley, R.A., Classification and overview of the genera *Streptococcus* and *Enterococcus, J. Appl. Microbiol.,* Sympos. Suppl., 83, 1S–11S, 1997.
27. Chenoweth, C., and Schaberg, D., The epidemiology of enterococci, *Eur. J. Clin. Microbiol. Infect. Dis.,* 9, 80–89, 1990.
28. Noble, C.J., Carriage of group D streptococci in the human bowel, *J. Clin. Pathol.,* 31, 1182–1186, 1978.
29. Murray, B.E., The life and times of the enterococcus, *Clin. Microbiol. Rev.,* 3, 46–65, 1990.
30. Collins, J.K., Thornton, G., and Sullivan, G.O., Selection of probiotic strains for human applications, *Int. Dairy J.,* 8, 487–490, 1998.
31. Salminen, S., von Wright, A., Morelli, L., Marteau, P., Brassart, D., de Vos, W.M., Fondén, R., Saxelin, M., Collins, K., Mogensen, G., Birkeland, S-E., and Mattila-Sandholm, T., Demonstration of safety of probiotics — a review, *Int. J. Food Microbiol.,* 44, 93–196, 1998.
32. Int'veld, J.H.J. and Havenaar, R., Selection criteria and application of probiotic microorganisms in man and animal, *Microecol. Ther.,* 26, 43–57, 1997.
33. Canganella, F., Paganini, S., Ovidi, M., Vettraino, A.M., Bevilacqua, L., Massa, S., and Trovatelli, L.D., A microbiological investigation on probiotic pharmaceutical products used for human health, *Microbiol. Res.,* 152, 171–179, 1997.
34. Franz, C.M.A.P., Holzapfel, W.H., and Stiles, M.E., Enterococci at the crossroads of food safety? *Int. J. Food Microbiol.,* 47, 1–24, 1999.
35. Lund, B., Edlund, C., Barkholt, L., Nord, C.E., Tvede, M., and Poulsen, R.L., Impact on human intestinal microflora of an *Enterococcus faecium* probiotic and vancomycin, *Scand. J. Infect. Dis.,* 32, 627–632, 2000.
36. Adams, M.R., The safety of lactic acid bacteria, *BNF Nutr. Bull.,* 22, 91–98, 1997.

37. van der Kamp, J.W., Safety considerations regarding probiotic strains. Conclusions of two special workshops, *IDF Nutr. Newsletter*, 5, 27–28, 1996.
38. Kawai, Y., Effect of cellular extracts of streptococci on hyperlipidemia in rats, rabbits and humans, *Microecol. Ther.*, 14, 109–126, 1984.
39. Grunewald, K.K., Serum cholesterol levels in rats fed skim milk fermented by *Lactobacillus acidophilus, J. Food Sci.,* 47, 2078–2079, 1982.
40. Payens, W., Rethans, E.J.M., and de Waard, H., The influence of consumption of a large quantity of yoghurt or milk on the serum cholesterol concentration (in German), *Milchwissenschaft,* 31, 525–530, 1976.
41. Zacconi, C., Bottazzi, V., Rebecchi, A., Bosi, E., Sarra, P.G., and Tagliaferri, L., Serum cholesterol levels in axenic mice colonized with *Enterococcus faecium* and *Lactobacillus acidophilus, Microbiologica,* 15, 413–418, 1992.
42. Mikeš, Z., Ferenčík, M., Jahnová, E., Ebringer, L., and Čižnár, I., Hypocholesterolemic and immunostimulatory effects of orally applied *Enterococcus faecium* M–74 in man, *Folia Microbiol.,* 40, 639–646, 1995.
43. Ebringer, L., Ferenčik, M., Lahitova, N., Kačani, L., and Michalkova, D., Antimutagenic and immuno-stimulatory properties of lactic acid bacteria, *World J. Microbiol. Biotechnol.,* 11, 294–298, 1995.
44. Akalin, A.S., Gönc, S., and Dűzel, S., Influence of yogurt and acidophilus yogurt on serum cholesterol levels in mice, *J. Dairy Sci.,* 80, 2721–2725, 1997.
45. Kawase, M., Hashimoto, H., Hosoda, M., Morita, H., and Hosono, A., Effect of administration of fermented milk containing whey protein concentrate to rats and healthy men on serum lipids and blood pressure, *J. Dairy Sci.,* 83, 255–263, 2000.
46. Sarkisov, K.G., Podrushnyak, E.P., Grigorov, Y.G., quoted in reference 47.
47. Agerbaek, M., Gerdes, L.U., and Richelsen, B., Hypocholesterolaemic effect of a new fermented milk product in healthy middle-aged men, *Eur. J. Clin. Nutr.,* 49, 346–352, 1995.
48. Friedewald, W.T., Levy, R.I., and Fredrickson, D.S., Estimation of the concentration of low-density lipoprotein cholesterol in plasma without use of the preparative ultracentrifuge, *Clin. Chem.,* 18, 499–502, 1972.
49. Richelsen, B., Kristensen, K., and Pedersen, S.B., Long-term (6 months) effect of a new fermented milk product on the level of plasma lipoproteins — a placebo-controlled and double blind study, *Eur. J. Clin. Nutr.,* 50, 811–815, 1996.
50. Durrington, P. N., Biological variation in serum lipid concentrations, *Scand. J. Clin. Invest.,* 198, S86–S91, 1990.
51. Bertolami, M.C., Faludi, A.A., and Batlouni, M., Evaluation of the effects of a new fermented milk product (Gaio) on primary hypercholesterolemia, *Eur. J. Clin. Nutr.,* 53, 97–101, 1999.
52. Agerholm-Larsen, L., Raben, A., Hansen, A.S., Manders, M., and Astrup, A., Effect of 8 week intake of probiotic milk products on risk factors for cardiovascular diseases, *Eur. J. Clin. Nutr.,* 54, 288–297, 2000.
53. Heinrich, J., Balleisen, L., Schulte, H., Assman, G., and van de Loo, J., Fibrinogen and factor VII in the prediction of coronary risk. Results from the PROCAM study in healthy men, *Arterioscler. Thromb.,* 14, 54–59, 1994.
54. Ridker, P., Cushman, M., Stampfer, M., Tracy, R., and Hennekens, C., Inflammation, aspirin, and the risk of cardiovascular disease in apparently healthy men, *N. Engl. J. Med.,* 336, 973–979, 1997.
55. Agerholm-Larsen, L., Bell, M.L., Grunwald, G.K., and Astrup, A., The effect of a probiotic milk product on plasma cholesterol: a meta-analysis of short term intervention studies, *Eur. J. Clin. Nutr.,* 54, 856–860, 2000.

56. Sessions, V.A., Lovegrove, J.A., Dean, T.S., Williams, C.M., Sanders, T.A.B., Macdonald, I.A., and Salter, A.M., The effects of a new fermented milk product on plasma cholesterol and apoliprotein B concentrations in middle-aged men and women, in *Functional Foods — the Consumer, the Product and the Evidence,* Sadler, M.J. and Saltmarch, M., Eds., The Royal Society of Chemistry, London, 1998, pp. 15–19.
57. Elmer, G.W., Probiotics: "living drugs," *Am. J. Health-Syst. Pharm.,* 58, 1101–1109, 2001.
58. Wunderlich, P.F., Braun, L., Fumagalli, I., D'Apuzzo, V.D., Heim, F., Karly, M., Lodi, R., Politta, G., Vonbank, F., and Zeltner, L., Double-blind report on the efficacy of lactic acid-producing *Enterococcus* SF68 in the prevention of antibiotic-associated diarrhoea in the treatment of acute diarrhoea, *J. Internat. Med. Res.,* 17, 333–338, 1989.
59. Mitra, A.K. and Rabbani, G.H., A double-blind, controlled trial of Bioflorin (*Streptococcus faecium* SF68) in adults with acute diarrhea due to *Vibrio cholerae* and enterotoxigenic *Escherichia coli, Gastroenterology,* 99, 1149–1152, 1990.
60. Belicová, A., Krajčovič, J., Dobias, J., and Ebringer, L., Antimutagenicity of milk fermented by *Enterococccus faecium, Folia Microbiol.,* 44, 513–518, 1999.
61. Gardiner, G.E., Ross, R.P., Wallace, J.M., Scanlan, F.P., Jägers, P.P.J.M., Fitzgerald, G.F., Collins, J.K., and Stanton, C. Influence of a probiotic adjunct culture of *Enterococcus faecium* on the quality of cheddar cheese, *J. Agric. Food Chem.,* 47, 4907–4916, 1999.
62. Dahlberg, A.C. and Kosikowski, F.W., The development of flavour in American Cheddar cheese made from pasteurized milk with *Streptococcus faecalis* starter, *J. Dairy Sci.,* 31, 275–284, 1948.
63. Jensen, J.P., Reinbold, G.W., Washam, C.J., and Vedamuthu, E.R., Role of enterococci in cheddar cheese: growth of enterococci during manufacture and curing, *J. Milk Technol.,* 36, 613–618, 1973.
64. Neviani, E., Muchetti, G., Contarini, G., and Carini, S., Ruolo delle enterococcacae nei formaggi italiani. I. Loro presenza in fromaggi di monte ed impiego in un innesto selezionato, *Il Latte,* 7, 722–728, 1982.

4 Kefir: A Fermented Milk Product

Edward R. Farnworth and Isabelle Mainville

CONTENTS

0-8493-1372-4/03/$0.00+$1.50

4.1 INTRODUCTION

Kefir is a beverage produced by the action of lactic acid bacteria (LAB), yeasts, and acetic acid bacteria on milk. This complex mixture of microorganisms produces a distinctive fermented milk product with unique properties.

The first fermented foods may have been produced by accident. However, fermentation of foods such as milk became a widespread method of preservation before refrigeration was introduced or preservation procedures such as canning and pasteurization were developed and used to extend shelf life. It would appear from the oral tradition of kefir that fermentation of milk in skin bags as a way of preserving milk led to the production of the first kefir grains and started the long tradition of producing kefir. Kefir has been produced using milk from cows, ewes, goats, and buffalo[1–3] and has been sold in Europe under a variety of names including kephir, kiaphur, kefyr, képhir, kéfer, knapon, kepi, and kippe.[4] (See Chapter 1 for more details on the history of kefir.)

Traditional home production of kefir has been joined by commercial production in many countries, and this has helped to increase the consumption of kefir and to promote its reputation as being good for health. Kefir is at the same time a functional food and a probiotic, and there is growing evidence that this unique fermented milk product may indeed be helpful in many disease/infection conditions.

Early reports on kefir quote original research that was carried out in the former Soviet Union.[2,5] The composition of this unique product has not been fully described, and it is evident that more research is required on the microbiology of kefir before the product can be fully understood from a scientific point of view.[6,7] Consumer acceptance may be slow because of the unique taste of kefir. The reported health properties of kefir and the desire of consumers to consume probiotic products may open new markets.[8]

4.2 KEFIR GRAINS

Kefir is produced by adding either a starter culture called kefir grains directly or a percolate of the grains to milk. Kefir grains are a mass of several different bacteria and yeasts embedded in a complex matrix of protein and carbohydrate. The microorganisms in the kefir grains ferment the milk, and the grains can be recovered at the end of the fermentation process. The grains have been described as resembling elastic small florets similar to cauliflower in shape, yellow/white in color, and 20 to 30 mm in size.[9,10] Figure 4.1 is a photograph of kefir grains. A crude analysis of the grains shows that they are a mass of bacteria, yeasts, polysaccharides, and proteins with a chemical composition of 890 to 900 g/kg water, 2 g/kg lipid, 30 g/kg protein, 60 g/kg sugars, and 7 g/kg ash.[3] A study of the proteins in kefir grains using SDS-PAGE on acrylamide gels indicated that the major grain proteins had a higher molecular weight than milk proteins, indicating that they were not proteolysis products.[11]

4.2.1 The Microorganisms in Kefir Grains and Kefir

The bacteria, yeasts, polysaccharides, and proteins in kefir grains added to milk produce kefir. Usually there is no pasteurization step after fermentation, and therefore

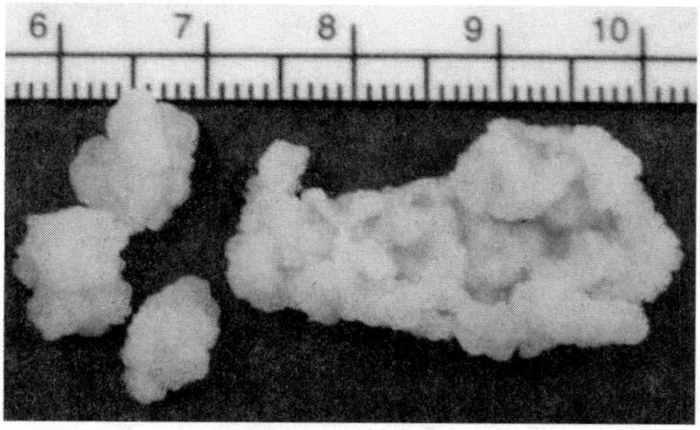

FIGURE 4.1 Kefir grains (scale in cm).

live bacteria and yeasts are found in the final product. In order to understand the fermentation process better and also to evaluate the health benefits of eating kefir, the microbiological profile of kefir has been studied by many researchers. The isolation and identification of microorganisms in kefir grains and kefir depends on the choice of suitable growth media, and more recently, sophisticated methods of identification have been used. It should be noted here that the identity of several of the microorganisms found in kefir has been revised over time as more definitive methods of identification have been used. In some cases, the nomenclature assigned to various bacterial species has also changed. The names of microorganisms used in this chapter are the names used in the original scientific articles cited.

Studies on the selection of a suitable selective growth medium for *Lactobacillus* species are numerous, but the media are often not suitable for the growth of some *Lactobacillus* species found in kefir or kefir grains.[12–14] Fujisawa et al.[15] observed that *Lactobacillus kefiranofaciens* grew on KPL agar at 30°C, but not on BL and MRS agars. Kojima et al.[16] found that Rogosa medium in 100% cheese–whey solution (Rogosa-CW) gave the best results for the isolation and cultivation of lactobacilli from kefir grains. Farnworth and Mainville[17] compared MRS, KPL, and Rogosa-CW to the lactic acid whey medium (LAW)[18] and found LAW to have the best recovery and growth rate for the lactobacilli present in kefir and kefir grains. This was especially true for *Lactobacillus kefirgranum* isolates.

Many studies have been carried out to identify the various bacteria and yeasts in kefir grains and in the final product.[19–23] Early studies revealed that many of the bacteria isolated are closely related and therefore difficult to isolate and identify.[20] As noted by Koroleva,[24] studying and monitoring kefir grains is difficult because when the various microorganisms are separated as pure cultures, they do not grow in milk or have decreased biochemical activity. Because of this, kefir has been cited as an example of symbiosis;[25] the growth and survival of individual strains are dependent on the presence of others. Toba[26] indicated several possible metabolic products that might contribute to the symbiotic relationship in kefir grains.

The growth of *Lactobacillus kefir* was enhanced when *Candida kefir* was added, either before or simultaneously, to the milk to be fermented.[27] Growth of the yeast, however, was not stimulated by the presence of the bacterium. When the two organisms were cultured together, the amounts of lactic acid, glycerol, and ethanol

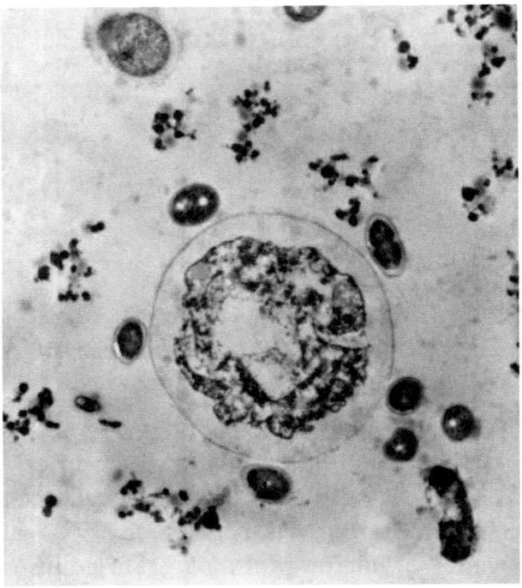

FIGURE 4.2 Electron micrograph of kefir showing symbiosis between bacteria and yeast. (Photo courtesy of D. Montpetit, Agriculture and Agri-food Canada.)

produced were increased. The growth of several bacteria isolated from kefir grains is improved when yeast extract is added to the growth medium,[17] indicating that the yeasts found in kefir grains are essential to maintain the integrity and viability of the microflora population. Vitamins, amino acids, and other essential growth factors for bacteria are produced by yeasts, while bacterial metabolic endproducts are used as energy sources by yeasts.[28] Figure 4.2 is an electron micrograph of kefir grains showing bacteria surrounding a yeast cell; the close proximity presumably indicates some sort of physical and chemical interaction.

The symbiosis found in the kefir grain microorganism population allows the grains to maintain uniformity so that throughout the year the microbiological profile of kefir grains and the kefir drink remain stable in spite of variations in milk quality and the presence of antibiotics and other inhibiting substances.[29] The microorganism profile of the final product does not necessarily parallel that of the grains because of conditions (pH and other) during the fermentation process. Also, the location of the microorganisms in the grains may be a factor. Yeasts are generally found in the interior of the grains, while lactococci are found on the exterior. Therefore, the numbers of yeasts found in the final product are lower than those counted in the grains themselves, while lactococci are numerous in the final drink. The complex microbiological composition of kefir grains produces kefir. Therefore, unlike the situation with yogurt, the drink kefir cannot be used as a starter to produce more kefir.

Initially, traditional selective media were used to isolate and identify individual bacterial strains. Recent studies dealing with the microflora identification of kefir grains and kefir drink show the importance of using the tools of molecular biology in order to clearly characterize the bacteria and yeasts present in these types of products.[30–32] Earlier reports on the composition of kefir list many different types of

LAB, yeasts, and acetic acid bacteria.[33–37] These identifications were based on phenotypic characteristics, which are often not sufficient to identify an organism to the species level.

Bacteria such as *Lactobacillus kefir* and *Lactobacillus brevis,* which are phenotypically very similar, are a good example to demonstrate the need to use molecular tools for identification. Some researchers have reported that they were able to differentiate the two species based on the fermentation of sucrose, trehalose, and xylose,[38] but results were not always obvious since some strains of *Lb. brevis* will not ferment one or more of these sugars.[39] Table 4.1 summarizes the most recent studies published on the identification of kefir microflora. Studies performed by Angulo et al.,[23] Rohm et al.,[40] Pintado et al.,[38] and Lin et al.[41] are based on phenotypic traits of the strains isolated, while papers published by Takizawa et al.,[30] Wyder and Puhan,[31] and Wyder et al.[32] used restriction length polymorphism (RFLP), DNA/DNA hybridization, and other molecular tools to characterize the strains they have isolated. One would not expect the list of bacteria and yeasts composing the grains to be very extensive, nor should it vary significantly from one part of the world to another if good care, similar growth conditions, and proper sanitary conditions are maintained. One could even assume that if a contaminant species were to come in contact with the kefir population, it would probably not survive or its growth would be inhibited due to the production of compounds such as bacteriocins by the symbiotic flora of kefir. However, over time and under different growing conditions, kefir grains may change their microbial makeup and fermentation properties.[42]

Most of the microbiological studies done on kefir and kefir grains have centered on the identification of the constituent bacteria. In many fermented milk products, yeasts are not desirable; they cause spoilage because the low pH provides a selective environment for their growth.[39] For kefir, yeasts play a key role in the fermentation process and, even though the number of yeasts in the final drink is less than in the grains,[43] it is important to maintain the balance of bacteria to yeasts. Yeasts contribute to the unique characteristics of kefir. Rosi[19] was one of the first researchers to study the yeast in kefir grains using electron microscopy. She found that the yeasts tended to be located in the center and along the peripheral channels of the grains. Using morphological and physiological characteristics, DNA base composition, and electrophoresis patterns, Iwasawa et al.[44] identified *Torulopsis holmii* in commercial kefir grains. Engel et al.,[21] in their survey of commercial and home-produced kefir, found both lactose-fermenting and nonlactose-fermenting yeasts in most products, although some commercial products called "kefir" contained no yeasts.

The stability and composition ratio of the microorganisms in kefir between the time of production and time of consumption is important to consumers who are eating kefir as a probiotic for its health-promoting properties. Figure 4.3 shows the changes in the total number of lactococci, lactobacilli, and yeasts in commercial kefir that had been stored at 4°C for 42 days.[45] The patterns for the three categories of microorganisms studied were similar in natural (unflavored) or strawberry (15% by weight) flavored kefir. Numbers of lactococci declined by almost 2 log units by the end of 42 days. Even with this decline during storage, the number of microorganisms in the drink remained high enough for the product to be considered a probiotic.

TABLE 4.1
Microorganisms Identified in Kefir and Kefir Grains

Lactic Acid Bacteria	Yeasts	Others	Sources	Ref.
Lb. kefiranum, Lb. kefieranofaciens, Lb. kefir, Lb. parakefir	N.D.	N.D.	Christian Hansen (Copenhagen); kefir grains	30
Lb. brevis, Lb. viridescens, Lb. kefir, Lb. fermentum, Lb. rhamnosus, Lb. casei ssp. *tolerans, Lb. casei* ssp. *pseudoplantarum, Lb. acidophilus, Lb. gasseri, L. lactis* ssp. *lactis, S. salivarius* ssp. *thermophilus, Leuconostoc* spp.	*Torulaspora delbrueckii, Saccharomyces cerevisiae, S. unisporus, Candida kefyr, C. holmii, C. friedrichii, Kluyveromyces lactis, Pichia fermentans*	*Pedioccus* spp., *Micrococcus* spp., *Bacillus* spp., *Acetobacter* spp., *Escherichia coli*	Eight sources from northwest region of Spain; kefir grains	23
N.D.	*Kluyveromyces marxianus, Pichia fermentans, Saccharomyces cerevisiae, S. dairensis*	N.D.	Nineteen sources from Austria; kefir grains and kefir drinks	40
Lb. helveticus, Leuconostoc mesenteroides	*Kluyveromyces marxianus, Pichia fermentans*	N.D.	Taiwan; kefir grains	41
N.D.	*Candida kefyr, Kluyveromyces marxianus, Candida colliculosa, Torulaspora delbrueckii, Saccharomyces unisporus, Brettanomyces anomalus, Saccharomyces turicensis*	N.D.	Five sources of different origins (Toni AG, Zürich, Switzerland; Biolacta-Texrl, Poland; private sources); kefir grains and kefir drinks	31, 32
Lb. kefir, L. lactis ssp. *lactis*	*Saccharomyces unisporus*	N.D.	Portugal; kefir grains	38

Note: N.D., not determined.

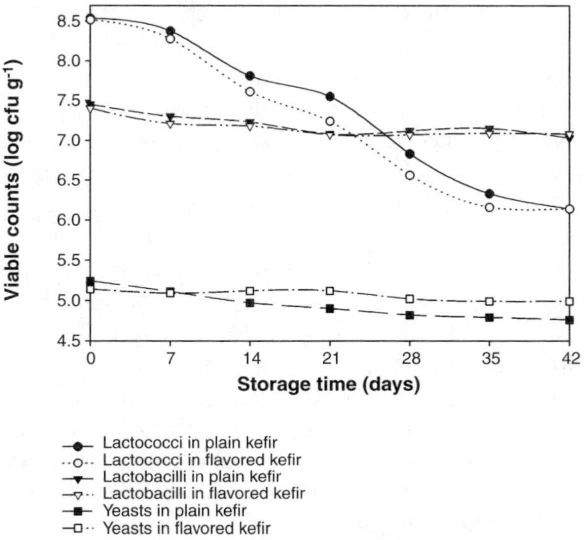

FIGURE 4.3 Changes in numbers of lactococci, lactobacilli, and yeasts during the storage of kefir.

4.2.2 Kefiran

Early observations of the structure of kefir grains noted that some of the bacteria were encapsulated by a polysaccharide.[33,36] La Riviére et al.[46] carried out the first studies of the polysaccharide and gave it the name kefiran. Kooiman[47] was able to show that the water-soluble polysaccharide consisted of approximately equal proportions of glucose and galactose. The chemical structure proposed is shown in Figure 4.4.

This structure was modified by Mukai et al.,[48] who used nuclear magnetic resonance (NMR) to show that 6-*O*-substituted galactose may also exist in the structure of kefiran. However, to date, no complete structural analysis with NMR confirmation has been published for kefiran.

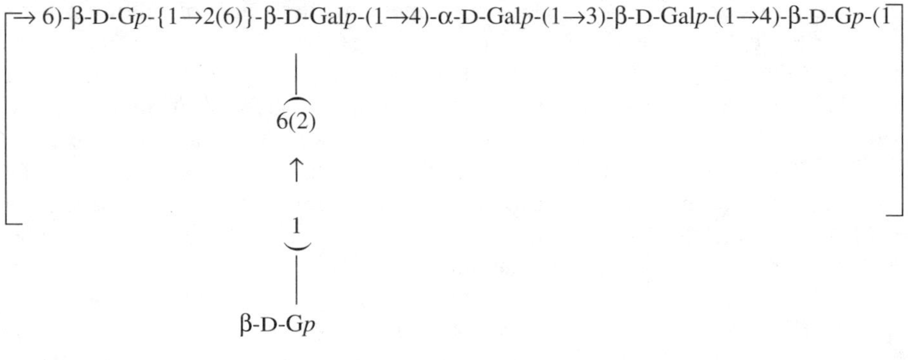

FIGURE 4.4 Chemical structure of kefiran. (From Kooiman, P., *Carbohyd. Res.*, 7, 200–211, 1968. With permission.)

It appears that kefir grains are the only source of kefiran, but it is not known whether more than one polysaccharide or various isomers of kefiran exist in kefir grains or kefir. Koroleva[24] quotes Russian references that claim kefir grains can contain up to 34% polysaccharide, while the final drink contains 0.2 to 0.7% polysaccharide.

The bacteria that produce kefiran have been the subject of additional research. La Riviére et al.[46] originally suggested that *Lb. brevis* was the bacterium producing kefiran. In 1983, Kandler and Kunath[49] identified *Lb. kefir* as the bacterium responsible. Toba et al.[50] isolated a polysaccharide-producing bacterium, which they named *Lactobacillus kefiranofaciens.*[15] *Lactobacillus* sp. KPB-167B was published by Yokoi et al.,[51] and *Lactobacillus kefirgranum* and *Lactobacilus parakefir* were reported by Takizawa et al.[30] as the responsible organisms. Yokoi et al.[52] were able to isolate five different polysaccharide-producing bacteria of the genus *Lactobacillus* from commercially available kefir grains. Yokoi et al. stated that this list of bacteria producing kefiran indicates the complexity of the taxonomic relationships of the bacteria isolated from kefir, and they implied that other bacteria found in kefir grains may be capable of producing kefiran-like polysaccharides, especially kefir grains originating from different geographical locations.

Toba's group tried to produce a new fermented milk drink using the kefiran-producing *Lb. kefiranofaciens* they had isolated from kefir grains, but the product had a ropy consistency and scored low on consumer acceptability criteria.[53] In spite of this, Toba et al.[53] stated that the health benefits of kefiran might be great enough to convince people to consume the product.

Several groups have tried to find media and growth conditions that would optimize the production of kefiran. Toba et al.[50,54] were able to produce 80 mg polysaccharide/l of medium using a whey-based growth medium that had increased glucose and added trypticase peptone compared to their standard whey-based medium. They also reported that the viscosity of the fermented drink was directly proportional to the polysaccharide content, and they used viscosity measures to gauge the efficiency of their different growth media.

The five different bacteria isolated by Yokoi et al.[52] were capable of producing 273 to 406 mg of total (supernatant + capsule) polysaccharide/l of medium. This group also used a whey-based growth medium. Recently, Yokoi and Watanabe[55] reported yields of 2.04 g polysaccharide/l of medium after 4 days of culturing, using a modified de Man, Rogusa, Sharp Lactose (MRSL) medium that contained 10% lactose. Micheli et al.[56] have been able to batch produce large quantities (2 g/l) of kefiran from a slime-forming rod-shaped *Lactobacillus* isolated from grains obtained from a local dairy in Italy, using the medium reported by Yokoi and Watanabe.[55]

It is not clear how many different bacteria synthesize kefiran or kefiran-like polysaccharides. Although kefiran is found in kefir grains, few reports have been published in which the concentration of kefiran in the final drink has been reported.[24] Kefiran may be contributing to the texture of kefir.

4.2.3 Electron Microscopy of Kefir Grains

Initially, light microscopy and later, transmission electron microscopy and scanning electron microscopy have been used extensively to study and describe the structure

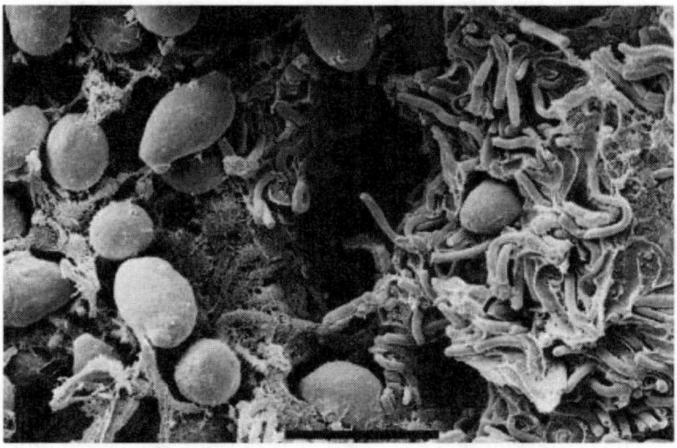

FIGURE 4.5 Electron micrograph of kefir grains showing bacteria and yeasts in carbohydrate/protein matrix. Magnification × 2555, bar indicates 10 μm. (Photo courtesy of M. Kalab, Agriculture and Agri-food Canada.)

and microbiology of kefir grains.[9,10,19,35,41,58–62] Figure 4.5 is a scanning electron micrograph of kefir grains obtained from the Moscow Dairy Institute that shows the matrix of bacteria, yeast, polysaccharide, and protein. Molska et al.[58] used electron micrographs to estimate the composition of kefir grains. They reported that their commercial grains from Poland contained 66% bacilli, 16% streptococci, and 18% yeasts. The microbiological population appears to be different on the surface of the grains than in the interior.[10] This may be due in part to differences in pH throughout the grain; the interior has been reported to have a very low pH that inhibits the growth of lactococci.[61] Generally, long rod-shaped bacteria predominate on the surface, while yeasts are concentrated in the interior of the grain.[41,57] This difference is important because the surface microorganisms are the ones that probably have the greatest impact on the fermentation process that produces kefir. Toba et al.[59] observed differences in the type of bacteria on the surfaces of propagable and nonpropagable grains and concluded that long rod-shaped bacteria with filamentous appendages were necessary for kefir grains to function. However, a large variation in the bacterial population occurs between grains and even within the same grain. Lin et al.[41] concluded that place of origin of the grains alone may explain the differences that have been reported in various electron microscopy studies.

A study of ways to preserve kefir grains showed that grains stored at –20°C for 120 days had the same microflora profile and produced kefir with the same rheological characteristics, acidity, and carbon dioxide content as kefir produced from nonstored grains. Grains stored at 4°C and then used to inoculate milk produced inferior quality kefir.[63] Lyophilizing kefir grains has also been used as a way of producing starter cultures for industrial production.[64]

Traditionally, kefir grains have been replicated in cows' milk. However, Abraham and de Antoni[11] have shown that the same grains that grow in cows' milk can grow and replicate in soybean milk. When soybean milk is used, the resulting grains are more compact and smaller in size, and they have a yellowish color — most likely due to the protein. When the grains are added to soybean milk, the pH is lower after 30 h than it is in cows' milk. After 20 subcultures, the soybean-grown grains had a

TABLE 4.2
Commercial Production of Kefir in Various Countries

Country	Year	Production	Comment	Ref.
USSR	1940	15,900 tons		3
USSR	1975	254,000 tons		3
Sweden	1985	17,000 tons		3
Poland	1982	22.5 million l		64
Poland	1988	35.3 million l		64
Poland	1997	255 million l	Kefir + buttermilk + flavored milk	71
Poland	1998	276 million l	Kefir + buttermilk + flavored milk	71
Poland	1999	327 million l	Kefir + buttermilk + flavored milk	71
Hungary	1996	116,337 tons	Fermented or acidified dairy products	72
Hungary	1997	119,255 tons	Fermented or acidified dairy products	72
Hungary	1998	138,986 tons	Fermented or acidified dairy products	72

higher protein content and a lower polysaccharide content than the milk-grown grains, although the microbial profiles of the two grains were similar.

4.3 COMMERCIAL KEFIR PRODUCTION

4.3.1 Size of Production

Kefir is a relatively new commercial product in North America, and this, combined with the fact that the majority of kefir is produced and consumed in eastern Europe, has made the gathering of statistics on kefir production and per capita consumption difficult.[65] In addition, many data gathering organizations do not differentiate between the various types of fermented (processed) milk products, and so kefir is often included with yogurt, buttermilk, and traditional Russian dairy products such as smetana, tvarog, blinchiki, and pelmeni. A sharp decrease in milk production and a substantial fall in the spending capacity of consumers have led to a significant decrease in the consumption of milk and milk products in the Ukraine and many eastern European nations.[66] However, the importance of kefir in the Russian diet is demonstrated by the fact that in 1992 it was included on the list of regulated commodities together with such other staples as milk, sugar, salt, and bread.[67] The industrial production of kefir has been large enough that Germany, Austria, Brazil, France, Luxembourg, Norway, Switzerland, Czechoslovakia, and Israel have regulations defining the composition and method of production of kefir.[68,69] In the United States, California has regulations concerning kefir milk.[70] Table 4.2 is a listing of production levels of kefir from various countries.

4.3.2 Methods of Production

Home production of kefir is practiced in many countries, and scientific studies have often used samples gathered from homes. The traditional method of kefir making is performed by adding kefir grains directly, as a starter, to pasteurized, cooled milk. In home production, fermentation temperature and time are not rigidly controlled. The final product cannot be used to inoculate new milk to produce kefir because the original balance of microorganisms in the grains has been disrupted, kefir grains are essential to the process.[73]

The properties of kefir (chemical, physical, and organoleptic) were initially difficult to duplicate in large-scale production. Some processes were developed where no grains were used, but the quality of the product was much different from that of kefir fermented with grains.

Large-scale production of kefir has been hindered by problems involved in reproducing the kefir grains and producing a consistent product. However, production of kefir on an industrial scale is common in many European countries, and patents have been taken out in several countries describing the process.[74–77] Several variations exist in the protocol used in the commercial production of kefir. Initially, the set method was used to produce kefir. In this procedure, inoculated milk was filled into bottles, fermented at a controlled temperature until a strong coagulum formed, and then cooled. However, the kefir produced was of low quality compared to that produced on a smaller scale using traditional methods. Today, kefir is produced by a stirred method where fermentation, coagulum formation, agitation, ripening, and cooling all occur in one vessel.[24] The composition (chemical, organoleptic, microbiological characteristics) of the final product depends on the type of milk used, the source of grains, the preparation of the mother culture (often produced by coarsely sieving the gains and using the percolate), the length of fermentation, the inclusion of a cooling step, and the inclusion of a maturation step.

A major difference in processing involves the choice of using either kefir grains or a kefir grain-free extract as the mother culture. Traditional artisan production of kefir involves inoculating milk with a quantity of grains (2 to 10%) and allowing the fermentation to proceed for approximately 24 h, to a predetermined pH or until a desired taste/texture is obtained. Fermentation is carried out at 20 to 25°C. A maturation step, carried out at 8 to 10°C for 15 to 20 h, is often added. The grains are then sieved out and can be used for a new fermentation or preserved (1 to 7 days) in fresh milk. Leaving the grains in the final product results in excessive acid production and inferior taste.[70,78,79]

For the production of "Russian style" kefir, a grain-free innoculum or mother culture is prepared by carrying out a traditional kefir fermentation, sieving the finished product, and using the percolate to inoculate the milk. About 1 to 3% mother culture is added to pasteurized milk. When lyophilized starter cultures are used, the mother culture is prepared by adding 1 g of lyophilized kefir grain starter to 3 liters of milk.[64] Koroleva[80,81] has suggested that the fermentation step should be followed by a slow cooling step (to a temperature of 8 to 10°C) and then a ripening (maturing) step that allows microorganisms to grow and taste and aroma to develop. Korovkina et al.[82] showed that as the temperature of the fermentation step increased (from 19 to 28°C), the pH of the kefir produced dropped and the viscosity increased. There was little change in the amounts of CO_2 and ethanol produced.

A third step, to produce what has been termed "industrial kefir," involves taking some of the Russian style kefir and adding it (2 to 3%) to a new source of milk, and again carrying out fermentation (8 to 20 h at 20–22°C) and maturation (12 h to 7 days at 8 to 16°C) steps.[79]

Koroleva[80,81] reported the effect of various variables on the production process and the composition of starters (mother culture) used to produce Russian style kefir. As the level of inoculum (as kefir grains) increased, the duration of the fermentation process became shorter. However, the changes to the microbiological profile (decreased levels of heterofermentative lactic acid streptococci and yeasts) were judged not to be favorable. The number of major groups of microorganisms increased as the quantity of kefir grains inoculated into the milk decreased. The pH of the starter was also dependent on the quantity of inoculum used: a starter pH of 3.6 to 3.8 was found using a ratio of 1:10 (starter to milk), and a pH of 4.4 to 4.6 with a ratio of 1:30 or 1:50. Garrote et al.[83] also showed that the drop in pH during fermentation was greater when the amount of grains added was increased, and that the ratio of grains to milk affected the viscosity of the final product. The apparent viscosity of kefir produced with 10 g grains/l of milk was highest; adding less or more grains per liter of milk reduced the viscosity. The numbers of yeasts and lactobacilli did not change in the final product produced with different ratios of grains to milk, but the number of lactococci decreased as the amount of grains added increased.

When starters were produced at 25°C, the numbers of homofermentive bacteria (particularly lactobacilli) were favored compared to preparation at 18 to 20°C. Higher temperature fermentation resulted in lower pH, which inhibited the growth of both homofermentive and heterofermentive lactic acid streptococci and yeasts. Agitation of the fermentation mix had the benefits of preventing the growth of molds on the starter surface and promoting the even distribution of microorganisms and metabolites. A tenfold increase in homofermentative lactic acid streptococci and yeasts followed agitation, but there was no effect on the numbers of heterofermentative lactic acid streptococci, thermophilic lactobacilli, or acetic acid bacteria or any changes in the quantities of volatile fatty acids in the finished starter. In some facilities, kefir grains are washed with water once a week. Data from Koroleva[80] showed that washing reduced the numbers of the main groups of starter microorganisms. When the washed grains were then used in the production process, fermentation times were increased and the final product had poor taste and consistency.

Various schematics have been published detailing the methods for producing kefir on an industrial scale.[8,79,84,85] Rossi and Gobbetti[86] outlined the details of a continuous process for producing kefir using a multistarter made up of bacteria and yeast isolated from kefir grains. Kefir was produced over a 30-day period; it was lower in viscosity and lacked some of the volatiles normally found in kefir.

Pettersson[87] and Pettersson et al.[88] listed several reasons for the need for a simpler starter culture to replace traditional kefir grains in the production process, including easier handling, uniformity of starter activity, reduced risk of infection and changes in the starter, final product uniformity, uniform yeast content, and improved storage characteristics. They developed a freeze-dried starter containing 75% homofermentative streptococci, 24% citric-fermenting streptococci, 0.5% lactobacilli, and 0.1% yeasts that produced a kefir with characteristics (pH, lactic acid production, viscosity) similar to those of kefir produced with traditional grains and that had higher organoleptic scores (yeast taste, overall aroma) than the kefir produced from grains. This

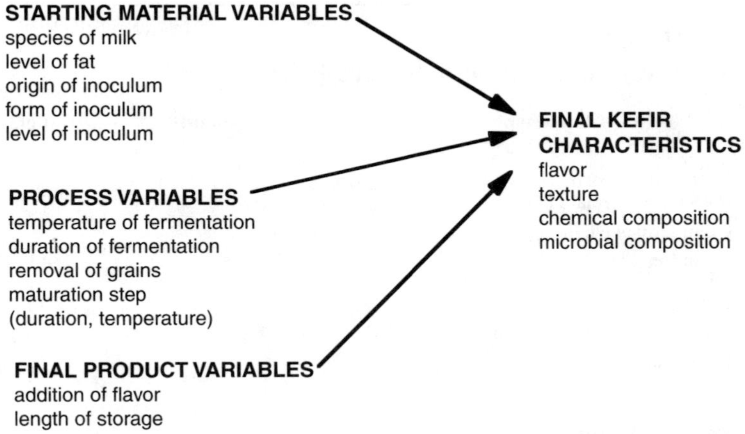

FIGURE 4.6 Processing factors affecting the characteristics of kefir.

freeze-dried starter was also found to produce a more stable bacterial population in prepared kefir than traditionally regenerated kefir grains did.

Other attempts have been made to produce a less complex starter culture that could replace kefir grains in the production of kefir.[89,90] By using two fermentation steps — *Lb. bulgaricus, Lb. acidophilus, Streptococcus thermophilus, Streptococcus lactis,* and *Leuconostoc* — followed by *Saccharomyces cerevisiae,* products were produced, but they had different characteristics (pH, viscosity, titratable acidity, and viable yeast counts) than the control kefir product. More recently, Rossi and Gobbetti[86] combined four lactic acid bacteria and two lactose-negative yeasts all isolated from kefir grains to produce a multistarter that allowed them to produce kefir using a continuous process.

No one to date has been successful in producing kefir grains themselves, and therefore new kefir grains can only be obtained by the propagation of existing grains. Few sources of kefir grains exist today, and although kefir drink can be found in many countries, grains are not commercially available. It has not been possible to establish whether all existing grains have a common origin. Koroleva[91] cautioned that any attempt to replace kefir grains by pure microorganism cultures would not be effective because the unique composition of kefir grains that has evolved over time cannot be replicated or replaced. Any symbiotic relationships between bacteria and yeasts that have developed over time would be difficult to replicate.

It is evident that the characteristics of kefir eaten by the consumer — chemical, microbiological, and sensory properties — are influenced by the starting material, various process variables, and manipulations of the final product. Figure 4.6 summarizes the factors that impact on the properties of the final product.

4.4 COMPOSITION OF KEFIR

The reported microbiological and chemical composition of kefir varies widely because the fermentation products in the final product are greatly influenced by the

TABLE 4.3
Chemical Composition of Various Kefirs

Component	Quantity	Ref.
CO_2		
Polish commercial kefir	24.74% (v/v)	77
Grain fermented, 24 h	1.33 g/l	43
Grain free, 24 h	0.65 g/l	43
Protein		
Taiwanese kefir from grains (cows' milk)	3.16–3.18%	62
Polish lab kefir	3.1%	78
Fat		
Taiwanese kefir from grains (cows' milk)	3.07–3.17%	62
Lactose		
Traditional kefir	2.5%	92
Eastern European stir type	3.7–3.8%	92
Irish kefir from grains (diluted milk)	1.8%	61
Taiwanese kefir from grains	2.81–3.13%	62
Ethanol		
Polish lab kefir	0.021–0.029%	78
Traditional kefir	0.5–1.5%	92
Eastern European commercial stir type	0.02–0.114%	92
Irish kefir from grains (diluted milk)	0.04%	61
Taiwanese kefir from grains	0.17–0.25%	62
Lactic Acid		
Traditional kefir	0.7–1%, 50% as L-(+) isomer	92
Eastern European commercial stir type	0.7–0.8%	92
Irish kefir from grains (diluted milk)	0.5%	61

source of kefir grains used during fabrication. In addition, different types of milk (various species, various levels of fat) and different production methods (commercial, artisan) can be used. Generally, the pH of kefir is between 4.2 and 4.6. The ingredients most commonly measured as indicators of quality are CO_2, protein, lipid (fat), lactose, ethanol, and lactic acid. Table 4.3 is a summary of the composition of kefir reported in the literature.

4.4.1 Carbon Dioxide Content

Yeasts and some heterofermentive lactic acid bacteria are responsible for the production of the CO_2 gas in kefir. The CO_2 content increases during fermentation as the pH drops. If the fermentation is carried on for longer than 24 hours, CO_2 production plateaus after 48 hours.[62] The gas produced leads to "fine flake" coagulum formation and also imparts a sparkling mouthfeel to kefir.[24] The presence of the gas bubbles in the drink has prompted some to refer to kefir as the champagne of fermented milk drinks.

Early research on yeasts in kefir showed that a yeast isolated from commercial kefir grains was able to ferment glucose, galactose, mannose, and sucrose, but not lactose. Fructose was only fermented after an inductive period.[42] The amount of CO_2 produced appeared to be dependent on the presence of constituent enzymes that Iwasawa et al.,[44] thought might be unique to this yeast. Clementi et al.[43] studied the production of CO_2 by two yeasts as a way to better understand the factors that control the CO_2 content of kefir because they felt that even with the possibility of producing a kefir from different starters (not grains), the CO_2 content had to be maintained. They were able to show that in 2.5 h, immobilized yeast cells added to lactic acid fermented milk could produce CO_2 levels comparable to those of traditional kefir that had been fermented for 24 h.

The production of CO_2 during the fermentation step and continuing on in the finished product presents a unique packaging problem that can cause bulging of containers and leakage of contents.[85,93] In some countries, kefir has been either produced or sold in glass bottles. However, polyethylene foil packs covered with aluminium foil or plastic containers with an aluminium foil cover are more flexible containers that can accommodate the gas produced after packaging.

4.4.2 Fat Content

The fat content of kefir can be altered based on the type of milk fermented. In the former Soviet Union, kefirs have been sold that vary from fat free up to 3.2% fat.[94] In her studies of various fermented milk products, Alm[95] found only minor differences in the fat content and composition (mono-, di-, and triglycerides, free fatty acids, and steroids) of kefir compared to the starting milk. The fact that free fatty acids were found in all the fermented milks analyzed (including kefir) indicated a disruption of the fat molecules in milk during fermentation, and Alm[95] speculated that this could result in an increase in the digestibility of milk fat in fermented products compared to other sources of fat.

4.4.3 Lactose/Lactic Acid Content

The lactose found in milk is degraded to lactic acid during fermentation; this causes the pH to drop and the milk to thicken. As much as 30% of the milk lactose is broken down during fermentation.[62] β-galactosidase is the enzyme that hydrolyzes lactose into glucose and galactose. β-galactosidase is found in yogurt and kefir grains, but β-galactosidase activity is very low in kefir.[17] The bacteria in kefir are able to break down glucose to lactic acid. Both D-(–)-lactic and L-(+)-lactic acid are produced from the milk lactose, and in kefir the L-(+) form tends to predominate. However, the ratio of L to D depends on the source and microbial composition of the grains.[61]

4.4.4 Ethanol Content

The production of ethanol in kefir is complex; both yeasts and heterofermentative bacteria produce ethanol. The quantity of ethanol produced is dependent on the fermentation process and the type of container used (tightly capped or not). Kefir produced in small dairies in the former Soviet Union in the early twentieth century contained alcohol levels between 1 and 2%. Present-day methods of production result in much lower levels of alcohol. This may be due in part to the fact that fermentations are stopped at higher pH levels than previously. The final alcohol concentration is determined for the most part by the number of yeasts present in the grains added to the milk and the duration of fermentation.[96] Ethanol appears to be produced toward the end of the fermentation process, and its formation can continue even when the pH has decreased to the point where the lactic acid bacteria in the product are no longer active.[93] The ethanol concentration can be increased in the final product by increasing the temperature during fermentation.[97] Kefir produced in the laboratory from grains had a higher ethanol content (0.040 to 0.30%) than kefir produced commercially in Germany (0.002 to 0.005%).[98]

Kwak et al.[93] studied ways to control the production of ethanol in kefir made from a defined starter culture and found that a two-stage fermentation was best. They used a nonlactose fermenting yeast — *Saccharomyces cerevisiae* — initially to ferment glucose added to milk and then to carry out a lactic acid bacterial fermentation using a mixture of lactobacilli, lactococci, leuconostocs, and propionibacteria obtained commercially. During the yeast fermentation stage, the pH remained stable; it only dropped when the lactic fermentation started. Adding 0.4 or 0.5% glucose to the starter milk resulted in ethanol production only during the yeast fermentation stage, yielding a final product with 0.07 or 0.08% ethanol. Storage experiments showed that kefir produced with the addition of 0.4% glucose was the most stable. When 1.0% glucose was added, the production of ethanol continued into the lactic fermentation and resulted in an ethanol concentration of 0.24% in the final product.

4.4.5 Amino Acids

Kefir has the same pattern of amino acids as milk. Kefir proteins are easily digested due to the action of acid coagulation and proteolysis of milk proteins. Free amino acids found in milk are consumed during the first hours of fermentation by selective bacteria. As the fermentation slows and the kefir enters the ripening stage, the proteolytic activity of other microorganisms such as acetic acid bacteria and yeasts causes more peptides and free amino acids to be formed in a manner seen in other fermented milk products. (See Chapters 5 and 7 for more details on the hydrolysis of milk proteins.)

4.4.6 Volatile Components

Minor volatile constituents have been studied because of their possible contribution to kefir's unique taste. Compositional data are inconsistent, however, because of the variety of sources of the kefir analyzed. Acetaldehyde and diacetyl are two important contributors to flavor, but propionaldehyde, 2-butanone, *n*-propanol, iso-amyl alcohol, acetic acid, and ethanol found in kefir may also influence the aroma.[99,100] Görner et al.[100] noted that the levels of volatiles found in kefir, particularly ethanol, changed during the course of the fermentation. Data from laboratory-produced kefir showed that maturation following fermentation lowered pH and acetaldehyde concentrations, but increased lactic acid and diacetyl levels.[83] Commercial kefir contains half as much orotic acid, twice as much pyruvic acid, nine times as much acetic acid, and about an equal amount of uric acid as does commercial yogurt. Some kefir also contains propionic acid.[17] Dousset and Caillet[97] followed the concentrations of seven organic acids during the fermentation process and found that propionic acid was produced only in the last stages of fermentation (pH 4.33 and below) and that when the temperature during fermentation was increased from 20 to 30°C, the concentrations of citric acid, lactic acid, acetic acid, propionic acid, and isobutyric acid increased in the final product, while pyruvic acid levels declined.

Guzel-Seydim et al.[101] followed the production of organic acids and volatiles during the fermentation of kefir produced in the laboratory from grains obtained from Turkey. Lactate production started slowly but the concentration of lactate rapidly climbed to 6000 μg/g by the end of the 22 h fermentation. Citrate, the next most abundant organic acid, declined during fermentation from 1760 to 1440 μg/g at the end of the fermentation. Pyruvate levels increased during fermentation to a final level of 18 μg/g. Levels of orotate and urate declined over the 22 h period; hippurate was totally consumed after 15 h. Acetic acid, propionic acid, and butyric acid were not found in any samples. Diacetyl was also absent. Acetaldehyde and acetoin levels increased as the fermentation progressed. However, ethanol production did not begin until after 5 h into the fermentation. Guzel-Seydim et al.[102] also studied the changes in organic acids and volatile flavor compounds in kefir stored at 4°C for up to 21 days. Lactate was the organic acid in highest concentration (>6000 μg/g), followed by citrate (1500 μg/g). Lactate increased slightly during storage. Pyruvate was found at day zero, but by day seven of storage it had disappeared. The conversion of pyruvate to ethanol and CO_2 may account for this observation. The stored kefir contained no hippuric acid, acetic acid, propionic acid, or butyric acid. Acetaldehyde levels doubled (to 11 μg/g) during storage, while acetoin levels decreased from 25 to 16 μg/g during the storage period. This group found no diacetyl in their kefir.

Linossier and Dousset[27] showed that the yeast *Candida kefir* was capable of producing malic acid, citric acid, and pyruvic acid in milk with *Lb. kefir*, but levels of lactic acid, fumaric acid, and butyric acid were low. Citric acid was not found after 70 h of fermentation. There appeared to be a symbiotic relationship between the *C. kefir* and the *Lb. kefir*, as the growth of the bacterium was enhanced by the presence of the yeast in the medium. The addition of *C. kefir* at levels as low as 0.5% of the total microbial population stimulated the growth of *Lb. kefir*.

4.5 THE TASTE OF KEFIR

The taste of unflavored kefir has been described as "yeasty," and the terms "prickling" and "sparkling" have been used to describe the mouthfeel of kefir caused by the liberation of trapped CO_2. Complaints from long-time consumers of kefir about the taste of traditional kefir occur only when the yeast taste is either very pronounced or absent.[87] However, the taste has not scored high in sensory evaluations by North American consumers.[10,88] Kefir that scores higher than unflavored kefir in consumer acceptability tests can be produced by adding flavor — either as as flavor itself or as fruit preserve. Starter cultures that contain bacteria that produce flavors can also be added to improve product acceptability.[103,104] Although flavored kefir may appeal to the North American consumer, the addition of fruit or other sources of sugars may cause unwanted fermentation by yeasts after packaging.[4]

In a sensory evaluation study carried out in Scotland, Muir et al.[105] showed that both the chemical composition and the scores given by sensory panelists differed between commercial traditional kefir made from kefir grains and modified kefir made from a defined blend of bacteria and yeasts. Traditional kefir had higher levels of lactic, acetic, and propionic acids. The sensory panelists judged the modified kefir as less acidic, with a creamier taste and less serum separation compared to the traditional kefir. This led the researchers to conclude that modified kefir might be more acceptable to Western European consumers than traditional kefir.

Kefir produced using buffalo milk and traditional grains was reported to have chemical properties similar to kefir produced from cows' milk, and it had acceptable organoleptic properties and was deemed suitable for Egyptian tastes.[1]

4.6 NUTRITIONAL VALUE OF KEFIR

The protein, fat, and mineral content of kefir is similar to that of the milk from which it is made, and therefore kefir has an inherently high nutritional value as a source of protein and calcium. Kefir also has a reputation as being palatable with a high digestibility, allowing for consumption of large quantities without intestinal disturbance.[106]

4.6.1 Digestibility

The fermentation process brings about denaturation of milk proteins and hydrolysis of some of the proteins, resulting in smaller structures that are more susceptible to digestion by gastric and intestinal juices. *In vitro* experiments carried out by Alm[107] showed that kefir developed a very small curd size when exposed to simulated gastric juice even after a three-hour exposure to acidic conditions. Kefir was widely used in hospitals and sanitaria in the former Soviet Union as part of the diet for patients with gastrointestinal and metabolic diseases, hypertension, ischemic heart disease, or allergies. Evenshtein[108] reported that kefir (250 ml/day) was used successfully to stimulate gastric secretions and acid formation in patients with pulmonary tuberculosis. Kefir is particularly recommended for babies and patients with malabsorption syndromes, presumably because of the small curd size that forms in the stomach and the observation that fermented dairy products in general are digested without the secretion of large amounts of gastric juices.[93] Mann[2] quotes two Russian publications that highlight the nutritional value of kefir especially for infants and indicate that products based on kefir were being produced for this market.

4.6.2 Protein Nutrition

Rat studies carried out by Vass et al.[109] confirmed that kefir had a superior biological value (protein efficiency ratio — PER) than milk, which could be explained as being due to its better protein digestibility. Also, as a result of bacterial metabolism, both the total nitrogen (TN) and the nonprotein nitrogen (NPN) in kefir are higher than those in the milk it is produced from. Schmidt et al.[110] reasoned that kefir, like yogurt, had a higher protein digestibility that contributed to its higher nutritional value and capacity to regenerate liver tissue in rats that had undergone partial (70%) hepatotectomy.

During fermentation and storage, the amounts of free amino acids in kefir increase, particularly lysine, proline, cystine, isoleucine, phenylalanine, and arginine.[111–113] The bacteria and yeasts used to produce kefir also result in the isomerization of the L-amino acids liberated from milk proteins to D-amino acids, particularly D-alanine, D-leucine, D-aspartic acid, and D-allo-isoleucine. D-amino acids are less common in yogurt.[114] The addition of sodium caseinate to milk results in a product that is even higher in protein content (up to 3.5%), and this has been proposed as a possible dietetic drink.[115]

4.6.3 Lactose Metabolism

Kefir contains a variety of microorganisms that have the potential to aid lactose digestion. People with lactose intolerance have an insufficient activity in their intestines of the enzyme β-galactosidase (*EC* 3.2.1.23) or lactase — the enzyme responsible for the hydrolysis of lactose into glucose and galactose. Fermented milk products offer hope to such people because some of the microorganisms used in the fermentation of these products possess lactase activity. Yogurt is often suggested as a dairy product that can be consumed by people with lactose intolerance because the lactase activity in yogurt microorganisms breaks down some of the milk lactose during yogurt production and storage and because lactase activity increases after consumption of unpasteurized yogurt due presumably to the presence of yogurt bacteria in the intestines that retain lactase activity.[116] Alm[117] studied the decrease in lactose content in several fermented milk products including kefir and found that compared to yogurt, acidophilus milk, and bifidus milk, the decrease in kefir lactose was less following fermentation and storage for up to 14 days. In their study with pigs, de Vrese et al.[118] showed that while kefir grains did have some β-galactosidase activity, kefir drink did not. Farnworth and Mainville[17] have found that the galactosidase activity in kefir is very low compared to that of yogurt containing live bacteria.

Lactic acid is found in all fermented dairy products as a result of the action of homo- and heterofermentative microorganisms. Lactic acid can exist in either the L-(+) or D-(–) isomeric forms and as a 50/50 DL racemic mixture. L-(+)-lactic acid is completely metabolized by the body, but D-(–)-lactic acid is used more slowly by the body, and excess D(–)-lactic acid can cause metabolic disturbances. Alm[119] quotes World Health Organization documents that suggest that infant nutrition products containing D-(–) or the DL mixture should be avoided, although it is admitted that more research needs to be done to verify this.[120] Kefir contains almost exclusively L-(+)-lactic acid; in yogurt the ratio of L(+)/D(–) is 58:42.[119]

4.6.4 Vitamin Content

Cows' milk is generally considered a good source of most water-soluble vitamins, except for ascorbic acid and vitamin B_{12}. Several investigators have measured the quantity of vitamins in kefir to determine whether fermentation changed levels compared to milk, but the results have not been consistent. An early study of the vitamin B_{12} content of kefir indicated that both during the fermentation and maturation stages, the vitamin B_{12} content went down.[121] Alm[122] used commercially available grains to produce kefir that had increased folic acid content compared to the starting milk but decreased concentrations of vitamin B_6, vitamin B_{12}, and biotin. Of the various fermented products studied by Alm, kefir had the lowest reduction (17%) in orotic acid (also referred to as vitamin B_{13}) as compared to the starting milk. Yogurt had a 47.8% reduction in orotic acid. Bossi et al.[112] reported declines in vitamin A, thiamin, riboflavin, nicotinamide, and vitamin C in laboratory-prepared kefir compared to the starting milk.

Kneifel and Mayer,[123] using ten different sources of grains collected from Austrian households, reported increases in pyridoxine (9/10), vitamin B_{12} (5/10), and folic acid (7/10) of laboratory-prepared kefir compared to the starting cows' milk. Most kefir samples had folic acid contents over 20% higher than those of the starting milk. Thiamin (8/10), riboflavin (10/10), niacin (8/10), pantothenic acid (7/10), and orotic acid or vitamin B_{13} (9/10) levels in the finished kefir were lower than in the starting milk. When different sources of milk were compared (cow, ewe, goat, mare), ewes' milk had the largest percent increases of thiamin, pyridoxine, and folic acid, while goats' milk had the largest percent losses of thiamin, riboflavin, vitamin B_{12}, niacin, pantothenic acid, and orotic acid. The source of the milk may influence the growth of particular microorganisms which, in turn, determine the final vitamin content of kefir.

Mann quotes Russian, Czech, and Polish references that describe attempts to increase the nutritional quality of kefir.[65] The total protein content of kefir can be increased by adding sodium caseinate to the starter milk. Specific bacteria can also be added to the initial mother culture to increase the levels of folic acid and vitamin B_{12}.

4.6.5 Kefir as an Infant Food

A Russian team studying premature infants found that a mixture of kefir and "Similac" type formula was well tolerated and produced adequate weight gain when fed to healthy premature infants. The blood fatty acid pattern in the infants was similar to that found in the kefir–Similac mixture that was fed.[124]

Kefir is used widely in Russian hospitals, particularly in neonatal wards, both for mothers and newborns. It is often recommended as the first food for babies after breastfeeding because of its high digestibility. Goncharova et al.[125] reported that kefir fed to premature infants did not restore their intestinal bifidobacteria nor reduce the number of disturbances of biocenosis. It appears that kefir was also used in the treatment of biliary tract diseases associated with diseases of the pancreas in children.[126] Ormisson and Soo[127] attempted to reduce the decline in buffer bases in children up to 2 years old suffering from acute pneumonia or acute bronchitis. Kefir and sour milk both shifted the acid–base balance to acidosis, which is not desirable in sick children. It appears that some kefir products were produced specifically for children.[128]

4.6.6 Other Nutritional Uses

Kefir has been used as part of a weight reduction program in the former Soviet Union, where obese patients are given only kefir to consume every second day of treatment. Kefir has also been used in the former Soviet Union as a vehicle for increasing the essential fatty acid (EFA) intake of patients with metabolic diseases and intestinal tract disorders.[94] This EFA-enriched kefir has been prescribed to patients with atherosclerosis, ischemic heart disease, obesity, peptic ulcers, and liver and gallbladder pathologies.

A Japanese patent has been granted for the formulation of a health food that contains kefir together with enzyme inhibitors such as lipase or α-amylase.[129] This product is purported to prevent and control obesity.

4.7 PHYSIOLOGICAL EFFECTS OF KEFIR CONSUMPTION

Kefir has a long tradition of being regarded as good for health in the countries where it is a staple in the diet. However, published human feeding trials to substantiate this view are not numerous.[7] Kefir, kefir grains, kefiran, and the bacteria found in kefir have been the subject of scientific studies to demonstrate beneficial effects on humans.

4.7.1 Kefir as a Probiotic

A probiotic is defined as a microbial preparation which contains live and/or dead cells including their metabolites, which is intended to improve the microbial or enzymatic balance at mucosal surfaces or to stimulate immune mechanisms.[130] Kefir contains many different bacteria and yeasts. To date, however, no scientific studies have been published showing health benefits of kefir microorganisms. The numbers of microorganims in kefir are large enough ($>10^7$ CFU/g) that kefir can be considered a probiotic.[131] During the fermentation process, microbial metabolites and/or degraded milk constituents may be produced that are also beneficial to human health.[132]

Studies have been published where kefir grains, kefir itself, or kefiran have been given to animals and humans to ameliorate a variety of conditions and diseases. The Russian literature contains articles describing the effects of kefir consumption, but these studies are not always readily available.

4.7.2 Antitumor Effect in Animals

The antitumor effect of kefiran was first reported by Shiomi et al.[133] They gave mice a polysaccharide isolated from kefir grains dissolved in drinking water for 7 days prior to injection with Ehrlich carcinoma (EC) cells and continuing for 24 days, or starting the same day as the injection of tumor cells and continuing for 23 or 24 days. Forty to 59% inhibition of tumor growth was found in the mice receiving 0.02 to 0.1% polysaccharide in their drinking water. The positive effect was observed for mice receiving the polysaccharide either starting before the administration of the tumor cells or at the same time as the tumor cell dose. In a second experiment, similar results (30 to 81% tumor growth inhibition) were found when mice were dosed with Sarcoma 180 (S180). Groups of mice were also given intraperitoneal injections of polysaccharide (0.05 to 2.0 mg/mouse). The polysaccharide was administered starting 7 days before or 1 day after tumor injection, and again the growth of both EC and S180 tumors was significantly reduced. Shiomi et al. carried out cytotoxicity tests in which they incubated EC and S180 cells with a solution of kefir polysaccharide (1 mg/ml) and showed no effect, and this led them to conclude that the antitumor effect was host mediated.

This same group carried out another study in order to define the mechanism of the antitumor effect of kefir polysaccharide.[134] They showed that when mice were given polysaccharide isolated from kefir grains either by gastric intubation or dissolved in the drinking water (on days 1 and 11 of the experiment), their cell-mediated immune response was increased as measured by the delayed-type hypersensitivity (DTH) test. As little as a single dose of 5 mg/kg body weight was sufficient to produce a significant effect. This effect was shown to be dependent on the total dose administered and not the duration or frequency of the dose. When these same mice were then inoculated with EC cells, mice receiving as low as 10 mg/kg body weight had significantly reduced tumor weights after 14 days. This relation between increased immune response and reduced tumor growth was shown to be active in tumor-bearing mice but not in intact mice. It was suggested that kefir polysaccharide could be useful against tumor growth when administered either before or after the initiation of the tumor. However, larger doses of polysaccharide might be necessary if administered after tumor inoculation.

In a third paper, Murofushi et al.[135] investigated the antibody response of mice given kefir polysaccharide. The polysaccharide was shown to significantly enhance the antisheep red blood cell (SRBC) response, but only at a low antigen dose (5×10^6 SRBC/mouse). A treatment of 100 mg polysaccharide/kg body weight was shown to be most effective, and the data indicated that the polysaccharide was exerting its effect in early events of the antibody response. A lack of response in nu/nu, nu/+ and conventional mice to 2,4-dinitrophenyl-alanylglycylglycyl-Ficoll (DNP-Ficoll) as a TI-2 antigen or trinitrophenyl $(TNP)_{2.3}$-lipopolysaccharide (TNP–LPS) prepared as a TI-1 antigen indicated that the kefir polysaccharide was perhaps enhancing only primary T-dependent B cells and not B cells responsive to TI-1 or TI-2 antigens. Using ^{3}H-labeled polysaccharide, they were able to show that the water soluble polysaccharide was absorbed into the body within 3 hours of oral administration. Radioactivity was found in all major organs, and it was concluded that intact polysaccharide reached the spleen or thymus to activate the immune system.

The Japanese literature contains a series of articles in which the antitumor properties of kefir itself are described. Furukawa et al.[136] found that mice given a subcutaneous inoculation of Lewis lung carcinoma had significantly smaller tumor weights and a 62% tumor inhibition rate 2 weeks after tumor inoculation. The mice received a gastric intubation of pasteurized kefir from day 1 to day 9 after tumor cell inoculation. Spleen weights and the number of leukocytes in tumor-bearing mice increased compared to control mice, but not in the mice receiving the kefir. This same group showed that feeding mice kefir (2 g/kg body weight) for 1 to 6 days after tumor inoculation resulted in an increase in delayed-type hypersensitivity response. However, no difference was found in the survival period of control mice vs. those receiving the kefir.[137] This group's data showed that kefir consumption significantly increased the number of leukocytes in normal mice and significantly decreased delayed-type hypersensitivity as measured by foot pad swelling in mice bearing Meth-A fibrinoma. Using similar protocols and tests, Kubo et al.[138] showed that oral administration of kefir (100 or 500 mg/kg body weight for 10 day starting 1 day after tumor inoculation) significantly reduced Ehrlich carcinoma tumor weight and inhibited tumor growth by up to 54%. When mice were given kefir together with mitomycin C, the average tumor weight was even further reduced.

Furukawa et al.[139] have also looked at antimetastic effects of two polysaccharide fractions of kefir grains. Both young (5 wk) and old (30 wk) female mice receiving the water-soluble fraction had significantly reduced pulmonary metastases of Lewis lung carcinoma. This fraction inhibited tumor growth when it was given to the young mice both before and after the tumor cell challenge. Mice given the insoluble fraction for 9 days before a challenge with B16 melanoma had an inhibition rate of 39% compared to the control mice. However, the water-soluble fraction was not protective against this highly metastatic cell line.[139]

Recent studies have looked at the antimutagenic properties of bacteria isolated from kefir in an attempt to understand their mechanism of action.[140] Starting with kefir manufactured in Mongolia, researchers isolated strains of *Streptococcus lactis* (5), *Str. cremoris* (3), *Str. faecalis* (1), *Lb. plantarum* (1), *Lb. brevis* (1), and *Leuconostoc dextranicum* (6). Using a binding assay in which the bacteria were incubated with mutagenic amino acid pyrolyzates, they showed that all the bacteria isolated from kefir had remarkable binding ability (>98.5%) to the mutagens 3-amino-1,4-dimethyl-5H-pyrido [4,3-b] indole and 3-amino-1-methyl-5H-pyrido [4,3-b] indole but a lesser binding ability to 2-amino-6-methyldipyrido [1,2-a:3',2',-d] imidazole. Hosono et al.[140] concluded that these findings, along with similar results obtained with bacteria isolated from yogurt, supported the idea that the consumption of fermented dairy products had a negative correlation with the risk for the development of colon cancer.

Miyamoto et al.[141] were able to isolate 31 strains of bacteria from kefir grains and kefir produced in western Europe, of which three strains of *Lactococcus lactis* ssp. *cremoris* had the strongest desmutagenic properties. These bacteria were also found to be slime-forming bacteria, and their antimutagenic properties appeared to be due to the binding affinity of AF-2 to bacterial cells. Cevikbas et al.[142] showed that intraperitoneal injection of 0.5 ml kefir/day for 20 days after the establishment of fusiform cell sarcomas significantly reduced the size of the tumors in mice and even resulted in the disappearance of tumoral necrosis in some animals.

Yoon et al.[143] used the Ames test to measure the antimutagenic properties of *Lactobacillus* spp. isolated from kefir and yogurt and were able to show that 36 out of 40 strains found in European kefir and yogurt protected *Salmonella typhimurium* TA 98 from the mutagen 2-nitrofluorene.

To date, no human studies have been carried out to verify the antitumor effect of kefir and kefiran in humans.

4.7.3 Antibacterial, Antifungal, and Antiviral Properties of Kefir

Recent studies looking at the production of bacteriocins by kefir microflora found different types of bacteriocins produced. A study on 33 sources of kefir grains from

Ireland showed the presence of at least three different types of bacteriocins produced by lactococcal isolates.[144] The first bacteriocin had a narrow spectrum, inhibiting only lactococci. A second bacteriocin inhibited strains of *Lb. casei, Lb. helveticus,* and *Pediococcus pentosaceus* of the strains tested. The third bacteriocin had a broad spectrum of effectiveness and inhibited all ten strains tested along with a number of other strains of *Lactococcus, Leuconostoc, Pediococcus, Streptococcus thermophilus,* and *Staphylococcus aureus.* This bacteriocin was named lacticin 3147 and is produced by *Lactococcus lactis* strain DPC3147.[145] Lacticin differs from nisin in that it is plasmid encoded. Lacticin 3147 and nisin have similar inhibitory properties, and like nisin, lacticin is heat stable. Lacticin could therefore be a key factor in maintaining the integrity of the microflora of kefir grains by inhibiting the growth of foreign organisms. A recent study by Morgan et al.[146] investigated the antimicrobial potential of 38 Irish kefir grains against the pathogens *Listeria innocua* DPC1770 and *Escherichia coli* 0157:H45. It was found that 18 grains were able to fully inhibit the growth of *L. innocua,* 13 could slightly inhibit growth, and seven grains had no inhibitory effect. Against *E. coli,* 34 of the grains completely inhibited growth, three slightly inhibited growth and one source of grains had no inhibitory activity. The inhibitory effect of the kefir starters against *L. innocua* was attributed to the production of bacteriocins by the microflora. However, the inhibitory effect on *E. coli* was attributed not to a bacteriocin, but to the combined effect of acid and hydrogen peroxide.

Bossi et al.[112] tested lab-produced kefir and various combinations of bacteria found in kefir for their inhibitory effect against several intestinal pathogenic microorganisms. The kefir inhibited the growth of eight pathogens but was not effective against two strains of *E. coli* and one strain of *Staphylococcus aureus.*

Kefir grains have antibacterial activity greater than kefir itself especially against Gram-positive cocci, including staphylococci, and Gram-positive bacilli. In addition, kefir was shown to have antifungal activity against a variety of fungi, yeasts, and molds. Cevikbas et al.[142] concluded that the antibacterial and antifungal activity of kefir helped to explain the wide use of kefir against infectious and neoplastic diseases. They quoted the work of Ormisson and Soo[127] as a human study that supports their conclusions. However, a translation of this latter article indicates that the use of kefir in children with pneumonia and acute bronchitis as a means of improving acid–base balance was not successful.

Serot et al.[147] were able to isolate and partially characterize two antimicrobial agents from bacteria found in kefir grains. These two substances were found to be effective against Gram-positive and Gram-negative bacteria, had molecular weights of approximately 1000 daltons, and were inhibited by some proteolytic enzymes. Zacconi et al.[148] showed that chicks challenged with *Salmonella kedougou* were protected by live kefir but not irradiated kefir, yogurt, or acidified milk. The treatment was most effective if the kefir was given at the same time as the *Salmonella* challenge, but it also offered protection if given 1 day or 6 days after infection.

The antimicrobial activity of fresh kefir differs from that of reconstituted lyophilized kefir. Fresh kefir had an intrinsic inhibitory power against *Staphylococcus aureus, Kluyveromyces lactis,* and *E. coli,* but not against *Saccharomyces cerevisiae* or *Candida albicans,* but kefir that had been lyophilized and then reconstituted in distilled water or reconstituted powdered milk had lost this intrinsic inhibitory factor. The addition of ribitol as a cryoprotective agent appeared the be the best way of preserving the inhibitory properties of kefir after lyophilization.[149]

The antibacterial properties of kefir may be due to a combination of factors including competition for available nutrients or the production of inhibitory metabolites, such as organic acids, during fermentation. Garrote et al.[150] observed that although the bacteria in kefir have been shown to produce a bacteriocin, lactic acid and acetic acid may also contribute to the antibacterial properties of kefir. Supernatant collected from kefir produced using grains collected from Argentinian households had inhibitory effects against Gram-positive and Gram-negative bacteria, although the effect was greater against Gram-positive bacteria. Milk had no effect. Analyses of the kefir supernatants showed that they contained both lactic acid and acetic acid. When milk supplemented with lactic acid and acetic acid was incubated with *E. coli* 3, the inhibitory effect was again observed, while nonsupplemented milk had no effect. Milk with its pH adjusted to that of fermented kefir or milk with added citric acid alone had no inhibitory effect against *E. coli* 3. Since the pH of the mixture determines the percentage of the acid in protonated or dissociated state, the pH of kefir supernatant (3.6 to 3.7) resulted in up to 1.5% of the lactic acid and 0.11% of the acetic acid being in the more powerfully inhibitory undissociated form. Garrote et al.[150] noted that previous research had shown that even these low levels were capable of inhibiting the growth of nonpathogenic *E. coli*.

Russian researchers have used a product called acipole that contains both *Lb. acidophilus* and kefir grains to manage antibiotic dysbacteriosis that is an adverse reaction to antibacterial therapy.[151] It was observed that patients suffering from pneumonia and chronic bronchitis treated with antibacterials often were susceptible to undesirable bacterial infections. Treating patients with an antibiotic and acipole lowered the frequency and severity of dysbacteriosis events.

Interferons initiate intracellular production of several kinds of antivirus proteins to indirectly induce the cell's virus-resistant state. Kefir has been shown to contain a sphingomyelin that is capable of enhancing the production of interferon-β from the human osteosarcoma cell line MG-63 treated with a chemical inducer poly I:poly C. Activity was found in the range of 2 to 100 μg/ml with a maximum secretion enhancement (14 times) occurring at 25 μg/ml.[152] However, insufficient experimental protocol details were included to allow a full evaluation of the credibility of these findings.

Recently, Besednova et al.[153] added to kefir a peptide isolated from nervous tissue of squid and found that this mixture, when fed to laboratory animals, stimulated their cellular and humoral immune systems. In mice that had originally been immune deficient, the mixture restored their immune function to normal.

4.7.4 Cholesterol Metabolism

The consumption of fermented milk products has long been proposed as a way of reducing serum cholesterol levels. Several human studies have been published in which yogurt was consumed. Taylor and Williams[154] summarized the results of 13 yogurt/cholesterol feeding trials and found that yogurt consumption lowered blood cholesterol in eight trials, had no effect in four trials, and increased blood cholesterol in one trial compared to control. Several biochemical mechanisms can be proposed that would predict that consumption of fermented milk will lower blood cholesterol levels.[155] It was argued that probiotic bacteria could cause an increase in the production of short-chain fatty acids, which would in turn decrease circulatory cholesterol levels either by inhibiting hepatic cholesterol synthesis or by redistributing cholesterol from the plasma to the liver. In addition, bile acid deconjugation could be increased in the large intestine. Deconjugated bile acids then would not be absorbed, but excreted from the body. This would stimulate an increase in bile acid synthesis, taking more cholesterol out of circulation.

Vujicic et al.[156] studied the ability of kefir grains to take up cholesterol from milk during a 24-h incubation at 20°C or after incubation and a 48 h storage at 10°C. Grains were obtained from Yugoslavia, Hungary, and the Caucasus. At the end of the fermentation, between 22 and 63% of the cholesterol in the starting milk had been assimilated; by the end of the 48 h storage period, between 41 and 84% had disappeared.

Recently, a randomized crossover feeding trial was carried out with 13 moderately hypercholesterolemic men (serum cholesterol levels 6.0 to 10 mmol/l). Subjects were fed 500 ml/day of kefir for 1 month or the same volume of milk and then the opposite after a 4-week washout period. By monitoring bacteria in fecal samples it was shown that 73% of the subjects were colonized by *Leuconostoc* sp. bacteria found in kefir (see Figure 4.7). Analyses of fecal samples after the washout period indicated that once the feeding of kefir was stopped, the microbial composition of feces returned to normal. Total cholesterol, low-density lipoprotein cholesterol, high-density lipoprotein cholesterol, and blood triglyceride levels were not changed during the period of kefir consumption.[157]

4.7.5 Other Uses

Kefir, like other probiotic products, may be most effective at influencing conditions in the gastrointestinal tract. Sukhov et al.[158] fed kefir (500 ml, 5 times/day) to 38 patients suffering from enteritis and found that there were no significant changes to the intestinal microflora population. In 1971, Russian researchers used kefir to treat patients with peptic ulcers in the stomach and duodenum.[159]

The abstract to a patent filed in 2000 claims that a dietary supplement consisting of milk fermented with two different sources of kefir grains (the first grains rich in bacteria and the second grains rich in yeasts) that had been filtered and then dried could be used to prevent osteoporosis.[160] It was stated that the supplement could be used in the prevention and treatment of osteoporosis and other diseases that result from calcium, magnesium, and potassium deficiencies. No data were included in the abstract to support this claim.

Kefir may also stimulate the mucosal immune system. It has been recently reported that young rats (6 months old) fed kefir *ad libitum* and then immunized intraduodenally with cholera toxin (CT) had significantly higher (86%) serum anti-CT antibodies compared to controls.[161] This response was attributed to an enhanced *in vitro* antibody secretion by cultured lymphocytes isolated from the Peyer's patches

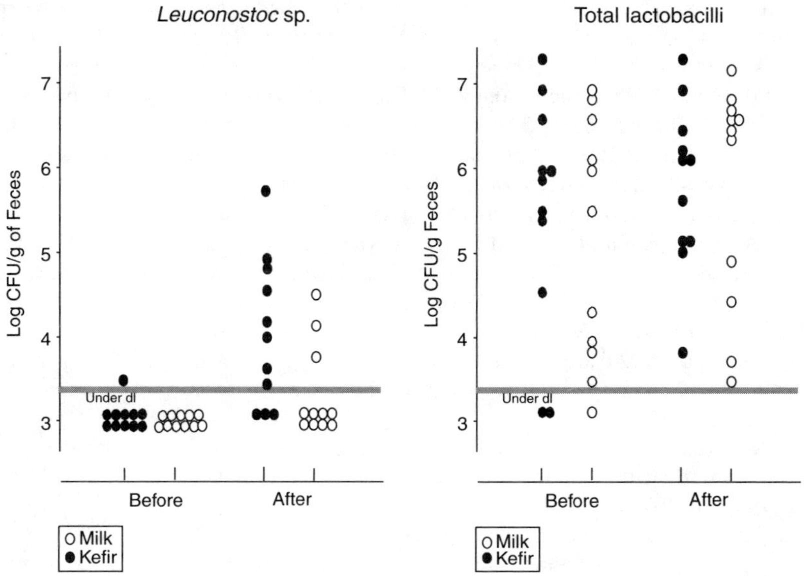

FIGURE 4.7 Colonization of subjects by bacteria during consumption of kefir.

and intestinal propria. The effect of kefir was not observed in older rats (26 months old) in the same experiment. Total serum IgA titers were also not changed due to kefir consumption, but in both young and old rats anti-CT IgG titers were lower in the kefir-fed animals. It was speculated that the immunomodulation effect of kefir may be due to bacterial wall components.

REFERENCES

1. Ismail, A.A., El-Nockrashy, S.A., and Khorshid, M.A., A beverage from separated buffalo milk fermented with kefir grains, *J. Soc. Dairy Technol.,* 36, 117–118, 1983.
2. Mann, E.J., Kefir and koumiss, *Dairy Indust. Int.,* 50, 11–12, 1985.
3. Zourari, A., and Anifantakis, E.M., Le kéfir. Caractères physico-chimiques, microbiologiques et nutritionnels. Technologie de production. Une revue, *Le Lait,* 68, 373–392, 1988.
4. Kemp, N., Kéfir, the champagne of cultured dairy products, *Cultured Dairy Prod. J.,* 19, 29–30, 1984.
5. Mann, E.J., Kefir and koumiss, *Dairy Indust. Int.,* 54, 9–10, 1989.
6. Farnworth, E.R., Properties and Characteristics of Kefir: A Fermented Milk Probiotic, Abs. No. PM332, 16th International Congress of Nutrition, Montreal, July 27–August 1, 1997.
7. Farnworth, E.R., Kefir: from folklore to regulatory approval, *J. Nutraceut. Funct. Med. Food,* 1, 57–68, 1999.
8. Tamime, A.Y., Muir, D.D., and Wszolek, M., Kefir, koumis, *Dairy Indust. Int.,* 64, 32–33, 1999.
9. Marshall, V., Cole, W.M., and Brooker, B.E., Observations on the structure of kefir grains and the distribution of the microflora, *J. Appl. Bacteriol.,* 57, 491–497, 1984.

10. Duitschaever, C.L., Kemp, N., and Smith, A.K., Microscopic studies of the microflora of kefir grains and of kefir made by different methods, *Milchwissenschaft,* 43, 479–481, 1988.
11. Abraham, A.G. and de Antoni, G.L., Characterization of kefir grains grown in cow's milk, *J. Dairy Res.,* 66, 327–333, 1999.
12. de Man, J.C., Rogosa, M., and Sharpe, M.E., A medium for the cultivation of lactobacilli, *J. Appl. Bacteriol.,* 23, 130–135, 1960.
13. Rogosa, M., Mitchell, J.A., and Wiseman, R.F., A selective medium for the isolation and enumeration of oral and fecal lactobacilli, *J. Bacteriol.,* 62, 132–133, 1951.
14. Mitsuoka, T., Sega, T., and Yamamoto, S., Eine verbesserte Methodik der qualitativen und quantativen Analyse der Darmflora von Menschen und Tieren, *Zentralbl. Bacteriol. Parasitenkd. Infektionskr. Hyg. Abt. 1 Orig. Reihe A.,* 195, 445–469, 1964.
15. Fujisawa, T., Adachi, S., Toba, T., Arihara, K., and Mitsuoka, T., *Lactobacillus kefiranofaciens* sp. nov. isolated from kefir grains, *Int. J. System. Bacteriol.,* 38, 12–14, 1988.
16. Kojima, S., Takizawa, S., Tamura, S., Fujinaga, S., Benno, Y., and Nakase, T., An improved medium for the isolation of *Lactobacilli* from kefir grains, *Biosci. Biotech. Biochem.,* 57, 119–120, 1993.
17. Farnworth, E.R., and Mainville, I., unpublished data, 2000.
18. American Type Culture Collection (ATCC), Medium 2025, http://www.atcc.org/SearchCatalogs/MediaFormulations.cfm.
19. Rosi, J.I., Microrganismi del kefir: gli acetobatteri, *Sci. Tecn. Lattiero-Casearia,* 29, 221–227, 1978.
20. Molska, I., Moniuszko, I., Komorowska, M., and Merilainen, V., Characteristics of bacilli of *Lactobacillus casei* species appearing in kefir grains, *Acta Alim. Pol.,* 9, 79–88, 1983.
21. Engel, V.G., Krusch, U., and Teuber, M., Mikrobiologische zusammensetzung von kefir. I. Helfen, *Milchwissenschaft*, 41, 418–421, 1986.
22. Rohm, H. and Lehner, M., Charakteristische eigenschften von österreichischem kefir, *Ernährung/Nutrition,* 14, 571–574, 1990.
23. Angulo, L., Lopez, E., and Lema, C., Microflora present in kefir grains of the Galician region (North-West of Spain), *J. Dairy Res.,* 60, 263–267, 1993.
24. Koroleva, N.S., Products prepared with lactic acid bacteria and yeasts, in *Therapeutic Properties of Fermented Milks,* Robinson, R.K., Ed., Elsevier Applied Science, London, 1991, pp. 159–179.
25. Margulis, L., From kefir to death, in *How Things Are,* Brockman, J. and Matson, K., Eds., William Morrow and Co., New York, 1995, pp. 69–78.
26. Toba, T., Fermented milks produced with mesophilic lactic acid bacteria: kefir and Scandinavian ropy sour milk, *Jpn. J. Dairy Food Sci.,* 36, 235–244, 1987.
27. Linossier, J.P. and Dousset, X., Stimulation de la croissance et du métabolisme de *Lactobacillus kefir* par *Candida kefir, Microbiol. Aliments Nutr.,* 12, 341–351, 1994.
28. Viljoen, B.C., The interaction between yeasts and bacteria in dairy environments, *Int. J. Food Microbiol.,* 69, 37–44, 2001.
29. Koroleva, N.S., Special products (kefir, koumyss, etc.), *Proc. XXI Internat. Dairy Congress,* Moscow, 1982, 2, pp.146–151.
30. Takizawa, S., Kojima, S., Tamura, S., Fujinaga, S., Benno, Y., and Nakase, T., *Lactobacillus kefirgranum* sp. nov. and *Lactobacillus parakefir* sp. nov.: two new species from kefir grains, *Int. J. Syst. Bacteriol.,* 44, 435–439, 1994.
31. Wyder, M-T. and Puhan, Z., A rapid method for identification of yeasts from kefyr at species level, *Milchwissenschaft,* 52, 3127–3330, 1997.

32. Wyder, M.-T., Meile, L., and Teuber, M., Description of *Saccharomyces turicensis* sp. nov., a new species from kefyr, *System. Appl. Microbiol.,* 22, 420–425, 1999.
33. Ottogalli, G., Galli, A., Resmini, P., and Volonterio, G., Composizione microbiologica, chimica et ultrastruttura dei ganuli di kefir, *Ann. Microbiol.,* 23, 109–121, 1973.
34. Vayssier, Y., Le kéfir: analyse qualitative et quantitative, *Rev. Lait. Fr.,* 361, 73–75, 1978.
35. Rosi, J. I., Microrganismi del kefir: I Lieviti, *Sci. Tech. Lattiero-Casaria,* 29, 59–67, 1978.
36. Rosi, J., and Rossi, J., I Microrganismi del kefir: I fermenti lattici, *Sci. Tech. Lattiero-Casearia,* 29, 291–305, 1978.
37. Kosikowski, F.V. and Mistry, V.V., *Cheese and Fermented Milk Foods,* Vol. 1, 3rd ed., F.V. Kosikowski LLC, Westport, CT, 1997, pp. 61–64.
38. Pintado, M.E., Lopes da Silva, J.A., Fernandes, P.B., Malcata, F.X., and Hogg, T.A., Microbiological and rheological studies on Portuguese kefir grains, *Int. J. Food Sci. Technol.,* 31, 15–26, 1996.
39. Mainville, I., unpublished data, 2001.
40. Rohm, H., Eliskases-Lechner, F., and Bräuer, M., Diversity of yeasts in selected dairy products, *J. Appl. Bacteriol.,* 72, 370–376, 1992.
41. Lin, C-W., Chen, H-L., and Liu, J-R., Identification and characterisation of lactic acid bacteria and yeasts isolated from kefir grains in Taiwan, *Aust. J. Dairy Technol.,* 54, 14–18, 1999.
42. Thompson, J.K., Johnston, D.E., Murphy, R.J., and Collins, M.A., Characteristics of a milk fermentation from rural Northern Ireland which resembles kefir, *Irish J. Food Sci. Technol.,* 14, 35–49, 1990.
43. Clementi, F., Gobbetti, M., and Rossi, J., Carbon dioxide synthesis by immobilized yeast cells in kefir production, *Milchwissenschaft,* 44, 70–74, 1989.
44. Iwasawa, S., Ueda, M., Miyata, N., Hirota, T., and Ahiko, K., Identification and fermentation character of kefir yeast, *Agric. Biol. Chem.,* 46, 2631–2636, 1982.
45. Farnworth, E.R., Unique problems in designing and testing probiotic foods, FoodInfo Online Features, http://foodsciencecentral.com/library.html#ifis/3803.
46. La Riviére, J.W.M., Kooiman, P., and Schmidt, K., Kefiran, a novel polysaccharide produced in the kefir grain by *Lactobacillus brevis, Arch. Mikrobiol.,* 59, 269–278, 1967.
47. Kooiman, P., The chemical structure of kefiran, the water-soluble polysaccharide of the kefir grain, *Carbohyd. Res.,* 7, 200–211, 1968.
48. Mukai, T., Toba, T., Itoh, T., and Adachi, S., Structural microheterogeneity of kefiran from kefir grains, *Jpn. J. Zootech.,* 59, 167–176, 1988.
49. Kandler, O. and Kunath, P., *Lactobacillus kefir* sp. nov., a component of the microflora of kefir, *System. Appl. Microbiol.,* 4, 286–294, 1983.
50. Toba, T., Abe, S., Arihara, K., and Adachi, S., A medium for the isolation of capsular bacteria from kefir grains, *Agric. Biol. Chem.,* 50, 2673–2674, 1986.
51. Yokoi, H., Watanabe, T., Fujii, Y., Mukai, T., Toba, T., and Adachi, S., Some taxonomical characteristics of encapsulated *Lactobacillus* sp. KPB-167B isolated from kefir grains and characterization of its extracellular polysaccharide, *Int. J. Food. Microbiol.,* 13, 257–264, 1991.
52. Yokoi, H., Watanabe, T., Fujii, Y., Toba, T., and Adachi, S., Isolation and characterization of polysaccharide-producing bacteria from kefir grains, *J. Dairy Sci.,* 73, 1684–1689, 1990.
53. Toba, T., Uemura, H., Mukai, T., Fujii, T., Itoh, T., and Adachi, S., A new fermented milk using capsular polysaccharide-producing *Lactobacillus kefiranofaciens* isolated from kefir grains, *J. Dairy Res.,* 58, 497–502, 1991.

54. Toba, T., Abe, S., and Adachi, S., Modification of KPL medium for polysaccharide production by *Lactobacillus* sp. isolated from kefir grain, *Jpn. J. Zootech. Sci.,* 58, 987–990, 1987.
55. Yokoi, H., and Watanabe, T., Optimum culture conditions for production of kefiran by *Lactobacillus* sp. KPB-167B isolated from kefir grains, *J. Fermen. Bioeng.,* 74, 327–329, 1992.
56. Micheli, L., Uccelletti, D., Palleschi, C., and Crescenzi, V., Isolation and characterization of a ropy *Lactobacillus* strain producing the exopolysaccharide kefiran, *Appl. Microbiol. Biotechnol.,* 53, 69–74, 1999.
57. Bottazzi, V., and Bianchi, F., A note on scanning electron microscopy of microorganisms associated with the kefir granule, *J. Appl. Bacteriol.,* 48, 265–268, 1980.
58. Molska, I., Kocon, J., and Zmarlicki, S., Electron microscopic studies on structure and microflora of kefir grains, *Acta Aliment. Pol.,* 6, 145–154, 1980.
59. Toba, T., Arihara, K., and Adachi, S., Distribution of microorganisms with particular reference to encapsulated bacteria in kefir grains, *Int. J. Food Microbiol.,* 10, 219–224, 1990.
60. Neve, H., Analysis of kefir grain starter cultures by scanning electron microscopy, *Milchwissenschaft,* 47, 275–278, 1992.
61. Rea, M.C., Lennartsson, T., Dillon, P., Drinan, F.D., Reville, W.J., Heapes, M., and Cogan, T.M., Irish kefir-like grains: their structure, microbial composition and fermentation kinetics, *J. Appl. Bacteriol.,* 81, 83–94, 1996.
62. Kuo, C-Y. and Lin, C-W., Taiwanese kefir grains: their growth, microbial and chemical composition of fermented milk, *Aust. J. Dairy Sci.,* 54, 19–23, 1999.
63. Garrote, G.L., Abraham, A.G., and de Antoni, G.L., Preservation of kefir grains, a comparative study, *Lebensm. Wiss. Technol.,* 30, 77–84, 1997.
64. Libudzisz, Z. and Piatkiewicz, A., Kefir production in Poland, *Dairy Indust. Int.,* 55, 31–33, 1990.
65. Mann, E.J., Kefir, *Dairy Indust. Int.,* May, 39,41,47, 1979.
66. Zimovetz, V.N. and Boyko R.L., 2000. — Ukraine, http://www.minagro.kiev.ua/english/Analitic/Milk/index-e.html.
67. Serova, E.V., 1997. Trade and market policy in Russia's agriculture: 1992–1996, http://www.iet.ru/personal/agro/berlin.htm.
68. Desez, M.A., Synthèse des réglementations connues sur les laits fermentés, *Ann. Bull. Int. Dairy Fed.,* Pt. III, 139–140, 1964.
69. Israel Standards Institution, 1987 Israel Standard S1 285:1987.
70. State of California, Food and Agricultural Code, Article 36, Kefir milk — a cultured milk, Sections 38871, 38872, 38873, 38874, 38875, 38876, 1991.
71. Anon., Dairying in Poland, 2000, http://www.imes.co.uk/articles/poland.html.
72. Hungarian Ministry of Economic Affairs, Milk and Dairy Products, http://www.ikm.iif.hu/english/economy/industr/tejfeld.htm
73. Kroger, M., Kefir, *Cultured Dairy Prod. J.,* 28(26), 28–29, 1993.
74. Klupsch, H.J., Method of Producing Kefir, German Patent Application DE3300122A1, 1984
75. Dmitrovskaya, G.P. et al., USSR Patent SU 1227 146 A, 1986
76. Tokumaru, S., et al., United States Patent US 4702923, 1987.
77. Kabore, P., Method for Preparing Kefir-type Stabilized Fermented Beverages, French Patent Application FR2665826A1, 1992.
78. Gawel, J. and Gromadka, M., Changements chimiques au cours de la fermentation et de la maturation du kéfir, *XX Internat. Dairy Cong. Brief Comm.,* Vol. E, 850, 1978.

79. Hallé, C., Leroi, F., Dousset, X., and Pidoux, M., Les kéfirs. Des associations bactéries lactique-levures, in *Bactéries Lactiques: Aspects Fondamentaux et Technologiques*, Vol. 2, de Roissart, H. and Luquet, F.M., Eds., Uriage, France, 1994, pp. 169–182.
80. Koroleva, N.S., Starters for fermented milks, *Bull. Internat. Dairy Fed.*, No. 227, 35–40, 1988.
81. Koroleva, N.S., Technology of kefir and kumys, *Bull. Internat. Dairy Fed.*, No. 227, 96–100, 1988.
82. Korovkina, L.N., Patkoul, G.M., and Maslov, A.M., Influence de la température d'acidification du lait sur les caractéristiques biochimiques et la consistance du kéfir, *Proc. 20e Congrés Internat. de Laiterie*, Vol. E, pp. 852–853, Paris, June 26–30, 1978.
83. Garrote, G.L., Abraham, A.G., and de Antoni, G.L., Characteristics of kefir prepared with different grain:milk ratios, *J. Dairy Res.*, 65, 149–154, 1998.
84. Glaeser, H. and Sigmaringen, F., Kefir: kulturen, herstellung, chemische zusammensetzung, ernährungsphysiologischer wert, *Ernährung-Umschau*, 28, 156–158, 1981.
85. Duitschaever, C.L., What is kefir and how can it be made? *Mod. Dairy*, 68, 18–19, 1989.
86. Rossi, J. and Gobbetti, M., Impiego di un multistarter per la produzione in continuo di kefir, *Ann. Microbiol.*, 41, 223–226, 1991.
87. Pettersson, H.-E., Freeze-dried concentrated starter for kefir, *IDF Bull.* 179, XVI–XVII, 1984.
88. Pettersson, H.E., Christiansson, A., and Ekelund, K., Making kefir without grains, *Scand. J. Dairy Technol. & Knowhow*, 2, 58–60, 1985.
89. Duitschaever, C.L., Kemp, N., and Emmons, D., Pure culture formulation and procedure for the production of kefir, *Milchwissenschaft*, 42, 80–82, 1987.
90. Duitschaever, C.L., Kemp, N., and Emmons, D., Comparative evaluation of five procedures for making kefir, *Milchwissenschaft*, 43, 343–345, 1988.
91. Koroleva, N.S., Technology of kefir and kumys, *Bull. Internat. Dairy Fed.*, 227, 96–100, 1988.
92. Puhan, Z., and Vogt, O., Hefehaltige Sauermilch Produkte-Technologie und Stuffwechsel, *Milchwirtschaftliche Berichte*, 78, 7–14, 1984.
93. Kwak, H.S., Park, S.K., and Kim, D.S., Biostabilization of kefir with a nonlactose-fermenting yeast, *J. Dairy Sci.*, 79, 937–942, 1996.
94. Samsonov, M. and Budagovskaya, V., The role of fermented dairy drinks as dietary products, *Proc. XXI Internat. Dairy Congr.*, Vol. 2, Moscow, 1982, pp. 140–145.
95. Alm, L., Effect of fermentation on milk fat of Swedish fermented milk products, *J. Dairy Sci.*, 65, 521–530, 1982.
96. Moulin, G., Arthaud, J.F., Ratomahenna, R., and Galzy, P., Remarques sur la production de kéfir, *Indust. Aliment. Agric.*, 94, 495–497, 1977.
97. Dousset, X. and Caillet, F., Aspects microbiologiques et biochimiques de la fermentation du kefir, *Microbiol. Aliments Nutr.*, 11, 463–470, 1993.
98. Glaeser, H. and Beiter, M., Enzymatische bestimmung des äthanolgehaltes in handelsublichem und selbst hergestelltem kefir, *Deut. Milchw.*, 30, 1865–1868, 1979.
99. Oberman, H., Fermented milks, in *Microbiology of Fermented Foods*, Vol. 1, Wood, B.J.B., Ed., Elsevier Applied Science, London, 1985, pp. 167–195.
100. Görner, F., Palo, V., and Seginova, M., Formation de composés hautement volatils au cours de la fermentation du kéfir, Abs. A.4.8, XVIII Congres Internat. de Laiterie, Vol. 1F, 738, 1970.
101. Guzel-Seydim, Z.B., Seydim, A.C., Greene, A.K., and Bodine, A.B., Determination of organic acids and volatile flavour substances in kefir during fermentation, *J. Food Comp. Anal.*, 13, 35–43, 2000.

102. Guzel-Seydim, Z., Seydim, A.C., and Green, A.K., Organic acids and volatile flavor components evolved during refrigerated storage of kefir, *J. Dairy Sci.,* 83, 275–277, 2000.
103. Marshall, V. and Cole, W.M., Methods for making kefir and fermented milks based on kefir, *J. Dairy Res.,* 52, 451–456, 1985.
104. Duitschaever, C.L., Toop, D.H., and Buteau, C., Consumer acceptance of sweetened and flavoured kefir, *Milchwissenschaft*, 46, 227–229, 1991.
105. Muir, D.S., Tamine, A.Y., and Wszolek, M., Comparison of the sensory profiles of kefir, buttermilk and yogurt, *Int. J. Dairy Technol.,* 52, 129–134, 1999.
106. Rosell, J.M., Yoghurt and kefir in their relation to health and therapeutics, *Can. Med. Assoc. J.,* March, 341–345, 1932.
107. Alm, L., Effects of fermentation on curd size and digestibility of milk proteins *in vitro* of Swedish fermented milk products, *J. Dairy Sci.,* 65, 509–514, 1982.
108. Evenshtein E.M., Use of kefir for stimulation of gastric secretion and acid-formation in patients with pulmonary tuberculosis (translated from Russian), *Probl. Tuberk.,* 2, 82–84, 1978.
109. Vass, A., Szakaly, S., and Schmidt, P.. Experimental study of the nutritional biological characters of fermented milks, *Acta Med. Hung.,* 41, 157–161, 1984.
110. Schmidt, P., Vass, A., and Szakaly, S., Effect of fermented milk diets on regeneration of rat liver, *Acta Med. Hung.,* 41, 163–169, 1984.
111. Alm, L., Effect of fermentation on proteins of Swedish fermented milk products, *J. Dairy Sci.,* 65, 1696–1704, 1982.
112. Bossi, M.G., Carminati, D., and Bettoni, L., Aspetti nutrizionali e terapeutici di latti fermentati, *L'industria Latte,* 22, 19–35, 1986.
113. Bottazzi, V., Zacconi, C., Sarra, P.G., Dallavalle, P., and Parisi, M.G., Kefir microbiologia, chimica, e tecnologia, *L'industria Latte,* 30, 41–62, 1994.
114. Brückner, H. and Hausch, M., D-Amino acids in dairy products: detection, origin and nutritional aspects. I. Milk, fermented milk, fresh cheese and acid curd cheese, *Milchwissenschaft*, 45, 357–360, 1990.
115. Bogdanova, E.A., Padaryan, I.M., Lavrenova, G.S., and Inozemtseva, V.F., Use of protein concentrates in the manufacture of dairy products, *20th Internat. Dairy Cong.,* Vol. 1E, 845, 1978.
116. Kotz, C.M., Furne, J.K., Savaiano, D.S., and Levitt, M.D., Factors affecting the ability of a high β-galactosidase yogurt to enhance lactose absorption, *J. Dairy Sci.,* 77, 3538–3544, 1994.
117. Alm, L., Effect of fermentation on lactose, glucose, and galactose content in milk and suitability of fermented milk products for lactose intolerant individuals, *J. Dairy Sci.,* 65, 346–352, 1982.
118. de Vrese, M., Keller, B., and Barth, C.A., Enhancement of intestinal hydrolysis of lactose by microbial β-galactosidase (*EC* 3.2.1.23) of kefir, *Br. J. Nutr.,* 67, 67–75, 1992.
119. Alm, L., Effect of fermentation on L(+) and D(-) lactic acid in milk, *J. Dairy Sci.,* 65, 515–520, 1982.
120. .Toxicological Evaluation of some Antimicrobials, Antioxidants, Emulsifiers, Stabilizers, Flour-treatment Agents, Acids, and Bases, FAO Nutrition Meetings. Report Series No. 40A,B,C, http://www.inchem.org/documents/jecfa/jecmono/40abcj44.htm.
121. Panic, B., Raisc, J., and Hristic, V., Variation de la teneur du kéfir en vitamin B_{12} pendant sa fabrication et sa maturation, *Le Lait,* 423/424, 148–154, 1963.
122. Alm, L., Effect of fermentation on B-vitamin content of milk in Sweden, *J. Dairy Sci.,* 65, 353–359, 1980.

123. Kneifel, W. and Mayer, H.K., Vitamin profiles of kefirs made from milks of different species, *Int. J. Food Sci. Technol.,* 26, 423–428, 1991.
124. Safonova, T.Y., Yatsyk, G.V., Yurkov, Y.A., and Volkova, L.D., Effect of varying types of feeding on fatty acid composition of blood serum in premature infants (in Russian, abstract only), *Vopr. Pitan,* 6, 44–49, 1979.
125. Goncharova, G.I., Semenova, L.P., Kozlova, E.P., Lyannaya, A.M., Labodo, K.S., Yatsyk, G.V., and Chistyakova, V.I., Effect of varying types of feeding infants on intestinal microbiocenosis (translated from Russian), *Vopr. Pitan,* 6, 49–53, 1979.
126. Bukhgalter F.L., Use of kefir in complex treatment of children with biliary tract diseases associated with diseases of the pancreas, *Pediatr. Akush. Ginekol.,* 6, 15–17, 1974.
127. Ormisson, A.A. and Soo, T.R., Effect of lactic acid milk and kefir on the indicators of acid–base equilibrium of arterial blood in healthy young children and patients with acute pneumonia and acute bronchitis (translated from Russian), *Pediatriia,* 10, 37–38, 1976.
128. Ivanova, L.N., Bulatskaya, A.N., and Silaev, A.E., Industrial production of kefir for children (in Russian), Molochn. Prom–st., 15–16 (DSA no. 106), 1981.
129. Tokumaru, S., Healthy Diets, Jpn. Kokai Tokkyo Koho, CODEN: JKXXAF JP 11346717 A2 19991221, 1999.
130. Reuter, G., Present and future of probiotics in Germany and in Central Europe, *Biosci. Microflor.,* 16, 43–51, 1997.
131. Ishibashi, N. and Shimamura, S., Bifidobacteria: research and development in Japan, *Food Technol.,* June, 126, 129, 130, 132–135, 1993.
132. Farnworth, E.R., Designing a proper control for testing the efficacy of a probiotic product, *J. Nutraceut. Functional Food,* 2, 55–63, 2000.
133. Shiomi M., Sasaki, K., Murofushi, M., and Aibara, K., Antitumor activity in mice of orally administered polysaccharide from kefir grain, *Jpn. J. Med. Sci. Biol.,* 35, 75–80, 1982.
134. Murofushi, M., Shiomi, M., and Aibara, K., Effect of orally administered polysaccharide from kefir grain on delayed-type hypersensitivity and tumor growth in mice, *Jpn. J. Med. Sci. Biol.,* 36, 49–53, 1983.
135. Murofushi, M., Mizuguchi, J., Aibara K., and Matuhasi T., Immunopotentiative effect of polysaccharide from kefir grain, KGF-C, administered orally in mice, *Immunopharmacology,* 12, 29–35, 1986.
136. Furukawa, N., Matsuoka, A., and Yamanaka, Y., Effects of orally administered yogurt and kefir on tumor growth in mice (in Japanese, abstract only), *J. Jpn. Soc. Nutr. Food Sci.,* 62, 579–585, 1991.
137. Furukawa, N., Matsuoka, A., Takahashi, T., and Yamanaka, Y., Effects of fermented milk on the delayed-type hypersensitivity response and survival day in mice bearing Meth-A (in Japanese), *Anim. Sci. Technol.,* 62, 450–453, 1991.
138. Kubo M., Odani T., Nakamura, S., Tokumaru, S., and Matsuda, H., Pharmacological study on kefir — a fermented milk product in Caucasus. I. On antitumor activity (1), (in Japanese, abstract only), *Yakugaku Zasshi.,* 112, 489–495, 1992.
139. Furukawa, N., Matsuoka, A., Takahashi, T., and Yamanaka, Y., Anti-metastatic effect of kefir grain components on Lewis lung carcinoma and highly metastatic B16 melanoma in mice, *J. Agri. Sci. Tokyo Nogyo Daigaku,* 45, 62–70, 2000.
140. Hosono, A., Tanabe, T., and Otani, H., Binding properties of lactic acid bacteria isolated from kefir milk with mutagenic amino acid pyrolyzates, *Milchwissenschaft,* 45, 647–651,1990.
141. Miyamoto, T., Morita, H., Nishioka, K., Kataoka, K., Izumimoto, M., and Kuyama, T., Constituent species of lactic acid bacteria from kefir and their desmutagenic properties (in Japanese), *Jpn. J. Dairy. Food Sci.,* 40, A111–A120, 1991.

142. Cevikbas, A., Yemni, E., Ezzedenn, F.W., Yardimici, T., Cevikbas, U., and Stohs, S.J., Antitumoural, antibacterial and antifungal activities of kefir and kefir grain, *Phytother. Res.*, 8, 78–82, 1994.
143. Yoon, Y.H., Cho, J.K., Baek, Y.J., and Huh, C.S., Antimutagenic activity of *Lactobacillus* spp. isolated from kefir and yogurt and non-starter strains (in Korean), *Kor. J. Anim. Sci.*, 41, 39–44, 1999.
144. Cogan, T.M., Barbosa, M., Beuvier, E., Bianchi-Salvadori, B., Cocconcelli, P.S., Fernandes, I., Gomez, J., Gomez, R., Kalantzopoulos, G., Ledda, A., Medina, M., Rea, M.C., and Rodriguez, E., Characterization of the lactic acid bacteria in artisanal dairy products, *J. Dairy Res.*, 64, 409–421, 1997.
145. Ryan, W.P., Rea, M.C., Hill, C., and Ross, R.P., An application in cheddar cheese manufacture for a strain of *Lactococcus lactis* producing a novel broad-spectrum bacteriocin, lacticin 3147, *Appl. Environ. Microbiol.*, 62, 612–619, 1996.
146. Morgan, S.M., Hickey, R., Ross, R.P., and Hill, C., Efficient method for the detection of microbially produced antibacterial substances from food systems, *J. Appl. Microbiol.*, 89, 56–62, 2000.
147. Serot, T., Dousset, X., Zucca, J., and Torcatis, N., Mise en évidence et purification partielle de substances antibacteriiennes produites par *Leuconostoc mesenteroides* et *Lactobacillus plantarum* isolés de grains de kéfyr, *Micro. Aliments Nutr.*, 8, 71–76, 1990.
148. Zacconi, C., Parisi, M.G., Sarra, P.G., Dallavalle, P., and Bottazzi, V., Competitive exclusion of *Salmonella kedougou* in kefir fed chicks, *Microbiol. Ailments Nutr.*, 12, 387–390, 1995.
149. Brialy, C., Rivalland, P., Coiffard, L., and de Roeck Holtzhauer, Y., Microbiological study of lyophilised dairy kefir, *Folia Microbiol.*, 40, 198–200, 1995.
150. Garrote, G.L., Abraham, A.G., and de Antoni, G.L., Inhibitory power of kefir: the role of organic acids, *J. Food Prot.*, 63, 364–369, 2000.
151. Oleinichenko, E.V., Mitrokhin, S.D., Nonikov, V.E., and Minaev, V.I., Acipole efficacy in prevention of enteric dysbacteriosis due to antibacterial therapy (in Russian, abstract only), *Anitibiot. Khimioter,* 44, 23–25, 1999.
152. Osada, K., Nagira, K., Teruya, K., Tachibana, H., Shirahata, S., and Murakami, H., Enhancement of interferon-β production with sphingomyelin from fermented milk, *Biotherapy,* 7, 115–123, 1994.
153. Besednova, N.N., Epshtein, L.M., Gazha, A.K., Borovskaia, G.A., Besednov, A.L., Rozhzhov, I.V., and Smolina, T.P., Therapeutic–prophylactic milk products with a new immunocorrector of natural origin (in Russian, abstract only), *Vopr. Pitan.*, 3, 31–34, 1997.
154. Taylor, G.R.J. and Williams, C.M., Effects of probiotics and prebiotics on blood lipids, *Br. J. Nutr.*, 80, S225–S230, 1998.
155. St-Onge, M.-P., Farnworth, E.R., and Jones, P.J.H., Fermented and non-fermented dairy product consumption: effects on cholesterol levels and metabolism, *Am. J. Clin. Nutr.*, 71, 674–681, 2000.
156. Vujicic, I.F., Vulic, M., and Könyves, T., Assimilation of cholesterol in milk by kefir cultures, *Biotechnol. Lett.*, 14, 847–850, 1992.
157. St-Onge, M-P., Farnworth, E.R., Savard, T., Chabot, D., Mafu, A., and Jones, P.J.H., Kefir consumption does not alter plasma lipid levels or cholesterol fractional synthesis rates relative to milk in hyperlipidemic men, *BMC Compl. Altern. Med. J.*, http://www.biomedcentral.com/1472–6882/2/1/, 2002.

158. Sukhov, S., Kalamkarova, L.I., Il Chenko, L.A., and Zhangabylov, A.K., Changes in the microflora of the small and large intestine in patients with chronic enteritis after dietary treatment with cultured milk products (in Russian, abstract only), *Vopr. Pitan.*, 4, 14–17, 1986.
159. Batinkov E.L., Use of milk and kefir in peptic ulcer of the stomach and duodenum (in Russian, abstract only), *Vopr. Pitan.*, 30(4), 89–91, 1971.
160. Weissmahr, J.A., Dietary Supplement Derived from Fermented Milks for Prevention of Osteoporosis, World Intellectual Property Organ, WO 00/60950 PCT/IT00/00130, 2000.
161. Thoreux, K. and Schmucker, D.L., Kefir milk enhances intestinal immunity in young but not old rats, *J. Nutr.*, 131, 807–812, 2001.

5 Yogurt and Immunity: The Health Benefits of Fermented Milk Products that Contain Lactic Acid Bacteria

Judy Van de Water

CONTENTS

0-8493-1372-4/03/$0.00+$1.50

5.1 HISTORY AND CULTURE RELATED TO FERMENTED FOOD

The history of fermented milk products is long and quite diverse culturally. While the exact origins are difficult to ascertain, they likely date back to more than 10,000 years ago. According to Persian tradition, Abraham owed his fecundity and longevity to the regular ingestion of yogurt. In the early 1500s, King Francis I of France was reportedly cured of a debilitating illness after eating yogurt made from goats' milk.[1] Scientific interest in the health benefits of yogurt was initiated by Élie Metchnikoff in the early 1900s. Metchnikoff proposed that the lactic acid microbes of fermentation must be antagonistic to the putrefying microbes of the gut, and once introduced into the intestine, they would prevent the breeding of the noxious microbes that required an alkaline environment. His hypothesis was stimulated by the fact that populations such as those living in the Balkans regularly ate yogurt and were noted for their longevity. He experimented on himself and reported that his health, which was generally poor, improved with regular ingestion of sour milk prepared with cultures of the Bulgarian lactic bacillus. Metchnikoff's enthusiasm about yogurt became more publicized, and doctors began recommending yogurt/sour milk as a hygienic food. Metchnikoff credited his relatively long life in part to the lactic bacilli in his diet, and hypothesized, "When people have learnt how to cultivate a suitable flora in the intestines of children as soon as they are weaned from the breast, the normal life may extend to twice my 70 years".[2] (See Chapter 1 for more details on the history of yogurt.)

Over the past several decades, the consumption of fermented milk products, especially yogurt, has greatly increased. The most dramatic increase occurred during the 1980s and 1990s, which is certainly in part due to increased knowledge regarding the health benefits of yogurt and other fermented milk products.[3] Moreover, the addition of fruit and sweeteners to yogurt has made it more widely palatable. However, it is likely the increasing knowledge regarding the health benefits of fermented foods, especially live-culture yogurt, that has driven the recent growth in consumption. The following sections of this chapter will discuss the fermentation process, the compositional changes of milk following fermentation, and health effects of consumption of fermented milk in both animals and man.

5.2 YOGURT PRODUCTION

Yogurt is produced using active cultures of bacteria to ferment cream and/or milk. Yogurt that is produced in the United States is made with two specific live and active cultures of lactic acid bacteria (LAB) — *Lactobacillus bulgaricus* (*Lb. bulgaricus*) and *Streptococcus thermophilus* (*S. thermophilus*). *Lb. bulgaricus* and *S. thermophilus* metabolize some of the milk sugar (lactose) in the milk into lactic acid. This action helps change the consistency of liquid milk into yogurt. The production of fermented milk, or yogurt, requires that the milk is first concentrated by the addition of dairy solids, evaporation, or membrane filtration. The mixture is then heated to destroy undesirable organisms and cooled. Then the starter cultures are added. Yogurt products may also have added ingredients such as sugar, sweeteners, fruits or vegetables, flavoring compounds, sodium chloride, coloring, stabilizers, and preservatives. In the United States, *Lb. bulgaricus* and *S. thermophilus* are required by U.S. Food and Drug Administration (FDA) standards in order for a product to be called yogurt. Other cultures may be added but are not required.

The fermentation process involves the inoculation of pasteurized milk that has been enriched in milk protein with concentrated cultures of bacteria; the milk is then incubated at 40 to 44°C for 4 to 5 h. During fermentation, lactic acid is produced from lactose by the yogurt bacteria, the population of which increases 100- to 10,000-fold to a final concentration of approximately 10^9/ml. The reduction in pH, due to the production of lactic acid, causes a destabilization of the micellar casein at a pH of 5.1 to 5.2, with complete coagulation occurring around pH 4.6. At the desired final pH, the coagulated milk is cooled quickly to 4 to 10°C to slow down the fermentation process.

Fermentation of milk with LAB leads to specific organoleptic characteristics (taste, aroma) of the final product. The metabolism of LAB and the interactions between the selected strains are responsible for the production of lactic acid, the coagulation of milk proteins, and the production of various compounds. Variables such as temperature, pH, the presence of oxygen, and the composition of the milk further contribute to the particular features of a specific product.[4–6] Fermented milks exhibit a wide variety of textures ranging from liquid drinks such as kefir, koumiss, and acidophilus milk to semisolid or firm products including yogurt, filmjölk, villi, dahi, and leben. (See Chapter 4 for more details on manufacture and properties of kefir.)

Certain strains of *S. thermophilus, Lb. bulgaricus,* and other LAB, such as *Lactococcus cremoris* and some species of *Leuconostoc,* produce exocellular polysaccharides which modify the texture of a fermented milk product, i.e., by increasing the viscosity or creating a "ropy" texture.[4,6–12] Lactic acid is also responsible for the slightly tart taste of the fermented milk product, while the other characteristic flavors and aromas are additional results of LAB metabolism. For example, acetaldehyde provides the characteristic aroma of yogurt, while diacetyl, produced by *Lc. diacetylactis* and *Leuconostoc cremoris,* imparts a buttery taste to some fermented milks. Acetoin, acetone, lactones, and volatile acids are other important flavor components that may be present in certain fermented milks as byproducts of bacterial metabolism.[8,9,12]

There is a symbiotic relationship, also known as "protocooperation," between *S. thermophilus* and *Lb. bulgaricus,* in which each species of bacteria stimulates the growth of the other. *Lb. bulgaricus* stimulates the growth of *S. thermophilus* by liberating amino acids and peptides from milk proteins; these substances enable *S. thermophilus* to grow faster in the early part of incubation. *S. thermophilus* in turn produces formic acid, which stimulates the growth of *Lb. bulgaricus,* resulting in a shortened fermentation time and a product with characteristics different from that of milk fermented with a single species.[4,8,9,12,13]

5.3 THE EFFECTS OF FERMENTATION ON MILK

5.3.1 Effects on Chemistry, Nutrient Content, and the Activity of Enzymes

One important result of the addition of the bacteria necessary for fermentation is the resulting proteolytic activity of the yogurt bacteria. Although this activity is slight, resulting in a breakdown of only 1 to 2% of milk protein,[14] it is essential to release small peptides and amino acids for the growth of the bacteria. *Lb. bulgaricus* is more proteolytic, but both yogurt bacteria contain peptidases that are necessary to hydrolyze large peptides into smaller peptides for transport into the cell. The principal substrate for such proteolysis is casein, but limited degradation of whey proteins may also occur.[15,16] The net effect of this proteolysis is that fermented milks have a higher content of peptides and free amino acids, especially valine, histidine, serine, and proline, than milk does.[17,18] Moreover, while the limited proteolytic action of yogurt bacteria does not significantly alter the nutritional value of milk proteins,[19] yogurt is significantly more digestible than the milk mixture from which it is made.[20] A study with rats demonstrated that feeding yogurt compared to the native milk from which it was prepared resulted in increased feed efficiency.[21] The increased digestibility of proteins in fermented milks may be related to the fine flocculation of caseins resulting from the joint action of proteolysis and acidification. (See Chapter 7 for more details on proteolysis of milk.)

With respect to the effects of fermentation on the carbohydrate fraction of milk, about 20 to 30% of the lactose in milk is fermented by LAB through different pathways (Figure 5.1). The bacteria used to produce yogurt are homofermentative, producing one major endproduct, lactic acid. This accounts for greater than 95% of the fermentation products found in yogurt. The final concentration of lactic acid in yogurt is 0.7 to 1.2%. This lactic acid is a mixture of both the L(+) and D(−) isomers. Although the quantity of each isomer present depends on the specific culture, the L(+) isomer generally represents between 50 and 70% of the total lactic acid.

It is the reduction in the lactose concentration coupled with the presence of a large number of viable bacteria containing ß-galactosidase that allows the consumption of yogurt by lactose-intolerant individuals. ß-Galactosidase is protected by the bacteria from denaturation by the acid in the stomach, thus allowing delivery of the enzyme to the intestine. Once in the small intestine, bile increases the permeability of the bacterial cells, thus facilitating the entry and subsequent hydrolysis of lactose by the bacteria.[22]

One study investigated the effects of acid milk and organic acids on the digestive tract in mice.[23] It was determined that lactic acid resulted in increased peristaltic movement in the duodenum, jejunum, ileum, cecum, and colon, but not in the stomach or rectum. In contrast, acetic acid stimulated movement only in the duode-

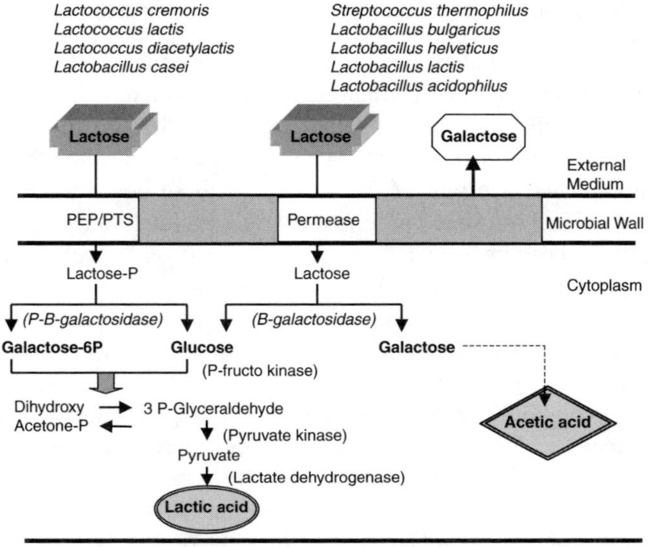

FIGURE 5.1 Major pathways of lactose metabolism by the lactic acid bacteria used for milk fermentation. PEP/PTS, phosphoenolpyruvate:sugar phosphotransferase system.

num and colon. The exact role of this acid-stimulated motility on digestion remains to be clarified.

With the exception of some B vitamins, changes in the vitamin and mineral content of fermented milk products are negligible.[24] Moreover, the pasteurization of milk prior to fermentation may destroy some vitamins such as B_6, B_{12}, and folic acid, while the level of thermostable vitamins (niacin and pantothenic acid) remains unchanged. Some lactic acid bacterial strains produce a net increase in B vitamins, notably folates, during fermentation, while others result in a net loss. Generally speaking, *Lb. bulgaricus* uses folic acid, whereas *S. thermophilus* produces it.[25] However, following fermentation, the levels of some vitamins, especially B_{12} and folic acid, decrease during cold storage.[26]

Yogurt, like milk, is an excellent source of calcium and phosphorus, both of which are essential for the development and maintenance of bones. In addition, yogurt also contains relatively large amounts of potassium and can be considered a good source of this mineral. Further, since yogurt is usually produced from milk that has been enriched either through concentration or the addition of milk powder, it is, on a unit-to-unit basis, a richer source of these minerals than milk is. This is especially important for those individuals who are lactose intolerant and must limit their intake of dairy products.

It should be noted that postfermentation heat treatment significantly alters some properties of fermented milks. Heat treatment above 65°C appreciably reduces the level of some thermosensitive vitamins.[26] In addition, enzymatic activity, notably the activity of ß-galactosidase, is markedly reduced.[27] This reduced activity dramatically lowers the ability of lactose intolerant individuals to tolerate the same amount of lactose that otherwise could be tolerated with live-culture yogurt.

5.3.2 Health Benefits of Fermented Milk Consumption

Numerous reports and studies regarding the health benefits of yogurt and other fermented milk products have been published. While the mechanisms behind such health claims are still being investigated, these benefits (on the immune and/or metabolic systems) appear to be real. Many of the data collected thus far indicate that it is through the ingestion of live LAB that these benefits are realized. The survival of bacteria administered in fermented milk products during passage through the human gut has been investigated intensely in recent years. Well-controlled, small scale studies on diarrhea in both adults and infants have shown that probiotics are beneficial and that they survive in sufficient numbers to affect gut microbial metabolism. Probiotics are nonpathogenic microorganisms that, when ingested, exert a positive influence on the health or physiology of the host.[28–32] They can influence intestinal physiology either directly or indirectly through modulation of the endogenous ecosystem or immune system. Survival rates have been estimated at 20 to 40% for selected strains; the main obstacles to survival are gastric acidity and the action of bile salts. Although it is believed that the maximum probiotic effect can be achieved if the organisms adhere to intestinal mucosal cells, there is no evidence demonstrating that exogenously administered probiotics do adhere to the mucosal cells. Instead, they seem to pass into the feces without having adhered or multiplied. Thus, to obtain a continuous exogenous probiotic effect, the probiotic culture must be ingested daily. Certain exogenously administered substances enhance the action of both exogenous and endogenous probiotics. Human milk contains many substances that stimulate the growth of bifidobacteria *in vitro,* especially in the small intestine of infants; however, it is unlikely that these substances function in the colon. Beneficial effects may thus accrue from exogenously administered probiotics, often administered with prebiotics (nondigestible food ingredients that benefit the host by selectively stimulating the growth or activity of one or a limited number of bacteria in the colon), or from endogenous bifidobacteria and lactobacilli, whose metabolic activity and growth may also be enhanced by the administration of prebiotics.[29]

Studies that have shown a sufficient level of proof to enable probiotics to be used as treatments for gastrointestinal disturbances include:

1. Increased tolerance of yogurt compared with milk in subjects with primary or secondary lactose maldigestion[33]
2. The use of *Saccharomyces boulardii, Lactobacillus,* and *Enterococcus faecium* SF 68 to prevent or shorten the duration of antibiotic-associated diarrhea[34–37]
3. The use of *S. boulardii* to prevent further recurrence of *Clostridium difficile*-associated diarrhea[38]
4. The use of fermented milks containing *Lb. rhamnosus* GG to shorten the duration of diarrhea in infants with rotavirus enteritis (and probably also in gastroenteritis of other causes)[39,40]

Additional situations in which probiotics may be of value include mitigation of diarrhea of miscellaneous causes; prophylaxis of gastrointestinal infections, including traveler's diarrhea; and immunomodulation.[41–43] Trials in gastrointestinal diseases that involve the ecosystem, e.g., *Helicobacter pylori* infections, inflammatory bowel disease, and colon cancer, are currently being performed.[44,45]

One of the most widely touted benefits of yogurt consumption is said to be the enhancement of the immune system. It has been proposed that LAB and fermented milks modulate certain parameters of both the nonspecific and specific immune responses. The link between these benefits and the immune system, however, has not been identified and the mechanisms involved are still unknown.

5.4 FERMENTED MILK AND THE IMMUNE SYSTEM

5.4.1 Overview of the Immune System

A brief overview will help to clarify the different branches of the immune system involved in the response to LAB and fermented foods.[30,46–49] The body has a number of defense systems used against the massive invasion of foreign matter it constantly receives. The nature of this response depends on two factors:

1. The type of foreign particles (viruses, bacteria, parasites, fungi, pollens, certain food proteins)
2. The route of entry (the skin, blood, the lungs, or the epithelium of the gastrointestinal tract)

The first line of defense involves physical–chemical barriers such as the skin and the mucus layers, e.g., in the nose and intestines, which indeed prevent the occurrence of most infectious diseases. The immune system represents the second line of defense against microorganisms, and its response involves a complex interrelation of its various components. The three principal steps in this response are:

1. Recognition of the foreign molecule
2. Destruction of the foreign matter
3. Regulation of the response through multiple feedback controls

The two major arms of the immune response are the innate or nonspecific response and the acquired or specific immune response. Generally, when one is faced with a microorganism, whether it be pathogenic or nonpathogenic, both arms of the immune system are involved in the response. The innate response consists of cells that are primarily phagocytic in nature and respond to the antigen by internalization. This in turn activates the responding cells and begins a cascade of events. Among these events are the presentation of the antigen to the acquired branch of the immune system and the activation of cells through cytokine production. Cytokines are proteins that are produced by cells and that affect other cells. Cytokines bind to a receptor on the target cell and can be either stimulatory or down-regulatory, depending on the cell type and the cytokine involved.

The specific or acquired immune response contains two general categories of reactivity, the cell-mediated and humoral responses. A complex interaction exists

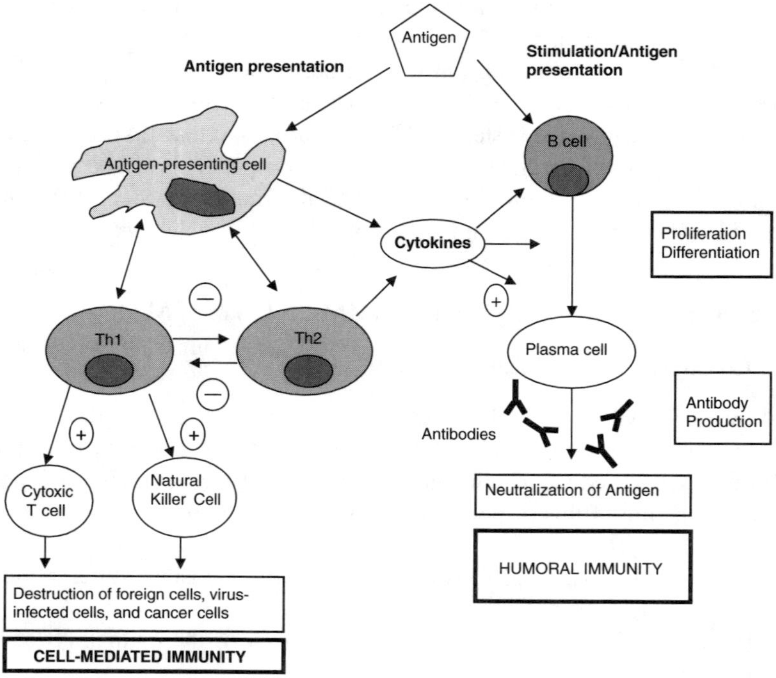

FIGURE 5.2 The acquired immune response.

between these two systems and between the acquired response and innate immunity (Figure 5.2). Which branch of the acquired immune system is involved in a response depends on the nature of the antigen and the route of exposure. Moreover, activation requires the identification and presentation of the antigen by the innate immune system (dendritic cells, monocytes/macrophages, and sometimes B cells).

The cell-mediated immune response involves T cells that directly kill pathogen-infected cells (T-cytotoxic cells, natural killer cells) or that regulate the immune response via cytokines (T-helper cells). As mentioned above, these mediators are multifunctional and can affect many organ systems. Examples of cytokines include interferons (IFN), interleukins (IL), colony-stimulating factors (CSF), and tumor necrosis factor (TNF).[50] T-helper cells are divided into three subgroups, Th1, Th2, and Th3. The Th1 cells produce the more inflammatory cytokines, such as IFN-γ and IL-2, while the Th2 cells produce the cytokines responsible for activation of the humoral response through the production of IL-4, IL-10, and IL-13. Th3 cells produce primarily TGF-β, a cytokine that is primarily responsible for the down-regulation of the immune response. There is also a feedback control system between the Th1 and Th2 cells for regulation of the response.[51]

The humoral response involves B cells that, upon stimulation by cytokines produced by helper T cells, divide and differentiate into plasma cells that secrete immunoglobulins or antibodies (Ab). Ab are capable of recognizing a specific segment of the antigen called the epitope. This can lead to neutralization of the antigen by blocking (e.g., viral attachment to cells), activation of the complement system (involved primarily with bacteria), and/or clearance (through the reticuloendothelial cell and gastrointestinal systems).[50,51]

5.4.2 The Immune System of the Gastrointestinal Tract

In addition to the systemic immune response, organ-specific lymphoid tissues have unique properties not found elsewhere. The most relevant of these, in the context of this chapter on fermented foods, is the gastrointestinal (GI) tract. The GI tract contains a specialized mucosa-associated lymphoid system (part of the mucosa-associated lymph tissue (MALT) system that also includes the lungs) called the gut-associated lymphoid tissue (GALT).

The GI tract encounters a myriad of antigens from the numerous microorganisms and foreign proteins derived from various foods that enter through the mouth. A first line of defense against pathogens is provided by the barrier effect of the gastrointestinal digestive juices, the intestinal flora, and the intestinal mucus layers, each of which prevents microorganisms from entering the body. The second defense mechanism is the GALT. It is composed of organized lymphoid tissue in the ileum (Peyer's patches), as well as a wide variety of immune cells (B and T lymphocytes, plasma cells, macrophages, mast cells, eosinophils, and basophils) that infiltrate the gastrointestinal mucosa. The GALT contains up to 60 billion cells and makes up 25% of the intestinal mucosa (Figure 5.3). The principal antibody synthesized by local plasma cells is secretory IgA (sIgA), which is found in other body secretions as well, such as in saliva, mucus, colostrum, and tears.[52] Secretory IgA is also part of the first line of defense for the mucosal barrier, where it binds to and blocks entry of a variety of intestinal pathogens.

The initial establishment of the intestinal microflora is essential for the development of a fully functional and balanced immune system, including the development and maturation of IgA plasmocytes and IgA production.[53,54] Part of the intestinal IgA production is directed against commensal bacteria themselves.[55,56] Much of this IgA is reportedly induced by T-cell independent pathways that appear to be

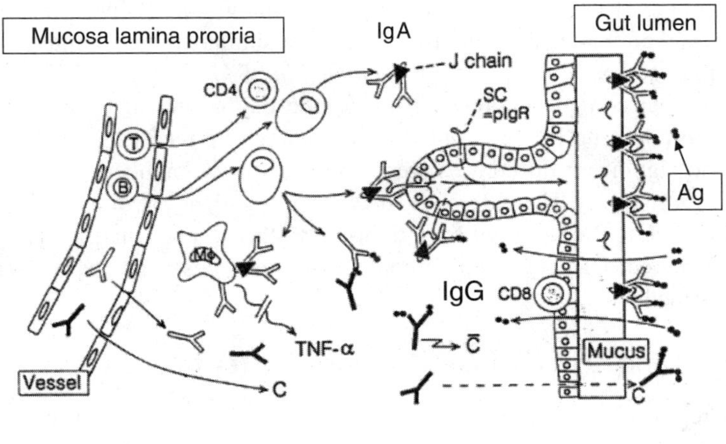

FIGURE 5.3 Gut-associated lymphoid tissue (GALT).

restricted to the intestinal mucosa, and no specific IgA was detected in serum.[56] Interestingly, in gnotobiotic rats colonized with a combination of *Escherichia coli* and *Lb. plantarum*, a large proportion of the *E. coli*-specific IgG and IgA antibodies cross-reacted with *Lb. plantarum.* Similar cross-reactivity was observed between *Lb. acidophilus* and *E. coli.*[57] It is conceivable that some of the increased secretory IgA responses reported after supplementation with various lactobacilli during viral infections may at least partially reflect such cross-reactivity.[58–60]

5.5 EFFECTS OF LACTIC ACID BACTERIA ON THE GASTROINTESTINAL SYSTEM

Supplementation with exogenous LAB can also affect IgA production. The addition of 3 ml of yogurt containing 2×10^8 cells/ml (*Lb. delbrueckii* ssp. *bulgaricus* and *S. salivarius* ssp. *thermophilus*) to the diet of Balb/c mice resulted in a significant increase in IgA-secreting cells in the small intestine after 7 days. However, this increase was no longer observed after 10 days of yogurt feeding.[61] Similarly, in protein–energy deficient mice given a protein-free diet, supplementation with 3 ml of the same yogurt preparation for only two consecutive days was associated with significantly higher numbers of IgA- and IgM-secreting cells in the ileum compared to animals supplemented with unfermented milk.[62] A group supplemented with *Lb. casei* in the diet exhibited intermediate responses. Yogurt was also most effective in improving the height of intestinal microvilli, enhancing mucus secretion, and preventing the translocation of intestinal bacteria, i.e., in improving the barrier functions of the intestine.

Swiss albino mice fed fresh yogurt for 7 days, then challenged with 20 LD_{50} of *Salmonella typhimurium,* exhibited significantly higher concentrations of *S. typhimurium*-specific IgA in their intestinal fluid than mice fed stored yogurt containing significantly lower numbers of live bacteria or control mice given nonfat milk.[63] An enhanced response in antigen-specific serum IgA following vaccination with attenuated *Salmonella typhi* Ty21a has also been reported in human volunteers consuming fermented milk containing *Lb. acidophilus* La1 (now *Lb. johnsonii* La1) and bifidobacteria (3×125 g/d, 10^7 to 10^8 colony forming units [CFU] per gram) before, during, and after vaccine administration.[64] In a similar study, volunteers were given the same vaccine and consumed *Lactobacillus* GG (ATCC 53103), *Lactocossus lactis,* or placebo for 7 days starting on the day of the first vaccine dose.[65] An increase in specific IgA was noted only in the group receiving *Lactobacillus* GG, while no significant differences were observed in the numbers of IgA-, IgG- and IgM-secreting cells. The differences between this and the previously discussed study could be due to the different bacteria used, to the different matrices in which they were delivered (fermented milk product versus capsules), and the different duration of LAB intake.

5.5.1 LACTIC ACID BACTERIA AND MUCOSAL HEALTH

It has repeatedly been observed that lactobacillus administration can shorten the duration of diarrhea,[66] including rotavirus-induced diarrhea in children.[58,59,67] Enzyme-linked immunospot (ELISPOT) assays of Ig-secreting cells and specific antibody-secreting cells (sASC) among circulating lymphocytes indicated that the nonspecific immune response was significantly greater in children receiving *Lactobacillus* GG than in those given a placebo.[58] In addition, although no difference existed in the number of IgA sASC during the acute phase, 90% of the lactobacillus-treated children and only 46% of the placebo-treated children had rotavirus-specific IgA sASC 3 weeks after recovery. Similarly, a higher rate of rotavirus IgA seroconversion (became positive for antibodies to rotavirus) was observed in infants given an oral rotavirus vaccine together with *Lb. casei* GG than in those who received the vaccine along with a placebo.[68] It is worth noting, however, that it is currently not clear what role, if any, IgA plays in the recovery from rotavirus infection and to what extent it provides protection from future recurrence.

In children with Crohn's disease, a disease often characterized by a relative deficiency in mucosal IgA, administration of *Lactobacillus* GG (10^{10} CFU twice daily) for 10 days was associated with a significant increase in cells secreting IgA specific for β-lactoglobulin and casein.[69] Such an increase was not observed for patients with juvenile chronic arthritis or healthy controls. The numbers of specific IgG- and IgM-secreting cells were not affected by *Lactobacillus* GG ingestion, nor was the number of nonspecific immunoglobulin-secreting cells of any isotype.

5.5.2 Lactic Acid Bacteria and Protection against Enteric Pathogens

One of the proposed mechanisms facilitating the health benefits of fermented milk products stems from the credo that feeding a live culture of LAB will inhibit the growth of pathogenic bacteria. Two of the most commonly found LAB in fermented milk products, *S. thermophilus* and *Lb. bulgaricus,* are normally isolated from green plant material and milk, respectively, and are not inhabitants of the intestinal tracts of humans and animals. These bacteria are not highly acid- and bile-resistant, with only 15% surviving the passage through the stomach and about 1% reaching the large intestine,[70] where they fail to colonize.[71] However, they may still exert an effect *in vivo* due to intracellular enzymes, cell surface antigenic receptors, or metabolites produced during fermentation. Moreover, while yogurt bacteria and other lactic acid bacteria have been shown to inhibit pathogenic bacteria *in vitro* by the production of organic acids and antibiotic-like substances, this interaction has not been clearly demonstrated *in vivo*.[72] It has been shown in studies with mice[47,73] that feeding yogurt results in an alteration in the intestinal flora, stimulating the growth of lactobacilli and bifidobacteria. Changes such as these in the intestinal microflora are thought to affect the intestinal transit time and may have an impact on nutrient absorption.

It has been suggested that while the numbers of colonizing bacteria are low, they do have an impact on intestinal health. For example, in one study, the effects of chronic consumption of yogurt with (fresh) or without (heated) live bacterial cultures (*Lb. bulgaricus* and *S. thermophilus*) on plasma glucose, insulin, triacylglycerols, cholesterol, fatty acids, and short-chain fatty acids were compared in two groups of healthy men, with or without lactose malabsorption. It was determined that in men with lactose malabsorption, chronic consumption of yogurt containing live bacterial cultures ameliorated the intolerance, as evidenced by lower breath hydrogen excretion, but increased propionate concentrations.[74]

5.5.3 Lactic Acid Bacteria, Nutrient Digestion, and Antigen Uptake

An additional vital function of LAB in the intestinal microflora is to aid in the absorption of otherwise indigestible nutrients, particularly carbohydrates (such as resistant starches, cellulose, hemicelluloses, pectins, and gums), through fermentation. Moreover, it was recently shown that colonization of adult germ-free mice with *Bacteroides thetaiotaomicron*, a major component of the intestinal microflora in mice and humans, was accompanied by marked changes in the transcription of a broad array of genes.[75] A majority of transcripts (95 of 118) increased, among them the mRNA of many proteins involved in nutrient absorption and metabolism. Similar changes, though generally of a lesser magnitude, were seen after colonization with *Escherichia coli* or *Bifidobacterium infantis* or complete conventional mouse microflora. This suggests that colonization can play an important role in increasing the efficiency of nutrient absorption and metabolism. Interestingly, higher nutrient efficiency has been reported in animals fed a yogurt-based diet as compared to animals fed a milk-based diet despite almost an identical nutrient composition of the diets.[76, 77]

In addition to enhancing nutrient absorption, LAB also have been reported to modulate antigen uptake. In suckling rats given cows' milk in addition to maternal milk, the absorption rate of degraded horseradish peroxidase (HRP) was greater in a group receiving cows' milk with *Lb. casei* GG than in a group receiving cows' milk alone, which did not differ from the controls.[78] In contrast, addition of *Lb. casei* GG to the hydrolysate decreased the absorption of degraded HRP. The extent to which HRP was degraded by the mucosa was similar in controls, mice supplemented with milk plus *Lb. casei* GG, and mice receiving the hydrolysate with or without *Lb. casei* GG, but was greater in all of these groups than in milk-supplemented animals.

When casein was first hydrolyzed with pepsin and trypsin, then additionally degraded by enzymes obtained from *Lb. casei* GG, the resulting products were found to consistently suppress concanavalin A (ConA)- or PHA-induced proliferation in peripheral blood mononuclear cells (PBMC) from healthy volunteers.[79] In contrast, only some of the molecules resulting from casein hydrolysis with pepsin and trypsin inhibited proliferation, while others stimulated it. In addition, casein degraded with *Lb. casei* GG was reported to decrease the anti-CD3-induced production of IL-4 by PBMC from atopic subjects, while not affecting interferon-γ (IFN-γ) secretion.[80] Intact casein stimulated the synthesis of both of these cytokines. In a similar experiment with T-cell-enriched PBMC from healthy subjects, *Lb. casei* GG degraded casein and suppressed IL-2 mRNA and protein synthesis, as well as IL-4 and IFN-γ production.[81] In addition, casein degraded with *Lb. casei* GG decreased protein kinase C translocation, another marker of T cell activation. Taken together, these results suggest that *Lb. casei* GG can digest casein, and possibly other macromolecules, into less antigenic and potentially immunomodulating peptides.

5.6 GUT-ASSOCIATED LYMPHOID TISSUE AND THE ESTABLISHMENT OF IMMUNE TOLERANCE

The GALT plays an important role in the establishment of systemic hyporesponsiveness to ubiquitous antigens such as food as well as commensal bacteria, in addition to acting as an aid in mucosal defense. The nature, dose, timing, and route of entry of the antigen all play important roles in determining the nature of the T cell responses and, thus, the outcome — sensitization or hyporesponsiveness. Cytokines produced by T cells upon antigenic stimulation are a major factor in directing the immune response.

Allergic diseases are thought to be a result of the skewing of an immune response to an allergen towards a Th2-type phenotype, i.e., an overproduction of IL-4 by $CD4^+$ helper T cells occurs along with a concomitant underproduction of IFN-γ. IL-4 promotes B cells to switch to IgE and IgG1 production and inhibits Th1-type responses, including the production of IFN-γ. Conversely, IFN-γ inhibits the proliferation of Th2 cells and the synthesis of IgG1 and IgE and instead induces secretion of IgG2a and IgG3. IL-12 is a potent inducer of IFN-γ production by natural killer (NK) and Th1 cells, but can also inhibit the development of a Th2-type response directly, i.e., independently of this IFN-γ induction.[51]

Interestingly, the immune responses of virtually all neonates to environmental allergens have been reported as skewed towards a Th2-type cytokine profile, whether or not they subsequently become atopic.[82,83] In the last decades, allergic diseases have increased dramatically in developed countries, and the "hygiene hypothesis" has been formulated as a possible explanation for this development. In somewhat oversimplified terms, this hypothesis states that partial deprivation of microbial stimuli due to increased hygiene results in an imbalance between Th1- and Th2-type immune responses, favoring the development of IgE-mediated allergies.[84,85] It was originally postulated that infections with measles and other airborne viruses played a major role in providing protection from atopic diseases. It is becoming increasingly recognized, however, that the earliest and largest exposure to microbial antigens occurs during intestinal colonization by bacteria starting at birth. Thus, the intestinal microflora may provide a major stimulus in directing immune responses away from the Th2 phenotype seen in neonates and infants towards the Th1-type response that predominates in nonatopic adults.

The results from several recent studies suggest that the indigenous intestinal microflora of allergic children differs from that of nonallergic children.[86–89] In particular, a significantly lower percentage of allergic children were reported to be colonized with lactobacilli.[87] In contrast, some aerobic microorganisms, such as coliforms and *Staphylococcus aureus,* were detected more frequently in allergic than in nonallergic children. In a prospective study,[89] children who had developed atopy by the age of 2 years were found to harbor significantly higher counts of clostridia and tended to have fewer bifidobacteria at the age of 3 weeks than children who did not develop atopic diseases. Further indications of an imbalance in the microflora of allergic children comes from observations that fecal short chain fatty acids differed between allergic and nonallergic children at the age of 12 months.[88]

Higher levels and proportions of L-caproic acid in the feces of allergic, as compared to nonallergic, children suggested that they harbored elevated numbers of *Clostridium difficile.*

5.7 INTESTINAL MICROFLORA AND ORAL TOLERANCE

Animal studies comparing germ-free and conventional mice provide further evidence for the importance of the intestinal microflora in the development of oral tolerance. In such experiments, oral feeding of an antigen before immunization with that antigen can induce oral tolerance, with the exact outcome depending on the nature and dose of the antigen, the dosing schedule, and the age and genetic background of the animals.[90] IgG unresponsiveness to ovalbumin (OVA) lasted longer in conventional C3H mice than in germ-free animals of the same strain when both groups were fed 20 mg OVA, then immunized three times intraperitoneally (i.p.) with 10 μg OVA adsorbed by alum.[91] No significant differences were observed in the duration of the IgE antibody response.

In contrast, Sudo et al.[92] reported that germ-free Balb/c mice exhibited significantly higher levels of total IgE as well as OVA-specific IgE and IgG1 than conventional mice 5 and 7 weeks after OVA feeding (5 mg/day for 4 days) followed by i.p. immunization with 1 μg OVA in alum every 2 weeks. OVA-specific IgG2 levels were comparable in the two groups. Oral tolerance induction in gnotobiotic (germ-free) mice monoassociated with *B. infantis* at the neonatal stage was similar to that seen in conventional mice, although their OVA-specific IgE levels were somewhat higher. In contrast, colonization of mice with *B. infantis* at the age of 5 weeks resulted in significantly higher OVA-specific IgE, IgG1, and IgG2a concentrations after OVA feeding before immunization compared to germ-free animals.

In the same study,[92] OVA-stimulated IL-4 synthesis was significantly higher in splenocytes from OVA-immunized germ-free animals without oral tolerance induction and was not down-regulated after oral tolerance induction, unlike in conventional animals. In contrast IFN-γ concentrations were similar in conventional and germ-free animals both with and without oral tolerance induction. IL-2 and TGF-β concentrations were also higher in germ-free than in conventional mice; they remained unaffected by oral tolerance induction in both groups. The authors suggested that, taken together with the data on IgG1 and IgG2a, Th1-mediated immune responses were abrogated, but Th2-mediated immune responses were unaltered by oral tolerance induction in germ-free mice.

The immune response to commensal bacteria is transient and is replaced by tolerance soon after the initial colonization.[54] It has therefore been proposed that, for the intestinal microflora to play a continued role in the protection from atopic diseases, a high turnover of bacterial genera, species, and strains may be required.[85,93] Supplementation with LAB may provide such renewed stimuli for the maintenance of predominant Th1-type responses. Findings from *in vitro* experiments and studies in which heat-killed LAB were fed suggest that intestinal colonization is not an absolute prerequisite for such stimulation.

For the *in vitro* studies, splenocytes from OVA-primed mice cocultured with OVA in the presence of heat-killed *Lb. casei* strain Shirota secreted significantly less total as well as OVA-specific IgE, whereas *Lb. johnsonii* JCM 0212 had no effect.[94] *Lb. casei* strain Shirota, but not *Lb. johnsonii*, also dose-dependently increased IFN-γ production, but inhibited IL-4 and IL-5 synthesis. In addition, *Lb. casei* strain Shirota induced IL-12 synthesis in splenic plastic-adherent cells (macrophages). Anti–IL-12 antibody abrogated the *Lb. casei* strain Shirota-induced suppression of IgE, IL-4, and IL-5 secretion in splenocytes, whereas IFN-γ antibody was only partially effective, suggesting that IL-12 did not solely act through the induction of IFN-γ.

Balb/c mice injected i.p. with OVA on day 0 and day 4 and fed a diet containing 0.1 or 0.05% heat-killed *Lb. casei* strain Shirota for 21 days starting on day 0 produced significantly lower levels of OVA-specific IgE and tended to have lower total serum IgE than controls not receiving *Lb. casei* strain Shirota.[95] This was associated with an interesting switch from a Th1-type to a Th2-type cytokine pattern in splenocytes isolated from these mice and restimulated with OVA, namely significant elevations in IFN-γ, IL-2, and IL-12 production and marked reductions in IL-4, IL-5, IL-6, and IL-10 synthesis.

DBA/2 mice fed a casein diet develop elevated casein-specific IgE in association with a Th2-like cytokine pattern. In such mice, i.p. injection of heat-killed *Lb. plantarum* L-137 inhibited casein-specific IgE production, although IgG1 levels actually increased, compared to saline-injected controls.[96] There was a concomitant increase in IL-12 p40 synthesis by unstimulated peritoneal macrophages from *Lb. plantarum*-injected mice and an even greater increase in macrophages restimulated with *Lb. plantarum.* Secretion of IL-4 by ConA-stimulated splenocytes was diminished in *Lb. plantarum*-injected mice, while IFN-γ production was not affected.

Perdigón et al. have also proposed that LAB induce a differential mucosal immune response (specific, nonspecific, or both) in mice. Moreover, they suggest that such behavior is due to the different sites of interaction of the LAB with the gut, resulting in the induction of these different immune responses. They have demonstrated that the LAB are present in different parts of the intestine and that the pathway of internalization of the strains used to make contact with the immune system is through either the Peyer's patch M cells or follicle-associated epithelium (FAE) cells, or the epithelial cells of the small and large intestine. These findings could explain the diversity of the mucosal immune response. It is well known that M cell interaction induces mainly a specific immune response but when internalized through the FAE cells, the response is nonspecific or inflammatory, even though these cells can later enhance a specific immune response. The interaction with epithelial cells can lead to the enhancement of local immunity or to tolerance by antigen clearance.[30,61,97]

5.8 CYTOKINES AND FERMENTED MILK

Some of the most compelling data with respect to the effects of fermented milk products on the immune system involve the production of various cytokines. Many LAB have been reported to induce the synthesis of IFN-γ *in vitro,* including *S.*

thermophilus, Lb. acidophilus, Lb. bulgaricus, and bifidobacteria, although inconsistent results have been obtained with *Lb. bulgaricus* and bifidobacteria (see Tables 5.1 and 5.2). In contrast, most LAB studied to date have little effect on IL-4 production *in vitro,* except for one study reporting a down-regulation of IL-4 synthesis in splenocytes from OVA-primed mice restimulated with OVA *in vitro* (Tables 5.1 and 5.2).[94] Studies on the effect of LAB on *ex vivo* cytokine production are summarized in Table 5.3.

In two separate animal feeding trials, the effects of three forms of yogurt followed by two different unheated or heat-treated yogurts fed to B6C3F1 mice on cytokine expression in spleen, mesenteric lymph nodes (MLN), or Peyer's patches (PP) were determined.[98] All yogurts were fermented with *Lb. bulgaricus* and *S. thermophilus;* in addition, some contained *Lb. acidophilus* and/or *Bifidobacterium* sp. The effects depended on the type of yogurt fed; the presence of *Lb. acidophilus* and/or *Bifidobacterium* sp. was found not to play a role. In addition, the duration of the feeding, the tissue, and the cytokine examined influenced the outcome. In the first feeding trial, yogurt feeding either had no significant effect or down-regulated cytokine expression in mice compared to nonfat dry milk. In particular, IL-4 mRNA was decreased in PP after 2 weeks, but not after 4 weeks, regardless of whether the yogurts contained live or heat-killed bacteria. IFN-γ mRNA was not affected in PP of any animals fed yogurt for either 2 or 4 weeks, but was down-regulated in MLNs by one type of live and three different heat-treated yogurts after 4 weeks of feeding. TNF-α expression was decreased after 4 weeks in spleen by all three live and heat-treated yogurts; all three heat-treated yogurts also reduced TNF-α expression in MLN after both 2 and 4 weeks, but only one live-culture yogurt did so after 4 weeks. In the second feeding trial, involving two other yogurts, the results were much more variable. Most notably, several yogurts increased IFN-γ mRNA levels in MLN; the increase was significant in the case of one heat-treated yogurt. The same yogurt, whether heat-treated or not, also increased IL-6 mRNA in MLN.

In additional mouse studies performed by Perdigón et al., the authors studied IL-4 and IFN-γ release by immunohistochemistry, due to the importance of these two cytokines in maintaining the balance between the populations Th2 and Th1, respectively. They also determined TNF-α as a measure of macrophage activity. It was determined in these studies that LAB were able to induce a diversity of responses mediated by the cytokines released, but different properties in the immunostimulation for each LAB assayed could not be determined. The authors interpreted these results as being due to the use of heterologous strains, of different origins, some of them from humans, tested in a mouse model. Therefore, they then compared the behavior of *Bifidobacterium* sp. and *Lb. animalis* (isolated from mice) with the corresponding human heterologous strains. These follow-up studies demonstrated that host specificity is important for the genus *Bifidobacterium,* but not for lactobacilli. However, this specificity allows a better regulation of the dose, thus avoiding dysregulation of the immune response.[30,97]

Several human studies also indicate that yogurt consumption may be associated with an increase in IFN-γ synthesis. Halpern et al.[99] reported an increase in IFN-γ production by T cells from subjects who consumed 450 g of live-culture yogurt per day for 4 months. This was determined by enzyme-linked immunosorbent assay

TABLE 5.1
Effect of Live LAB on the Production of Various Cytokines *In Vitro*

Bacteria and Cell Type	TNF-α	IL-1β	IL-2	IL-4	IL-6	IL-10	IL-12	IFN-γ	Ref.
Lb. bulgaricus E585 stimulation of PBMC	↑↑	↑↑	—	No effect	↑	↑	↑	No effect	128
Lb. bulgaricus stimulation of PBMC	↑	↑	No effect	—	—	—	—	↑	129
Lb. acidophilus stimulation of PBMC	↑	↑	No effect	—	—	—	—	↑	129
Lb. acidophilus stimulation of PBMC	—	↑	—	—	↑	—	—	↑	102
Lb. acidophilus E 507 stimulation of PBMC	↑	—	—	—	↑	↑ only in responders	—	—	130
Lb. acidophilus stimulation of epithelial cell/leukocyte co-cultures	—	—	—	—	—	No effect	↑	↑	131
Lb. casei stimulation of PBMC	—	↑	—	—	↑	—	—	No effect	102
Lb. rhamnosus	↑	↑	—	No effect	↑	↑	↑	↑	128
Lb. sakei	—	—	—	—	—	No effect	↑	↑	131
Bifidobacterium sp.	↑	↑	No effect	—	—	—	—	↑	129
Bifidobacterium	—	—	↑	—	↑	—	—	No effect	102
S. thermophilus stimulation of PBMC	—	↑	—	—	↑	—	—	↑	102
S. thermophilus stimulation of PBMC	↑	↑	No effect	—	—	—	—	↑	129

TABLE 5.2
Effect of Heat-Killed Lactic Acid Bacteria on Cytokine Production

Bacteria	TNF-α	IL-1β	IL-2	IL-4	IL-5	IL-6	IL-10	IL-12	IFN-γ	Ref.
Lb. acidophilus La1 (*Lb. johnsonii* La1) in RAW 264.7 cells	↑	—	—	—	—	↑	—	—	—	132
Lb. bulgaricus 1489 NCK 231 in RAW 264.7 cells	↑	—	—	—	—	↑	—	—	—	132
Lb. bulgaricus (2 strains)										133
In RAW264.7	↑	—	—	—	—	↑	—	—	—	
In PMA-stimulated EL4.IL-2 cells	—	—	↑	—	—	—	—	—	—	
Lb. johnsonii JCM 0212 in mouse splenocytes from OVA-primed mice	—	—	—	No effect	No effect	—	—	No effect	No effect	94
Lb. casei in x-rayed mouse splenocytes (i.e., monocytes)	—	—	—	—	—	—	—	↑	↑ ConA-induced IFN-γ; no effect alone	127
Lb. casei in mouse splenocytes from OVA-primed mice restimulated with OVA	—	—	—	↓	↓	—	—	↑	↑↑	94
Lb. casei B ATCC 11578 cell wall										134
In rat peritoneal macrophages	↑	↑	—	—	—	—	↑	—	—	
In splenocytes	—	↑	—	No effect	—	—	↑	—	—	
B. bifidum										133
In RAW 264.7 cells	↑	—	—	—	—	↑	—	—	—	
In PMA-stimulated EL4.IL-2 cells	—	—	↑	—	—	—	—	—	—	
B. bifidum in RAW 264.7 cells	↑	—	—	—	—	↑	—	—	—	132
S. thermophilus St-133 in RAW 264.7 cells	↑	—	—	—	—	↑	—	—	—	132
S. thermophilus (4 different strains)										133
In RAW264.7 cells	↑	—	—	—	—	↑	—	—	—	
In PMA-stimulated EL4.IL-2 cells	—	—	↑	—	—	—	—	—	—	

TABLE 5.3
Effects of LAB Consumption on *ex Vivo* Cytokine Production

Bacteria	Treatment	Effect on Cytokine Production	Ref.
Bifidobacterium (Bf)	B6C3F1 mice received a single dose of 10^9 cells	Bf decreased PMA-induced IL-6 & IFN-γ	135
Lb. acidophilus (La)	Peritoneal cells were cultured in the absence or presence of LPS or PMA	La induced IL-6 and IL-12 p40, enhanced PMA-induced IFN-γ	135
Lb. bulgaricus		No effect on IL-6 or IFN–γ	
Lb. casei		Induced IL-6 and IL-12 p40	
S. thermophilus		No effect on IL-6, IFN-γ, or IL-12 p40	
Lb. acidophilus	Balb/c mice were gavaged with 10^9 CFU for 10 or 28 d	↑ ↑IFN-γ compared to milk ± IL-4	116
Lb. rhamnosus	Their splenocytes were cultured with ConA		
Lb. casei strain Shirota (LcS)	Balb/c mice received 1.3 × 10^9 LcS orally for 7 d; their splenocytes were cultured with Con A	↑ ↑IFN-γ ± IL-4 and IL-5[a]	127
Lb. casei strain Shirota	Collagen-induced arthritis (CIA) was induced in DBA/1 mice; some received 2.5 or 5 × 10^8 LcS orally for 5 d; splenocytes were restimulated with denatured collagen	↓IFN-γ compared to water-treated[a] controls	126
Lb. casei strain Shirota	Balb/c mice were i.p. injected with OVA; some were fed a diet containing 0.05% (wt/wt) heat-killed LcS for 21 d; splenocytes were restimulated with OVA	↓IL-4 , IL-5, IL-6, IL-10 ↑ IL-2, IL-12, IFN-γ	95
Lb. casei ATCC 393	Female Balb/c were immunized with Chikungunya virus and fed *L. casei* for 4 alternate d; on d 7, immunohistochemically cytokine-positive cells in the gut villi were counted	↑ TNF-α ($P < 0.05$ compared to $NAHCO_3$–treated controls); IL-1α, IL-1β, IL-2, IL-10, and IFN-γ not significantly different	60

[a] Note that *Lb. casei* strain Shirota appears to have opposite effects in healthy mice and mice in the early stages of collagen-induced arthritis.

(ELISA) analysis of supernatants collected from PHA-stimulated PBMC. In a similar study by the same group of researchers, the health of a college-age population during chronic yogurt consumption was followed. Subjects were asked to eat 200 g of plain yogurt every day for a year; one group ate heat-inactivated yogurt, and a group that ate no yogurt served as controls. The effects of year-long daily consumption of 200 g of yogurt on IFN-γ levels were measured in the two subject populations by an ELISA assay on PHA-stimulated PBMC. While the data were quite variable, when the data were analyzed according to subject age, there was a significant decrease in IFN-γ production in the older group of subjects.[99,100,101] Clearly, further studies involving dose kinetics would be beneficial.

In another human study, subjects eating yogurt of their own choice (i.e., with unknown numbers and types of live or inactive bacteria) for 2 weeks showed a significant increase in 2–5A synthetase activity.[102] This enzyme is reportedly a specific marker for the production of interferons. Significant up-regulation of 2–5 A synthetase activity was also reported in volunteers who consumed 100 g yogurt containing 10^9 *Lb. bulgaricus*/g and 10^9 *S. thermophilus*/g compared to volunteers consuming milk.[103] Enzyme activity was further increased after volunteers ate 250 g yogurt per day for 15 days.

In a double-blind crossover experiment by Wheeler et al., the effect of 450 g/day live-culture yogurt with or without *Lb. acidophilus* was studied in adult patients with moderate asthma.[104] After two 1-month crossover test periods, no significant changes were noted in peripheral cell counts, IgE, IL-2, or IL-4 when the two diets were compared to each other. In addition, ConA-stimulated lymphocytes from patients who consumed yogurt containing *Lb. acidophilus* produced borderline elevated interferon gamma levels ($P = .054$). No differences were noted in mean daily peak flows or changes in spirometric values, and the quality of life indices were unchanged. The authors concluded that the live-culture yogurt generated trends toward an increase in IFN-γ and decreased eosinophilia.[104]

5.9 LACTIC ACID BACTERIA AND IMMUNE CELL FUNCTION

Highly variable results have been reported from both *in vitro* and *ex vivo* experiments assessing the effects of various LAB on lymphocyte proliferation. Whether LAB enhance or inhibit proliferation appears to depend not only on the specific bacterial strain investigated and the type and concentration of the mitogen used to induce proliferation, but also on the activation state of the immune system of the experimental animals or humans.

In vitro, incubation of human PBMC with live *Lb. johnsonii* or *Lb. sakei* for 5 days resulted in a strong proliferative response.[105] The effect obtained with *Lb. johnsonii* was significantly greater than that of *Lb. sakei* and was of a magnitude similar to that observed with PHA (10 μg/ml). Heat-killed bacteria produced similar results. Homogenates of *Lb. rhamnosus* GG, *B. lactis, Lb. acidophilus, Lb. delbrueckii* ssp. *bulgaricus*, and *S. thermophilus* all suppressed proliferation of human PBMC induced with a very high concentration (125 μg/ml) of PHA.[106] This effect was significantly reduced, but not completely eliminated, by heat inactivation. Cytoplasmic extracts, whether unheated or heated, significantly inhibited PHA-induced proliferation, whereas cell wall extracts had no effect.

In mice fed yogurt for 10 months, splenocyte proliferation in response to ConA and PHA was significantly increased compared to milk-fed controls.[76] In contrast, after a shorter feeding period (4 weeks), the ConA-, PHA-, and LPS-induced proliferation of splenic and intestinal lymphocytes from DBA/2J mice fed yogurt or dried milk powder as 50% of their energy for 4 weeks did not differ significantly.[77] When, however, such mice were challenged with an LD_{50} dose of *Salmonella typhimurium,* proliferation of splenocytes after ConA stimulation was significantly higher in the yogurt-fed than in the milk-fed group, while the response to PHA or LPS was similar in the two groups. The mitogenic response of intestinal lymphocytes to ConA and LPS was also significantly higher in yogurt-fed than in milk-fed mice. The proliferative response to PHA, though greater in yogurt-fed animals, was not significantly different from that in milk-fed animals. Similarly, the proliferative response to ConA and LPS was significantly increased in splenocytes from *S. typhimurium*-challenged Balb/c mice fed a diet supplemented with 30 g of yogurt (8×10^8 LAB/g) for 12 days compared to powdered milk-supplemented controls.[107] A slight, statistically nonsignificant increase in proliferation was also observed in mice fed heat-treated yogurt. In another study by the same group, supplementing the diet of unchallenged Balb/c mice with 20% heat-treated yogurt resulted in similar proliferative responses of PP to PHA as did yogurt containing live LAB (*S. thermophilus* and *Lb. bulgaricus*) after 14 or 21 days.[108] Only yogurt containing live LAB, however, increased the LPS-induced proliferation of PP after 7, 14, and 21 days. These enhancing effects of yogurt consumption on the mitogenic response were transient and were no longer observed after 28 days of supplementation with yogurt containing live or heat-killed LAB.

Basal and LPS-stimulated proliferation was increased in splenic lymphocytes from mice fed 10^9 viable *Lb. acidophilus* per kg body weight for 7 days compared to saline-treated controls.[109] In contrast, the same number of bacteria from the strains *Lb. casei, Lb. gasseri,* and *Lb. rhamnosus* inhibited basal proliferation and proliferation stimulated with supraoptimal concentrations of LPS or ConA. In a similar study by the same authors, however, oral supplementation with *Lb. rhamnosus* for 7 days did not affect basal proliferation, while administration for 14 days enhanced it.[110] Unlike in the previous study, an increase in proliferation stimulated with the optimal concentrations of ConA or LPS after 7 days was also observed.

Intraperitoneal administration of *Lb. casei* to MRL/lpr mice (a model for systemic lupus erythematosus) was also reported to significantly inhibit the ConA-, LPS-, or pokeweed mitogen-induced proliferation of splenocytes compared to saline-injected controls.[111] In mesenteric lymph node cells, however, *Lb. casei* injection did not significantly affect the proliferative response to these mitogens.

5.9.1 Innate Immune Responses

5.9.1.1 Phagocytic Activity of Macrophages and Granulocytes

Peritoneal macrophages from mice supplemented with milk fermented with *Lb. acidophilus, Lb. casei,* or both (both isolated from human feces) for 8 days exhibited significantly increased phagocytic activity compared to the peritoneal macrophages from nonsupplemented mice.[112] Colloidal carbon clearance as a measure of *in vivo* phagocytic activity was similarly enhanced. In a similar study by the same group of researchers,[113] oral administration of *Lb. casei* enhanced *in vitro* and *in vivo* phagocytosis, whereas *Lb. bulgaricus* had little effect on *in vitro* phagocytosis but nonetheless significantly increased the *in vivo* colloidal carbon clearance rate. Macrophage activation has also been observed in tumor-bearing mice fed yogurt mixed into their diets.[114,115] In mice immunized with cholera toxin or tetanus vaccine, supplementation with 10^9 CFU of *Lb. acidophilus* (HN017), *Lb. rhamnosus* (HN001), or *B. lactis* for 10 or 28 days was associated with a significant increase in phagocytic activity of both PBL and peritoneal macrophages when compared with controls given skim milk without LAB.[116]

In a study where human volunteers consumed milk for 3 weeks, followed by fermented milk providing either 7×10^{10} CFU *Lb. acidophilus* La1 or 1×10^{10} CFU *B. bifidum* strain Bb 12 for another 3 weeks, there was a highly significant increase in leukocyte phagocytic activity in both groups, with granulocytes showing a greater increase than monocytes (measured as uptake of opsonized *E. coli*).[117] A significant elevation in phagocytic activity could still be detected 6 weeks after cessation of fermented milk intake, even though fecal counts of bifidobacteria and lactobacilli, respectively, had returned to baseline levels by day 12 after the end of bacterial supplementation. Another study by the same group, using a similar protocol, compared the effects of supplementation with 150 ml/day of a milk fermented with *S. thermophilus* to those seen after consumption of the same product supplemented with *Lb. johnsonii* La1 either fresh or stored for 21 to 28 days to yield a tenfold lower bacterial count.[118] Despite the somewhat lower daily dose compared to the previous study (1.5×10^9 CFU *Lb. johnsonii* per day), significant increases in leukocyte phagocytic activity and respiratory burst were observed in the group receiving fresh fermented milk with *Lb. johnsonii* La1. Similar trends were seen in the group consuming stored fermented milk with *Lb. johnsonii* La1, but this did not reach statistical significance for either parameter. The authors concluded that the minimal effective dose for modulating granulocyte/monocyte activities was 10^9 CFU per day. It should be noted, however, that storage of fermented milk products results in changes other than the decrease in viable bacteria due to the accumulation of metabolites generated by these bacteria. Hence, whether the lack of a significant effect in the group receiving stored fermented milk with *Lb. johnsonii* La1 was due to the lower cell count or to other factors (or a combination of the two) is not clear. In order to establish a true dose response, it will be necessary to test products in which differences in the number of viable bacteria are obtained by the appropriate changes in manufacture, rather than through prolonged storage.

In healthy subjects, consumption of milk supplemented with 2.6×10^8 CFU *Lactobacillus* GG resulted in a significant up-regulation of the expression of phagocytosis receptors (CR1, CR3, FcγRI, and IgαR) on neutrophils, whereas the increases on monocytes were not significant, compared to the period prior to milk consumption without *Lactobacillus* GG.[119] Milk alone did not affect receptor expression. In contrast, in milk-hypersensitive subjects, there was considerably higher expression of all of these receptors on monocytes and neutrophils during milk consumption.

When, however, *Lactobacillus* GG was added to the milk, receptor expression did not significantly differ from that seen before milk challenge, suggesting that *Lactobacillus* GG down-regulated the milk-induced phagocyte activation.

The cytokines TNF-α and IL-1β are predominantly produced by activated macrophages. As summarized in Tables 5.1 and 5.2, numerous *in vitro* studies have reported an up-regulation of the synthesis of these cytokines when a variety of cell types, including those used for the production of yogurt or commonly added to it, were incubated with live or heat-killed LAB. This further suggests that numerous LAB have stimulatory effects on macrophage activities.

Interestingly, growth phase, heat treatment, and interactions between the two were found to influence the ability of three different lactobacilli to induce TNF-α production in human peripheral blood monocytes.[120] Live bacteria from the logarithmic growth phase stimulated considerably higher amounts of TNF-α secretion than did heat-killed cells. In contrast, heat-killed bacteria harvested at the stationary phase induced 4- to 5-fold higher levels of TNF-α than live cells did.

5.9.1.2 Natural Killer Cell Activity

Natural killer (NK) cells are thought to play an important role in inhibiting carcinogenesis. IL-12 is a strong inducer of NK cells, and — as discussed and summarized in Tables 5.1 and 5.2 — a variety of LAB, including *Lactobacillus casei* strain Shirota (LcS),[94,95] have been shown to stimulate IL-12 production. The enhancement of NK activity and a significant reduction in tumor incidence have been reported after LcS supplementation in 3-methylcholanthrene-induced carcinogenesis in C3H/HeN mice.[121,122] Such inhibition was not observed in NK cell-deficient beige mice,[122] suggesting that a major mechanism by which LcS inhibits carcinogenesis in this model is through enhancement of NK cytotoxicity.

In mice, oral administration of *Lb. rhamnosus* resulted in a significant increase in NK cell activity, while the increase after *Lb. acidophilus* or *B. lactis* supplementation was not significant.[116] Daily intake for 3 weeks of *B. lactis* in low-fat milk by human volunteers, however, was associated with significantly higher NK cytotoxicity compared to consumption of low-fat milk alone.[123]

Perdigón et al. studied the mechanisms by which yogurt was able to inhibit the growth of a chemically induced intestinal tumor.[124] They demonstrated that yogurt can down-regulate the inflammatory response induced by the carcinogen by:

1. The increase of IgA+ producing cells and CD4+ T cells
2. The diminution of T cells, CD8+, and cytotoxic activity
3. The increase of cellular apoptosis of infiltrating immune cells
4. Apoptosis, which was favored by an increase in the levels of TNF release induced by yogurt
5. Induction of IL-10 release, which plays an important role in the mechanisms of down-regulation

The authors concluded that the mechanisms involved in the immunostimulation by LAB or yogurt are multiple. These findings indicate that it will be very difficult to establish a reliable test for selection of LAB with immunostimulatory capacity. However, the knowledge of the mechanisms controlling the response to an individual strain or yogurt will allow a better choice for therapeutic purposes.[124]

5.9.1.3 Immunostimulating vs. Immunosuppressive Effects

TNF-α and IL-1β are proinflammatory cytokines, and stimulation of their production suggests that LAB may have proinflammatory activities. Such activities would be undesirable in situations already characterized by inflammation, such as allergy, asthma, and several autoimmune diseases. LAB therapy has been examined in several of these conditions, and to date no exacerbation of inflammation has been observed. Instead, in children with atopic eczema and cows' milk allergy, consumption of extensively hydrolyzed whey formula supplemented with *Lactobacillus* GG, but not the same formula alone, was associated with a significant decrease in fecal TNF-α, although TNF-α release by ConA-stimulated PBMC was similar before and after treatment.[125] This may be a strain-dependent effect; it may, however, also reflect the potential of LAB to have differential effects in health and disease.

Indeed, it has repeatedly been reported that various LAB differentially affect the immune responses of healthy animals or volunteers and those whose immune system is activated due to pathogen challenge or autoimmune disease. For example, *Lactobacillus* GG down-regulated phagocyte receptor expression induced by milk challenge in milk-hypersensitive subjects but stimulated receptor expression in healthy subjects.[119] Administration of LcS was associated with increased ConA-stimulated IFN-γ production by splenocytes from healthy Balb/c mice, but had the opposite effect in DBA/1 mice in the early stages of collagen-induced arthritis.[126,127] Mitogen-induced proliferation of splenic lymphocytes from DBA/2J was not enhanced after yogurt feeding.[77] When, however, such mice were challenged with an LD_{50} dose of *Salmonella typhimurium,* the proliferative response to ConA was significantly higher in the yogurt than in the milk group. Nonetheless, because of the species and strain dependence of the immunomodulation by LAB, thorough assessment of LAB *in vitro* and in animal models will be crucial before they can be used to modulate diseases characterized by immune deviation.

5.10 CONCLUSIONS

An aura has surrounded the association of fermented milk products, especially yogurt, and health for more than a century. The pivotal finding within this association has been the direct effect of yogurt upon the immune system. Although this area has been studied in both humans and mice for some time, a significant amount of research is still required to address the fundamental basis for the mechanisms of yogurt's biological consequences. It is perhaps the complexity of the immune system coupled to an equally complex microbial physiology *in vivo* that limits rapid progress within this area. Despite these limitations, the yogurt–immunity connection remains an exciting and attractive area of research for a variety of disciplines. Within the foreseeable future, it is envisioned that these disciplines will collectively develop a core of paradigms regarding the science of yogurt and immunity that will allow rigorous examination and verification with respect to human health. Moreover, through these efforts, new methodology, procedures, microbial strains, and research findings will consolidate the variation in studies extant today and drive the development of improved fermented milk products.

REFERENCES

1. Deeth, H.C. and Tamine, A.Y., Yogurt: nutritive and therapeutic aspects, *J. Food Prot.*, 44, 78–86, 1981.
2. Metchnikoff, O., *Life of Elie Metchnikoff,* Houghton Mifflin Co., Boston, 1921.
3. Dannon, *The History of Yogurt,* The Dannon Institute, Tarrytown, NY, 2002.
4. Roissart, H.D. and Luquet, F.M., *Bactéries Lactiques: Aspects Fondamentaux et Techonologiques,* Vols. 1 and 2, Lorica, Uriage, 1994.
5. Friend, B.A., Fiedler, J.M., and Shahani, K.M., Influence of culture selection on the flavor, antimicrobial activity, beta-galactosidase, and B-vitamins of yogurt, *Milchwissenschaft,* 38, 133–136, 1983.
6. Nakazawa, Y. and Hosono, A., *Functions of Fermented Milk: Challenges for the Health Sciences,* Elsevier Applied Science, London, 1992, p. 518.
7. Salminen, S. and Von Wright, A., *Lactic Acid Bacteria*, Marcel Dekker, New York, 1993, pp. 1–95.
8. Hartley, D. and Denariaz, G., The role of lactic acid bacteria in yogurt fermentation, *Int. J. Immunother.,* 9, 3–17, 1993.
9. Zourari, A., Accolas, J.P., and Desmazeaud, M.J., Metabolism and biochemical characteristics of yogurt bacteria: a review, *Lait,* 72, 1–34, 1992.
10. Cerning, J., Exocellular polysaccharides produced by lactic acid bacteria, *FEMS Microbiol. Rev.,* 7, 113–130, 1990.
11. Kurmann, J.A., Rasic, J.L., and Kroger, M. *Encyclopedia of Fermented Fresh Milk Products: An International Inventory of Fermented Milk, Cream, Buttermilk, Whey, and Related Products*, Van Nostrand Reinhold Co., New York, 1992, p. 368.
12. Marshall, V.M., Starter cultures for milk fermentation and their characteristics, *J. Soc. Dairy Technol.,* 46, 49–56, 1993.
13. Veringa, H.A., Galesloot, T.E., and Davelaar, H., Symbiosis in yogurt II. Isolation and identification of a growth factor for *Lactobacillus bulgaricus* produced by *Streptococcus thermophilus, Neth. Milk Dairy J.,* 22, 114, 1968.
14. Rasic, J.L. and Kurmann, J.A., *Yogurt,* Technical Dairy Publishing House, Copenhagen, 1978.
15. Chandan, R.C., Argyle, P.J., and Mathison, G.E., Action of *Lactobacillus bulgaricus* proteinase preparations on milk proteins, *J. Dairy Sci.,* 65, 1408–1413, 1982.
16. Khalid, N.M., El-Soda, M., and Marth, E.H., Peptide hydrolases of *Lactobacillus helveticus* and *Lactobacillus delbruechkii* sub-sp. *bulgaricus, J. Dairy Sci.,* 74, 29–45, 1991.
17. Tamime, A.Y. and Deeth, H.C., Yogurt: technology and biochemistry, *J. Food Prot.*, 43, 939–977, 1980.
18. Vaitheeswaran, N.I. and Bhat, G.S., Influence of lactic cultures in denaturation of whey proteins during fermentation of milk, *J. Dairy Res.,* 55, 443–448, 1988.
19. Hewitt, D. and Bancroft, H.J., Nutritional value of yogurt, *J. Dairy Res.,* 52, 197–207, 1985.
20. Breslaw, E.S. and Kleyn, D.H., *In vitro* digestibility of protein in yogurt at various stages of processing, *J. Food Sci.,* 38, 1016–1021, 1973.

21. Mcdonough, F.E., Hitchins, A.D., and Wong, N.P., Effect of yogurt and freeze-dried yogurt on growth stimulation rate, *J. Food Sci.,* 47, 1463–1465, 1982.
22. Gilliland, S.E., Properties of yogurt, in *Therapeutic Properties of Fermented Milks,* Robinson, R.K., Ed., Elsevier Applied Science, London, 1991, p. 75.
23. Yokokura, T., Effect of organic acid on gastrointestinal motility of rat *in vitro, Life Sci.,* 21, 59–61, 1977.
24. Ilic, D.B. and Ashoor, S.H., Stability of vitamins A and C in fortified yogurt, *J. Dairy Sci.,* 71, 1492–1498, 1988.
25. Kneifel, W., Kaufmann, M., Fleischer, A., and Ulberth, F., Screening of commercially available mesophilic dairy starter cultures: biochemical, sensory and microbiological properties, *J. Dairy Sci.,* 75, 3158–3166, 1992.
26. Reddy, K.P., Shahani, K.M., and Kulkarni, S.M., B-complex vitamins in cultured and acidified yogurt, *J. Dairy Sci.,* 59, 191–195, 1976.
27. Savaiano, D.A., Abouelanouar, A., Smith, D.E., and Levitt, M.D., Lactose malabsorption from yogurt, pasteurized yogurt, sweet acidophilus milk, and cultured milk in lactase-deficient individuals, *Am. J. Clin. Nutr.,* 40, 1219–1223, 1984.
28. Elmer, G.W., Probiotics: "living drugs," *Am. J. Health-Syst. Pharm.,* 58, 1101–1109, 2001.
29. Bezkorovainy, A., Probiotics: determinants of survival and growth in the gut, *Am. J. Clin. Nutr.,* 73, 399S–405S, 2001.
30. Perdigón, G., Alvarez, S., Rachid, M., Agüero, G., and Gobbato, N., Immune system stimulation by probiotics, *J. Dairy Sci.,* 78, 1597–1606, 1995.
31. Heller, K.J., Probiotic bacteria in fermented foods: product characteristics and starter organisms, *Am. J. Clin. Nutr.,* 73, 374S–379S, 2001.
32. Roberfroid, M.B., Prebiotics and probiotics: are they functional foods? *Am. J. Clin. Nutr.,* 71, 1682S–1687S, 1688S–1690S, 2000.
33. De Vrese, M., Stegelmann, A., Richter, B., Fenselau, S., Laue, C., and Schrezenmeir, J., Probiotics: compensation for lactase insufficiency, *Am. J. Clin. Nutr.,* 73, 421S–429S, 2001.
34. Gotz, V., Romankiewicz, J.A., Moss, J., and Murray, H.W., Prophylaxis against ampicillin-associated diarrhea with a lactobacillus preparation, *Am. J. Hosp. Pharm.,* 36, 754–757, 1979.
35. Clements, M.L., Levine, M.M., Black, R.E., Robins-Browne, R.M., Cisneros, L.A., Drusano, G.L., Lanata, C.F., and Saah, A.J., *Lactobacillus* prophylaxis for diarrhea due to enterotoxigenic *Escherichia coli, Antimicrob. Agents Chemother.,* 20, 104–108, 1981.
36. Tankanow, R.M., Ross, M.B., Ertel, I.J., Dickinson, D.G., Mccormick, L.S., and Garfinkel, J.F., A double-blind, placebo-controlled study of the efficacy of Lactinex in the prophylaxis of amoxicillin-induced diarrhea, *DICP,* 24, 382–384, 1990.
37. Elmer, G.W., Surawicz, C.M., and Mcfarland, L.V., Biotherapeutic agents. A neglected modality for the treatment and prevention of selected intestinal and vaginal infections, *JAMA,* 275, 870–876, 1996.
38. Surawicz, C.M., Mcfarland, L.V., Greenberg, R.N., Rubin, M., Fekety, R., Mulligan, M.E., Garcia, R.J., Brandmarker, S., Bowen, K., Borjal, D., and Elmer, G.W., The search for a better treatment for recurrent *Clostridium difficile* disease: use of high-dose vancomycin combined with *Saccharomyces boulardii, Clin. Infect. Dis.,* 31, 1012–1017, 2000.
39. Raza, S., Graham, S.M., Allen, S.J., Sultana, S., Cuevas, L., Hart, C.A., Kaila, M., Isolauri, E., Saxelin, M., Arvilommi, H. et al., *Lactobacillus* GG in acute diarrhea, *Indian Pediatr.,* 32, 1140–1142, 1995.

40. Raza, S., Graham, S.M., Allen, S.J., Sultana, S., Cuevas, L., and Hart, C.A., *Lactobacillus* GG promotes recovery from acute nonbloody diarrhea in Pakistan, *Pediatr. Infect. Dis. J.,* 14, 107–111, 1995.
41. Oksanen, P.J., Salminen, S., Saxelin, M., Hamalainen, P., Ihantola-Vormisto, A., Muurasniemi-Isoviita, L., Nikkari, S., Oksanen, T., Porsti, I., Salminen, E. et al., Prevention of travellers' diarrhoea by *Lactobacillus* GG, *Ann. Med.,* 22, 53–56, 1990.
42. Heyman, M., Effect of lactic acid bacteria on diarrheal diseases, *J. Am. Coll. Nutr.,* 19, 137S–146S, 2000.
43. Dunne, C., O'Mahony, L., Murphy, L., Thornton, G., Morrissey, D., O'Halloran, S., Feeney, M., Flynn, S., Fitzgerald, G., Daly, C., Kiely, B., O'Sullivan, G.C., Shanahan, F., and Collins, J.K., *In vitro* selection criteria for probiotic bacteria of human origin: correlation with *in vivo* findings, *Am. J. Clin. Nutr.,* 73, 386S–392S, 2001.
44. Michetti, P., Dorta, G., Wiesel, P.H., Brassart, D., Verdu, E., Herranz, M., Felley, C., Porta, N., Rouvet, M., Blum, A.L., and Corthesy-Theulaz, I., Effect of whey-based culture supernatant of *Lactobacillus acidophilus* (*johnsonii*) La1 on *Helicobacter pylori* infection in humans, *Digestion,* 60, 203–209, 1999.
45. Marteau, P.R., De Vrese, M., Cellier, C.J., and Schrezenmeir, J., Protection from gastrointestinal diseases with the use of probiotics, *Am. J. Clin. Nutr.,* 73, 430S–436S, 2001.
46. Borchers, A.T., Keen, C.L., and Gershwin, M.E., The influence of yogurt/*Lactobacillus* on the innate and acquired immune response, *Clin. Rev. Allergy Immunol.,* 22, 207–230, 2002.
47. De Simone, C., Vesely, R., Negri, R., Bianchi Salvadori, B., Zanzoglu, S., Cilli, A., and Lucci, L., Enhancement of immune response of murine Peyer's patches by a diet supplemented with yogurt, *Immunopharmacol. Immunotoxicol.,* 9, 87–100, 1987.
48. Meydani, S.N. and Ha, W.K., Immunologic effects of yogurt, *Am. J. Clin. Nutr.,* 71, 861–872, 2000.
49. Wheeler, J.G., Bogle, M.L., Shema, S.J., Shirrell, M.A., Stine, K.C., Pittler, A.J., Burks, A.W., and Helm, R.M., Impact of dietary yogurt on immune function, *Am. J. Med. Sci.,* 313, 120–123, 1997.
50. Roitt, I.M., Brostoff, J., and Male, D.K., *Immunology,* Mosby, London, 1993, pp. 1.1–1.12, 12.12–12.20.
51. Janeway, C.A. and Travers, P, *Immunobiology: The Immune System in Health and Disease*, Garland Pub., London, 1997.
52. Kraehenbuhl, J.P. and Neutra, M., Molecular and cellular basis of immune protection of mucosal surfaces, *Am. Physiol. Soc.,* 72, 853–879, 1992.
53. Moreau, M.C., Ducluzeau, R., Guy-Grand, D., and Muller, M.C., Increase in the population of duodenal immunoglobulin A plasmocytes in axenic mice associated with different living or dead bacterial strains of intestinal origin, *Infect. Immun.,* 21, 532–539, 1978.
54. Shroff, K.E., Meslin, K., and Cebra, J.J., Commensal enteric bacteria engender a self-limiting humoral mucosal immune response while permanently colonizing the gut, *Infect. Immun.,* 63, 3904–3913, 1995.
55. Cebra, J. J., Influences of microbiota on intestinal immune system development, *Am. J. Clin. Nutr.,* 69, 1046S–1051S, 1999.
56. Macpherson, A.J., Gatto, D., Sainsbury, E., Harriman, G.R., Hengartner, H., and Zinkernagel, R.M., A primitive T cell–independent mechanism of intestinal mucosal IgA responses to commensal bacteria, *Science*, 288, 2222–2226, 2000.
57. Herías, M.V., Midtvedt, T., Hanson, L.Å., and Wold, A.E., Increased antibody production against gut-colonizing *Escherichia coli* in the presence of the anaerobic bacterium *Peptostreptococcus, Scand. J. Immunol.,* 48, 277–282, 1998.

58. Kaila, M., Isolauri, E., Soppi, E., Virtanen, E., Laine, S., and Arvilommi, H., Enhancement of the circulating antibody secreting cell response in human diarrhea by a human *Lactobacillus* strain, *Pediatr. Res.*, 32, 141–144, 1992.
59. Kaila, M., Isolauri, E., Saxelin, M., Arvilommi, H., and Vesikari, T., Viable versus inactivated lactobacillus strain GG in acute rotavirus diarrhoea, *Arch. Dis. Child.*, 72, 51–53, 1995.
60. Maassen, C.B.M., Van Holten-Neelen, C., Balk, F., Den Bak-Glashouwer, M.-J. H., Leer, R.J., Laman, J.D., Boersma, W.J.A., and Claassen, E., Strain-dependent induction of cytokine profiles in the gut by orally administered *Lactobacillus* strains, *Vaccine*, 18, 2613–2623, 2000.
61. Perdigón, G., Rachid, M., De Budeguer, M.V., and Valdez, J.C., Effect of yogurt feeding on the small and large intestine associated lymphoid cells in mice, *J. Dairy Res.*, 61, 553–562, 1994.
62. Agüero, G., Sanchez, S., Fernandez, S., Allori, C., De Ruiz Holgado, A.P., and Perdigón, G., Administration of yoghurt or *Lactobacillus casei* to malnourished mice: comparative effect of lymphoid cells and mucosal reconditioning of the intestine, *Food Agric. Immunol.*, 8, 229–238, 1996.
63. Perdigón, G., Alvarez, S., Medici, M., Vintiñi, E., De Giori, G.S., De Kairuz, M.N., and De Ruiz Holgado, A.P., Effect of yogurt with different storage periods on the immune system in mice, *Milchwissenschaft,* 50, 367–371, 1995.
64. Link-Amster, H., Rochat, F., Saudan, K.Y., Mignot, O., and Aeschlimann, J.M., Modulation of a specific humoral immune response and changes in intestinal flora mediated through fermented milk intake, *FEMS Immunol. Med. Microbiol.*, 10, 55–63, 1994.
65. Fang, H., Elina, T., Heikki, A., and Seppo, S., Modulation of humoral immune response through probiotic intake, *FEMS Immunol. Med. Microbiol.*, 29, 47–52, 2000.
66. Marteau, P.R., De Vrese, M., Cellier, C.J., and Schrezenmeir, J., Protection from gastrointestinal diseases with the use of probiotics, *Am. J. Clin. Nutr.*, 73, 430S–436S, 2001.
67. Isolauri, E., Juntunen, M., Rautanen, T., Sillanaukee, P., and Koivula, T., A human *Lactobacillus* strain (*Lactobacillus* GG) promotes recovery from acute diarrhea in children, *Pediatrics,* 88, 90–97, 1991.
68. Isolauri, E., Joensuu, J., Suomalainen, H., Luomala, M., and Vesikari, T., Improved immunogenicity of oral D × RRV reassortant rotavirus vaccine by *Lactobacillus casei* GG, *Vaccine*, 13, 310–312, 1995.
69. Malin, M., Suomalainen, H., Saxelin, M., and Isolauri, E., Promotion of IgA immune response in patients with Crohn's disease by oral bacteriotherapy with *Lactobacillus* GG, *Ann. Nutr. Metab.*, 40, 137–145, 1996.
70. Isolauri, E., Majamaa, H., Arvola, T., Rantala, I., Virtanen, E., and Arvilommi, H., *Lactobacillus casei* strain GG reverses increased intestinal permeability induced by cow milk in suckling rats, *Gastroenterology*, 105, 1643–1650, 1993.
71. Takahashi, T., Oka, T., Iwana, H., Kuwata, T., and Yamamoto, Y., Immune response of mice to orally administered lactic acid bacteria, *Biosci. Biotech. Biochem.*, 57, 1557–1560, 1993.
72. Garrote, G.L., Abraham, A.G., and De Antoni, G.L., Inhibitory power of kefir: the role of organic acids, *J. Food Prot.*, 63, 364–369, 2000.
73. Halpern, G.M., Vruwink, K.G., Van De Water, J., Keen, C.L., and Gershwin, M.E., Influence of long-term yoghurt consumption in young adults., *Int. J. Immunotherapy,* 7, 205–210, 1991.

74. Rizkalla, S.W., Luo, J., Kabir, M., Chevalier, A., Pacher, N., and Slama, G., Chronic consumption of fresh but not heated yogurt improves breath-hydrogen status and short-chain fatty acid profiles: a controlled study in healthy men with or without lactose maldigestion, *Am. J. Clin. Nutr.,* 72, 1474–1479, 2000.
75. Hooper, L.V., Wong, M.H., Thelin, A., Hansson, L., Falk, P.G., and Gordon, J.I., Molecular analysis of commensal host-microbial relationships in the intestine, *Science*, 291, 881–884, 2001.
76. Puri, P., Mahapatra, S.C., Bijlani, R.L., Prasad, H.K., and Nath, I., Feed efficiency and splenic lymphocyte proliferation response in yogurt- and milk-fed mice, *Int. J. Food Sci. Nutr.,* 45, 231–235, 1994.
77. Puri, P., Rattan, A., Fbijlani, R.L., Mahapatra, S.C., and Nath, I., Splenic and intestinal lymphocyte proliferation response in mice fed milk or yogurt and challenged with *Salmonella typhimurium, Int. J. Food Sci. Nutr.,* 47, 391–398, 1996.
78. Pessi, T., Sutas, Y., Marttinen, A., and Isolauri, E., Probiotics reinforce mucosal degradation of antigens in rats: implications for therapeutic use of probiotics, *J. Nutr.,* 128, 2313–2318, 1998.
79. Sütas, Y., Soppi, E., Korhonen, H., Syvaoja, E.-L., Saxelin, M., Rokka, T., and Isolauri, E., Suppression of lymphocyte proliferation *in vitro* by bovine caseins hydrolyzed with *Lactobacillus casei* GG-derived enzymes, *J. Allergy Clin. Immunol.,* 98, 216–224, 1996.
80. Sütas, Y., Hurme, M., and Isolauri, E., Down-regulation of anti-CD3 antibody-induced IL-4 production by bovine caseins hydrolysed with *Lactobacillus* GG-derived enzymes, *Scand. J. Immunol.,* 43, 687–689, 1996.
81. Pessi, T., Isolauri, E., Sütas, Y., Kankaanranta, H., Moilanen, E., and Hurme, M., Suppression of T cell activation by *Lactobacillus rhamnosus* GG-degraded bovine casein, *Int. Immunopharmacol.,* 1, 211–218, 2001.
82. Prescott, S.L., Macaubas, C., Holt, B.J., Smallacombe, T.B., Loh, R., Sly, P.D., and Holt, P.G., Transplacental priming of the human immune system to environmental allergens: universal skewing of initial T cell responses toward the Th2 cytokine profile, *J. Immunol.,* 160, 4730–4737, 1998.
83. Prescott, S.L., Macaubas, C., Smallacombe, T., Holt, B.J., Sly, P.D., and Holt, P.G., Development of allergen-specific T-cell memory in atopic and normal children, *Lancet,* 353, 196–200, 1999.
84. Erb, K.J., Atopic disorders: a default pathway in the absence of infection? *Immunol. Today,* 20, 317–322, 1999.
85. Matricardi, P.M. and Bonini, S., High microbial turnover rate preventing atopy: a solution to inconsistencies impinging on the hygiene hypothesis? *Clin. Exp. Allergy,* 30, 1506–1510, 2000.
86. Sepp, E., Julge, K., Vasar, M., Naaber, P., Björkstén, B., and Mikelsaar, M., Intestinal microflora of Estonian and Swedish infants, *Acta Paediatr.,* 86, 956–961, 1997.
87. Björkstén, B., Naaber, P., Sepp, E., and Mikelsaar, M., The intestinal microflora in allergic Estonian and Swedish 2-year-old children, *Clin. Exp. Allergy,* 29, 342–346, 1999.
88. Böttcher, M.F., Nordin, E.K., Sandin, A., Midtvedt, T., and Björkstén, B., Microflora-associated characteristics in faeces from allergic and nonallergic infants, *Clin. Exp. Allergy,* 30, 1590–1596, 2000.
89. Kalliomäki, M., Kirjavainen, P., Eerola, E., Kero, P., Salminen, S., and Isolauri, E., Distinct patterns of neonatal gut microflora in infants in whom atopy was and was not developing, *J. Allergy Clin. Immunol.,* 107, 129–134, 2001.

90. Weiner, H.L., Oral tolerance: immune mechanisms and treatment of autoimmune diseases, *Immunol., Today,* 18, 335–343, 1997.
91. Moreau, M.-C., and Gaboriau-Routhiau, V., The absence of gut flora, the doses of antigen ingested and aging affect the long-term peripheral tolerance induced by ovalbumin feeding in mice, *Res. Immunol.,* 147, 49–59, 1996.
92. Sudo, N., Sawamura, S., Tanaka, K., Aiba, Y., Kubo, C., and Koga, Y., The requirement of intestinal bacterial flora for the development of an IgE production system fully susceptible to oral tolerance induction, *J. Immunol.,* 159, 1739–1745, 1997.
93. Wold, A.E., The hygiene hypothesis revised: is the rising frequency of allergy due to changes in the intestinal flora? *Allergy,* 53, 20–25, 1998.
94. Shida, K., Makino, K., Morishita, A., Takamizawa, K., Hachimura, S., Ametani, A., Sato, T., Kumagai, Y., Habu, S., and Kaminogawa, S., *Lactobacillus casei* inhibits antigen-induced IgE secretion through regulation of cytokine production in murine splenocyte cultures, *Int. Arch. Allergy Immunol.,* 115, 278–287, 1998.
95. Matsuzaki, T., Yamazaki, R., Hashimoto, S., and Yokokura, T., The effect of oral feeding of *Lactobacillus casei* strain Shirota on immunoglobulin E production in mice, *J. Dairy Sci.,* 81, 48–53, 1998.
96. Murosaki, S., Yamamoto, Y., Ito, K., Inokuchi, T., Kusaka, H., Ikeda, H., and Yoshikai, Y., Heat-killed *Lactobacillus plantarum* L-137 suppresses naturally fed antigen-specific IgE production by stimulation of IL-12 production in mice, *J. Allergy Clin. Immunol.,* 102, 57–64, 1998.
97. Perdigon, G., Nader de Macias, M.E., Alvarez, S., Oliver, G., and Pesce de Ruiz Holgado, A.A., Enhancement of immune response in mice fed with *Streptococcus thermophilus* and *Lactobacillus acidophilus, J. Dairy Sci.,* 70, 919–926, 1987.
98. Ha, C.L., Lee, J.H., Zhou, H.R., Ustunol, Z., and Pestka, J.J., Effects of yogurt ingestion on mucosal and systemic cytokine gene expression in the mouse, *J. Food Prot.,* 62, 181–188, 1999.
99. Halpern, G.M., Vruwink, K.G., Van De Water, J., Keen, C.L., and Gershwin, M.E., Influence of long-term yogurt consumption in young adults, *Int. J. Immunother.,* 7, 205–210, 1991.
100. Trapp, C.L., Chang, C.C., Halpern, G.M., Keen, C.L., and Gershwin, M.E., The influence of chronic yogurt consumption of populations of young and elderly adults, *Int. J. Immunother.,* 9, 53–64, 1993.
101. Van De Water, J., Keen, C.L., and Gershwin, M.E., The influence of chronic yogurt consumption on immunity, *J. Nutr.,* 129, 1492S–1495S, 1999.
102. Aattouri, N. and Lemonnier, D., Production of interferon induced by *Streptococcus thermophilus:* role of CD4+ and CD8+ lymphocytes, *Nutr. Biochem.,* 8, 25–31, 1997.
103. Solis-Pereyra, B., Aattouri, N., and Lemonnier, D., Role of food in the stimulation of cytokine production, *Am. J. Clin. Nutr.,* 66, 521S–525S, 1997.
104. Wheeler, J.G., Shema, S.J., Bogle, M.L., Shirrell, M.A., Burks, A.W., Pittler, A., and Helm, R.M., Immune and clinical impact of *Lactobacillus acidophilus* on asthma, *Ann. Allergy Asthma Immunol.,* 79, 229–233, 1997.
105. Haller, D., Blum, S., Bode, C., Hammes, W.P., and Schiffrin, E.J., Activation of human peripheral blood mononuclear cells by nonpathogenic bacteria *in vitro:* evidence of NK cells as primary targets, *Infect. Immun.,* 68, 752–759, 2000.
106. Pessi, T., Sütas, Y., Saxelin, M., Kallioinen, H., and Isolauri, E., Antiproliferative effects of homogenates derived from five strains of candidate probiotic bacteria, *Appl. Environ. Microbiol.,* 65, 4725–4728, 1999.

107. De Simone, C., Tzantzoglou, S., Baldinelli, L., Di Fabio, S., Bianchi-Salvadori, B., Jirillo, E., and Vesely, R., Enhancement of host resistance against *Salmonella typhimurium* infection by a diet supplemented with yogurt, *Immunopharmacol. Immunotoxicol.,* 10, 399–415, 1988.
108. De Simone, C., Vesely, R., Negri, R., Bianchi Salvadori, B., Zanzoglu, S., Cilli, A., and Lucci, L., Enhancement of immune response of murine Peyer's patches by a diet supplemented with yogurt, *Immunopharmacol. Immunotoxicol.,* 9, 87–100, 1987.
109. Kirjavainen, P.V., El-Nezami, H.S., Salminen, S.J., Ahokas, J.T., and Wright, P.F.A., The effect of orally administered viable probiotic and dairy lactobacilli on mouse lymphocyte proliferation, *FEMS Immunol. Med. Microbiol.,* 26, 131–135, 1999.
110. Kirjavainen, P.V., Elnezami, H.S., Salminen, S.J., Ahokas, J.T., and Wright, P.F.A., Effects of orally administered viable *Lactobacillus rhamnosus* GG and *Propionibacterium freudenreichii* subsp. *shermanii* JS on mouse lymphocyte proliferation, *Clin. Diagn. Lab. Immunol.,* 6, 799–802, 1999.
111. Mike, A., Nagaoka, N., Tagami, Y., Miyashita, M., Shimada, S., Uchida, K., Nanno, M., and Ohwaki, M., Prevention of B220+ T cell expansion and prolongation of lifespan induced by *Lactobacillus casei* in MRL/lpr mice, *Clin. Exp. Immunol.,* 117, 368–375, 1999.
112. Perdigón, G., De Macias, M.E., Alvarez, S., Oliver, G., and De Ruiz Holgado, A.P., Systemic augmentation of the immune response in mice by feeding fermented milks with *Lactobacillus casei* and *Lactobacillus acidophilus, Immunology,* 63, 17–23, 1988.
113. Perdigon, G., De Macias, M.E., Alvarez, S., Oliver, G., and De Ruiz Holgado, A.A., Effect of perorally administered lactobacilli on macrophage activation in mice, *Infect. Immun.,* 53, 404–410, 1986.
114. Valdez, J.C., Rachid, M., Bru, E., and Perdigón, G., The effect of yoghurt on the cytotoxic and phagocytic activity of macrophages in tumour-bearing mice, *Food Agric. Immunol.,* 9, 299–308, 1997.
115. Perdigón, G., Valdez, J.C., and Rachid, M., Antitumour activity of yogurt: study of possible immune mechanisms, *J. Dairy Res.,* 65, 129–138, 1998.
116. Gill, H.S., Rutherford, K.J., Prasad, J., and Gopal, P.K., Enhancement of natural and acquired immunity by *Lactobacillus rhamnosus* (HN001), *Lactobacillus acidophilus* (HN017) and *Bifidobacterium lactis* (HN019), *Br. J. Nutr.,* 83, 167–176, 2000.
117. Schiffrin, E.J., Rochat, F., Link-Amster, H., Aeschlimann, J.M., and Donnet-Hughes, A., Immunomodulation of human blood cells following the ingestion of lactic acid bacteria, *J. Dairy Sci.,* 78, 491–497, 1995.
118. Donnet-Hughes, A., Rochat, F., Serrant, P., Aeschlimann, J.M., and Schiffrin, E.J., Modulation of nonspecific mechanisms of defense by lactic acid bacteria: effective dose, *J. Dairy Sci.,* 82, 863–869, 1999.
119. Pelto, L., Isolauri, E., Lilius, E.M., Nuutila, J., and Salminen, S., Probiotic bacteria down-regulate the milk-induced inflammatory response in milk-hypersensitive subjects but have an immunostimulatory effect in healthy subjects, *Clin. Exp. Allergy,* 28, 1474–1479, 1998.
120. Haller, D., Bode, C., and Hammes, W.P., Cytokine secretion by stimulated monocytes depends on the growth phase and heat treatment of bacteria: a comparative study between lactic acid bacteria and invasive pathogens, *Microbiol. Immunol.,* 43, 925–935, 1999.

121. Takagi, A., Matsuzaki, T., Sato, M., Nomoto, K., Morotomi, M., and Yokokura, T., Inhibitory effect of oral administration of *Lactobacillus casei* on 3-methylcholanthrene-induced carcinogenesis in mice, *Med. Microbiol. Immunol. (Berlin),* 188, 111–116, 1999.
122. Takagi, A., Matsuzaki, T., Sato, M., Nomoto, K., Morotomi, M., and Yokokura, T., Enhancement of natural killer cytotoxicity delayed murine carcinogenesis by a probiotic microorganism, *Carcinogenesis*, 22, 599–605, 2001.
123. Chiang, B.L., Sheih, Y.H., Wang, L.H., Liao, C.K., and Gill, H.S., Enhancing immunity by dietary consumption of a probiotic lactic acid bacterium (*Bifidobacterium lactis* HN019): optimization and definition of cellular immune responses, *Eur. J. Clin. Nutr.,* 54, 849–855, 2000.
124. Perdigon, G., Valdez, J.C., and Rachid, M., Antitumour activity of yogurt: study of possible immune mechanisms, *J. Dairy Res.,* 65, 129–138, 1998.
125. Majamaa, H. and Isolauri, E., Probiotics: a novel approach in the management of food allergy, *J. Allergy Clin. Immunol.,* 99, 179–185, 1997.
126. Kato, I., Endo-Tanaka, K., and Yokokura, T., Suppressive effects of the oral administration of *Lactobacillus casei* on type II collagen-induced arthritis in DBA/1 mice, *Life Sci.,* 63, 635–644, 1998.
127. Kato, I., Tanaka, K., and Yokokura, T., Lactic acid bacterium potently induces the production of interleukin-12 and interferon-g by mouse splenocytes, *Int. J. Immunopharmacol.,* 21, 121–131, 1999.
128. Miettinen, M., Matikainen, S., Vuopio-Varkila, J., Pirhonen, J., Varkila, K., Kurimoto, M., and Julkunen, I., Lactobacilli and streptococci induce interleukin-12 (IL-12), IL-18, and gamma interferon production in human peripheral blood mononuclear cells, *Infect. Immun.,* 66, 6058–6062, 1998.
129. Solis Pereyra, D.L., Induction of human cytokines by bacteria used in dairy foods, *Nutr. Res.,* 13, 1127–1140, 1993.
130. Miettinen, M., Vuopio-Varkila, J., and Varkila, K., Production of human tumor necrosis factor alpha, interleukin-6, and interleukin-10 is induced by lactic acid bacteria, *Infect. Immun.,* 64, 5403–5405, 1996.
131. Haller, D., Bode, C., Hammes, W.P., Pfeifer, A.M., Schiffrin, E.J., and Blum, S., Nonpathogenic bacteria elicit a differential cytokine response by intestinal epithelial cell/leucocyte co-cultures, *Gut,* 47, 79–87, 2000.
132. Tejada-Simon, M.V. and Pestka, J. J., Proinflammatory cytokine and nitric oxide induction in murine macrophages by cell wall and cytoplasmic extracts of lactic acid bacteria, *J. Food Prot.,* 62, 1435–1444, 1999.
133. Marin, M.L., Tejada-Simon, M.V., Lee, J.H., Murtha, J., Ustunol, Z., and Pestka, J.J., Stimulation of cytokine production in clonal macrophage and T-cell models by *Streptococcus thermophilus:* comparison with *Bifidobacterium* sp. and *Lactobacillus bulgaricus, J. Food Prot.,* 61, 859–864, 1998.
134. Simelyte, E., Isomäki, P., Rimpiläinen, M., Zhang, X., and Toivanen, P., Cytokine production in arthritis susceptible and resistant rats: a study with arthritogenic and nonarthritogenic *Lactobacillus* cell walls, *Scand. J. Immunol.,* 53, 132–138, 2001.
135. Tejada-Simon, M.V., Ustunol, Z., and Pestka, J.J., *Ex vivo* effects of lactobacilli, streptococci, and bifidobacteria ingestion on cytokine and nitric oxide production in a murine model, *J. Food Prot.,* 62, 162–169, 1999.

6 Health Properties of Milk Fermented with *Lactobacillus casei* Strain Shirota (LcS)

Takeshi Matsuzaki

CONTENTS

6.1 HISTORY AND BACKGROUND

Throughout history, humans have made use of lactic acid bacteria (LAB), which are distributed widely in nature. LAB have traditionally been employed to produce fermented milk products, including yogurt, leiben, dahi, kefir, and koumiss. Currently, they are also used to produce many processed foods, such as fermented meat products, brewed products, Japanese pickles, and bread, as well as silage.[1]

LAB were first discovered by Pasteur in 1857.[2] In 1878, Lister reported the isolation of LAB from rancid milk.[2] Subsequently, these bacteria were also isolated from the intestinal tract. *Bifidobacterium* species were discovered by Tissier[3] in 1889, and *Lactobacillus acidophilus* was discovered by Moro[4] in 1890. Soon afterwards, attempts were made to classify these new species of LAB. Although there was some confusion for a while, Orla-Jensen[5] laid the foundation for an overall classification of LAB in 1919.

The genus *Lactobacillus* comprises Gram-positive, nonsporing, noncatalase-producing facultative anaerobic rods that commonly produce lactic acid as their major metabolite. *Lactobacilli* are generally isolated from fermented milk and from the intestinal tract of humans and other animals.[6] Major *Lactobacillus* species isolated from the human intestinal tract include *Lb. gasseri, Lb. crispatus, Lb. johnsonii, Lb. salivarius, Lb. reuteri, Lb. casei, Lb. ruminis, Lb. itulinus, Lb. plantarum,* and *Lb. brevis.*[7] *Lb. acidophilus* is rarely isolated from the human intestinal tract.

Lb. casei strain Shirota (LcS) is one of the *Lb. casei* strains isolated from the human intestine. LcS was first cultured stably in 1930 by the late Minoru Shirota (1899 to 1982) at the Microbiological Laboratory of the Faculty of Medicine at Kyoto University, Kyoto, Japan. This organism was isolated from a healthy human and is resistant to gastric and bile acids, so live bacteria can reach the lower intestine after oral administration. Dr. Shirota developed "Yakult," a dairy product manufactured using LcS. Yakult was first produced in 1935 and is based on his hypothesis that daily oral intake of LAB promotes intestinal health and prevents diseases, thereby prolonging the life span of humans. Yakult that contains LcS has been sold all over the world for over 60 years. The production of products that contain LcS is about 7 hundred million and 15 hundred million units/day in Japan and in other countries, respectively. Subsequent studies of LcS have focused on the taxonomy, genetics, and physiology of *Lactobacillus* species and other LAB as well as the nutritional value of fermented milk products, their beneficial effects on fecal properties, and their preventive action against gastrointestinal infection. Additional recent studies have included the investigation of the beneficial effects of fermented milk products as well as the effects of cellular constituents and metabolic products of LAB used to manufacture these products on host homeostasis, e.g., prevention of carcinogenesis, beneficial effects on lipid metabolism and blood pressure, and immunomodulatory effects. The results obtained from these studies have shown that LcS and dairy products manufactured using this lactobacillus strain have a range of biological activities. Recently, this and other beneficial bacterial strains and their products have been termed "probiotics" and have attracted the attention of many investigators.

The present chapter summarizes various findings on the biological activities of LcS that can impact favorably on human health. This review presents information from studies performed both within and outside the Yakult Institute[8] (Yakult Central Institute for Microbiological Research, Tokyo, Japan).

6.2 GENERAL PROPERTIES OF LcS

6.2.1 Morphology and Structure of LcS

LcS is an average-sized bacillus. Each cell is about 1 to 2.5 μm long and has a diameter of about 0.5 μm as observed under the electron microscope (Figure 6.1). In general, the size of bacterial cells changes slightly during culture. In the case of LcS, cells longer than 4 μm sometimes appear at the late stage of culture, when cell division ceases. LcS is normally roundish and has a smooth surface without any of the flagella or cilia possessed by *Escherichia coli* (Figure 6.2).

Each LcS organism is surrounded by a cell wall about 20 to 30 nm thick, which is composed of peptidoglycan (murein), teichoic acid, and proteins, as is the case for other Gram-positive bacteria. After fixation by the Ryter–Kellenberger method,[9] the standard fixation method for bacteria, LcS appears to have a polysaccharide-like

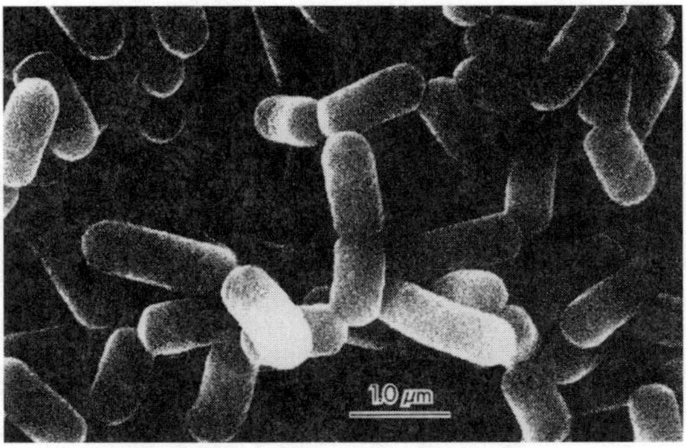

FIGURE 6.1 Scanning electron micrograph of *L. casei* strain Shirota (LcS), showing bacilli about 0.5 μm in diameter and about 1.5 μm in length (× 17,000). (From Lactobacillus casei *strain Shirota — Intestinal Flora and Human Health,* Yakult Central Institute for Microbiological Research, Tokyo, 1999, p. 23. With permission.)

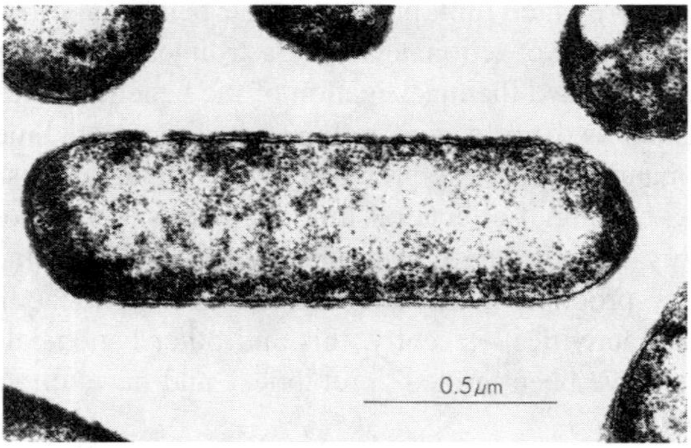

FIGURE 6.2 Electron micrograph of an ultrathin section of LcS, showing a relatively smooth cell surface (× 66,000). (From Lactobacillus casei *strain Shirota — Intestinal Flora and Human Health,* Yakult Central Institute for Microbiological Research, Tokyo, 1999, p. 24. With permission.)

coat outside the cell wall. This coating may not be an intrinsic structure like a capsule, but instead may be a layer of metabolic products adhering to the cell wall, since no such coat is observed after double fixation with glutaraldehyde and osmium tetroxide or after rapid freezing and substitution fixation. LcS does not have a regular array layer composed of fine protein granules.[10]

Chemical analysis shows that LcS is surrounded by a cell wall composed of peptidoglycans to which polysaccharides are bound as accessory polymers, while lipoteichoic acid and other polymers extend toward the cell wall from the cell membrane as in other Gram-positive bacteria.[11] Some other Gram-positive bacteria have a capsule as the outermost structure or have M protein or S layers on the cell wall.[12] However, no such structures have been detected in LcS. Thus, the major constituents of the surface of LcS appear to be lipoteichoic acid, a peptidoglycan layer, and polysaccharides attached to the cell wall (Figure 6.3).

6.2.2 Energy Metabolism of LcS

In LcS, the phosphoenolpyruvic acid-dependent phosphotransferase system (PTS) may facilitate cellular uptake of lactose. When grown in a medium with glucose or lactose as the major carbohydrate source, LcS produces lactic acid as the predominant product of fermentation. Thus, this strain belongs to the homofermentative group of LAB. Homofermentative LAB produce two moles of lactic acid and two moles of adenosine triphosphate (ATP) from one mole of glucose. The efficiency of ATP formation in the process of homolactic acid fermentation is twice that of heterolactic acid fermentation. Therefore, the efficiency with which LcS obtains energy is higher than that of heterofermentative lactic acid bacteria.[13,14]

LAB are generally classified as homofermentative or heterofermentative. Homofermentative bacteria are further classified into obligate homofermentatives (which are always homofermentative under any culture conditions) and facultative homofermentatives (which can switch to the heterofermentative pattern with a change of

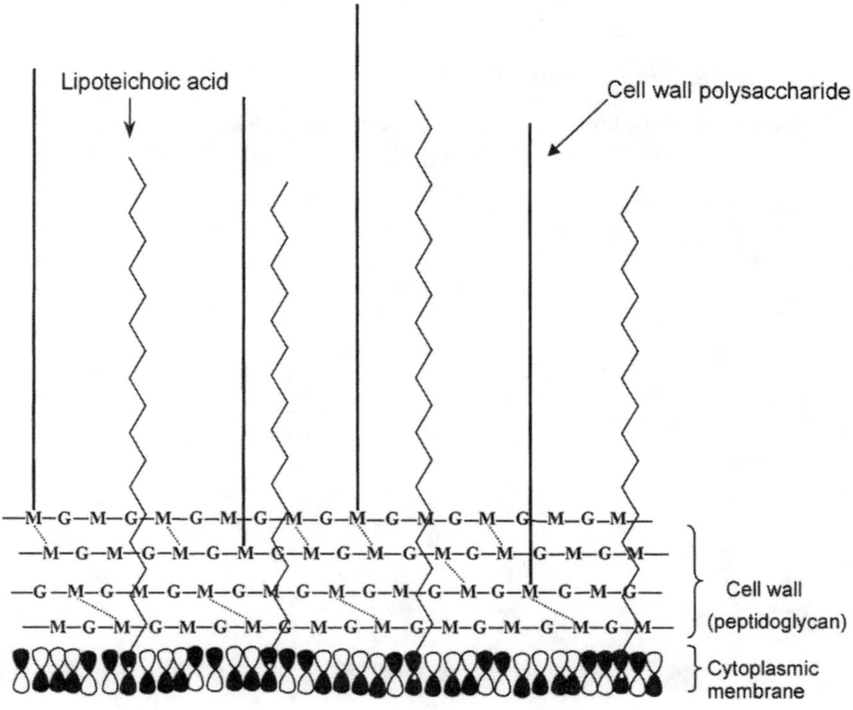

Diagram of the cell surface structures of Gram-positive bacteria.
M: *N*–acetylmuramic acid, G: *N*–acetylglucosamine,
------ : Interpeptide bridge

FIGURE 6.3 Diagram of the cell surface structures of Gram-positive bacteria. M: *N*-acetylmuramic acid, G: *N*-acetylglucosamine, -----: Interpeptide bridge. (From Lactobacillus casei *strain Shirota — Intestinal Flora and Human Health,* Yakult Central Institute for Microbiological Research, Tokyo, 1999, p. 31. With permission.)

carbohydrate source and/or culture conditions). Since LcS also produces trace amounts of acetic acid and ethanol, this strain is a facultative homofermentative in the strict sense.[15,16]

6.2.3 Nutritional Requirements of LcS

The nutritional requirements of LcS are very complicated, as are those of many other LAB (Table 6.1). To achieve the maximum growth of LcS in a synthetic culture medium, as many as 12 amino acids and 4 vitamins are required.[17] Several additional amino acids, vitamins, and nucleic acid bases have also been found to promote the growth of this strain. LcS is auxotrophic for valine, glutamic acid, nicotinic acid, and pantothenic acid, which are required for the growth of almost all species of LAB. Various minerals are also required by LcS; the concentration of manganese required is particularly high.

TABLE 6.1
Nutritional Requirements of LcS

Nutrients Eliminated from the Basal Synthetic Medium	Viability	Nutrients Eliminated from the Basal Synthetic Medium	Viability
Amino Acids		**Vitamins**	
Alanine	+	Thiamin	+
Arginine	–	Riboflavin	–
Aspartic acid	–	Pyridoxal	–
Cysteine	±	Biotin	+
Glutamic acid	–	Pantothenic acid	–
Glycine	+	Nicotinic acid	–
Histidine	+	Folic acid	±
Isoleucine	–	*p*-Aminobenzoic acid	+
Leucine	–		
Lysine	–	**Purines and Pyrimidines**	
Methionine	–	Adenine	±
Phenylalanine	±	Cytosine	+
Serine	+	Guanine	+
Proline	–	Thymine	+
Threonine	–	Uracil	±
Tryptophan	–	Xanthine	±
Tyrosine	–		
Valine	–		

Note: After incubation at 37°C for 16 h in Rogosa medium, bacteria were harvested and washed twice with saline. Then the organisms were incubated into liquid basal synthetic medium containing all but the specified nutrient and were incubated at 37°C for 3 days. During incubation, the turbidity of each culture was examined at 24-h intervals and graded as + (good growth), ± (slight growth), or – (no growth).

Source: Morishita, T., Fukuda, T., Shirota, T., and Yura, T., *J. Bacteriol.*, 120, 1078–1084, 1974. With permission.

6.3 FERMENTATION PROCESS OF LcS

6.3.1 Cultivation of LcS

LcS has been shown to have various biological effects, which will be described in the following sections. It should be noted that the cell components and biological activities may change with different culture conditions. Therefore, it is important to study the growth characteristics of this strain in detail, and to optimize the culture conditions for mass production of the cells with the desired biological activity. The most commonly used media for cultivating LAB are MRS medium[18] and Rogosa medium.[19] For experimental use, LcS is cultured for 48 h at 37°C in Rogosa's medium. After cultivation, cells are collected by centrifugation, washed with sterile distilled water, and lyophilized.

On the other hand, for industrial use, corn steep liquor (CSL) medium, which is composed of glucose and corn steep liquor, is more popular because it is less expensive and ensures favorable bacterial growth.

6.3.2 Basic Growth Aspects of LcS

LcS is a homofermentative organism that derives energy from glucose by the following metabolic reaction:

$$C_6H_{12}O_6 + 2ADP + 2Pi \rightarrow 2C_3H_6O_3 + 2ATP + 2H_2O$$

The metabolic product, lactic acid, accumulates in the medium, and the pH of the medium is decreased. When the lactic acid concentration in the medium increases to 16 g/l, the pH is decreased to nearly 4, and cell growth ceases due to the high acidity (static culture). The cell concentration when growth ceases is 5×10^9/ml (82.2 g-cell/l). When the pH of the medium is constantly adjusted to about 7.0 with an alkaline agent such as ammonia or sodium hydroxide (constant pH culture), the cells continue to grow until the cell concentration plateaus at 1 to 2×10^{10}/ml (5.0 g-cell/l).

6.3.3 Optimization of Culture Temperature and pH

The parameters that most significantly affect the growth of LAB are pH and temperature. As the pH continually changes in static culture, it is difficult to optimize the conditions for growth.

To determine the optimum culture conditions in constant pH culture, batch culture was performed with the two major variables, pH and temperature, changed within the range of 5.0 to 8.0 and 25 to 45°C, respectively. The relationships of these variables with the growth time (T h), final cell concentration (X g-cell/l), cell yield of glucose (Yx g-cell/g-glucose), and cell productivity (Px g-cell/h/l) were determined. X and Yx remained relatively constant independent of both variables, at 4.8 to 5.0 g-cell/l and 0.11 to 0.12 g-cell/g-glucose, respectively. In contrast, T and Px were significantly affected by both temperature and pH. Thus, the optimum pH and temperature were found to be 6.5 and 35°C, respectively, for maximum mass production of LcS.[20]

6.3.4 Effect of Temperature on the Survival of LcS in Dairy Products

Strains of LAB used to manufacture fermented milk or sour milk beverages usually have a high proliferative activity, show high LAB production in a specific medium, and create no metabolites that adversely affect the flavor of the product. On the other hand, the strains used are also required to create little or no lactic acid in the final product during storage. During transportation of the final product to the point of sale, the ambient temperature must be controlled strictly to suppress lactic acid production by the bacterial strain and thus maintain product quality.

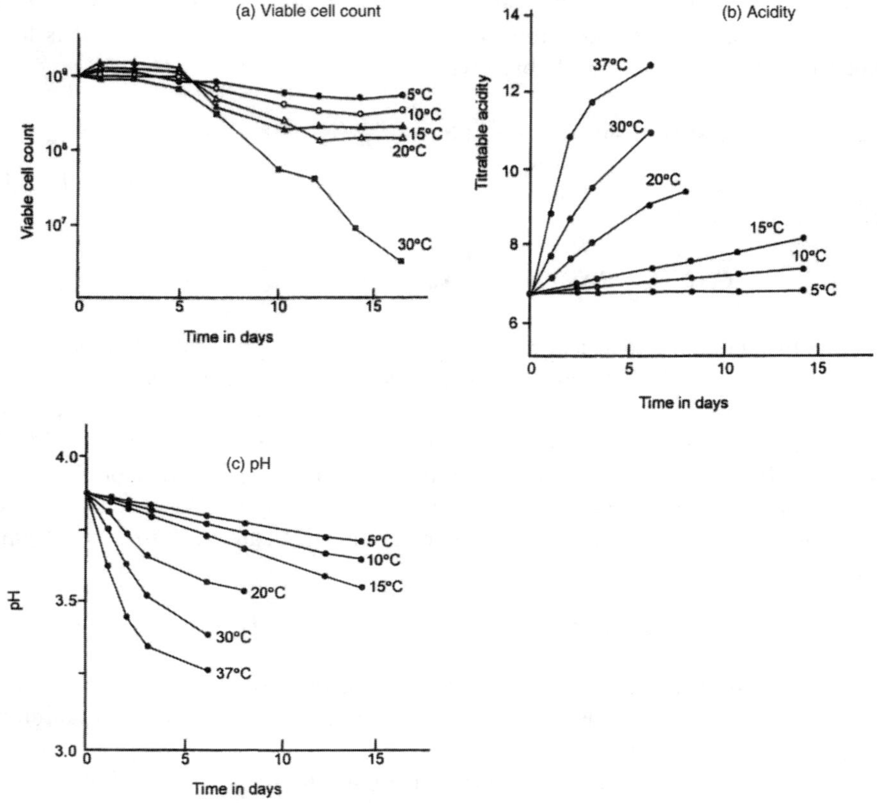

FIGURE 6.4 Effect of storage temperature on the viable cell count (a), titratable acidity (b), and pH (c) of LcS fermented milk (Yakult). Sample batches were serially diluted tenfold with sterile saline to obtain bacterial growth of 30 to 300 colonies per plate. A 1-ml aliquot of the dilution obtained in this way was mixed well with 15 ml of heated medium in a Petri dish. After solidification, the plates were inverted and cultured at 35 to 37°C for 72 hours. At the end of incubation, the yellow colonies on each plate were counted. Titratable acidity: volume of 0.1 *N* NaOH required to neutralize a 10-ml sample. (From Lactobacillus casei *strain Shirota — Intestinal Flora and Human Health,* Yakult Central Institute for Microbiological Research, Tokyo, 1999, p. 91. With permission.)

Figure 6.4 shows the results of a study on the effects of storage temperature on the viability of LcS in a fermented milk product together with effects on acidity and pH of the product. The viable cell count was determined by culture on agar plates containing bromocresol purple. This culture medium is officially sanctioned by the Ministry of Health and Welfare in Japan and available commercially.[21,22]

With a rise of storage temperature, activation of bacterial metabolism results in an increase in acidity and a decrease in pH, causing a decrease in the number of live bacterial cells. At a storage temperature above 15°C, these changes become marked. Such changes may impair the balance of metabolites that determines the product's flavor and may reduce product quality with respect to taste, aroma, and flavor.

It was concluded that fermented milk and sour milk beverages should be stored at 10°C or lower to maintain their quality for a specified period after manufacture. In general in Japan, the expiration date of such dairy products is set at 14 days after manufacture, provided that the products are stored at 10°C or lower. During storage at 10°C or lower, the viable cell count in the product shows only a slight decrease in number, which is not associated with any noticeable changes in taste or flavor.

6.4 MODIFICATION OF GASTROINTESTINAL FUNCTION BY LcS

Digestion and absorption of nutrients are the major functions of the stomach and small intestine. It has long been considered that fermented milk and yogurt prepared using lactobacilli promote lactose digestion and absorption and may be of nutritional benefit for individuals with low lactase activity.[23]

Promotion of lactose absorption by fermented milk and yogurt has been explained by prolongation of the gastric emptying time and small intestinal transit time as well as by hydrolysis of lactose via the lactase activity in these products. The former effect may be attributed to the physicochemical properties and fat content of the fermented milk. Gastric emptying functions independently for liquids and solids, with liquids being transported more rapidly. Transit of liquid gastric contents is controlled by mechanical stimulation through changes of the intragastric pressure or chemical stimulation caused by lipids, carbohydrates, or acids, which induce contraction of the pyloric sphincter via mechanoreceptors or chemoreceptors present in the duodenal mucosa.[24]

Studies on the function of the stomach and small intestine and gastrointestinal kinetics of absorption were carried out in rats using a fermented milk drink containing LcS; L-lactic acid, glucose, sucrose, and lactose concentrations were measured. Following intragastric administration of the fermented drink to rats, the amounts of these ingredients that remained unabsorbed in each segment of the small intestine were determined. L-lactic acid was not absorbed in the stomach, reached the ileum in 15 minutes after administration, and was absorbed in the small intestine without passage through the ileocecal valve. Glucose and sucrose were absorbed in the jejunum, while some lactose remained undigested and passed through the ileocecal valve. In an experiment involving intestinal perfusion of anesthetized rats, L-lactic acid showed a more rapid absorption in the ileum than in the jejunum or colon, which suggests that a specific mechanism exists for L-lactic acid absorption in the ileum. The gastric emptying time and the small intestinal transit time after oral administration of the fermented milk prepared using LcS were compared with these same parameters in rats after administration of a solution with the same carbohydrate composition and after administration of unfermented milk. The gastric emptying time was longest after administration of the fermented milk, followed by unfermented milk and then the carbohydrate solution, while the small intestinal transit time was similar in all three cases (Figure 6.5).

In normal and gnotobiotic rats, the gastric emptying time was longer after oral administration of the fermented milk prepared using LcS than after administration

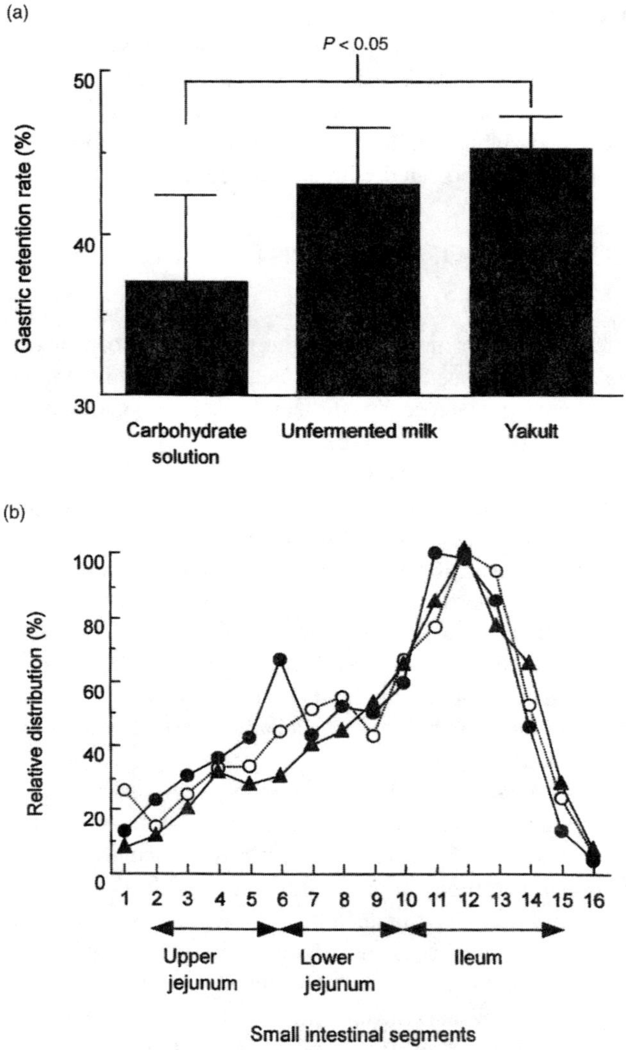

FIGURE 6.5 Gastric retention rate (a) and distribution of phenol red (a marker of unabsorbed substances) in 16 small intestinal segments of equal length (b) at 30 min after administration of 2 ml of Yakult (●), a solution with the same carbohydrate composition (○), or unfermented milk (▲) in normal rats fasted for 24 h. Data on the distribution of phenol red are expressed as a percentage of the maximum value. (From Ohashi, Y. and Umesaki, Y., *Dig. Absorp.*, 20, 119–123, 1997. With permission.)

of a control carbohydrate solution.[25] This suggests that lactic acid or other fermentation products of LcS act as chemical stimuli in the upper gastrointestinal tract. LcS may help increase carbohydrate absorption in individuals who have impaired absorption in the small intestine. Administration of LcS has also been reported to improve colonic function by increasing the frequency of bowel movements and relieving abdominal symptoms associated with constipation (unpublished data).

Although the exact mechanism of these effects is not clear, LcS may improve colonic motility by reestablishing a normal intestinal flora.[26]

6.5 EFFECTS OF LcS ON EXPERIMENTAL ANIMALS

6.5.1 Antitumor Activity of LcS

In 1981, Yokokura et al.[27] screened 26 strains of 14 species of LAB for *in vivo* antitumor activity against sarcoma 180, a transplantable mouse tumor, and found that some strains had potent antitumor activity. Among them, LcS had an especially high potency. The antitumor effect of LcS administered at various times and via various routes of administration was assessed using transplantable mouse tumors (Table 6.2). Intravenous or intraperitoneal administration of LcS caused a dose-dependent inhibition of the growth of subcutaneously implanted sarcoma 180 in ICR mice as well as Meth A fibrosarcoma and another methylcholantherene-induced tumor (MCA K-1) in syngeneic BALB/c mice.[28] LcS also exerted a potent antitumor effect on Lewis lung carcinoma and B16 melanoma, as well as on highly metastatic variants of B16 melanoma (B16-BL6 and B16-F10) in syngeneic C57BL/6 mice.

Intrapleural administration of LcS has been shown to markedly improve the survival of mice with carcinomatous pleurisy. Mice inoculated intrapleurally with Meth A fibrosarcoma cells eventually die of massive pleural effusion associated with tumor cell proliferation. The pattern of proliferation of tumor cells in the pleural cavity is quite similar to that of carcinomatous pleurisy in humans. Therefore, this system may provide an appropriate animal model of carcinomatous pleurisy. When administered intrapleurally, LcS showed a significantly greater survival benefit than other bacterial preparations — OK-432, *Corynebacterium parvum*, and bacillus Calmette-Guerin (BCG) — and has been established as a useful therapy for malignant pleural effusion.[29] Also, it has been reported that oral administration of LcS effectively inhibited methylcholantherene-induced carcinogenesis in mice.[30]

LcS has been reported to augment the antitumor effect of a combination chemotherapy with agents such as doxorubicin, mitomycin-C, cyclophosphamide, bleomycin, and 5-fluorouracil. Significant synergism between LcS administered via various routes (intrapleural, intraperitoneal, intravenous, and subcutaneous) and these cytotoxic drugs has been observed in mice with various transplantable tumors, including mouse Meth A fibrosarcoma, L1210 leukemia, and Lewis lung carcinoma.[31]

6.5.2 Augmentation of Host Immune Cells by LcS

The initial response induced by administration of LcS in a host is an activation of neutrophils, macrophages, or natural killer (NK) cells. In particular, LcS has a strong potential to augment host NK activity as well as the activation of macrophages.[29,32] Even in mice inoculated with Meth A fibrosarcoma cells, administration of LcS induced a high level of NK activity in splenocytes. As shown in Table 6.3, lymphocytes in pleural exudate exhibited a strong NK activity from 3 days after administration onward and still remained active at 7 days.[29] Also, it was reported that oral

TABLE 6.2
Antitumor Effect of LcS Administered via Various Routes

Route of Administration	Dose (mg/kg) × Number of Doses	Transplantable Tumor	Inoculation Site	Antitumor Effect
Intrapleural (i.pl.)	4 × 5	Meth A	i.pl.	T/C 250
Intraperitoneal (i.p.)	5 × 5	Meth A	i.p.	T/C > 157 (1/10)
	10 × 3	L1210	i.p.	T/C 138
	2 × 5	C57AT1	i.p.	T/C 134
	2 × 5	Sarcoma 180	i.p.	T/C > 209 (1/9)
	10 × 5	Meth A	s.c.	I.R. 87.6
	10 × 5	MCA K-1	s.c.	I.R. 61.3
Intravenous (i.v.)	10 × 5	Meth A	s.c.	I.R. 88.2
	2 × 5	MCA K-1	s.c.	I.R. 60.4
	10 × 5	Sarcoma 180	s.c.	I.R. 75.9
	10 × 4	3LL	s.c.	T/C > 132 (2/7)
	10 × 5	B16	s.c.	T/C > 150 (1/10)
	10 × 5	B16-F10	i.v.	T/C 134
	10 × 10	AH 130 (rats)	i.v.	T/C 181
	10 × 10	AH 66 (rats)	i.v.	T/C 159
	10 × 10	AH 7974 (rats)	i.v.	T/C 139
	10 × 10	AH 41C (rats)	i.v.	T/C > 178 (2/6)
Subcutaneous (s.c.)	30 × 7	Meth A	s.c.	I.R. 72.9
Intratumoral (i.t.)	4 × 5	Meth A	s.c.	I.R. 88.2
	4 × 5	Meth A	s.c.	T/C > 152 (2/10)
	4 × 5	K234	s.c.	T/C > 162 (2/10)
	10 × 4	B16-BL6	s.c.	I.R. 81.9
	10 × 4	B16-BL6	s.c.	T/C 156
	10 × 5	B16-F10	s.c.	T/C 142 (5/6)
	4 × 4	Line-10 (guinea pigs)	Intradermal	

Note: T/C: Survival benefit (%) = mean survival time (days) of treated animals/mean survival time (days) of control animals × 100; I.R.: Inhibition rate (%) = (1 – mean tumor weight in treated animals/mean tumor weight in control animals) × 100. Animals used were BALB/c mice (Meth A fibrosarcoma, MCA K-1 sarcoma, K 234 sarcoma), DBA/2 mice (L1210 leukemia), C57BL/6 mice (C57AT1 virus-induced tumor, B16 [B16-F10 and B16-BL6] melanoma, Lewis lung carcinoma), ICR mice (Sarcoma 180), Donryu rats (AH 130, AH 66, AH 7974, and AH 41C ascitic hepatoma), and strain-2 guinea pigs (Line-10 hepatoma). Allogeneic tumors are underlined. Numbers in parentheses represent the percentage of animals surviving for 40 days showing complete tumor regression.

Source: Yokokura, T., Kato, I., and Mutai, M., in *Intestinal Flora and Carcinogenesis,* Mitsuoka, T., Ed., Japan Scientific Societies Press, Tokyo, 1981, pp. 125–137. With permission.

TABLE 6.3
NK Activity of Pleural Exudate Cells Harvested after Intrapleural Administration of LcS

Days after Administration of LcS[a]	Cytotoxicity (%) Against YAC-1 Lymphoma[b] 100:1[c]	50:1
Untreated control	1.8 ± 1.3	1.7 ± 1.8
1	2.6 ± 1.8	1.3 ± 0.6
3	52.4 ± 3.0	47.0 ± 7.7
5	47.0 ± 9.2	34.6 ± 5.0
7	57.0 ± 7.0	50.6 ± 6.9

[a] Pleural exudate cells were obtained after intrapleural administration of LcS (100 μg) to BALB/c mice.
[b] Cytotoxicity was determined by the ^{51}Cr-release method.
[c] Effector/target ratio.

Source: Matsuzaki, T., Yokokura, T., and Mutai, M., *Cancer Immunol. Immunother.,* 26, 209–214, 1988. With permission.

administration of LcS significantly enhanced splenic NK activity in a murine carcinogenesis model[33] (Figure 6.6). Furthermore, oral administration of LcS significantly delayed the onset of carcinogenesis in mice treated with 3-methylcholanthrene and recovered the decreased T-cell responses such as concanavalin A-induced T-cell proliferation and interleukin-2 production[34] (Figure 6.7). Important antitumor mechanisms induced by LcS include enhancement of such nonspecific antitumor activity and induction of cell-mediated immunity through subsequent activation of T cells. LcS was shown to be potent in inducing cell-mediated immunity.[26] In BALB/c mice, tumor-specific immunity was induced by intraperitoneal or subcutaneous inoculation of a mixture of LcS and tumor cells.[35] Induction of tumor-specific antitumor immunity was observed in some strain-2 guinea pigs, which showed complete tumor regression after intratumoral administration of LcS.[36] Splenic and peritoneal T cells obtained from these animals exhibited cytostatic activity against tumor cells in a neutralization test. In C57BL/6 mice, priming with LcS augmented the antitumor and antimetastatic effects of LcS. This may have resulted from enhancement of tumor-specific effector cells by various cytokines produced by LcS-specific helper T cells.[37]

6.5.3 Protection against Infection by LcS

LcS has been reported to protect a host against infection by the pathogens *Listeria monocytogenes, Mycobacterium fortunum, Pseudomonas aeruginosa, Escherichia coli, Salmonella typhimurium, Staphylococcus aureus,* murine cytomegalovirus, and herpes simplex virus. Prior intraperitoneal administration of LcS increased the resistance of mice to infection induced by intraperitoneal inoculation of a lethal amount of *P. aeruginosa,* compared with untreated mice. The protective effect of

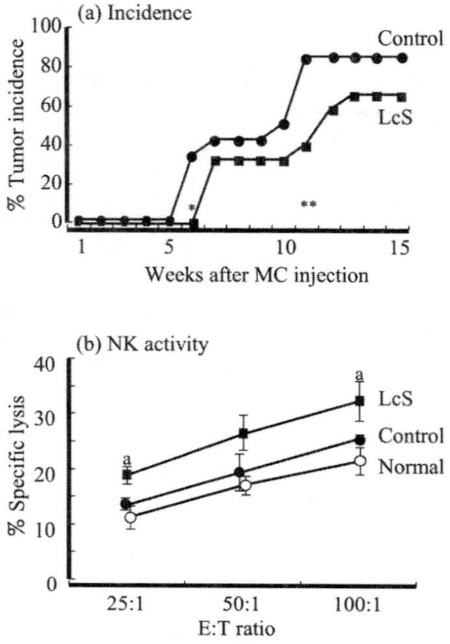

FIGURE 6.6 (a) Inhibition of 3-methylcholanthrene (MC)-induced carcinogenesis and (b) enhancement of natural killer activity by LcS. * $P < 0.05$ at 6 weeks, ** $P < 0.05$ up to 11 weeks in the cumulative incidence; E:T, effector cell to target cell; (a) $P < 0.05$ vs. control. (From Takagi, A., Matsuzaki, T., Sato, M., Nomoto, K., Morotomi, M., and Yokokura, T., *Carcinogenesis,* 22, 599–605, 2001. With permission.)

LcS against *P. aeruginosa* infection was also noted in mice with severe neutropenia following whole-body x-ray irradiation, while no such effect was observed in mice treated with carrageenan, which causes a specific inactivation of macrophages.[38] These findings suggested that macrophages can act as effector cells in protection against infection.

Listeria monocytogenes, a Gram-positive bacillus forming short rods, is an intracellular parasite that is widely distributed in nature. This bacterium has been isolated from numerous species of mammals, birds, fish, crustaceans, and insects. Because of its widespread occurrence, *L. monocytogenes* has many opportunities to enter food production and processing environments. Recently, one outbreak of gastroenteritis with fever but without progression to invasive disease was linked to the consumption of milk highly contaminated by *L. monocytogenes,* adding *Listeria* to the causes of food poisoning.[39] However, the pathogenesis of *L. monocytogenes* in human infection is unclear. The protective effect of LcS against *L. monocytogenes* infection was assessed in mice[40] (Table 6.4). Administration of LcS caused sustained activation of macrophages. Mice inoculated intravenously with a lethal dose of *L. monocytogenes* were protected if they had received LcS for 3 weeks prior to the lethal dose. In the experiment some mice were given *Corynebacterium parvum* (1 week prior to the lethal dose) as a possible protection, but some of these mice died.

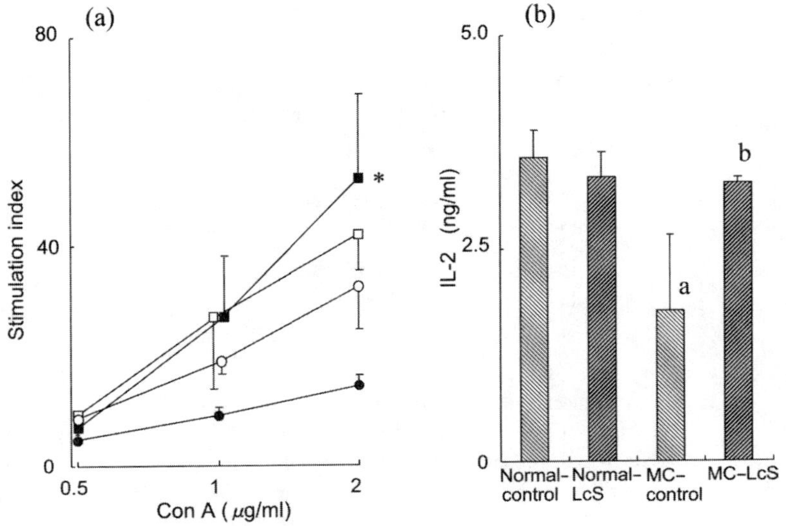

FIGURE 6.7 Improvement of methylcholantherene (MC)-induced suppression of proliferative responses of spleen cells to concanavalin (Con) A by LcS. The spleen cells obtained from the mice at week 16 after MC treatment were cultured with Con A, and the splenic cell proliferation (a) and the release of interleukin (IL)–2 (b) were measured. (a) $P < 0.01$ vs. normal-control; □, non(MC)-treated; ■, MC-treated; ○, non(MC)-treated; ●, MC-treated. (b) $P < 0.05$ vs. MC-control. (From Takagi, A., Matsuzaki, T., Sato, M., Nomoto, K., Morotomi, M., and Yokokura, T., *Med. Microbiol. Immunol.,* 188, 111–116, 1999. With permission.)

In this experimental system, the number of viable bacteria in the liver was markedly reduced in mice pretreated with LcS compared with untreated mice and mice pretreated with *C. parvum.* In mice treated with LcS, there was increased production of bactericidal lysosomal enzymes (e.g., β-glucuronidase) and bactericidal oxygen radicals (superoxide radical; $\cdot O_{2-}$) by peritoneal macrophages and Kupffer cells in the liver, and this change was more sustained than that caused by *C. parvum.*

6.5.4 Antihypertensive Effect of Oral Administration of LcS

The antihypertensive effect of LcS was assessed in spontaneously hypertensive rats (SHR), an animal model commonly used for antihypertensive drugs.[41] A single oral administration of LcS (100 mg/kg) had no effect on blood pressure in normotensive Wistar-Kyoto rats (WKY), while the same dose significantly decreased the blood pressure of SHR. At this dose, LcS did not affect the heart rate of WKY or SHR. In addition, long-term administration of LcS at 100 or 1000 mg/kg/day suppressed the elevation of systolic blood pressure (SBP) in SHR (Figure 6.8).

A water-soluble fraction of LcS was obtained with high yield by autolysis at 55°C and pH 7.0 for 2 h followed by heating at 100°C for 10 min. This soluble fraction was lyophilized and termed LEx, and the antihypertensive effect of LEx was studied. This soluble fraction was composed of sugar (about 20% by the phenol–sulfuric acid method), protein (about 45% by Lowry's method), nucleic acid

TABLE 6.4
Protective Effect against *Listeria monocytogenes* Infection of Pretreatment with LcS or *Corynebacterium parvum* in Mice

Treatment[b]	Survival Rate (%)[a]	
	LcS	*C. parvum*
–7	9/9 (100)[c]	9/9 (100)
–14	9/9 (100)	0/8 (0)
–21	9/9 (100)	0/9 (0)
–28	1/8 (12.5)	0/10 (0)

[a] After inoculation of 5×10^4 CFU of *L. monocytogenes* into the tail vein, survival was monitored for 14 days. The survival rate of untreated control mice was 0.0% (0/15 mice).

[b] Each mouse received an intravenous dose of 1 mg of LcS (10^9 cells) or *C. parvum* at 7, 14, 21, or 28 days before inoculation with *L. monocytogenes*.

[c] Number surviving/number treated (% survival).

Source: Nomoto, K., Miake, S., Hashimoto, S., Yokokura, T., Mutai, M., Yoshikai, Y., and Nomoto, K., *J. Clin. Lab. Immunol.*, 17, 91–97, 1985. With permission.

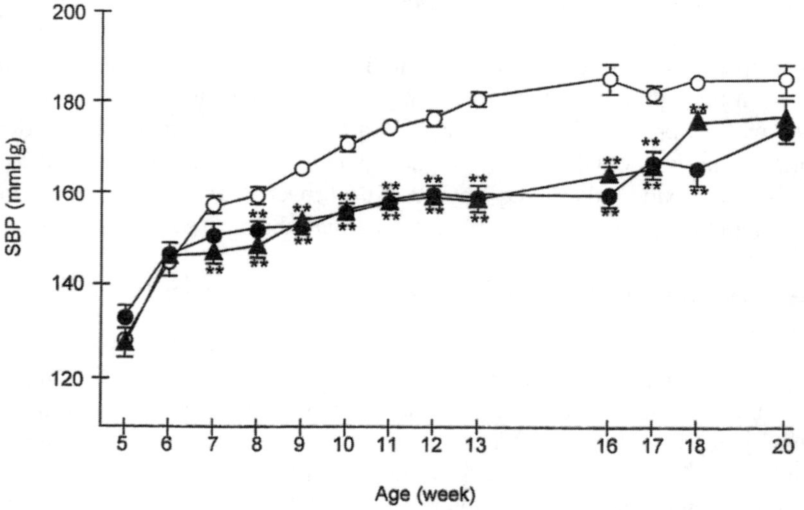

FIGURE 6.8 Effect of long-term oral administration of LcS on SBP in SHR. LcS was dispersed in distilled water (DW) at appropriate concentrations that resulted in oral doses of 100 (▲) and 1000 (●) mg/kg/day delivered in a volume of 0.5 ml/100 g body weight; (○) control rats. Administration was performed daily for 11 weeks from 5 to 16 weeks of age. ** $P < 0.01$ vs. control. (From Furushiro, M., Sawada, S., Hirai, K., Motoike, M., Sansawa, H., Kobayashi, S., Watanuki, M., and Yokokura, T., *Agric. Biol. Chem.*, 54, 2193–2198, 1990. With permission.)

(about 10%), and ash (about 10%). A single oral administration of LEx at a dose of 10 mg/kg lowered the SBP in SHR. Moreover, long-term oral administration of 1 mg/kg suppressed the development of high SBP.[42]

6.5.5 Inhibitory Effect on Immunoglobulin E Production

It has been reported that some probiotics exert a regulatory effect on the immune response, resulting in the inhibition of immunoglobulin E (IgE) production by murine spleen cells *in vitro*.[43] Recently, it has been demonstrated that LcS has a preventive effect on IgE production in BALB/c mice *in vivo*.[44] The mice were immunized by intraperitoneal injection of ovalbumin (OVA) and $Al(OH)_3$ on days 0 and 14. Seven days after final immunization, blood was collected from all mice and assayed for OVA-specific serum IgE, while spleen cells were prepared for assays of OVA-specific IgE production and OVA-induced cytokine production. The level of OVA-specific IgE in serum in each group is shown in Figure 6.9. In the group fed a diet containing 0.05% (w/w) LcS, inhibition of OVA-specific IgE production was evident at 3 weeks of feeding of the LcS-containing diet compared with the response of the control group. Also, in the mice fed LcS, the level of production of Th1-associated cytokines, such as IFN-γ and IL-2, by spleen cells was higher than that in the control group (Figure 6.10a). In contrast, the level of production of Th2-associated cytokines, such as IL-4, IL-5, and IL-6, by spleen cells from the mice fed LcS was lower than that of the control group (Figure 6.10b). Furthermore, the level of production of IL-12, which augments IFN-γ production, by spleen cells from the mice fed LcS was also higher than that of the control group (Figure 6.10c). These results indicate that LcS induces a Th1 response rather than a Th2 response. Therefore, it may be concluded that functional augmentation of Th1 cells and inhibition of Th2 cells by LcS (or LcS components) are probably critical in the inhibition of IgE production in mice.

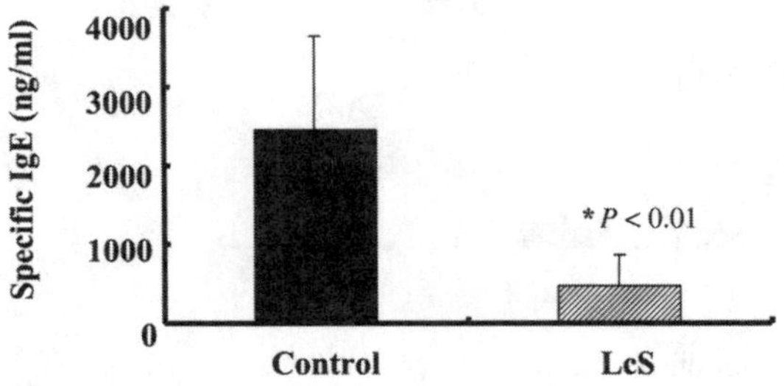

FIGURE 6.9 Effect of LcS on ovalbumin (OVA)-specific IgE production. BALB/c mice were injected i.p. on day 0 and 14 with 20 μg of OVA and 2 mg of $Al(OH)_3$ in a total volume of 0.2 ml. The mice were then fed a diet containing (w/w) 0.05% LcS for 21 days. Control mice were injected with OVA only. Sera were collected on day 21, and the level of OVA-specific IgE was determined by passive cutaneous anaphylaxis reaction. Bars represent mean values (± standard deviation) for five mice. Significant difference from control: * $P < 0.01$. (From Matsuzaki, T., Yamazaki, R., Hashimoto, S., and Yokokura, T., *J. Dairy Sci.*, 81, 48–53, 1988. With permission.)

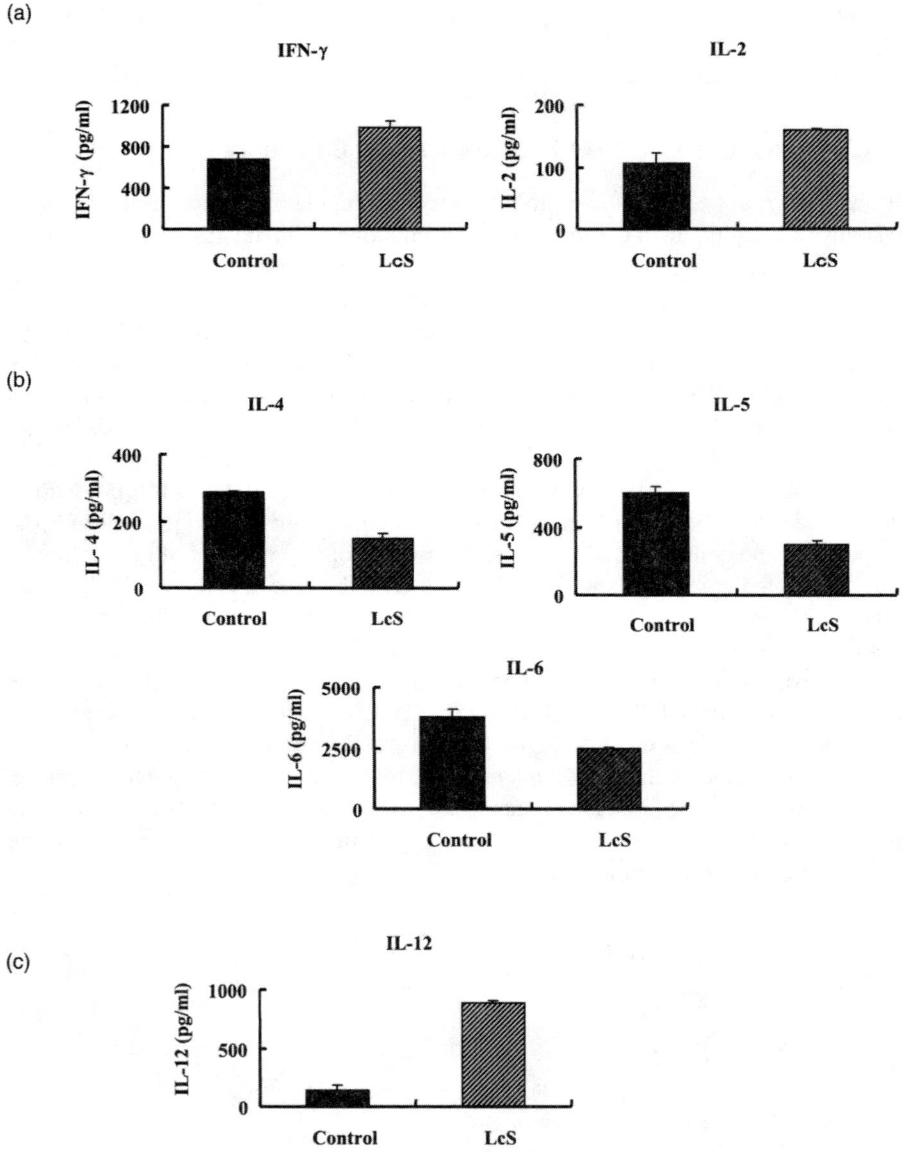

FIGURE 6.10 Production of interferon (IFN)-γ, interleukin (IL)-2 (a), IL-4, IL-5, IL-6 (b), and IL-12 (c) by spleen cells *in vitro*. BALB/c mice were intraperitoneally injected on day 0 and 14 with 20 μg of OVA and 2 mg of $Al(OH)_3$ in a total volume of 0.2 ml. The mice were then fed a diet containing (wt/wt) 0.05% LcS for 21 days. Spleen cells were collected on day 21 and were co-cultured with OVA (final concentration, 100 μg/ml) for 24 h. The amounts of IFN-γ, IL-2, IL-4, IL-5, IL-6, and IL-12 in the supernatant were measured by ELISA. (From Matsuzaki, T., Yamazaki, R., Hashimoto, S., and Yokokura, T., *J. Dairy Sci.*, 81, 48–53, 1988. With permission.)

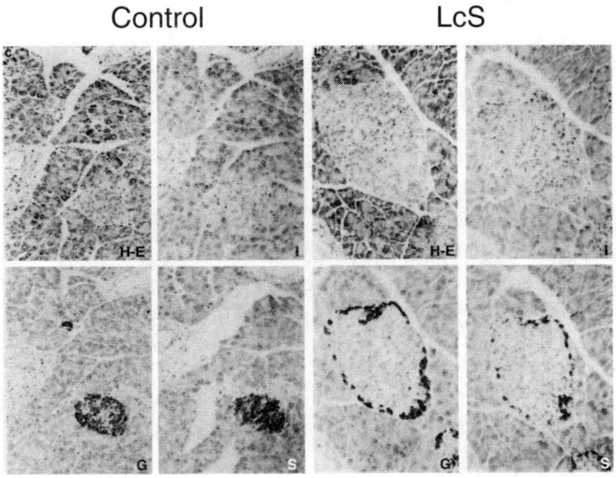

FIGURE 6.11 Photomicrographs of islets of Langerhans from NOD mice. Hematoxylin (H-E) staining and immunoperoxidase staining (I: insulin, G: glucagon, S: somatostatin) (× 120). (From Matsuzaki, T., Nagata, Y., Kado, S., Uchida, K., Kato, I., Hashimoto, S., and Yokokura, T., *APMIS,* 105, 643–649, 1997. With permission.)

6.5.6 Impact of LcS on Autoimmune Diseases

It has been reported that oral administration of LcS effectively inhibits the onset of diabetes in an insulin-dependent diabetes mellitus model, the nonobese diabetic (NOD) mouse.[45] In this study, 4-week-old female NOD mice were fed a diet of either standard laboratory chow or the same chow containing 0.05% (by weight) LcS, and the onset of diabetes was thereafter recorded. The incidence of diabetes was significantly higher in the control group (10/12) than in the LcS-treated group (3/12) ($P < 0.01$). Pathological analysis in the LcS-treated group revealed strong inhibition of the disappearance of insulin-secreting β cells in Langerhans islets caused by autoimmune disease (Figure 6.11).

It was also shown that oral administration of LcS significantly prolonged the life span of MRL/lpr mice, which develop an autoimmue disease resembling human systemic lupus erythematosus. Also, it was demonstrated that LcS accelerates macrophage recruitment and prevents the expansion of $B220^+$ T cells without affecting the functions of T cells in MRL/lpr mice.[46]

Kato et al.[47] reported the effect of the oral administration of LcS on the development of type II collagen (CII)–induced arthritis (CIA) in DBA/1 mice. It was shown that the LcS treatment significantly reduced the incidence and the development of CIA and the levels of antibody to CII in serum compared with the control group. The CII-specific IgG2a and IgG2b antibodies in serum were also down-regulated in the LcS-treated group. It was also demonstrated that LcS inhibited the delayed-type hypersensitivity response to CII in DBA/1 mice immunized with CII and complete Freund's adjuvant and suppressed the CII-specific secretion of IFN-γ from splenocytes. Taken together, this suggests that LcS has the potential to ameliorate or prevent autoimmune diseases through modification of the humoral and cellular immune response in the host.

6.6 EFFECTS OF LcS IN HUMAN TRIALS

6.6.1 Survival of LcS in the Gastrointestinal Tract and Modification of Intestinal Flora

6.6.1.1 Survival of LcS in Infants

Aritaki and Ishikawa[48] studied the effect of a daily intake of 30 ml of a fermented milk containing LcS (1×10^7 to 3×10^8 cells/ml given as three doses of 10 ml each) on the intestinal flora of 10 healthy formula-fed infants aged 2 to 5 months.[40] The intestinal flora were assessed by classifying 50 to 200 colonies grown on a smear of an appropriate dilution of feces by Gram staining and morphological observation. The organisms were classified as lactobacilli, bifidobacteria, Gram-positive bacilli, Gram-negative bacilli, or cocci, and the percentage of colonies in each group was calculated. The mean percentage of *Lactobacillus* species increased from 4.5% before drinking fermented milk to 20.9% after 3 days and to 53.8% after 7 days. After cessation of consumption of the fermented milk, the percentage of *Lactobacillus* species remained high, at 52.5 and 32.2% at 3 and 6 days afterwards, respectively. The percentage of Gram-negative bacilli decreased from 43.1% before administration to 29.4 and 14.7% after 3 and 7 days of administration, respectively. The percentage of this bacterial group remained lower after cessation of administration. Yamagishi et al.[49] also studied the effect of a fermented milk containing LcS on intestinal flora in 12 infants (aged 4 to 19 months) hospitalized in a nursing home. Each infant was given 65 ml of the drink daily for 60 consecutive days, and lactobacilli were determined. The results were similar to those reported by Aritaki and Ishikawa. After administration of fermented milk, the numbers of lactobacilli in the feces and the number of infants excreting lactobacilli both increased.

6.6.1.2 Survival of LcS in Children

Shirota et al.[50] studied the fecal recovery of LcS during the long-term intake of a fermented milk containing LcS and the effect of habitual intake of this fermented milk on the intestinal flora in 30 children aged 2 to 6 years (Figure 6.12). The children were divided into two equal groups following randomization with respect to age, sex, and living environment. One group drank 50 ml of a fermented milk containing live LcS (1×10^8 to 2×10^8 cells/ml) together with 180 ml of milk every morning. The other group drank 50 ml of a fermented milk sterilized at 80°C for 30 min together with 180 ml of milk daily for the same period. The fecal recovery of LcS increased rapidly after the start of daily intake of fermented milk containing live LcS. The number of bacteria recovered per gram of feces was in the range 10^6 to 10^8 (mean: 2.0×10^8) after 1 week of administration and 6×10^7 to 1×10^9 (mean: 2.0×10^8) after 2 weeks. Constant fecal recovery of the bacteria continued thereafter. Following cessation of drinking of the fermented milk containing live LcS, the fecal recovery of LcS did not decrease rapidly and remained at 10^6 to 10^7 cells/g in four of the 15 children after 2 weeks. After 3 weeks, however, no LcS cells were detected in the feces of any subject. In both groups receiving either live or killed bacteria, 10^6 to 10^9 (mean: 10^8) cells of Enterobacteriaceae were detected per gram of feces

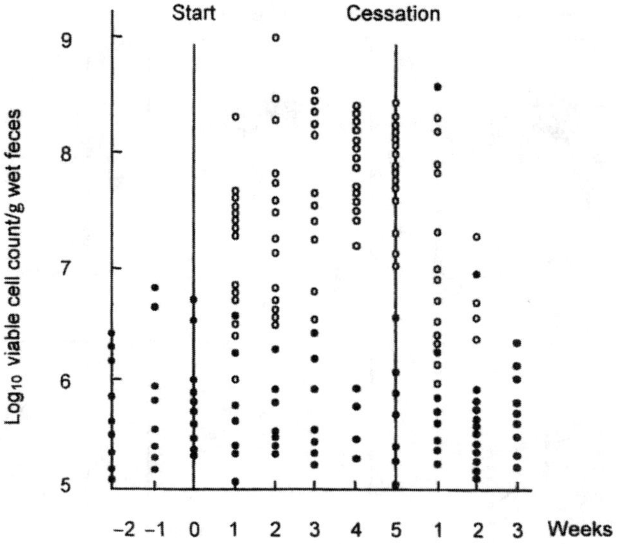

FIGURE 6.12 Changes of fecal Enterobacteriaceae in the Yakult group and sterilized Yakult group. ○,Yakult group; ●, sterilized Yakult group. (From Shirota, M., Aso, K., and Iwabuchi, A., *Jpn. J. Bacteriol.,* 21, 274–283, 1966. With permission.)

at baseline. In the children drinking fermented milk containing LcS, the number of Enterobacteriaceae decreased to 10^4 to 10^7 (mean: 10^6) cells/g after 2 to 3 weeks. However, the number of bacteria rapidly returned to the baseline values after cessation of the intake of live LcS. In the control group given sterilized fermented milk, fecal Enterobacteriaceae remained unchanged. The number of enterococci in the feces decreased daily in a similar manner after the start of intake of fermented milk with LcS; values gradually returned to baseline values after cessation of LcS intake. Other microbial species, such as anaerobes (e.g., bifidobacteria), staphylococci, and yeasts, were not affected by fermented milk with live LcS.

Hanada et al.[51] also assessed the effect of daily intake of a fermented milk (30 ml/day, containing 1×10^7 to 2×10^8 cells/ml of LcS) for 10 weeks on fecal lactobacilli, Enterobacteriaceae, group A enterococci (e.g., *Enterococcus* sp.), and group B enterococci (e.g., *Staphylococcus* sp.) in a controlled study involving 20 children aged 4 to 7 years. In children drinking the fermented milk with LcS, LcS was detected in the feces after 1 to 2 weeks. Fecal recovery transiently decreased around the fifth week and increased again from the sixth week. During the period that fermented milk containing LcS was ingested, Enterobacteriaceae, *Enterococcus* spp., and *Staphylococcus* spp. decreased.

6.6.1.3 Survival of LcS in Adults

Tanaka et al.[52] orally administered live LcS (1×10^{10} cells) together with 200 ml of milk to five healthy adults (aged 25 to 32 years) daily for 5 weeks. They found that the feces of these subjects contained 10^7 to 10^8 cells of LcS/g. Lactobacilli increased in all five subjects, while bifidobacteria increased in four of them (Figure 6.13).

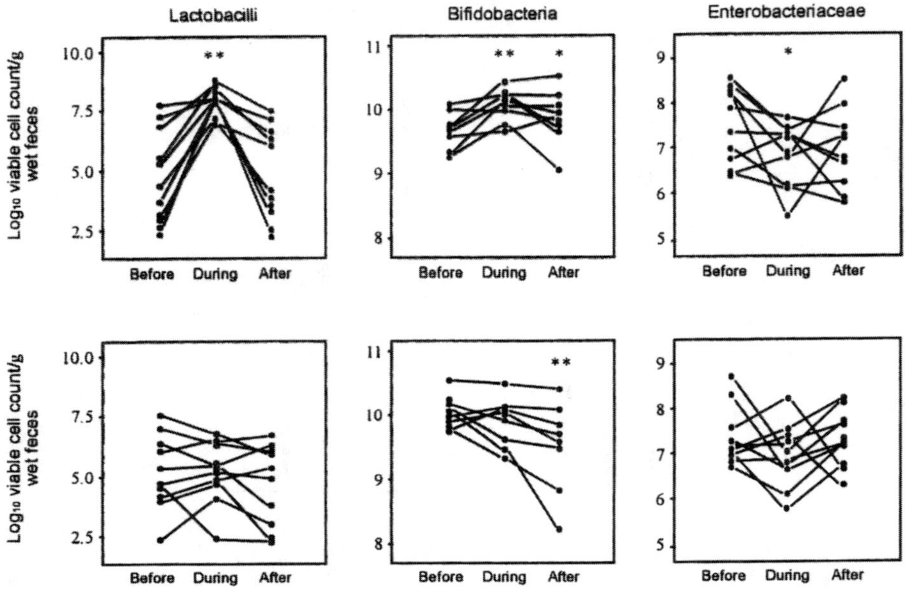

FIGURE 6.13 Changes of fecal microflora in healthy adults orally administered 10^{10} cells LcS (top) and unfermented milk (bottom) groups. Significantly different from baseline; * $P < 0.05$, ** $P < 0.01$. (From Tanaka, R., Tohyama, K., Morotomi, M., Takayama, H., Nanno, M., Kuroshima, T., and Mutai, M., in *Proceeding of I RIKEN Symposium on Intestinal Flora: Intestinal Flora and Cancer,* Mitsuoka, T., Ed., Japan Scientific Societies Press, Tokyo, 1981, pp. 79–103. With permission.)

Tanaka and Ohwaki[53] also performed a double-blind controlled study to confirm this finding. Twenty adult men consumed either 80 ml of a fermented milk (containing 7.5×10^8 cells of LcS/ml) or nonfermented milk three times daily for 4 consecutive weeks. In the fermented milk group, *Lactobacillus* spp. significantly increased, and LcS was the predominant strain of this bacterial group in the fecal flora. Bifidobacteria also increased, while Enterobacteriaceae decreased.

Spanhaak et al.[54] performed a similar study to assess the effect of a fermented milk containing LcS in European individuals. There were two 2-week untreated control periods before and after the treatment period. Ten subjects drank 100 ml of a fermented milk (containing 10^9 cells of LcS/ml) three times daily for 4 weeks, while 10 control subjects drank unfermented control milk in the same manner. During the two control periods, all subjects received 100 ml of sterilized milk three times daily. None of the fermented milk recipients complained of any adverse effects during the treatment period. More than 10^7 cells of LcS were recovered per gram of feces, and bifidobacteria also increased.

6.6.2 Modification of Bowel Movements by LcS

LAB have long been used and have an established efficacy for treatment of defecation disorders (e.g., diarrhea and constipation) and various associated abdominal symptoms.[55] Kawamura et al.[56] evaluated the efficacy of *Lactobacillus* powder (containing 1×10^{10} viable cells of LcS/g) in 30 patients with a long history of defecation disorders (diarrhea and constipation) or unidentified abdominal complaints, including abdominal pain and bloating. The patients were given 1 g of the *Lactobacillus* preparation orally once daily for 5 to 82 days (mean: 14.4 days). The number of bowel movements, the properties of the feces, and the severity of abdominal pain, bloating, anorexia, and nausea were measured during the treatment period. The patients were rated as markedly improved when all complaints improved, moderately improved when more than half of the complaints improved, and slightly improved when fewer than half of the complaints improved. A rating of unchanged was assigned when none of the complaints improved. Fifteen patients (50%) showed some response, with the rating being markedly improved for eight patients, moderately improved for three patients, and slightly improved for four patients. A response was obtained in 10 of the 15 patients without organic disease (e.g., those with irritable bowel syndrome or habitual constipation) and in two of the five patients with organic disease (chronic colitis, Crohn's disease, and diverticular disease of the colon). When stratified by symptoms, the highest response rate (45%; 9/20) was obtained for bloating, followed by abnormal bowel movements (9/27), abnormal fecal properties (10/29), abdominal pain (6/22), and anorexia (3/8).

6.6.3 Suppression of Intestinal Putrefaction by LcS in Healthy Adults

To evaluate the effect of LcS on intestinal putrefaction in humans, seven healthy adult men were given 10^{10} CFU of LcS daily together with 200 ml of milk for 5 weeks (Figure 6.14).[57] Throughout the experimental period, the urine excreted after rising (morning urine specimen) was collected and assayed to determine indoxyl sulfate, *p*-cresol, and phenol excretion. The mean urinary concentrations of indoxyl sulfate and *p*-cresol during the treatment period were significantly lower than those before the administration when only milk was fed to the subjects ($P < 0.05$).

Five subjects were enrolled for a second similar study and also received 10^{10} CFU of LcS daily for 5 weeks. In this study, only the urinary concentration of phenol decreased significantly ($P < 0.05$), whereas the urinary concentrations of indoxyl sulfate and *p*-cresol showed tendencies toward reduction.

The data obtained from 12 subjects in the two studies were combined and analyzed together. As shown in Figure 6.14, the mean urinary concentrations of indoxyl sulfate ($P < 0.01$), *p*-cresol ($P < 0.05$), and total phenol ($P < 0.05$) during treatment with LcS were significantly lower than the levels in the pretreatment period when only milk was fed to the subjects. Moreover, on withdrawal of LcS, the urinary concentrations of indoxyl sulfate and *p*-cresol returned to their respective pretreatment levels ($P < 0.05$). Although the suppression of putrefaction by LcS shown in these studies may be regarded as a combined effect with milk, these data suggest that LcS tends to suppress the formation of putrefactive substances by gut bacteria in the human intestine.

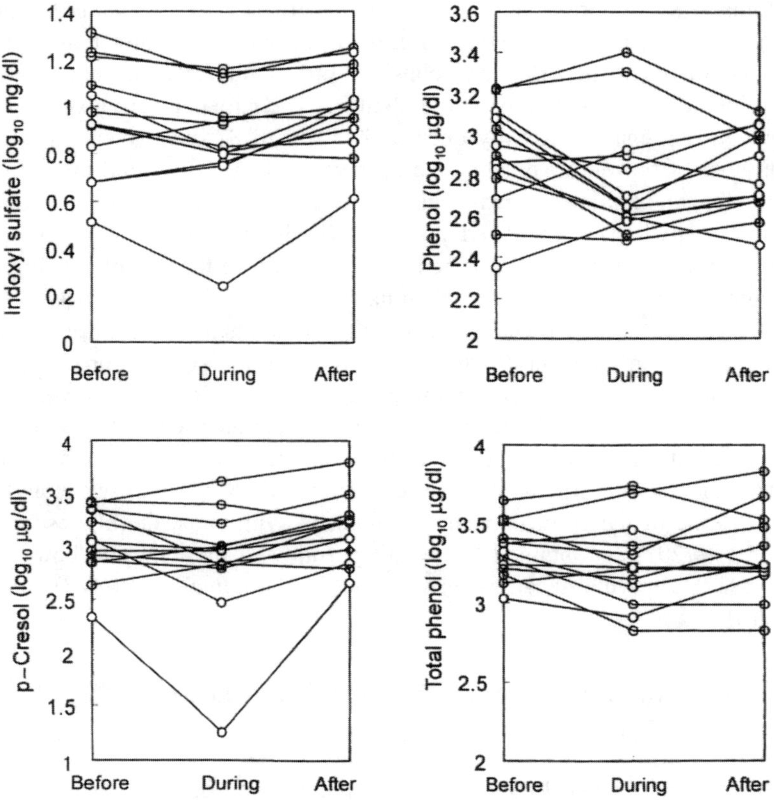

FIGURE 6.14 Suppression of urinary excretion of indoxyl sulfate and phenols in morning urine specimens from healthy adults receiving treatment with LcS. (From Tohyama, K., Kobayashi, S., Kan, T., Yazawa, K., Terashima, T., and Mutai, M., *Microbiol. Immunol.,* 25, 101–112, 1981. With permission.)

6.6.4 Antitumor Effects in Humans

6.6.4.1 Antitumor Activity of Heat-Killed Cells of LcS

As the impact and the safety of LcS have been confirmed in several experimental tumor models, the clinical efficacy of subcutaneous administration of heat-killed cells of LcS (LC9018) combined with radiotherapy has been assessed. In a late Phase II study and a Phase III study involving patients with cervical cancer, LC9018 combined with radiotherapy caused significantly greater tumor regression and was of greater survival benefit than radiotherapy alone.[58]

Since intrapleural administration of LC9018 significantly suppressed pleural effusion and inhibited tumor growth in mice with experimental carcinomatous pleurisy, its clinical use for the treatment of carcinomatous pleurisy was assessed in 47 patients with malignant pleural effusion secondary to primary lung cancer. When combined with chemotherapy agents, LC9018 administered at doses of 0.5, 0.2, and 0.1 mg achieved response rates of 90.0, 83.3, and 67.7%, respectively. Intrapleural administration of LC9018 was found to be effective as an antitumor agent for controlling malignant pleural effusion, while improving the quality of life (QOL) of cancer patients.[59,60]

6.6.4.2 Preventive Effect of LcS on the Recurrence of Bladder Cancer

Aso et al.[61] conducted two clinical studies to assess the preventive effect of a LcS preparation biolactis powder (BLP), which contained LcS at 10^{10} cells/g) on the recurrence of bladder cancer after transurethral resection. The first study was a randomized controlled study designed to determine the preventive effect of BLP on the recurrence of superficial bladder cancer after transurethral resection. Patients assigned to the BLP group received 1 g of the preparation three times daily for 1 year or until tumor recurrence. Following surgery, all patients underwent cystoscopy and cytological examination of the bladder every 3 months to detect cancer recurrence. The time to the first recurrence was compared between patients treated with BLP and those not treated with BLP (control group). The 50% recurrence-free interval was 352 days in the BLP group compared with 208 days in the control group. This represented a 2.4-fold higher risk of tumor recurrence in the control group than in the BLP group.

Subsequently, a double-blind placebo-controlled study was conducted to obtain more objective evidence of the efficacy of BLP against superficial bladder cancer (Figure 6.15). In this study, the 50% recurrence-free interval was 688 days in the BLP group compared with 543 days in the placebo group. Based on these results, Aso et al.[61] concluded that BLP significantly prolonged the time to recurrence of bladder cancer after transurethral resection. Furthermore, recently the efficacy of dairy products containing LcS on the recurrence of superficial bladder cancer was confirmed epidemiologically.[62]

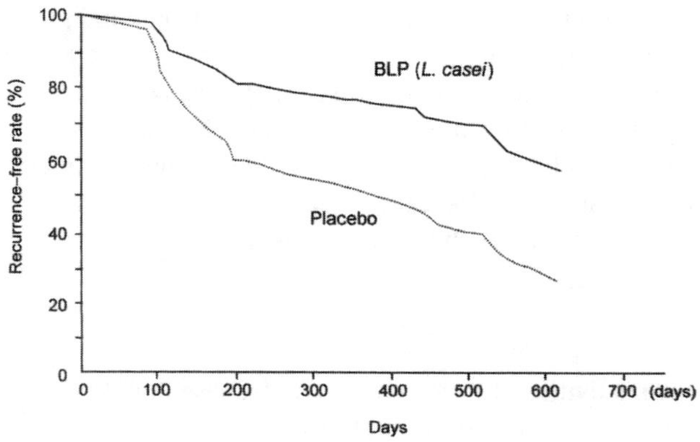

FIGURE 6.15 Corrected cumulative recurrence-free rates for patients treated with BLP containing LcS and placebo recipients in a double-blind placebo-controlled study. (From Aso, Y., Akaza, H., and the BLP Study Group, *Urol. Int.*, 49, 125–129, 1992. With permission.)

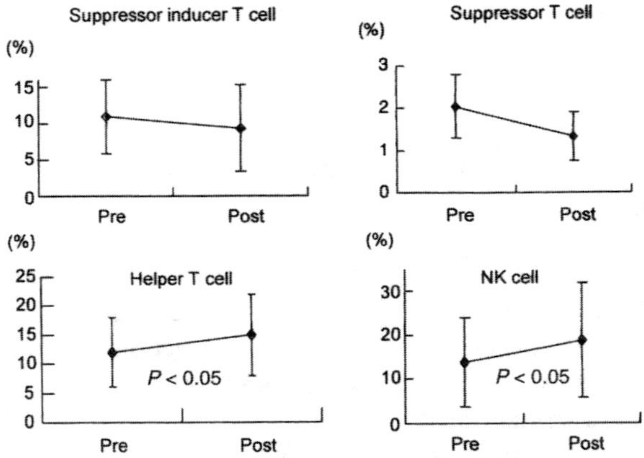

FIGURE 6.16 Effect of oral administration of BLP containing LcS on peripheral blood lymphocyte subsets in patients with colon cancer. (From Aso, Y., Akaza, H., Tsukamoto, T., Imai, K., Naito, S., and the BLP Study Group, *Eur. Urol.*, 27, 104–109, 1995. With permission.)

6.6.4.3 Augmentation of Host Immune Parameters

Sawamura et al.[63] determined the natural killer (NK) cell activity and the response to phytohemagglutinin of lymphocytes obtained from the peripheral blood and regional lymph nodes before and after tumor resection in patients with Dukes A colon cancer who were treated with BLP. Flow cytometric analysis of lymphocyte subsets showed an increase of helper T cells and NK cells and a decrease of suppressor T cells in the peripheral blood of patients with BLP (Figure 6.16). Among the lymphocytes in the regional lymph nodes obtained from the treated patients, suppressor T cells were slightly decreased and suppressor/inducer T cells were decreased. Because treatment with BLP increased helper T cells and NK cells and also decreased suppressor T cells in the regional lymph nodes as well as in the peripheral blood, BLP seems to have a systemic immunopotentiating effect. Recently, it was reported that the intake of a dairy product containing LcS augments host NK cells in the peripheral blood.[64] This suggests that it is possible to prevent disorders such as cancer, autoimmune diseases, etc. by taking some kind of dairy product, because it was demonstrated that medium and high levels of NK activity of peripheral blood lymphocytes are associated with a reduced cancer risk, whereas low activity is associated with increased cancer risk, suggesting a role for natural immunological host defense mechanisms against cancer.[65]

6.7 CONCLUSIONS

Various biological activities of LcS were discussed based on data obtained not only *in vitro* but also *in vivo* (Table 6.5). *Lactobacilli* including LcS have traditionally been used for food processing and have been ingested orally as live cells in yogurt and other fermented foods. Therefore, the modulating effect on intestinal function of lactic acid bacteria and their metabolites has been thought to be their primary benefit for human health. In recent years, the importance of probiotics has been

TABLE 6.5
Positive Effect of LcS in Several Models

Disease/Condition	Treatment/Dose	Effect of LcS	Ref.
Animal			
Bowel movement	p.o.	Normalization	25, 26
Tumors	i.p., i.v. /100~250 μg	Inhibition	27, 28, 29
Immune modulation	i.p., i.v., i.pl., s.c., p.o. / 100~250 μg	Regulation (augmentation)	30, 31, 33–35
Metastases	i.v., i.l. /100~250 μg	Inhibition	32, 36, 37
Infection	i.p., s.c. /100~250 μg	Inhibition	38, 40
Immunoglobulin E	p.o. /0.05%-containing diet (0.05%-LcS)	Inhibition of IgE and augmentation of cytokines	43, 44
Hypertension	p.o. /10 mg/kg	Inhibition of increased blood pressure	42
Diabetes	p.o. /0.05%-LcS	Inhibition of onset	45
Autoimmune	p.o. /0.05%-LcS	Inhibition of onset	46
Arthritis	p.o. /100~250 μg	Inhibition of onset	47
Human			
Normalization of intestinal flora	LcS 10^7~10^{10}	Normalization	48–53
Constipation	1 g/day, 2 weeks	Decreases constipation	56
Uterine/lung cancer	s.c., i.d., i.pl.	Prolongation of survival	57–59
Bladder cancer	p.o.	Prevention of recurrence	60, 61
Immune modulation	p.o.	Augmentation of NK activity	4, 63

Note: i.v., intraveneous injection; i.p., intraperitoneal injection; i.l., intralesional injection; i.pl., intrapleural injection; s.c., subcutaneous injection; i.d., intradermal injection; p.o., per oral administration.

Source: Lactobacillus casei *strain Shirota — Intestinal Flora and Human Health,* Yakult Central Institute for Microbiological Research, Tokyo, 1999. With permission.

recognized in many countries. With recent technological advances, this field will show even more progress in the twenty-first century.

ACKNOWLEDGMENTS

I summarized all the data on LcS obtained so far on behalf of the Yakult Central Institute for Microbiological Research. All the data cited in this chapter come from Lactobacillus casei *strain Shirota — Intestinal Flora and Human Health,* edited by Yakult Central Institute for Microbiological Research, 1999. I thank all members of the Yakult Institute who gave me data and advice.

REFERENCES

1. Salminen, S. and von Wright, A., Eds., *Lactic Acid Bacteria,* Marcel Dekker, Inc., New York, 1993.

2. Stanier, R.Y., Doudoroff, M., and Adelberg, E.A., *The Microbial World*, Prentice-Hall, Inc., New York, 1957.
3. Tissier, H., *Recherches sur La Flore Intestinale. Normale et Pathologique du Nourrisson.*, Georges Carre et C. Naud, Paris, 1900.
4. Moro, E., Uber die nach Gram farbbaren Bacillen des Sauglingsstuhles, *Wien. Klin. Wochenschr.,* 13, 114–115, 1900.
5. Orla-Jensen, S., *The Lactic Acid Bacteria,* Host, Copenhagen, 1919, pp. 1–196.
6. Skerman, V.B.D., McGowan, V., and Sneath, P.H.A., Approved list of bacterial names, *Int. J. Syst. Bacteriol.*, 30, 225–420, 1980.
7. Mitsuoka, T., Intestinal flora and animal models, in *Proceedings of 2nd Symposium on Intestinal Flora: Intestinal Flora and Cancer,* Part 2, Mitsuoka, T., Ed., Japan Scientific Societies Press, Tokyo, 1995, pp. 3–27.
8. Lactobacillus casei *strain Shirota — Intestinal Flora and Human Health,* Yakult Central Institute for Microbiological Research, Tokyo, 1999.
9. Ryter, A. and Kellenberger, E., Etude au microscope electronique de plasmas contenant de l'acide desoxyribonucleique. I. Les nucleoides des bacteries en croissance active. *Z. Naturforsch.*, 136, 597, 1958.
10. Amako, K., Fine structure of bacteria, in *Electron Micrographs of Bacteria,* Amako, K., Ed., Nanzando, Tokyo, 1983, pp. 16–89.
11. Nagaoka, M., Muto, M., Nomoto, K., Matsuzaki, T., Watanabe, T., and Yokokura, T., Structure of polysaccharide–peptidoglycan complex from the cell wall of *Lactobacillus casei* YIT9018, *J. Biochem.,* 108, 568–571, 1990.
12. Smith, J., Protein surface layers of bacteria, in *Bacterial Outer Membranes as Model Systems,* Inoue, M., Ed., John Wiley and Sons Ltd., New York, 1986, pp. 343–376.
13. Kaneko, T., Morita, H., Abe, T., Uchida, K., Yamamoto, N., and Morishita, T., Nutrition and metabolism of lactic acid bacteria, in *Science and Technology of Lactic Acid Bacteria,* Japanese Scientific Societies Press, Tokyo, 1996, pp. 89–132.
14. Shimizu-Kadota, M., Properties of lactose plasmid pLY101 in *Lactobacillus casei, Appl. Envron. Microbiol.,* 53, 2987–2991, 1991.
15. Yamachita, T., Takahashi, Y., Usami, A, Takahashi, H., and Aso, K., Flavor produced during lactic acid fermentation: part 1. Analysis of flavor factors, *Annual Report of Yakult Central Institute for Microbiological Research,* 2, 37–47, 1971.
16. Chang, I.S., Kim, B.H., and Shin, P.K., Use of sulfite and hydrogen peroxide to control bacterial contamination in ethanol fermentation, *Appl. Environ. Microbiol.,* 63, 1–6, 1997.
17. Morishita, T., Fukuda, T., Shirota, T., and Yura, T., Genetic basis of nutritional requirements in *Lactobacillus casei, J. Bacteriol.,* 120, 1078–1084, 1974.
18. De Man, J.C., Rogosa, M., and Sharpe, M.E., Medium for lactobacilli, *J. Appl. Bacteriol.,* 23, 130–135, 1960.
19. Efthymiou, C. and Hansen, P.A., An antigenic analysis of *Lactobacillus acidophilus, J. Infect. Dis.,* 110, 258–267, 1962.
20. Sansawa, H., in Lactobacillus casei *strain Shirota — Intestinal Flora and Human Health*, Yakult Central Institute for Microbiological Research, Tokyo, 1999, p. 94.
21. Kawami, K., Stability of Yakult beverage during storage, Reports on Yakult Beverage and Yakult Strain, No. 13, Yakult Co., Tokyo, 1980.
22. Shelef, L.A. and Firstenberg-Eden, R., Novel selective and nonselective optical detection of microorganisms, *Lett. Appl. Microbiol.,* 25, 202–206, 1997.
23. Kolars, J.C., Yogurt: an autodigesting source of lactose, *N. Engl. J. Med.,* 310, 1–3, 1984.

24. Mayer, E.A., The physiology of gastric storage and emptying, in *Physiology of the Gastrointestinal Tract,* 3rd ed., Johnson, L.R., Ed., Raven Press, New York, 1994, pp. 929–976.
25. Ohashi, Y. and Umesaki, Y., Kinetics of *L. casei* strain Shirota fermented milk in upper intestinal tract of the rat model, *Dig. Absorp.,* 20, 119–123, 1997.
26. Ohashi, Y., Inoue, R., Tanaka, K., Matsuki, T., Umesaki, Y., and Uchida, K., *Lactobacillus casei* strain Shirota–fermented milk stimulates indigenous *Lactobacilli* in the pig intestine, *J. Nutr. Sci. Vitaminol.,* 47, 172–176, 2001.
27. Yokokura, T., Kato, I., and Mutai, M., Antitumor effect of *Lactobacillus casei* (LC9018), in *Intestinal Flora and Carcinogenesis,* Mitsuoka, T., Ed., Japan Scientific Societies Press, Tokyo, 1981, pp. 125–137.
28. Kato, I., Kobayashi, S., Yokokura, T., and Mutai., M., Antitumor activity of *Lactobacillus casei* in mice, *Gann,* 72, 517–523, 1981.
29. Matsuzaki, T., Yokokura, T., and Mutai, M., Antitumor effect of intrapleural administration of *Lactobacillus casei* in mice, *Cancer Immunol. Immunother.,* 26, 209–214, 1988.
30. Takagi, A., Matsuzaki, T., Sato, M., Nomoto, K., Morotomi, M., and Yokokura, T., Inhibitory effect of oral administration of *Lactobacillus casei* on 3-methylcholanthrene-induced carcinogenesis in mice, *Med. Microbiol. Immunol.,* 188, 111–116, 1999.
31. Matsuzaki, T., Tsukio, K., Yokokura, T., and Mutai, M., Effect of subcutaneous injection of a lactobacilli preparation, LC9018, and antitumor drugs on murine tumor, *Biotherapy,* 2, 440–445, 1988.
32. Matsuzaki, T., Yokokura, T., and Azuma, I., Antimetastatic effect of *Lactobacillus casei* YIT9018 (LC 9018) on a highly metastatic variant of B16 melanoma in C57BL/6J mice, *Cancer Immunol. Immunother.,* 24, 99–105, 1987.
33. Takagi, A., Matsuzaki, T., Sato, M., Nomoto, K., Morotomi, M., and Yokokura, T., Enhancement of natural killer cytotoxicity delayed murine carcinogenesis by a probiotic microorganism, *Carcinogenesis,* 22, 599–605, 2001.
34. Yasutake, N., Kato, I., Ohwaki, M., Yokokura, T., and Mutai, M., Host-mediated antitumor activity of *Lactobacillus casei* in mice, *Gann,* 75, 72–80, 1984.
35. Yasutake, N., Ohwaki, M., Mutai, M., Koide, Y., and Yoshida, T., Anti-tumor effect of humoral and cellular immunities mediated by a bacterial immunopotentiator, *Lactobacillus casei,* in mice, *Cancer Immunol. Immunother.,* 20, 109–116, 1985.
36. Matsuzaki, T., Yokokura, T., and Azuma, I., Antitumor activity of *Lactobacillus casei* on Lewis lung carcinoma and line-10 hepatoma in syngeneic mice and guinea pigs, *Cancer Immunol. Immunother.,* 20, 18–22, 1985.
37. Matsuzaki, T., Shimizu, Y., and Yokokura, T., Augmentation of antimetastatic effect on Lewis lung carcinoma (3LL) in C57BL/6 mice by priming with *Lactobacillus casei, Med. Microbiol. Immunol.,* 179, 161–168, 1990.
38. Miake, S., Nomoto, K., Yokokura, T., Yoshikai, Y., Mutai, M., and Nomoto, K., Protective effect of *Lactobacillus casei* on *Pseudomonas aeruginosa* infection in mice, *Infect. Immun.,* 48, 480–485, 1985.
39. Dalton, C.B., Austin, C.C., Sobel, J., Hayes, P.S., Bibb, W.F., Graves, L.M., Swaminathan, B., Proctor, M.E., and Griffin, P.M., An outbreak of gastroenteritis and fever due to *Listeria monocytogenes* in milk, *N. Engl. J. Med.,* 336, 100–105, 1997.
40. Nomoto, K., Miake, S., Hashimoto, S., Yokokura, T., Mutai, M., Yoshikai, Y., and Nomoto, K., Augmentation of host resistance to *Listeria monocytogenes* infection by *Lactobacillus casei, J. Clin. Lab. Immunol.,* 17, 91–97, 1985.

41. Furushiro, M., Hashimoto, S., Hamura, M., and Yokokura, T., Mechanism of the antihypertensive effect of a polysaccharide–glycopeptide complex from *Lactobacillus casei* in spontaneously hypertensive rats (SHR), *Biosci. Biotech. Biochem.,* 57, 978–981, 1993.
42. Furushiro, M., Sawada, S., Hirai, K., Motoike, M., Sansawa, H., Kobayashi, S., Watanuki, M., and Yokokura, T., Blood pressure-lowering effect of extract from *Lactobacillus casei* in spontaneously hypertensive rats (SHR), *Agric. Biol. Chem.,* 54, 2193–2198, 1990.
43. Shida, K., Makino, K., Morishita, A., Takamizawa, K., Hachimura, S., Ametani, A., Sato, T., Kumagai, Y., Habu, S., and Kaminogawa, S., *Lactobacillus casei* inhibits antigen-induced IgE secretion through regulation of cytokine production in murine splenocytes cultures, *Int. Arch. Allergy Immunol.,* 115, 278–287, 1998.
44. Matsuzaki, T., Yamazaki, R., Hashimoto, S., and Yokokura, T., Effect of oral feeding of *Lactobacillus casei* strain Shirota on immunoglobulin E production in mice, *J. Dairy Sci.,* 81, 48–53, 1988.
45. Matsuzaki, T., Nagata, Y., Kado, S., Uchida, K., Kato, I., Hashimoto, S., and Yokokura, T., Prevention of onset in an insulin-dependent diabetes mellitus model, NOD mice, by oral feeding of *Lactobacillus casei, APMIS,* 105, 643–649, 1997.
46. Mike, A., Nagaoka, N., Tagami, Y., Miyashita, M., Shimada, S., Uchida, K., Nanno, M., and Ohwaki, M., Prevention of B220^{+} T cell expansion and prolongation of lifespan induced by *Lactobacillus casei* in MRL/lpr mice, *Clin. Exp. Immunol.,* 117, 368–375, 1999.
47. Kato, I., Endo-Tanaka, K., and Yokokura, T., Suppressive effects of the oral administration of *Lactobacillus casei* on type II collagen-induced arthritis in DBA/1 mice, *Life Sci.,* 63, 635–644, 1998.
48. Aritaki, S. and Ishikawa, S., Application of milk fermented by *L. acidophilus* in the pediatric field, *J. Jpn. Pediatr. Soc.,* 66, 125–130, 1962.
49. Yamagishi, T., Serikawa, T., Morita, R., Takahashi, K., and Nishida, S., Effect of a *Lactobacillus* product administration on the anaerobic intestinal flora of aged adults, *Jpn. J. Microbiol.,* 18, 211–216, 1974.
50. Shirota, M., Aso, K., and Iwabuchi, A., Studies on microflora of the human intestine. I. Alteration of the constitution of intestinal flora by oral administration of *L. casei* strain Shirota to healthy infants, *Jpn. J. Bacteriol.,* 21, 274–283, 1966.
51. Hanada, K., Kondo, F., Kitahara, R., Tokushige, H., Ueno, T., Kutani, T., and Ochiai, K., Effect of daily administration of *Lactobacillus* fermented milk on children, *J. Kobe Inst. Health,* 55–60, 1956.
52. Tanaka, R., Tohyama, K., Morotomi, M., Takayama, H., Nanno, M., Kuroshima, T., and Mutai, M., Effect on the fecal flora and urinary metabolites of the administration of *Lactobacillus casei* and *Bifidobacterium breve,* in *Proceeding of I RIKEN Symposium on Intestinal Flora: Intestinal Flora and Cancer,* Mitsuoka, T., Ed., Japan Scientific Societies Press, Tokyo, 1981, pp. 79–103.
53. Tanaka, R. and Ohwaki, M., A controlled study of the effects of the ingestion of *Lactobacillus casei*–fermented milk on the intestinal microflora, its microbial metabolism and the immune system of healthy humans, in *Proceeding of XII RIKEN Symposium on Intestinal Flora: Intestinal Flora and Diet,* Mitsuoka, T., Ed., Japan Scientific Societies Press, Tokyo, 1993, pp. 85–104.
54. Spanhaak, S., Havenaar, R., and Schaafsma, G., The effect of consumption of milk fermented by *Lactobacillus casei* strain Shirota on the intestinal microflora and immune parameters in humans, *Eur. J. Clin. Nutr.,* 52, 899–907, 1998.

55. Winkelstein, A., *Lactobacillus acidophilus* tablets in the therapy of various intestinal disorders: a preliminary report, *Am. Proct. Dig. Treat.,* 1, 1637–1639, 1955.
56. Kawamura, T., Ohnuki, K., and Ichida, F., Clinical study of BLG-01 containing *Lactobacillus casei;* Effect on irregular bowel movements and abdominal discomfort, *Jpn. Pharmacol. Ther.,* 9, 4361–4370, 1981.
57. Tohyama, K., Kobayashi, S., Kan, T., Yazawa, K., Terashima, T., and Mutai, M., Effect of lactobacilli on urinary indican excretion in gnotobiotic rats and in man, *Microbiol. Immunol.,* 25, 101–112, 1981.
58. Okawa, T., Kita, M., Arai, T., Iida, K., Dokiya, T., Takegawa, Y., Hirokawa, Y., Yamazaki, K., and Hashimoto, S., Phase II randomized clinical trial of LC9018 concurrently used with radiation in the treatment of carcinoma of the uterine cervix. Its effect on tumor reduction and histology, *Cancer,* 64, 1769–1776, 1989.
59. Masuno, T., Kishimoto, S., Ogura, T., Honma, T., Niitani, H., Fukuoka, M., and Ogawa, A., Comparative trial of LC 9018 plus doxorubicin and doxorubicin alone for the treatment of malignant pleural effusion secondary to lung cancer, *Cancer,* 68, 1495–1500, 1991.
60. Okawa, T., Niibe, H., Arai, T., Sekiba, K., Noda, K., Takeuchi, S., Hashimoto, S., and Ogawa, N., Effect of LC 9018 combined with radiation therapy on carcinoma of the uterine cervix, *Cancer,* 72, 1949–1954, 1993.
61. Aso, Y., Akaza, H., and the BLP Study Group, Prophylactic effect of a *Lactobacillus casei* preparation on the recurrence of superficial bladder cancer, *Urol. Int.,* 49, 125–129, 1992.
62. Aso, Y., Akaza, H., Tsukamoto, T., Imai, K., Naito, S., and the BLP Study Group, Preventive effect of a *Lactobacillus casei* preparation on the recurrence of superficial bladder cancer in a double-blind trial, *Eur. Urol.,* 27, 104–109, 1995.
63. Sawamura, A., Yamaguchi, Y., Tohge, T., Nagata, N., Ikeda, H., Nakanishi, K., and Asakura, S., Enhancement of immuno-activities by oral administration of *Lactobacillus casei* in colorectal cancer patients, *Biotherapy,* 8, 1567–1572, 1994.
64. Nagao, F., Nakayama, M., Muto, T., and Okumura, K., Effects of fermented milk drink containing *Lactobacillus casei* strain Shirota on the immune system in healthy human subjects, *Biosci. Biotechnol. Biochem.,* 64, 2706–2708, 2000.
65. Imai, K., Matsuyama, S., Miyake, S., Suga, K., and Nakachi, K., Natural cytotoxic activity of peripheral-blood lymphocytes and cancer incidence: an 11-year follow-up study of a general population, *Lancet,* 356, 1795–1799, 2000.

7 Biologically Active Peptides Released in Fermented Milk: Role and Functions

Chantal Matar, Jean Guy LeBlanc, Luc Martin, and Gabriela Perdigón

CONTENTS

7.1 INTRODUCTION

Recent years have seen a growing interest in the health enhancement effects of lactic acid bacteria (LAB).[1] More and more evidence indicates that probiotics — foods that contain live bacteria — might constitute a valuable therapeutic and preventive

0-8493-1372-4/03/$0.00+$1.50

tool against a number of diseases in humans and animals. Probiotics can confer on their hosts multiple beneficial effects including prevention and treatment of diarrhea, induction of protective immunity against pathogens and tumors, prevention of allergies, control of inflammatory diseases, modulation of gastrointestinal functions, and alleviation of lactose intolerance and hypertension.[2] Numerous health and functional attributes of fermented dairy products are ascribed to the microorganisms that induce physical and/or chemical modifications of milk components. The mechanism responsible for health benefits of LAB is multifactorial and probably relates to the complex interaction between milk components, LAB and their constituents, and the intestinal mucosa.

Recent studies have reported the particular role of the metabolic products derived from milk fermentation by probiotics. An important metabolic activity that occurs during milk fermentation is proteolysis. Proteolysis ensures LAB growth in the medium and consequently might influence the potential release of physiologically active peptides. These peptides are encrypted in the milk protein sequences in a latent state and might be released during food processing or after degradation by digestive enzymes. They have been shown to possess opiate, antithrombotic, antihypertensive, immunomodulating, antibacterial, antigastric, and mineral carrier properties. Hence, milk-related bioactive peptides and other derived metabolites from fermentation might play an important role in health enhancement and reducing the incidence of many diseases. Physiologically active peptides might particularly contribute to the phenomenom of probiosis due to their hormone-like activities. Many of these sequences can be found in the milk of a large number of mammalian species. Some multifunctional bioactive sequences present in overlapping sequence, in the β-caseins for example, are endowed with multiple physiological activities (morphinomimetic and immunomodulating). Bioactive sequences are also found in plant and animal proteins as immunomodulating peptides in soybean[3,4] and rice albumin.[5]

Several peptides derived from milk proteins that have effects on behavioral, neurological, physiological, and vasoregulatory responses have been identified. Table 7.1 lists the major peptides identified in cows' milk and their physiological activities.

This chapter will review the occurrence of biologically active peptides in fermented milk and their functional and potential physiological activities, specifically opioid, antihypertensive, and immunomodulating activities.

7.2 RELEASE OF BIOLOGICALLY ACTIVE PEPTIDES PRODUCED DURING FERMENTATION

The amino acids present in milk have limited nutritional support for LAB growth, and therefore LAB rely on a complex proteolytic system to ensure optimal growth in milk. The proteolytic system is composed of proteinases, peptidases, and transport systems. These LAB cells possess:

1. Proteinases located in the microbial cell envelope that permit the degradation of caseins into oligopeptides

TABLE 7.1
Bioactive Peptides Derived from Milk Proteins

Name	Protein Precursor	Physiological Activity
Casomorphins	α-, β-Casein	Opioid agonist
α-Lactorphin	α-Lactalbumin	Opioid agonist
β-Lactorphin	β-Lactoglobulin	Opioid agonist
Lactoferroxins	Lactoferrin	Opioid antagonist
Casoxins	κ-Casein	Opioid antagonist
Casokinins	α-, β-Casein	ACE inhibitor
Lactokinins	β-Lactalbumin, β-lactoglobulin, serum albumin	ACE inhibitor
Antihypertensive peptides	β-casein	Antihypertensive
Immunopeptides	α-, β-Casein	Immunomodulator
Lactoferricin	Lactoferrin	Antimicrobial
Casocidin	α_{S2}-Casein	Antimicrobial
Isracidin	α_{S1}-Casein	Antimicrobial
Casoplatelins	κ-Casein	Antithrombotic
Phosphopeptides	α-, β-Casein	Mineral binding

Source: Adapted from Meisel, H. and Bockelmann, W., *Antonie Van Leeuwenhoek,* 76, 207–215, 1999; Clare, D.A. and Swaisgood, H.E., *J. Dairy Sci.,* 83, 1187–1195, 2000. With permission.

2. Peptide transport systems that allow the internalization of the released oligopeptides
3. Intracellular peptidases that hydrolyze the oligopeptides into peptides or into amino acids to be used by the cell.[8–11]

The first step in milk protein breakdown involves proteinases, which are responsible for the release of peptides from caseins. Then, the peptides released are hydrolyzed by peptidases to release amino acids and small peptides, which are taken up in the cell by transport systems.

7.2.1 Proteinases

Most of the proteinases are located in the cell envelope. Lactococcal proteinases, which have been extensively studied (for a review, see Law and Handrikmann[12]), are very large proteins with a molecular weight of 140,000 kDa and pH optima of 5.5 to 6.5. They are classified as serine proteinase inhibitors and have a conserved active site triad consisting of aspartic acid, histidine, and serine. Proteinases can be classified as PI, when β-casein is the principal protein degraded by the enzyme, or PIII, when a different specificity toward the β-casein is shown in addition to the α_{S1}-casein hydrolysis.[13] When purified, these enzymes can strongly degrade caseins and release a large number of oligopeptides. Approximately 20% of the oligopeptides released are small enough to be taken up by the transport system of *Lactococcus lactis.*[14]

Proteinase genes were found to be encoded on a plasmid. Molecular cloning of proteinase genes in *Lactococcus* and its transfer into a Prt– *L. lactis* ssp. *lactis* strain resulted in its conversion to the Prt+ phenotype.[15] Proteinase genes prtP and prtM encode respectively a 200 kDa protein and a 32 kDa lipoprotein. The latter is implicated in the conversion of the inactive proteinase to its active form during or after translocation across the membrane.[16,17] The prtP gene codes for a preproteinase with a signal peptide at the N-terminus and a membrane anchoring sequence at the C-terminus. The C-terminus of the protein consists of membrane spanning domains with a hydrophobic α-helix. Many basic amino acids constitute the C-terminal region of the helix, and it is believed that their role is related to the interaction with charged phosphate groups of the phospholipid bilayer acting then as a stop-transfer signal.[12,18] *Lactobacillus helveticus* and *Lb. delbrueckii* ssp. *bulgaricus* were shown to possess stronger enzymatic activity than lactoccocci. The activity of a cell wall proteinase from *Lb. helveticus* L89 was studied by Martín-Hernandéz et al.[19] Foucaud and Juillard[20] demonstrated that a high peptide content is noticed after a 24 h of fermentation of milk. At the end of growth, proteinase activity still occurs at low pH, but peptides cease to be taken up by the cell because the transport system does not operate at low pH.

Proteinase activity, which differs markedly among species of LAB, may also have a major impact on the peptide pattern. Each probiotic enzyme possesses a unique enzymogram, which differs from one species to another as well as from one strain to another within the same species, contributing to the release of a different pool of potentially bioactive peptides. Thus, proteinases, by their specificities and activities, play a primary role in the generation of the peptide pool in fermented milk and potentially the release of bioactive peptides.

7.2.2 Peptidases

Lactic acid bacteria possess a large spectrum of proteolytic enzymes including endopeptidases, aminopeptidases, tripeptidases, and dipeptidases. The biochemical characterization of these peptidases reflects their intracellular location. An endopeptidase PepO was isolated by Tan et al.[21] The enzyme is an intracellular metalloproteinase, capable of hydrolyzing oligopeptides such as bradykinin. A specific endopeptidase pepX degrades oligopeptides by releasing dipeptides containing proline.[22] Aminopeptidases act on oligopeptides released by a proteinase by cleaving a single N-terminal amino acid residue. Aminopeptidases designated PepN and PepC were found to possess a broad specificity toward peptide bonds. Others, such as the PepA that releases the amino-terminal glutamate residue, are less specific.

Dipeptides and tripeptides can also be cleaved by specific peptidases. Some of these enzymes have a broad specificity, cleaving all dipeptides and tripeptides except those containing proline. The presence of proline-specific peptidases is necessary for optimal growth of LAB because of their ability to degrade proline-rich oligopeptides from caseins. Proline aminopeptidases PIP and PepI, specific for di- and tripeptides with N-terminal proline residues, were isolated from two strains of LAB.[23,24]

7.2.3 Peptide Transport

The oligopeptides and small peptides are taken up by the cell by a dependent oligopeptide transport system (Opp) and a di- and tripeptide transport system (DtpT) and DtpP. The Opp system is comprised of two ATP-binding proteins, OppD and OppF, two integral membrane proteins, OppB and OppC, and a substrate binding protein, OppA.[25] Only a limited number of peptides resulting from the proteinase action are transported by the oligopeptide transport system. Foucaud and Juillard[20] noted that <25% of casein-derived peptides were used to sustain LAB growth. Most of these oligopeptides do not promote growth of LAB because they are not transported into the cell. In addition, the uptake of small peptides via Opp is limited by the size of the peptides. The transport system of LAB enables them to internalize oligopeptides up to 18 amino acids in length, although only peptides smaller than 11 amino acid residues were found to be transported.[10] Thus, the nontransported peptides left in the medium constitute an important pool of potential biologically active peptides.

7.2.4 Biopeptides from Fermented Milk and Cheese

Cell wall-bound proteinases of LAB, as well as enzymes from the endogenous microflora of milk, including digestive enzymes, have the capacity to release bioactive peptides from within the sequences of milk proteins. However, microbial proteolysis is highly specific, leading to the release of very potent bioactive peptides. Microbial proteolysis also has an impact on the hydrolytic pattern of milk protein prior to the action of digestive enzymes. It was shown from high-pressure liquid chromatography (HPLC) analysis of the gastrointestinal digests of human subjects that the release of peptides is greater after ingestion of yogurt than with nonfermented milk.[26] The peptidic profile of milk proteins is significantly different after microbial fermentation (Figure 7.1), suggesting that microbial proteolysis can be a potential source of bioactive peptides.[27] The degradation of milk proteins produced by proteolysis with *Lb. helveticus* was confirmed by size-exclusion HPLC. The fermentation of milk by this LAB increased the amount of smaller weight protein-derived compounds from the degradation of milk proteins. This is shown by the appearance of peaks caused by the proteolytic system of the bacteria after milk fermentation (Figure 7.1b) that were not present before fermentation (Figure 7.1a). In addition to the release of biologically active peptides after fermentation, microbial proteolysis may also expose the inner protein bonds, favoring the action of digestive enzymes and the release of potentially active peptides. α_{S1}-casein and β-casein are major proteins found in large quantities in milk; they are very susceptible to proteolysis. With 199 and 209 amino acid residues in their sequences, respectively, these proteins have the capacities to release more than 20,000 different peptides.[28]

It has been shown that milk fermented by *Lb. helveticus* R389, a bacterium that has strong protease and peptidase activities compared to other lactic acid bacteria,[28] is capable of exerting an antimutagenic effect, while milk fermented by its protease-deficient derivative is not.[29] Other studies have shown that a proteinase from *Lb. helveticus* CP790 was able to release an antihypertensive peptide from casein hydrolysates.[30,31] Bioactive sequences from whey-soluble proteins might also be released, after fermentation, by *Kluyveromyces marxianus* var. *marxianus*[32] and from milk fermented by *Lb.* GG and digested by pepsin and trypsin.[33]

Intense proteolysis during cheese ripening was found to affect the presence of various bioactive peptides in a variety of cheeses (for a review, see Smacchi and

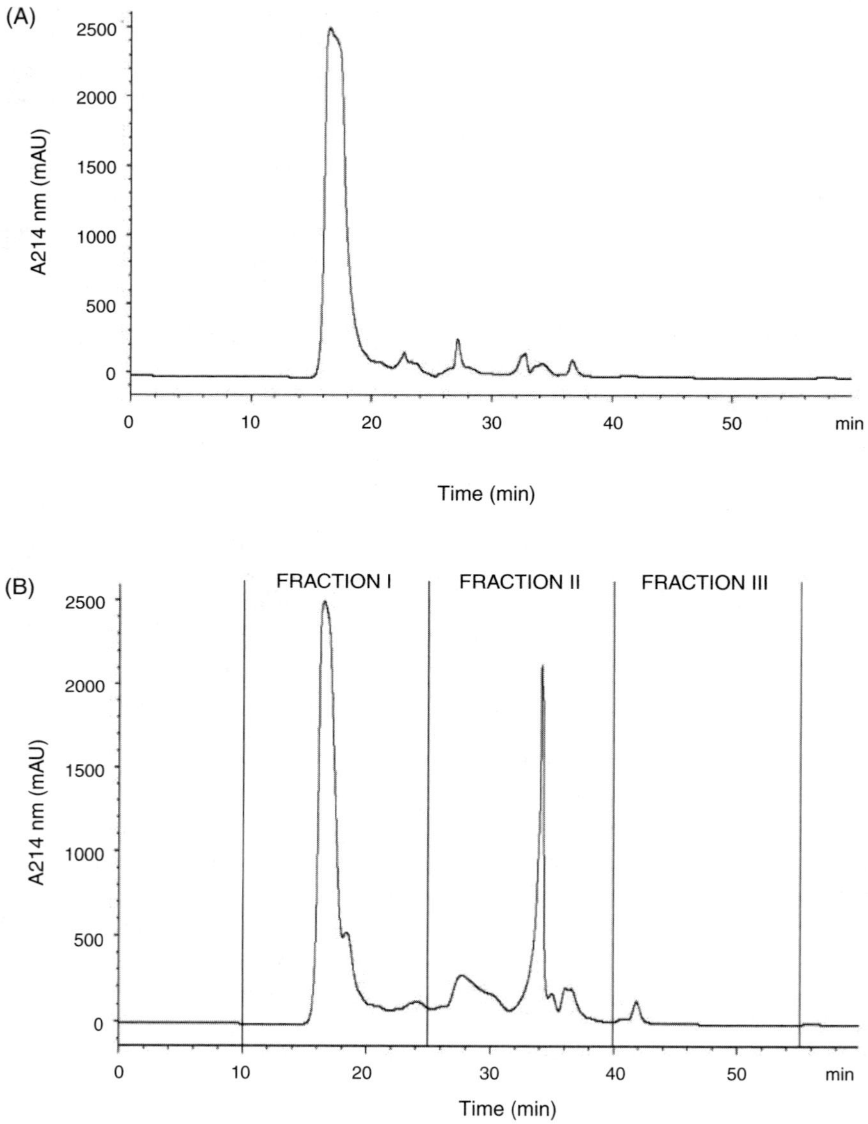

FIGURE 7.1 Size-exclusion HPLC profile of unfermented milk (a) and milk fermented with *Lactobacillus helveticus* R389 for 24 h (b). Samples were passed on a LKB TSK-G2000SW gel filtration column (600 × 7.5 mm TosoHaas, United States). The flow rate was 0.7 ml/min, and the eluted proteins were monitored at 214, 220, 224, and 280 nm using a HP1100 diode array detector. Fractions were collected with a Gilson FC104 fraction collector (Gilson, U.S.), then pooled and concentrated using an Automatic Environmental SpeedVac® System (AES1010, Savant, United States) and stored at 4°C until their use during *in vivo* studies. (From Matar, C., Nadathur, S.S., Bakalinsky, A.T., and Goulet, J., *J. Dairy Sci.,* 80, 1965–1970, 1997. With permission.)

Gobbetti[34]). These peptides are found after a short or medium ripening period. Casein phosphopeptides were produced during the ripening of Comté and Grana Padano cheeses.[35,36] Angiotensin-I converting enzyme (ACE) inhibiting peptides have also been isolated from several Italian cheeses.[39] The antihypertensive peptides were also found in medium aged Gouda[38] and in Parmesan cheese.[38] It seems that a long ripening period for cheese could inactivate these peptides due to further proteolysis. Smacchi and Gobbetti[34] have reported that bioactive peptides might selectively inhibit proteolytic enzymes of LAB as well as the dairy spoilage enzymes in cheese, since those peptides, when exhibiting their physiological activities (immunostimulatory, antithrombotic, and antihypertensive), utilize a common mechanism based on the inhibition of the target proteolytic enzymes.[34] The authors postulated that a similar mechanism may influence the food ecosystem. Bioactive peptides with an inhibitory effect on intracellular peptidases of LAB have been isolated from different types of cheese.[37,40,41]

7.3 EFFECT OF IMMUNOPEPTIDES ON MUCOSAL IMMUNITY

Intake of milk fermented by LAB has led to a significant increase in various immune responses, such as sIgA-producing cells, macrophage activity, and specific antibody responses during infections.[42–45] The stimulation of the immune system by LAB depends on the survival of the bacteria in the gastrointestinal tract, resistance to gastric acid, and ability to adhere to the mucosal surface.[46] However, studies have shown that components of fermented milks other than bacteria also contribute to the immunostimulating effects. Specifically:

- The dialysate and anion exchange fraction of yogurt showed significant inhibitory action against tumors in mice *in vivo*.[47]
- The soluble compounds produced by LAB during milk fermentation can be used to prevent gastrointestinal disorders and cancer.[48]
- The supernatant of fermented milk cultured with *Lb. casei* and *Lb. acidophilus* increased the immune response independent of the presence of lactobacilli.[49]
- Filtered yogurt increases interferon (IFN)-γ production and natural killer (NK) activity of human peripheral blood lymphocytes.[50]

Ng and Griffiths[51] suggested that soluble substances released during fermentation by proteolytic LAB might influence the activity of macrophages. They studied the macrophage cytokine release by cell-free fractions of fermented milk in an *in vitro* model. *Lb. helveticus*-fermented milk and its cell-free fractions enhanced the cytokine release by macrophages. The cell-free supernatant from LAB-fermented milk along with LPS (lipopolysaccharides) demonstrated an effect on interleukin (IL)-6 production.[51] Peptides released during the fermentation of κ-casein by *Lb.* GG, followed by pepsin/trypsin treatment, were found to enhance the mitogen responsiveness of *in vitro* cell cultures of human lymphocytes.[52] β-Casein permeate medium fermented with *Lb. helveticus* is able to affect the lymphocyte proliferation *in vitro* of human peripheral blood lymphocytes.[53]

7.3.1 Immunomodulatory Peptides from Caseins

Many studies have reported the presence of immunomodulatory sequences within milk proteins. Several whey and casein-derived peptides, as well as peptides derived from lactoferrin, may play a role in the modulation of the immune system. Bovine κ-caseinoglycopeptide (CGP) is obtained by the chymosin digestion of κ-casein. CGP has been shown to down-regulate the immune system by suppressing the proliferation of murine splenic lymphocytes *in vitro* by lipopolysaccharides or phytohemagglutinin.[54] The dipeptide Tyr-Gly from κ-casein increases cellular proliferation of human peripheral blood lymphocytes activated with concanavalin A.[55,56] Due to its small size, this peptide can, in principle, pass across the intestine in quantitatively significant amounts to reach local lymphocytes.[57] κ-Casein treated with pepsin/trypsin upregulated IL-4 and IFN-γ production, whereas *Lb.* GG–degraded casein downregulated IL-4 production with no effect on IFN-γ.[58] A hexapeptide isolated from the trypsin digest of human casein, Val-Glu-Pro-Ile-Pro-Tyr, when intravenously injected into mice, improved resistance to *Klebsiella pneumoniae. In vitro,* this hexapeptide also stimulated the phagocytosis of sheep red blood cells by peritoneal macrophages of mice.[59]

Enzymatic digests of αs1-casein inhibit the proliferation responses of murine splenic lymphocytes and rabbit Peyer's patch cells[60] and cause the release of the peptide Thr-Thr-Met-Pro-Leu-Tyr. This peptide has been shown to promote antibody formation and phagocytosis *in vitro* and reduce *K. pneumoniae* infection in mice.[61,62] Peptide fragments from β-casein (PGPIPM and LLY) have been demonstrated to stimulate phagocytosis of sheep red blood cells by peritoneal macrophages and protect against *K. pneumoniae* infections, and the LLY peptide also enhances antigen-dependent T cell proliferation.[63] Depending on the concentration, hydrolysis of β-casein by pepsin/trypsin either stimulated or inhibited the proliferation of human peripheral blood lymphocytes; however, pepsin/chymosin digests directly stimulated the proliferation of rat lymphocytes *in vitro*.[56] Pancreatin/trypsin digests of β-casein inhibit murine splenic lymphocyte proliferation and Peyer's patch cells *in vitro*.[60] Coste et al.[64] have obtained evidence that the peptide 193–209 from bovine β-casein can enhance rat lymphocyte proliferation.

7.3.2 Immunomodulatory Peptides from Minor Proteins in Milk

Not only casein-derived peptides exert immunomodulating activities. Many studies have shown that peptides derived from whey or lactoferrin also exhibit important physiological activities. A pepsin-generated hydrolysate of lactoferrin has been shown to contain immunostimulating peptides, which can enhance the proliferation of spleen cells[65] and stimulate the phagocytic activity of human neutrophils.[66] Peptides obtained by tryptic hydrolysis of bovine α-lactoglobulin have been shown to induce oral tolerance in mice.[67] Hydrolyzed α-lactalbumin enhances murine humoral responses to sheep and human red blood cells caused by the modulation of both B lymphocyte and T helper cell activities.[68] Commercial hydrolyzed α-lactalbumin stimulates B lymphocytes in the absence of T cell cooperation due to an enhanced immune response to the T-cell-independent antigen TNP-Ficoll.[69] Bovine lactoferricin B, a peptide released from the hydrolysis of lactoferrin, has been found to suppress IL-6 production by human monocytic cells in response to LPS stimulation.[70] Lactoferricin has also been shown to be able to stimulate the release of neutrophil-activating chemokine IL-8 from human polymorphonuclear leucocytes.[71]

Bioactive peptides might exert an indirect effect on the immune system. Opioid peptides, such as β-endorphins, enhance lymphocyte proliferation, NK activity and neutrophil locomotion.[72,73] These effects could be explained by the presence of opioid μ receptors on T lymphocytes and human phogocytic leucocytes. β-casokinin inhibits ACE enzymes, causing a decrease of blood pressure and aldosterone and acts on bradykinin, a hormone with immune enhancing effects. Bradykinin is able to stimulate macrophages, enhance lymphocyte migration, and increase secretion of lymphokines.[74,75] Thus, this chain of events indirectly produces an overall immunostimulating response.[7]

7.4 EFFECT OF MILK PEPTIDES ON TUMOR GROWTH

Lactic acid bacteria, through the mechanism of fermentation, may release compounds that react with the immune system parameters and induce protective immunity against infections and some tumors.[76] *In vivo* experiments showed that potentiation of the immune system and inhibition of tumor development is likely related to the proteolysis that occurs during milk fermentation.[77,78] Milk fermented with *Lb. helveticus* R389 increased the number of IgA+ B cells in the small intestine and bronchial tissues (Table 7.2). Increases in the levels of secretory IgA and activation of the B cells to enter the IgA cycle were not noticeable when milk was fermented by the nonproteolytic variant of this strain (Table 7.3).[77]

Perdigon et al.[79] suggested that the increased numbers of cells secreting IgA in the large intestine of mice given yogurt could contribute to decreasing the tissue-damaging consequences of a permanent inflammatory response, which occurs during the development of tumors and neoplasia. IgA is considered to be an immune barrier in colonic neoplasia. The increase of the mucosal immunity and the enhancement of cellular mobilization of IgA+ cells are properties that can be exploited clinically to prevent infections and development of some tumors in the mucosal network. Those results could be therapeutically used to redirect the immunologic memory and favor a response other than the T helper subset 2 cell, which could prevent an allergic inflammatory disease.[80] *Lb. helveticus,* in our laboratory, was able to hydrolyze milk proteins and cause the release of peptides, as was shown by the level of proteolysis and the HPLC elution pattern of milk after fermentation (Figure 7.1). This same fermented milk increased the phagocytic index of peritoneal macrophages in mice, which was directly correlated with a regression of fibrosarcoma.[77] Further *in vivo* studies designed to analyze the particular effect of the different peptidic fractions in milk after fermention by *Lb. helveticus* have led to the identification of a bioactive fraction. This fraction increased the number of IgA secreting cells in the GALT (gut-associated lymphoid tissue) and caused the regression of subcutaneously implanted fibrosarcomas, an immunodependent tumor (Figure 7.2).[78] These studies

TABLE 7.2
Effect of Oral Administration of Milk Fermented (12 h) by *Lb. helveticus* Wild Type on IgA Cell Numbers in the Intestine and the Bronchial-Associated Lymphoid Tissues in Mice

	Number of IgA+ B Cells			
	Intestine		Bronchus	
Feeding Period	Nonfermented Milk[a]	Milk Fermented by *Lb. helveticus* Wild Type	Nonfermented Milk[a]	Milk Fermented by *Lb. helveticus* Wild Type
3 days	80 ± 5.0	180[b] ± 3.5	18 ± 4.0	57[b] ± 5.2
5 days	86 ± 3.2	141[b] ± 1.3	20 ± 3.4	38[b] ± 5.2
7 days	85 ± 2.5	81 ± 3.5	19 ± 1.3	27[c] ± 3.2

Note: IgA^+ B cells were measured by an immunofluorescence test using a monospecific antibody after 3, 5, and 7 days of feeding. Values are means for $n = 4$ ± standard deviation.

[a] Controls were animals given uninoculated milk.
[b] Significantly different from the corresponding value for controls, $P < 0.01$.
[c] Significantly different from the corresponding value for controls, $P < 0.05$.

Source: Matar, C., Valdez, J.C., Medina, M., Rachid, M., and Perdigon, G., *J. Dairy Res.,* 68, 601–609, 2001. With permission.

TABLE 7.3
Effect of Oral Administration of Milk Fermented (12 h) by *Lb. helveticus* Protease (–) (a Nonproteolytic Variant) on IgA Cell Numbers in the Intestine and the Bronchial-Associated Lymphoid Tissues in Mice

	Number of IgA Cells			
	Intestine		Bronchus	
Feeding Period	Nonfermented Milk (+ 0.4% Yeast Extract)[a]	Milk (+ 0.4% Yeast Extract) Fermented by *Lb. helveticus* Protease (–)	Nonfermented Milk (+ 0.4% Yeast Extract)[a]	Milk (+ 0.4% Yeast Extract) Fermented by *Lb. helveticus* Protease (–)
3 days	76.0 ± 3.0	71 ± 6.0	25.3 ± 3.1	23 ± 6.0
5 days	92.5 ± 1.0	81 ± 1.4	22.3 ± 2.0	15 ± 1.15
7 days	75.5 ± 1.5	80.5 ± 5.0	22.66 ± 2.2	18 ± 6.0

Note: IgA^+ B cells were measured by an immunofluorescence test using a monospecific antibody after 3, 5, and 7 days of feeding. Values are means for $n = 4$ ± standard deviation.

[a] Controls were animals given uninoculated milk supplemented with 0.4% yeast extract. Addition of yeast extract allowed comparable growth in wild type and protease (–) strains.

Source: Matar, C., Goulet, J., Bernier, R.L., and Brochu, E., in *Probiotics 3: Immunomodulation by the Gut Microflora and Probiotics,* Fuller, R., Ed., Kluwer Academic Publishers, Dordrecht, The Netherlands, 2000. With permission.

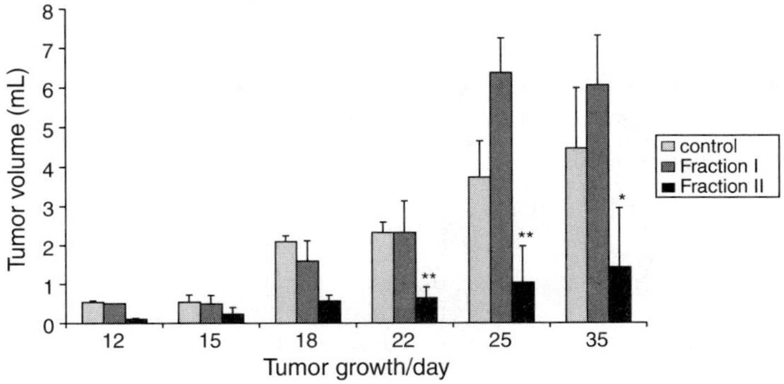

FIGURE 7.2 Effect of previous 7 days feeding of mice with fractions I or II on fibrosarcoma volume (ml). The volume and the growth of the methylcholanthrene-induced fibrosarcomas were recorded on days 12, 15, 18, 22, 25, and 35. Mice were given 50 µg/day of milk protein isolated from fermented milk after size exclusion separation on HPLC. Fraction I is the amount of proteins separated on HPLC during the first 25 minutes, and Fraction II is the amount of proteins separated under the same conditions after 25 minutes. Chromatography conditions are explained in the caption to Figure 7.1. Fractions I and II were given to mice 7 days prior to the inoculation of 5×10^5 tumor cells. Values are means for $n = 5 \pm$ SD. Values were significantly different from the corresponding values for control, ** $P < 0.01$ and * $P < 0.05$. (From Matar, C., Valdez, J.C., Medina, M., Rachid, M., and Perdigon, G., *J. Dairy Res.,* 68, 601–609, 2001. With permission.)

thus confirm that peptides released by bacterial proteolysis might have important implications in modulation of the host's immune response and have an impact on tumor growth.

7.5 ANTIHYPERTENSIVE PEPTIDES

Antihypertensive agents can be classified in five different categories depending on their modes of action. They are categorized as diuretics, sympatholytic agents, angiotensin conversion enzyme (ACE) inhibitors, calcium channel blockers, and arterial vasodilators. Antihypertensive peptides are considered, by their mode of action, to be ACE inhibitors. Antihypertensive peptides are one class of bioactive peptides having an inhibitory effect on the ACE.

ACE, found in lungs, blood vessels, and mucosa wall, is an important component of the renin–angiotensin system, implicated in the regulation of arterial tension, blood volume, and the balance of electrolytes. Renin, an aspartic proteinase of the renin–angiotensin system (RAS), hydrolyzes angiotensinogen, releasing angiotensin I, a decapeptide. Then, ACE hydrolyzes angiotensin I to angiotensin II, an octapeptide, by removing His-Leu from its C-terminal. Angiotensin II is a vasoconstrictor — an inhibitor of bradykinin (a vasodilator). This hormone increases the production of aldosterone, which affects ion retention and the excretion of fluids and therefore causes hypertension. Angiotensin II directly influences blood pressure. Consequently, any peptide that has an inhibitory action on ACE can be considered an antihypertensive peptide.

TABLE 7.4
Some Bioactive Peptides with ACE Inhibition Activity

Peptide	Source	Preparation	IC_{50} (μ*M*)[a]	Ref.
FFVAPFPEVFGK	α_{S1}-Casein	Trypsin	77	101
FFVAP	α_{S1}-Casein	Proteinases	6	74
TTMPLW	α_{S1}-Casein	Trypsin	16	102
PLW	α_{S1}-Casein	Synthesis	36	102
LW	α_{S1}-Casein	Synthesis	50	102
VAP	α_{S1}-Casein	Synthesis	2	74
FVAP	α_{S1}-Casein	Synthesis	10	74
AYFYPE	α_{S1}-Casein	*Lb. helveticus* proteinase	106	31
KYPVQPFTESQSLTL	β-Casein	*Lb. helveticus* proteinase	93	31
SVLSLSESKVLPVPE	β-Casein	*Lb. helveticus* proteinase	39	31
PPQSVLSLSESKVLPVPE	β-Casein	*Lb. helveticus* proteinase	25	31
RDMPIQAF	β-Casein	*Lb. helveticus* proteinase	209	31
YQQPVLGPVRGPFPIIV	β-Casein	*Lb. helveticus* proteinase	101	31
LPQNIPPLTQTPVVVPPFLQPEVMGVSK	β-Casein	*Lb. helveticus* proteinase	144	31
LLYQQPVLGPVRGPFPIIV	β-Casein	*Lb. helveticus* proteinase	21	31
DELQDKIHPFATQSLVYPFPGPIHNS	β-Casein	*Lb. helveticus* proteinase	4	31
AVPYPQR	β-Casein	*Lb. helveticus* proteinase	15	31
IPP	β-Casein	Fermentation	5	90
VPP	β-Casein	Fermentation	9	90
KVLPVP	β-Casein	Synthesis	5	30
YKVPQL	β-Casein	*Lb. helveticus* proteinase	22	30
WLAHK	α-Lactalbumin	Trypsin	77	103
IVY	Wheat germ	Hydrolyzate	0.48	104
SAYPGQITSN	Zein	Trypsin	7	86
QVSLNSGYY	Hordein	Trypsin	23	86

[a] Concentration of ACE inhibitor needed to inhibit 50% of the ACE activity.

Various peptides having ACE inhibition activity *in vitro* have been isolated from different proteins including α_{S1}- and β-casein (CN);[74,81] whey proteins;[82,83] soy proteins;[84] gelatin, fish proteins, and maize;[85] and gluten, zein, and hordein,[86] but milk proteins are the principal sources for such bioactive peptides (Table 7.4).

Different kinds of fermented milk have shown a high hemodynamic regulatory activity. Antihypertensive peptides derived from fermented milk have been studied *in vivo*.[87] LAB have extracellular proteinases that hydrolyze casein to release ACE inhibitory peptides.[88] Some antihypertensive peptides are released from casein by a purified extracellular proteinase from *Lb. helveticus* CP790 (Table 7.4).[31] Moreover, a dodecapeptide derived from casein enzymatic hydrolysate has been found to have antihypertensive properties *in vivo*.[89] Two other antihypertensive peptides, Val-Pro-Pro and Ile-Pro-Pro, were purified from a sour milk called Calpis fermented with *Lb. helveticus* and *Saccharomyces cerevisiae*.[90] Their amino acid sequences can be found at three positions in bovine caseins, β-CN(f74–76), β-CN(f84–86) and κ-CN(f108–110). As shown in Table 7.4, these peptides have a low IC_{50} and therefore a strong ACE inhibitory activity.

Several low molecular weight bioactive peptides with ACE-inhibitory activity have also been isolated from ripened cheeses.[56] An antihypertensive casein fragment, β-CN f58–72, which contains the β-casomorphin-7 sequence, has been isolated from Crescenza[37] and cheddar cheeses.[41] Two bioactive peptides have been isolated from the fermentation of whey by *Kluyveromyces marxianus*.[32] The two oligopeptides obtained were the result of the proteolysis of β-lactoglobulin in fragments f98–132 and f4–31. These oligopeptides contain the peptidic sequence YLLF, known to have antihypertensive properties, and are found in β-lactorphin. Consequently, these peptides could have some effects on hypertension. *In vivo* studies to evaluate antihypertensive activity of peptides are usually carried out on spontaneously hypertensive rats (SHR). The antihypertensive effect is observed by a drop in systolic blood pressure after oral administration of the antihypertensive peptide. Milk fermented by *Lb. helveticus* showed strong antihypertensive activity after oral administration to SHR. However, milk fermented by *Lb. helveticus* CP791, defective for proteinase activity, showed no significant antihypertensive activity after oral administration to SHR.[31] Gobbetti et al.[91] used two selected strains of LAB, *Lb. delbruecki* ssp. *bulgaricus* SS1 and *Lactococcus lactis* ssp. *cremoris* FT4, to produce two types of fermented milk that contained ACE-inhibitory peptides. The casein-derived ACE-inhibitory peptides liberated by these two strains were derived mainly from β-casein. The purified crude fractions that showed the highest ACE-inhibitory activity in milk fermented by *Lb. delbruecki* ssp. *bulgaricus* SS1 contained a mixture of peptides such as β-CN f6–14, f7–14, f73–82, f74–82, and f75–82.[91]

In general, ACE-inhibitory peptides that have been found to have antihypertensive activity in SHR have IC_{50} values lower than 150 μ*M*. However, in some cases, the extent of ACE inhibitory activity of the peptide is not correlated with the antihypertensive activity.[92] Some peptides show strong antihypertensive activity at a low dose even though they possess a low ACE inhibitory activity.[30] For example, Tyr-Pro Pro found in β-CN, κ-CN, and α_{S1}-casein, purified from a yogurt-like product fermented by *Lb. helveticus*, showed a significant antihypertensive activity after oral administration to SHR, even though it has a very low ACE inhibitory activity (IC_{50} of 720 μ*M*).[93,94] The antihypertensive activity of Tyr-Pro was measured by observing a drop of the systolic blood pressure in SHR after oral administration. Although Tyr-Pro has a strong antihypertensive activity, it has a low inhibitory activity against ACE, suggesting another possible way of controlling hypertension. Chymase, a major angiotensin-II forming enzyme, plays a role in the development of hypertension in the vessels of the heart.[95] The Tyr-Pro peptide could have an inhibitory effect on that enzyme. In addition, antihypertensive peptides also display an effect on bradykinin levels by inhibition of ACE.[96] Further digestion of the precursors of the bioactive peptides themselves by digestive enzymes could affect the activity of the peptides by enhancing or inactivating their physiological activity. The liberation of C-terminal amino acid residues of the peptide β-casein f169–175 after *in vitro* pancreatin digestion increases tremendously the ACE-inhibitory activity and consequently the antihypertensive effect *in vivo*.[30] It has been demonstrated that

ACE-inhibitory peptides, formed in cheese, decreased when proteolysis exceeded a certain level during the storage period.[97]

Most antihypertensive peptides obtained from food proteins have not yet been studied in humans. However, it has been demonstrated that *Lb. casei* cell extract had an antihypertensive effect on hypertensive patients.[98] Val-Pro-Pro and Ile-Pro-Pro have also been tested on humans by daily ingestion of sour milk for 8 weeks. Systolic blood pressure decreased significantly after 4 weeks of treatment.[87] Moreover, no side effects, such as cough and serum lipid metabolism problems usually observed with ACE inhibitors such as Captopril, were observed in patients treated with the tripeptides.[99]

Many peptides with elevated ACE-inhibitory activity are short and possess proline residues at the C-terminus. Another common structural feature of these peptides is the presence of C-terminal aromatic amino acids. Trp, Tyr, and Phe and imino acid Pro residues have more affinity for binding to the active site of ACE.[100] The preferable structure for optimal activity must contain principally hydrophobic residues at the three C-terminal positions. The presence of a hydrophobic residue within the ACE-inhibitory sequence is noteworthy. Gobbetti et al.[91] pointed out that ACE-inhibitory peptides derived from caseins contain a high proportion of hydrophobic peptides (>60%).

7.6 β-CASOMORPHINS

7.6.1 Antidiarrheal Effect

Oral rehydration solutions are the remedy of choice for the treatment of diarrhea, but increasing numbers of investigations are showing the beneficial uses of opioid receptor ligands for this symptom. The production of opioid receptor ligand peptides in milk-derived products could be a cost-effective remedy for the treatment of diarrhea in malnourished children in developing countries. Opioid receptors (μ, δ, and κ) are located in the nervous, endocrine, and immune systems and in the gastrointestinal tract. These receptors can interact with their endogenous ligands as well as with exogenous opioids and opioid antagonists such as milk-derived opioid peptides.[105] Many opioid receptor binding peptides have been obtained by the hydrolysis of milk proteins or by artificial production using peptidic sequences encoded by these proteins, such as β-casomorphins (f60–70 β-casein), α-casomorphin (αs1-casein), α-lactorphin (α-lactalbumin), and β-lactorphins.[106–108] The major opioid peptides, called β-casomorphins, are fragments of the bovine β-casein sequence 60–70 (YPFPGPIPASL) and have been characterized as μ-type ligands,[109] thus exerting a morphine-like activity.[109–111] Opioid peptides are designated opioid agonists. The opioid antagonist peptides called casoxins and lactoferroxins are also found in milk proteins. Their action consists of antagonizing the activities of the opioids, such as the inhibition of gut motility.[62,75]

Opioid peptides are characterized by a particular structure that facilitates their interaction with the binding site of opioid receptors. They are characterized by the presence of a tyrosine residue at the amino terminal end, followed by a proline and another aromatic residue, normally tyrosine or phenylalanine, in the third or fourth

TABLE 7.5
Examples of Opiod Peptides Derived from Bovine Milk Proteins

Bioactive Peptide	Sequence	Source (Fragment)	Release Protease	Ref.
Opioid Agonists				
β-Casomorphin-4	YPFP	β-Casein (f60–63)	LAB protease	26
β-Casomorphin-5	YPFPG	β-Casein (f60–64)	Chyme	106
β-Casomorphin-7	YPFPGPI	β-Casein (f60–66)	Chyme	106
β-Casomorphin-11	YPFPGPIPNSL	β-Casein (f60–70)	Chyme	106
Exorphin	RYLGYLE	α_{S1}-Casein (f90–96)	Pepsin	130
α-Lactorphin	YGLF	α-Lactalbumin (f50–53)	Trypsin	132
β-Lactorphin	$YLLF-NH_2$	β-Lactoglobulin (f102–105)	Trypsin	83
Opioid Antagonists				
Lactoferroxin A	YLGSGY	Lactoferrin (f318–323)	Pepsin	88
Casoxin A	YPSYGLNY	κ-Casein (f53–42)	Trypsin	111
Casoxin B	YPYY	κ-Casein (f58–61)	Trypsin	111
Casoxin C	YIPIQYVLSR	κ-Casein (f25–34)	Trypsin	131
Casoxin D	YVPFPPF	αs_1-Casein (f158–164)	Pepsin–chymotrypsin	111

Source: Meisel, H. and FitzGerald, R.J., *Br. J. Nutr.,* 84, S27–S31, 2000. With permission.

position (Table 7.5). The removal of the tyrosine residue results in an absence of bioactivity.[112] The proline residue is crucial for maintaining the proper orientation of the aromatic residue side chains.[113]

Opioid peptides have a large spectrum of activity including analgesic activity[114,115] and stimulation of endocrine response.[116] The mechanism of their antidiarrheal properties is due to a potent antisecretory effect at the intestinal mucosa and a potent inhibitory effect on gastrointestinal motility. It has already been demonstrated that opioid peptides such as the β-casomorphins are able to alter intestinal electrolyte and fluid movement in the *in vitro* stripped rabbit ileum.[117–120] Also, it has been shown in *in vivo* studies in rats that these peptides can prolong gastrointestinal transit time by inhibiting intestinal peristalsis and motility.[121] A reduction of intestinal motility has also been shown with the administration of casein in dogs.[122] The effect of casein was shown to be due to the presence of opioid receptor binding peptides, since treatments with naloxone, a potent opioid receptor antagonist, inhibited this effect. The action of casomorphins at the gastrointestinal level is important, since it can retard the rate of passage of the digesta through interaction with other components.[123]

It has been suggested that the brush border membrane could contain specific binding sites for β-casomorphins.[124] The β-casomorphins could thus act directly at the brush border membrane, affecting electrolyte transport. It is also noteworthy that the control of intestinal electrolyte and water transport involves a variety of regulatory sites including enteroendocrine cells, the nervous network, and the immune system. β-Casomorphins, due to their structure and their physicochemical characteristics, could interact with the various control sites altering intestinal electrolyte and fluid transport, and not just exert their effects via opioid receptors.[125] The antidiarrheal effect of casomorphins could thus be multifactorial. Much more research is necessary to understand the complete biochemical mechanisms of their actions in this field.

7.6.2 β-Casomorphins in Fermented Milk

Some reports have noted the presence of casomorphins and their derivatives in fermented dairy products. Fermented milks containing β-casomorphins resulting from the incomplete proteolysis of caseins by mutants of lactobacilli deficient in proline-specific peptidases were reported by Matar and Goulet.[27] Milk fermented with a x-prolyldipeptidyl aminopeptidase (XPDAP)-deficient mutant of *Lb. helveticus* was able to liberate β-casomorphin-4, since this LAB had lost its ability to hydrolyze the peptidic bond next to the proline residues.[26] Since β-casomorphins contain two proline residues (Table 7.5), they would be digested by the proline-specific peptidases widely present in LAB, if these enzymes were not inhibited or eliminated by mutations.[91] β-Casomorphin-4 amide showed a high potency for endogenous receptors.[112] β-Casomorphin immunoreactive material has been identified in milk incubated with various bacterial species.[126] This milk was incubated with caseolytic bacteria, such as *Bacillus cereus* and *Pseudomonas aeruginosa.* The presence of β-casomorphins in cheese was studied, and it was shown that they can be degraded by the enzymes of *Lactococcus lactis* ssp. *cremoris.*[127] It has been demonstrated that trypsin/pepsin hydrolysis of fermented milk by *Lb.* GG led to the release of opioid peptide sequences from caseins.

7.7 HEALTH BENEFITS OF MILK-BASED BIOACTIVE PEPTIDES

It is now accepted that metabolites generated during milk fermentation have more impact on the enhancement of the *in vivo* response and bioactivities than do the cellular components themselves (cell wall, peptidoglycan, or cytoplasm fraction). Physiologically active peptides are among the most important metabolites that have proven to influence many health parameters in humans and animals. Various strategies have been adopted to develop functional foods containing novel biopeptide preparations. Despite the proven *in vivo* biological activities of many biopeptides, more emphasis should be placed on the specific functionality of these peptides before adding them as single ingredients in food preparations. More research is needed to better understand the interactions of biopeptides with each other, the interaction of biopeptides with other nutrients, the attainment of optimum effects and doses, the mechanisms of action, and their bioavailability. Many observations tend to demonstrate the alteration of biological activities of these compounds when ingested singly.

Schanbacher et al.,[123] in a review on bioactive peptides in milk, noted the importance of the interaction and synergism of the peptides among themselves and with other nonpeptide components. The authors pointed to the fact that peptides, such as casomorphins, might influence the kinetics and the dynamics of other bioactivities in the intestine. Casomorphins, by slowing the passage of the digesta through the gut, increase the time available for the bioactive agents in milk to assert their action on target cells, bacteria, or organs. In addition, the synergistic action of bioactive peptides with nonpeptide (lipid, glycolipid, and oligosaccharide) or peptide agents in milk is necessary for the expression of the bioactivity.[128,129] The major bioactivities in milk are dose dependent and act principally at the intestinal site. When a single bioactive peptide was administered, even if its concentration was increased, the biological effect attributed to the single ingredient was still lower than the effect observed with complete fermented milk. The dose and the mode of administration of bioactive substances play tremendous roles in determining the target functionality. An excessive dose might be more harmful than beneficial to the host. The question remains, "Should probiotic preparations or *de novo* generated peptides be administered continuously or periodically?" Continuous administration of immunoregulatory substances might initiate downregulatory signals at the level of the immune system and result in an increase in inflammatory response.[45]

Use of fermentation organisms with enhanced proteolytic activities or engineered probiotic organisms to favor the overproduction of peptides is one option to enrich a food. Applications of bioactive peptides to nutraceutical and pharmaceutical formulations should be submitted to more scientific scrutiny to ascertain whether or not the specific health benefits are achieved. Many concerns should be taken into consideration when processing such products, such as interaction of peptides with their receptors *in vivo*, the influence of microflora, which differs among individuals, and the dose/effect. The presence of both agonist and antagonist, as is the case for opioid peptides, may have physiological importance and should also be taken into consideration when formulating enriched bioactive peptide products.

7.8 CONCLUSIONS

The role of peptides derived from milk proteins has been reviewed. The multifunctional properties and potential activities on several biological functions of the host, such as effects on the tumor and mucosal immunity, antidiarrheal effects, and antihypertensive activity of peptides, were analyzed. The identification of several bioactive sequences in dietary proteins has contributed to the understanding of the role of these molecules as precursors of bioregulating peptides at the intestinal level. During the fermentation process, LAB use a large spectrum of proteolytic enzymes (endopeptidases, aminopeptidases, tripeptidases, and dipeptidases) to fulfill their nitrogen need for growth. During the fermentation of milk, caseins undergo proteolysis, generating potentially bioactive peptides.

LAB themselves can modulate the immune response. However, more and more evidence now links the health benefits of fermented products to specific enzymatic activities of LAB and the peptides released during the fermentation process. The mechanisms responsible for the improvement of health are probably multifactorial and involve complex interactions between peptides from milk proteins, LAB, and intestinal cells.

It is very difficult to demonstrate *in vivo* the effect of different LAB administered with the diet or to show the effects of individual peptides on the gastrointestinal tract, especially related to mucosal immunostimulation. Casein-derived proteins such as human β-casein, bovine κ-casein, and human- and bovine-α lactalbumin enhance immune functions (macrophage activity, lymphocyte proliferation, protection against infections, etc.). Thus, consumption of fermented milk products may help to prevent intestinal or nonintestinal tumors.

Recent studies[76,77] have shown protection of the mucosal immune system and inhibition of the development of a nonintestinal tumor (fibrosarcoma) by peptides released from milk fermented with a *Lb. helveticus* proteolytic strain, and since similar results were not observed with a milk fermented with a nonproteolytic *Lb. helveticus*, a biological peptide obtained using a proteolytic strain is believed to be responsible.

Antihypertensive peptides can be released from casein using a purified extracellular proteinase from *Lb. helveticus*. The ACE inhibitory activity of this peptide is not correlated with the antihypertensive effect. The presence of hydrophobic residues at the C-terminal positions is necessary for optimal activity.

Peptides derived from milk proteins may play a significant role in the reported beneficial effects of fermented milk consumption. To improve human health, further studies involving molecular biology are needed to better understand the complex network of interactions between food microorganisms and the digestive system.

REFERENCES

1. Marteau, P., Seksik, P., and Jian, R., Probiotics and intestinal health effects: a clinical perspective, *Br. J. Nutr.,* 88, S51–S57, 2002.
2. Fuller, R., Ed., *Probiotics 3: Immunomodulation by the Gut Flora and Probiotics,* Kluwer Academic Press, Dordrecht, The Netherlands, 2000.
3. Yoshikawa, M., Fujita, H., Matoba, N., Takenaka, Y., Yamamoto, T., Yamauchi, R., Tsuruki, H., and Takahata, K., Bioactive peptides derived from food proteins preventing lifestyle-related diseases, *Biofactors,* 12, 143–146, 2000.
4. Yamauchi, F. and Suetsuna, K., Immunological effects of dietary peptide derived from soybean protein, *J. Nutr. Biochem.,* 4, 450–457, 1993.
5. Takahashi, M., Moriguchi, S., Ikeno, M., Kono, S., Ohata, K., Usui, H., Kurahashi, K., Sasaki, R., and Yoshikawa, M., Studies on the ileum-contracting mechanisms and identification as a complement C3a receptor agonist of oryzatensin, a bioactive peptide derived from rice albumin, *Peptides,* 17, 5–12, 1996.
6. Meisel, H. and Bockelmann, W., Bioactive peptides encrypted in milk proteins: proteolytic activation and thropho-functional properties, *Antonie Van Leeuwenhoek,* 76, 207–215, 1999.
7. Clare, D.A. and Swaisgood, H.E., Bioactive milk peptides: a prospectus, *J. Dairy Sci.,* 83, 1187–1195, 2000.
8. Juillard, V., Guillot, A., Le Bars, D., and Gripon, J.C., Specificity of milk peptide utilization by *Lactococcus lactis, Appl. Environ. Microbiol.,* 64, 1230–1236, 1998.
9. Kunji, E.R., Mierau, I., Hagting, A., Poolman, B., and Konings, W.N., The proteolytic systems of lactic acid bacteria, *Antonie Van Leeuwenhoek,* 70, 187–221, 1996.
10. Kunji, E.R., Fang, G., Jeronimus-Stratingh, C.M., Bruins, A.P., Poolman, B., and Konings, W.N., Reconstruction of the proteolytic pathway for use of beta-casein by *Lactococcus lactis, Mol. Microbiol.,* 27, 1107–1118, 1998.
11. Mierau, I., Kunji, E.R., Venema, G., and Kok, J., Casein and peptide degradation in lactic acid bacteria. *Biotechnol. Gen. Engin. Rev.,* 14, 279–301, 1997.

12. Law, J. and Handrikmann, A., Proteolytic enzymes of lactic acid bacteria, *Int. Dairy J.*, 7, 1–15, 1997.
13. Visser, S., Exterkate, F.A., Slangen, C.J., and de Veer, G.J.C.M., Comparative study of action of cell wall proteinases from various strains of *Streptococcus cremoris* on bovine α_{S1}-, β-, and κ-casein, *Appl. Environ. Microbiol.*, 52, 1162–1166, 1986.
14. Kunji, E.R., Smid, E.J., Plapp, R., Poolman, B., and Konings, W.N., Di-tripeptides and oligopeptides are taken up via distinct transport mechanisms in *Lactococcus lactis, J. Bacteriol.*, 175, 2052–2059, 1993.
15. Kok, J., van der Vossen, J.M.B.M., and Venema, G., Construction of plasmid cloning vectors for lactic streptococci which are replicate in *Bacillus subtilis* and *Escherichia coli, Appl. Environ. Microbiol.*, 48, 726–731, 1984.
16. Haandrikman, A.J., Kok, J., Laan, H., Soemitro, S., Ledeboer, A.M., Konings, W.N., and Venema, G., Identification of a gene required for the maturation of an extracellular serine proteinase, *J. Bacteriol.*, 171, 2789–2795, 1989.
17. Vos, P., Simons, G., Siezen, R.J., and de Vos, W.M., Primary structure and organisation of the genes for procaryotic, cell envelope–located serine proteinase, *J. Biol. Chem.*, 264, 13579–13585, 1989.
18. von Heijne, G., Membrane proteins. The amino acid composition of membrane penetrating segments, *Eur. J. Biochem.*, 120, 275–278, 1981.
19. Martín-Hernández, M.C., Alting, A.C., and Exterkate, F.A., Purification and characterization of the mature, membrane-associated cell-envelope proteinase of *Lactobacillus helveticus* L89, *Appl. Microbiol. Biotechnol.*, 40, 828–834, 1994.
20. Foucaud, C. and Juillard, V., Accumulation of casein-derived peptides during growth of proteinase-positive strains of *Lactococcus lactis* in milk: their contribution to subsequent bacterial growth is impaired by their internal transport, *J. Dairy Res.*, 67, 233–240, 2000.
21. Tan, P.S.T., Pos, K.M., and Konings, W., Purification and characterization of an endopeptidase from *Lactococcus lactis* subsp. *cremoris* Wg2, *Appl. Environ. Microbiol.*, 57, 3593–3599, 1991.
22. Booth, M., Ní Fhaolain, I., Jennings, P.V., and O'Cuinn, G., Purification and characterization of a post-proline dipeptidyl aminopeptidase from *Streptococcus cremoris* AM2, *J. Dairy Res.*, 57, 89–99, 1990.
23. Baankreis, R. and Exterkate, F.A., Characterization of a peptidase from *Lactococcus lactis* ssp. *cremoris* HP that hydrolyses di- and tripeptides containing proline or hydrophobic residues at the aminoterminal amino acid, *System. Appl. Microbiol.*, 14, 317–323, 1991.
24. Atlan, D., Laloi, P., and Portalier, R., X-prolyldipeptidyl aminopeptidase of *Lactobacillus delbrueckii* subsp. *bulgaricus:* characterization of the enzyme and isolation of deficient mutants, *Appl. Environ. Microbiol.*, 56, 2174–2179, 1990.
25. Tynkkynen, S., Buist, G., Kunji, E., Kok, J., Poolman, B., Venema, G., and Haandrikman, A., Genetic and biochemical characterization of the oligopeptide transport system of *Lactococcus lactis*: demonstration of the essential role of oligopeptide transport in the utilization of casein, *J. Bacteriol.*, 175, 7523–7532, 1993.
26. Chabance, B., Marteau, P., Rambaud, J.C., Migliore-Samour, D., Boynard, M., Perrotin, P., Guillet, R., Jolles, P., and Fiat, A.M., Casein peptide release and passage to the blood in humans during digestion of milk or yogurt, *Biochimie*, 80, 155–165, 1998.
27. Matar, C. and Goulet, J., β-Casomorphin 4 from milk fermented by a mutant of *Lactobacillus helveticus*, *Int. Dairy J.*, 6, 383–397, 1996.

28. Yamamoto, N. and Takano, T., Antihypertensive peptides derived from milk proteins, *Nahrung,* 43, 159–164, 1999.
29. Matar, C., Nadathur, S.S., Bakalinsky, A.T., and Goulet, J., Antimutagenic effects of milk fermented by *Lactobacillus helveticus* L89 and a protease-deficient derivative, *J. Dairy Sci.,* 80, 1965–1970, 1997.
30. Maeno, M., Yamamoto, N., and Takano, T., Identification of an antihypertensive peptide from casein hydrolysate produced by a proteinase from *Lactobacillus helveticus* CP790, *J. Dairy Sci.,* 79, 1316–1321, 1996.
31. Yamamoto, N., Akino, A., and Takano, T., Antihypertensive effect of the peptides derived from casein by an extracellular proteinase from *Lactobacillus helveticus* CP790, *J. Dairy Sci.,* 77, 917–922, 1994.
32. Belem, M.A.F., Gibbs, B.F., and Lee, B.H., Proposing sequence for peptides derived from whey fermentation with potential bioactive sites, *J. Dairy Sci.,* 82, 486–493, 1999.
33. Rokka, T., Syvaoja, E.L., Tuominen, J., and Korhonen, H., Release of bioactive peptides by enzymatic proteolysis of *Lactobacillus* GG fermented UHT milk, *Milchwissenschaft,* 52, 675–678, 1997.
34. Smacchi, E. and Gobbetti, M., Bioactive peptides in dairy products: synthesis and interaction with proteolytic enzymes, *Food Microbiol.,* 17, 129–141, 2000.
35. Roudot-Algaron, F., LeBars, D., Kerhoas, L., Einhorn, J., and Gripon, J.C., Phosphopeptides from Comté cheese: nature and origin, *J. Food Sci.,* 59, 544–547, 1994.
36. Pellegrino, L., Battelli, G., Resmini, P., Ferranti, P., Barone, F., and Addeo, F., Alkaline phosphatase inactivation during Grana Padano cheese-making and related effects on cheese characterization and ripening, *Lait,* 77, 217–228, 1997.
37. Smacchi, E. and Gobbetti, M., Peptides from Italian cheeses inhibitory to proteolytic enzymes of lactic acid bacteria, *Pseudomonas fluorescens* ATCC 948 and to the angiotensin-I converting enzyme, *Enzyme Microb. Technol.,* 22, 687–694, 1998.
38. Meisel, H., Goepfert, A., and Günther, S., ACE-inhibitory activities in milk products, *Milchwissenschaft,* 52, 307–311, 1997.
39. Addeo, F., Chianese, L., Salzano, A., Sacchi, R., Cappuccio, U., Ferranti, P., and Malorni, A., Characterization of the 12% tricholoroacetic acid-insoluble oligopeptides of Parmigiano–Reggiano cheese, *J. Dairy Res.,* 59, 401–411, 1992.
40. Gobbetti, M., Stepaniak, L., Fox, P.F., Sorhaug, T., and Tobiassen, R., Inhibition of endo- and amino-peptidase activities in cytoplasmic fractions of *Lactococcus, Lactobacillus* and *Propionibacterium* by peptides from different cheeses, *Milchwissenschaft,* 50, 565–570, 1995.
41. Stepaniak, L., Fox, P.F., Sørhaug, T., and Grabska, J., Effect of peptides from the sequence 58–72 of β-casein on the activity of endopeptidase, aminopeptidase and X-prolyl-dipeptidyl aminopeptidase from *Lactoccocus, J. Agric. Food Chem.,* 43, 849–853, 1995.
42. Perdigon, G., Alvarez, S., Rachid, M., Aguero, G., and Gobbato, N., Immune system stimulation by probiotics, *J. Dairy Sci.,* 78, 1597–1606, 1995.
43. Perdigon, G., Vintini, E., Alvarez, S., Medina, M., and Medici, M., Study of the possible mechanisms involved in the mucosal immune system activation by lactic acid bacteria, *J. Dairy Sci.,* 82, 1108–1114, 1999.
44. Perdigon, G. and Holgado, A., Mechanisms involved in the immunostimulation by lactic acid bacteria, in *Probiotics 3: Immunomodulation by the Gut Microflora and Probiotics,* Fuller, R., Ed., Kluwer Academic Publishers, Dordrecht, The Netherlands, 2000, pp. 213–229.

45. Perdigon, G., Fuller, R., and Raya, R., Lactic acid bacteria and their effect on the immune system, *Curr. Issues Intest. Microbiol.,* 2, 27–42, 2001.
46. Schiffrin, E.J., Brassart, D., Servin, A.L., Rochat, F., and Donnet-Hughes, A., Immune modulation of blood leukocytes in humans by lactic acid bacteria: criteria for strain selection, *Am. J. Clin. Nutr.,* 66, 515S–520S, 1997.
47. Ayebo, A.D., Shahani, K.M., Dam, R., and Friend, B.A., Ion exchange separation of the antitumor component(s) of yogurt dialyzate, *J. Dairy Sci.,* 65, 2388–2390, 1982.
48. Biffi, A., Coradini, D., Larsen, R., Riva, L., and Di Fronzo, G., Antiproliferative effect of fermented milk on the growth of a human breast cancer cell line, *Nutr. Cancer,* 28, 93–99, 1997.
49. Perdigon, G., de Marcia, M.E.N., Alvarez, S., Medici, M., Oliver, G., and de Ruiz Holgado, A.P., Effect of a mixture of *Lactobacillus casei* and *Lactobacillus acidophilus* administered orally on the immune system in mice, *J. Food Prot.,* 49, 986–989, 1986.
50. de Simone, C., Slavadori, B.B., Negri, R., Ferrazzi, M., Bladinelli, L., and Vesely, R., The adjuvant effect of yogurt on production of gamma-interferon by con a-stimulated human peripheral blood lymphocytes, *Nutr. Rep. Int.,* 33, 419–433, 1986.
51. Ng, K.-Y. and Griffiths, M.W., Enhancement of macrophage cytokine release by cell-free fractions of fermented milk, *Milchwissenschaft,* 57, 66–70, 2002.
52. Sutas, Y., Soppi, E., and Korhonen, H., Suppression of lymphocyte proliferation *in vitro* by bovine caseins hydrolyzed with *Lactobacillus casei* GG-derived enzymes, *J. Allergy Clin. Immunol.,* 98, 216–224, 1996.
53. Laffineur, E., Genetet, N., and Leonil, J., Immunomodulatory activity of beta-casein permeate medium fermented by lactic acid bacteria, *J. Dairy Sci.,* 79, 2112–2120, 1996.
54. Otani, H., Monnai, M., Kawasaki, Y., Kawakami, H., and Tanimoto, M., Inhibition of mitogen-induced proliferation responses of lymphocytes by bovine κ-caseinoglycopeptides having different carbohydrate chains, *J. Dairy Res.,* 62, 349–357, 1995.
55. Kayser, H. and Meisel, H., Stimulation of human peripheral blood lymphocytes by bioactive peptides derived from bovine milk proteins, *FEBS Lett.,* 383, 18–20, 1996.
56. Gill, H.S., Doull, F., Rutherford, K.J., and Cross, M.L., Immunoregulatory peptides in bovine milk, *Br. J. Nutr.,* 84, S111–S117, 2000.
57. Sutas, Y., Hurne, M., and Isolauri, E., Down-regulation of CD3 antibody-induced IL-4 production by bovine caseins hydrolysed with *Lactobacillus* GG–derived enzymes, *Scand. J. Immunol.,* 43, 687–689, 1996.
58. Parker, F., Migliore-Samour, D., and Floc'h, F., Immunostimulating hexapeptide from human casein, amino acid sequence, synthesis and biological properties, *Eur. J. Biochem.,* 145, 677–682, 1984.
59. Otani, H. and Hata, I., Inhibition of proliferation responses of mouse spleen lymphocytes and rabbit Peyer's patch cells by bovine milk caseins and their digests, *J. Dairy Res.,* 62, 339–348, 1995.
60. Jolles, P., Fiat, A.-M., Migliore-Samour, D., Drouet, L., and Caen, J.P., Peptides from milk proteins implicated in antithrombosis and immunomodulation, in *New Perspectives in Infant Nutrition, Symposium Antwerp,* Renner, B. and Sawatzki, G., Eds., Thieme Medical Publishers, New York, 1992, pp. 160–172.
61. Schlimme, E. and Meisel, H., Bioactive peptides derived from milk proteins. Structural, physiological and analytical aspects, *Nahrung,* 39, 1–20, 1995.
62. Migliore-Samour, D., Floch, F., and Jolles, P., Biologically active casein peptides implicated in immunomodulation, *J. Dairy Res.,* 56, 357–362, 1989.

63. Meisel, H., Biochemical properties of regulatory peptides derived from milk proteins, *Biopolymers,* 43, 119–128, 1997.
64. Coste, M., Rochet, V., Leonil, J., Molle, D., Bouhallab, S., and Tome, D., Identification of C-terminal peptides of bovine beta-casein that enhance proliferation of rat lymphocytes, *Immunol. Lett.,* 33, 41–46, 1992.
65. Miyauchi, H., Kaino, A., Shinoda, I., Fukuwatari, Y., and Hayasawa, H., Immunomodulatory effect of bovine lactoferrin pepsin hydrolysate on murine splenocytes and Peyer's patch cells, *J. Dairy Sci.,* 80, 2330–2339, 1997.
66. Miyauchi, H., Hashimoto, S., Nakajima, M., Shinoda, I., Fukuwatari, Y., and Hayasawa, H., Bovine lactoferrin stimulates the phagocytic activity of human neutrophils: identification of its active domain, *Cell Immunol.,* 187, 34–37, 1998.
67. Pecquet, S., Bovetto, L., Maynard, F., and Fritsche, R., Peptides obtained by tryptic hydrolysis of bovine beta-lactoglobulin induce specific oral tolerance in mice, *J. Allergy Clin. Immunol.,* 105, 514–521, 2000.
68. Bounous, G. and Kongshavn, P.A.L., Differential effect of dietary protein type on the B cell and T cell response in mice, *J. Nutr.,* 115, 1403–1408, 1985.
69. Bounous, G. and Kongshavn, P.A.L., Influence of dietary proteins on the immune system of mice, *J. Nutr.,* 112, 1747–1755, 1982.
70. Mattsby-Baltzer, I., Roseanu, A., Motas, C., Elverfors, J., Engberg, I., and Hanson, L.A., Lactoferrin or a fragment thereof inhibits the endotoxin-induced interleukin-6 response in human manocytic cells, *Pediatric Res.,* 40, 257–262, 1996.
71. Shinoda, I., Takase, M., Fukuwatari, Y., Shimamura, S., Koller, M., and Konig, W., Effects of lactoferrin and lactoferricin on the release of interleukin 8 from human polymorphonuclear leucocytes, *Biosci. Biotech. Biochem.,* 60, 521–523, 1996.
72. Migliore-Samour, D. and Jolles, P., Casein, a prohormone with an immunomodulating role in the newborn? *Experientia,* 44, 188–193, 1988.
73. Elitsur, Y. and Luk, G.D., Beta-casomorphin (BCM) and human colonic lamina propria lymphocyte proliferation, *Clin. Exp. Immunol.,* 85, 493–497, 1991.
74. Maruyama, S., Mitachi, H., Awaya, J., Kurono, M., Tomizuka, N., and Suzuki, H., Angiotensin I–converting enzyme inhibitory activity of the C-terminal hexapeptide of alpha S1 casein, *Agric. Biol. Chem.,* 51, 2557–2561, 1987.
75. Fiat, A.M., Migliore-Samour, D., Jolles, P., Drouet, L., Sollier, C.B., and Caen, J., Biologically active peptides from milk proteins with emphasis on two examples concerning antithrombotic and immunomodulating activities, *J. Dairy Sci.,* 76, 301–310, 1993.
76. Matar, C., Goulet, J., Bernier, R.L., and Brochu, E., Bioactive peptides from fermented foods: their role in the immune system, in *Probiotics 3: Immunomodulation by the Gut Microflora and Probiotics,* Fuller, R., Ed., Kluwer Academic Publishers, Dordrecht, The Netherlands, 2000.
77. Matar, C., Valdez, J.C., Medina, M., Rachid, M., and Perdigon, G., Immunomodulating effects of milks fermented by *Lactobacillus helveticus* and its non-proteolytic variant, *J. Dairy Res.,* 68, 601–609, 2001.
78. LeBlanc, J.G., Matar, C., Valdez, J.C., LeBlanc, J., and Perdigon, G., Identification of a peptidic fraction in fermented milk that enhances the immune system response and regression of fibrosarcoma, *J. Dairy Sci.,* (submitted).
79. Perdigon, G., Valdez, J.C., and Rachid, M., Antitumour activity of yogurt: study of possible immune mechanisms, *J. Dairy Res.,* 65, 129–138, 1998.
80. Isolauri, E., Probiotics in human disease, *Am. J. Clin. Nutr.,* 73, 1142S–1146S, 2001.

81. Maruyama, S., Mitachi, H., Awaya, J., Kurono, M., Tomizuka, N., and Suzuki, H., Angiotensin I-converting enzyme inhibitory activity of the C-terminal hexapeptide of alpha S1 casein, *Agric. Biol. Chem.,* 51, 2557–2561, 1987.
82. Mullally, M.M., Meisel, H., and FitzGerald, R.J., Synthetic peptides corresponding to alpha-lactalbumin and beta-lactoglobulin sequences with angiotensin-I-converting enzyme inhibitory activity, *Biol. Chem. Hoppe Seyler,* 377, 259–260, 1996.
83. Mullaly, M., Meisel, H., and FitzGerald, R.J., Angiotensin-I-converting enzyme inhibitory activities of gastric and pancreatic proteinase digests of whey proteins, *Int. Dairy J.,* 7, 299–303, 1997.
84. Rhyu, M., Nam, Y., and Lee, H., Screening of angiotensin-I-converting enzyme inhibitors in cereals and legumes, *Food Biotechnol.,* 5, 334, 1996.
85. Ariyoshi, Y., Angiotensin-converting enzyme inhibitors derived from food proteins, *Trends Food Sci. Technol.,* 4, 139–144, 1993.
86. Gobbetti, M., Smacchi, E., Corsetti, A., and Bellucci, M., Inhibition of proteolytic enzymes from *Pseudomonas fluorescens* ATCC 948 and angiotensin-I converting enzyme by peptides from zein, hordein and gluten hydrolysates, *J. Food Prot.,* 60, 499–504, 1997.
87. Hata, Y., Yamamoto, M., Ohni, M., Nakajima, K., Nakamura, Y., and Takano, T., A placebo-controlled study of the effect of sour milk on blood pressure in hypertensive subjects, *Am. J. Clin. Nutr.,* 64, 767–771, 1996.
88. Yamamoto, N., Akino, A., and Takano, T., Purification and specificity of a cell-wall-associated proteinase from *Lactobacillus helveticus* CP790, *J. Biochem.,* 114, 740–745, 1993.
89. Sugai, R., ACE-inhibitors and functional foods, abstract, *Int. Dairy Fed. Wk.,* Mar. 9–11, 1998.
90. Nakamura, Y., Yamamoto, N., Sakai, K., Okubo, A., Yamazaki, S., and Takano, T., Purification and characterization of angiotensin I-converting enzyme inhibitors from sour milk, *J. Dairy Sci.,* 78, 777–783, 1995.
91. Gobbetti, M., Ferranti, P., Smacchi, E., Goffredi, F., and Addeo, F., Production of angiotensin-I-converting-enzyme-inhibitory peptides in fermented milks started by *Lactobacillus delbrueckii* subsp. *bulgaricus* SS1 and *Lactococcus lactis* subsp. *cremoris* FT4, *Appl. Environ. Microbiol.,* 66, 3898–3904, 2000.
92. Yamamoto, N. and Takano, T., Antihypertensive peptides derived from milk proteins, *Nahrung,* 43, 159–164, 1999.
93. Yamamoto, N., Ono, H., Maeno, M., Ueda, Y., Takano, T., and Momose, H., Classification of *Lactobacillus helveticus* strains by immunological differences in extracellular proteinases, *Biosci. Biotechnol. Biochem.,* 62, 1228–1230, 1998.
94. Yamamoto, N., Maeno, M., and Takano, T., Purification and characterization of an antihypertensive peptide from yogurt-like product fermented by *Lactobacillus helveticus* CPN4, *J. Dairy Sci.,* 82, 1388–1393, 1999.
95. Urata, H., Boehm, K.D., Philip, A., Kinoshita, A., Gabrovsek, J., Bumpus, F.M., and Husain, A., Cellular localization and regional distribution of an angiotensin II–forming chymase in the heart, *J. Clin. Invest.,* 91, 1269–1281, 1993.
96. Berthou, J., Migliore-Samour, D., Lifchitz, A., Delettre, J., Floc'h, F., and Jolles, P., Immunostimulating properties and three-dimensional structure of two tripeptides from human and cow caseins, *FEBS Lett.,* 218, 55–58, 1987.
97. Ryhanen, E.-L., Pihlanto-Leppala, A., and Pahkala, E., A new type of ripened, low-fat cheese with bioactive properties, *Int. Dairy J.,* 11, 441–447, 2001.

98. Hata, Y., Yamamoto, M., Ohni, M., Nakajima, K., Nakamura, Y., and Takano, T., A placebo-controlled study of the effect of sour milk on blood pressure in hypertensive subjects, *Am. J. Clin. Nutr.,* 64, 767–771, 1996.
99. Takano T., Milk derived peptides and hypertension reduction, *Int. Dairy J.,* 8, 375–381, 1998.
100. Kawakami, A. and Kayahara, H., Synthesis of Leu-Lys-Tyr derivatives and their interaction with angiotensin-I converting enzyme, *J. Jpn. Soc. Nutr. Food Sci.,* 46, 425–428, 1993.
101. Maruyama, S. and Suzuki, H., A peptide inhibitor of angiotensin-I converting enzyme in the tryptic hydrolysate of casein, *Agric. Biol. Chem.,* 46, 1393, 1982.
102. Maruyama, S., Nakagomi, K., Tomizuka, N., and Suzuki, H., Angiotensin I–converting enzyme inhibitor derived from an enzymatic hydrolysate of casein. II. Isolation and bradykinin-potentiating activity on the uterus and ileum of rats, *Agric. Biol. Chem.,* 49, 1405, 1985.
103. Pihlanto-Leppälä, A., Koskinen, P., Piilola, K., Tupasela, T., and Korhonen, H., Angiotensin I–converting enzyme inhibitory properties of whey protein digests: concentration and characterization of active peptides, *J. Dairy Res.,* 67, 53–64, 2000.
104. Matsui, T., Li, C.-H., and Osajima, Y., Preparation and characterization of novel bioactive peptides responsible for angiotensin I-converting enzyme inhibition from wheat germ, *J. Peptide Sci.,* 5, 289–297, 1999.
105. Teschemacher, H., Koch, G., and Brandt, V., Milk protein derived opioid peptides and correlated compounds with antagonist activity, in *β-Casomorphins and Related Peptides: Recent Developments*, Brandt, V. and Teschemacher, H., Eds., VCH-Weinheim, Germany, 1994, pp. 9–14.
106. Meisel, H. and FitzGerald, R.J., Opioid peptides encrypted in intact milk protein sequences, *Br. J. Nutr.,* 84, S27–S31, 2000.
107. Antila, P., Paakkari, I., Jarvinen, A., Mattila, M.J., Laukkanen, M., Pihlanto-Leppälä, A., Mantsala, P., and Hellman, J., Opioid peptides derived from *in vitro* proteolysis of bovine whey proteins, *Int. Dairy J.,* 1, 215–229, 1991.
108. Teschemacher, H., Koch, G., and Brantl, V., Milk protein-derived opioid receptor ligands, *Biopolymers,* 43, 99–117, 1997.
109. Brantl, V., Teschemacher, H., Henschen, A., and Lottspeich, F., Novel opioid peptides derived from casein (beta-casomorphins). I. Isolation from bovine casein peptone, *Hoppe Seyler's Physiol. Chem.,* 360, 1211–1216, 1979.
110. Henschen, A., Lottspeich, F., Brandt, V., and Teschemacher, H., Novel opioid peptides derived from casein, β-casomorphins. II. Structure of active components from bovine casein peptone, *Hoppe Seyler's Physiol. Chem.,* 360, 1217–1224, 1979.
111. Yoshikawa, M., Tani, F., Ashikaga, T., Yoshimura, T., and Chiba, H., Opioid peptides from milk proteins, *Agric. Biol. Chem.,* 50, 2419–2421, 1986.
112. Chang, K.J., Killian, A., Hazum, E., and Cuatrecasas, P., Morphiceptin (NH_4-Tyr-Pro-Phe-Pro-$CONH_2$): a potent and specific agonist for the morphine (μ) receptors, *Science,* 212, 75–77, 1981.
113. Mierke, D.F., Nobner, G., Shiller, P.W., and Goodman, M., Morphiceptin analogs containing 2-aminocyclopentane carboxylic acid as a peptidomimetic for proline, *Int. J. Peptide Res.,* 35, 34–45, 1990.
114. Matthies, H., Stark, H., Hartrodt, B., Ruethrich, H.L., Spieler, H.T., Barth, A., and Neubert, K., Derivatives of beta-casomorphins with high analgesic potency, *Peptides,* 5, 463–470, 1984.
115. Paroli, E., Opioid peptides from food (the exorphins), *World Rev. Nutr. Diet.,* 55, 58–97, 1988.

116. Meisel, H. and Schlimme, E., Milk proteins: precursors of bioactive peptides, *Trends Food Sci. Technol.,* 1, 41–43, 1990.
117. Hautefeuille, M., Brandt, V., Dumontier, A.-M., and Desjeux, J.-F., *In vitro* effects of β-casomorphins on ion transport in rabbit ileum, *Am. J. Physiol.,* 250, G92–G97, 1986.
118. Tomé, D., Dumontier, A.M., Hautefeuille, M., and Desjeux, J.F. Opiate activity and transepithelial passage of intact β-casomorphins in rabbit ileum, *Am. J. Physiol.,* 253, G737–G744, 1987.
119. Mansour, B.A., Tomé, D., Rautureau, M., Bisalli, A., and Desjeux, J.-F., Luminal anti-secretory effects of a casomorphin analogue on rabbit ileum treated with cholera toxin, *Pediatr. Res.,* 24, 751–755, 1988.
120. Mahé, S., Tomé, D., Dumontier, A.M., and Desjeux, J.-F., Absorption of intact β-casomorphins (β-CM) in rabbit ileum *in vitro*, *Reprod. Nutr. Dévelop.,* 29, 725–732, 1989.
121. Daniel, H., Vohwinkel, M., and Rehner, G., Effect of casein and beta-casomorphins on gastrointestinal motility in rats, *J. Nutr.,* 120, 252–257, 1990.
122. Defilippi, C. and Gomez, E., Effect of casein and casein hydrolysate on small bowel motility and D-xylose absorption in dogs, *Neurogastroenterol. Motil.,* 7, 229–234, 1995.
123. Schanbacher, F.L., Talhouk, R.S., Murray, F.A., Gherman, L.I., and Willett, L.B., Milk-borne bioactive peptides, *Int. Dairy J.,* 8, 393–403, 1998.
124. Ermisch, A., Brust, P., and Brandsch, M., β-Casomorphins alter the intestinal accumulation of L-leucine, *Biochim. Biophys. Acta,* 982, 79–84, 1989.
125. Erll, G., Hahn, A., Brandt, V., and Daniel, H., β-Casomorphins and intestinal net fluid transport *in vivo*, in *β-Casomorphins and Related Peptides: Recent Developments,* Brandt, V. and Teschemacher, H., Eds., VCH, Weinheim, 1994, pp. 143–149.
126. Hamel, U., Kielwein, G., and Teschemacher, H., Beta-casomorphin immunoreactive materials in cow's milk incubated with various bacterial species, *J. Dairy Res.,* 52, 139–148, 1985.
127. Muehlenkamp, M.R. and Warthesen, J.J., β-Casomorphins: analysis in cheese and susceptibility to proteolytic enzymes from *Lactococcus lactis* ssp. *cremoris*, *J. Dairy Sci.,* 79, 20–26, 1996.
128. Nuijens, J.H., van Berkel, P.H.C., and Schanbacher, F.L., Structure and biological actions of lactoferrin, *J. Mammary Gland Biol. Neoplasia,* 1, 285–294, 1996.
129. Schanbacher, F.L., Talhouk, R.S., and Murray, F.A., Biology and origin of bioactive peptides in milk, *Livestock Product. Sci.,* 50, 105–123, 1997.
130. Loukas, S., Varoucha, D., Zioudrou, C., Straty, R.A., and Klee, W.A., Opioid activities and structures of alpha casein–derived exorphins, *Biochemistry,* 22, 4567–4573, 1983.
131. Chiba, H., Tani, F., and Yoshikawa, M., Opioid antagonist peptides derived from κ-casein, *J. Dairy Res.,* 56, 363–366, 1989.
132. Fiat, A.M. and Jolles, P., Caseins of various origins and biologically active casein peptides and oligosaccharides: structural and physiological aspects, *Mol. Cell Biochem.,* 87, 5–30, 1989.

8 Cheese and Its Potential as a Probiotic Food

Knut J. Heller, Wilhelm Bockelmann, Juergen Schrezenmeir, and Michael deVrese

CONTENTS

0-8493-1372-4/03/$0.00+$1.50

8.1 INTRODUCTION

For the production of cheese from milk, two key steps are essential:

1. Concentration of the milk casein and fat through coagulation of the casein by proteolytic enzymes and/or lactic acid
2. Drainage of the whey after mechanical disruption of the coagulated casein

Starting with this simple basic technique, more than a thousand cheese varieties are produced today.[1] Variety is brought about by altering different aspects of cheese manufacture: type of starter culture, additional cultures, fermentation conditions, renneting, cutting the curd, scalding, drainage of whey, forming of green cheese, salting and addition of spices, and ripening. The vast variety of cheeses makes cheese production profitable as a small cottage industry as well as for national and multinational food producers, since both gourmet and commodity markets may be served. The variety of the market and the steady growth in cheese production provide opportunities to adopt new food trends. One important food trend in developed countries has been the introduction of health and functional foods during recent years. Especially in the dairy sector, the introduction of the probiotic concept has been successfully accomplished. Although most of the probiotic products developed are based on fermented milks, a few examples are visible in the cheese sector.

In this chapter, the production of the different varieties of cheese and their potential to serve as carriers for probiotic bacteria are discussed. Available data on the beneficial effects of cheeses containing probiotic microorganisms are also presented.

8.2 HISTORY

It is commonly believed that cheese evolved in the area of the former Assyria some seven to eight thousand years ago. For people of that time, and probably for those who lived during the following centuries, the most important incentive for cheese production was that cheese constituted a highly nutritious, high-energy food with a much longer shelf life than liquid milk. With the increasing knowledge of cheese production and the influence of acidification, salt dehydration, spices, and ripening on shelf life and taste, very different cheese varieties were developed. Whereas some of our present-day cheeses with international recognition were described more than 1000 years ago, others are rather recent developments of the last three to four centuries. Table 8.1 lists some major cheese varieties together with their first recorded date of production.

Today cheese is recognized as having very high nutritional value due to its generally high content of protein, calcium, riboflavin, and vitamins A and D.[3] Its reputation has not always been that positive, however. In the oldest existing paper dealing exclusively with milk and milk products,[4] the second century physician Galen from Pergamonisis stated: "There is no cheese I can praise except Oxigalactinus, which is made from sour milk. All others are hard to digest, cause heartburn, and fill the stomach with flatulence." (See Chapter 1 for more details on the history of cheese.) Today, adverse effects due to consumption of cheese are no longer

TABLE 8.1
Some Major Cheese Varieties and Their First Recorded Date[2]

Gorgonzola	897	Cheddar	1500	Stilton	1785
Roquefort	1070	Parmesan	1579	Camembert	1791
Grana	1200	Gouda	1697	St. Paulin	1816

Source: Scott, R., *Cheese Making Practice,* Elsevier Applied Science Publishers, London, 1986. With permission.

observed, and reports from developed countries on hygiene problems in cheese are extremely rare. However, consumption of some cheese varieties with high fat and salt content may not be recommended for people who have other risk factors such as high blood pressure, smoking, obesity, diabetes, and heredity for a variety of modern day diseases.

8.3 SIZE OF PRODUCTION

Cheese is one of the most important products of the dairy industry. World cheese production increased by about 10% from 1992 to 1999 to reach ca. 15.4×10^6 tonnes in the latter year. Europe is by far the largest producing block, followed by North America (Table 8.2).

In the European Union, ca. 50% of the cheeses produced are hard/semi hard cheeses, followed by fresh cheese, which accounts for ca. 30% of the cheese produced (Table 8.3).

Especially in Europe, dairy companies are very heterogeneous with respect to the amount of milk processed; the largest multinational companies process several million liters, the smallest cottage companies just process a few hundred liters per day. The large number of small companies is one of the major reasons for the great variety of cheese produced.

8.4 CHEESE GROUPS AND FERMENTATION PROCESSES

8.4.1 Cheese Groups

Cheese manufacture essentially involves concentrating the fat and casein of milk by coagulating the casein enzymatically (rennet cheeses) or by acidic pH (quarg, acid-curd cheeses). Cows' milk is used predominantly in industrial cheesemaking. In Mediterranean countries, sheep's and goats' milks are used for cheesemaking to a large extent. Outside Europe, the milk of water buffaloes, camels, and mares is often used for cheesemaking. Over the centuries, cheese production has led to the existence of an extensive range of cheese varieties. Although more than one thousand individual varieties exist, some of these differ only in size, packaging, place of origin, or name.[5,6]

TABLE 8.2
World Cheese Production Trends, 1992 to 1999[a]

Thousands of Tonnes	1992	1993	1994	1995	1996	1997	1998	1999	% Change 1992–1999
United States and Canada	3590	3594	3693	3807	3938	4002	4085	4126	14.9
South and Central America	803	828	820	814	832	903	914	876	9.1
European Union (15 countries)	5879	5993	6083	6295	6409	6448	6415	6417	9.2
Other Europe	1904	1898	1859	1662	1684	1685	1719	1735	–8.9
Africa	456	474	558	593	575	602	636	640	40.4
Asia	963	966	990	950	957	974	974	984	2.2
Australia and New Zealand	335	335	427	432	503	552	571	602	79.7
World total	13,936	14,107	14,429	14,550	14,897	15,167	15,313	15,378	10.3

[a] Includes cheese made from cows' milk, buffaloes' milk, goats' milk, etc.

Source: The world market for cheese, *Bull. Int. Dairy Fed.*, 359, 12, 2001. With permission.

TABLE 8.3
Cheese Production Trends in the European Union, by Cheese Types, 1992 to 1999[a]

Thousands of Tonnes	1992	1993	1994	1995	1996	1997	1998	1999	% Change 1992–1999
Hard/semi-hard	2418	2619	2671	2720	2842	2847	2841	2828	17.0
Soft	829	878	869	871	844	863	878	885	6.8
Blue	93	93	98	94	100	102	106	108	16.1
Fresh	1339	1355	1407	1472	1478	1539	1547	1574	17.6
Not specified	93	98	112	126	138	139	146	132	41.9
Processed	446	458	460	470	472	473	475	471	5.6
Total	4772	5043	5157	5283	5402	5490	5518	5527	15.8

[a] Except figures for Italy and Portugal.

Source: The world market for cheese, *Bull. Int. Dairy Fed.*, 359, 12, 2001. With permission.

The classification of cheeses includes an indication of the process used in the manufacture.[5,7] Apart from the moisture content, the method of coagulation of the milk is also used for classification. The main difficulty in using classification schemes is that large moisture ranges are allowed for the major cheese groups. According to moisture levels, at least three cheese types are generally defined: hard cheeses (moisture 20 to 45%), semi-hard/semi-soft cheeses (moisture 45 to 55%), and soft cheeses (moisture >55%). All three types are consumed after a certain ripening period in contrast to fresh cheeses (moisture >70%), which are consumed after draining (Table 8.4).

8.4.2 Handling of Milk

Of particular significance for cheese manufacture is the variation of fat and casein levels in milk. The relative proportions of casein and fat in milk for cheesemaking are standardized to minimize seasonal variation in milk composition and to facilitate the production of cheeses that comply with specific regulations that ensure the uniform aroma and texture required for a single cheese variety. Homogenization is not routinely used in cheese manufacture, but it can be used to improve the texture of cheeses. Homogenization is used mainly in the manufacture of unripened fresh cheeses.[6]

In general, cold storage of milk is necessary before cheese manufacture. The main problems with this step are growth and enzyme activities of psychrotrophic microorganisms, including Gram-negative bacteria (e.g., *Pseudomonas, Enterobacter*) and Gram-positive spore forming bacilli (e.g., *Bacillus cereus*). Psychrotrophic microorganisms have been shown to produce heat-stable proteinases, which may be responsible for decreasing cheese yields, and lipases, which can cause rancid off-flavors in aged cheeses.[5]

Most cheeses are made from pasteurized milk (72°C for 15 sec). Pasteurization does not affect the physicochemical parameters of the milk significantly, but it

TABLE 8.4
Major Cheese Groups

Fresh Cheeses (Quarg Type Cheeses, Cottage Cheese, Etc.)

Cheeses with 14–30% dry matter (D.M.) and 0–75% fat
Acid dominated coagulation (16–48 h)
High initial heat treatment (82–88°C)
Spontaneous draining by optional slicing or molding
Acidification and renneting at 18–28°C (mesophilic lactic starters)
For some varieties, more rennet and higher temperatures are used (28–32°C), resulting in cheeses with a DM of 25–33%
Salting: mixing with salt
No ripening, cheeses are consumed after packaging

Soft Cheeses (Camembert, Brie, Chaumes, Romadour)

Cheeses with 40–50% DM and 25–60% fat i.d.
Produced from thermized or pasteurized milk
Acidification and coagulation (30–90 min); balanced action of rennet and lactic starters
Temperatures 32–35°C favor the action of rennet
Spontaneous draining promoted by optional slicing, molding, or pressing
Dry salting or brining (NaCl 1.6–2%)
Short ripening period (14 d)
Very heterogeneous cheese group; appearance and taste depending mainly on the surface flora (molds or red smear)

Pressed Cheeses (Semisoft, Semihard)

Uncooked: No or Slight Heating During Vat Mixing (Gouda, Edam, Cheddar)
44–55% DM, 30–55% fat i.d.
Acidification and coagulation rennet dominant (32–37°C, 30–60 min)
Draining by mechanical treatments: slicing, mixing, prepressing, pressing
Salting to 1.5–2% NaCl by brining
Ripening: 12–60 d (or longer: 6–12 months)
Characteristics: homogenous dough with few small "eyes"

Pressed Cheeses (Semihard, Hard Varieties)

Cooked: Emmental, Gruyère, Comté, Parmesan, Sbrinz
Cheeses with 58–64% DM, 30–55 fat i.d.
Cheeses with or without propionic acid fermentation
Acidification and coagulation rennet dominant (33–38°C, 12–30 min)
Heating during vat mixing (52–55°C)
Mesophilic and thermophilic starters (propionic starters)
Use of thermized milk, often raw milk
Draining by mechanical treatments: slicing, mixing (heating), pressing
Surface with or without microflora
Ripening 6 weeks to >12 months
Characteristics: firm and smooth dough with or without "eyes"

destroys most of the pathogenic and spoilage bacteria contaminating the milk. Some nonstarter lactic acid bacteria (*Lactobacillus* spp., *Streptococcus* spp.) and, if present, sporeforming bacteria (*Clostridium, Bacillus*) can survive pasteurization and can affect cheese ripening. Heating of milk before cold storage (63 to 65°C, 15 to 20 sec) may be used for prolonged storage; however, the milk is still pasteurized before cheesemaking.[6]

8.4.3 Starter Cultures

Acid production during cheesemaking is essential in the formation of the gel from casein. Starter cultures promote rapid acid development during curd manufacture and contribute to distinct textural and flavor properties in the cheese after ripening.[5,8] Bacteria used as starter cultures generally belong to the genera *Lactococcus, Streptococcus, Leuconostoc,* and *Lactobacillus.* The typical mesophilic starter culture consists of *Lactococcus lactis* ssp. *lactis* (*L. lactis*) and *L. lactis* ssp. *cremoris* (*L. cremoris*). These starters are used for cheeses with moderate scald temperatures (<40°C, Gouda, Edam, etc.). For cheeses requiring higher scald temperatures, thermophilic starter cultures are used (*Streptococcus thermophilus*, *Lactobacillus helveticus, Lactobacillus delbrueckii* ssp. *bulgaricus,* etc.). When "eyes" are required, the gas-forming species *L. lactis* ssp. *lactis* biovar. *diacetylactis, Leuconostoc mesenteroides,* or *Propionibacterium shermanii* can be used.

8.4.4 Clotting Methods

Most of the major international cheese varieties are rennet cheeses. A small percentage of cheeses are acid curd cheeses (cottage cheese, quarg, feta, Harz cheese) or heat/acid coagulated cheeses (ricotta). For rennet-type cheeses, the gel is formed at a higher pH than achieved by acid alone. The rennet gel is more elastic than an acid gel and shrinks and expels moisture in the presence of heat and acid. Rennet is used in cheese manufacture primarily to coagulate milk. However, residual rennet, which is kept in the curd, also plays an important role in the generation of flavor compounds during cheese ripening. In the past, the principal coagulant used was calf rennet, mainly consisting of chymosin and some pepsin or some microbial proteinases from different mold species. Today a pure recombinant chymosin is increasingly used. Some well known trade names are "Chymogen," "Maxiren," and "Chy-Max."[5,9,10]

8.4.5 Removal of Whey and Salting

After cooking or scalding, the curd, containing the caseins and milk fat, is separated from the whey, containing lactose, whey proteins, and minerals. In the case of Cheddar cheese, the curd may be textured, milled, and salted. The curd of other cheese varieties may be hooped directly into perforated forms, and then an appropriate pressure is applied for draining. After whey removal, cheeses may be salted by applying dry salt to the cheese surface or by immersing cheese blocks in brines containing 18% or more NaCl. Brining is performed on small-sized soft cheeses for a few hours up to 24 h, and for longer periods of time for larger cheeses such as Gouda, Edam, or Tilsit (>2.5 kg). Aged brines develop a typical salt- and acid-tolerant microflora (10^4 to 10^6 colony-forming units [CFU]/ml), often with *Debaryomyces hansenii* and *Staphylococcus equorum* as the predominant species. An influence of the brine microflora on cheese ripening is significant for surface ripened cheeses, especially for smear cheese varieties.[5–7,10,11]

8.5 CHANGES IN THE COMPOSITION OF CHEESES RESULTING FROM FERMENTATION AND RIPENING

8.5.1 Initial Changes during Production

When the casein in milk is coagulated enzymatically or by acidic pH, the fat and casein of the milk are concentrated 6- to 12-fold. Cows' milk contains about 3.3% protein (caseins, whey proteins). The primary proteolysis is mediated by chymosin, which specifically hydrolyzes the peptide bond Phe_{105}–Met_{106} of κ-casein. Upon cleavage, κ-casein loses its micelle-stabilizing properties. The solubility of the micelles is reduced, an aggregation occurs, and a coagulum is formed. The texture of the coagulum is dependent on the protein content of the milk, the pH, and the level of calcium ions in the milk. The gel starts to shrink (syneresis) when the coagulum is cut, and the water content of 87% is reduced to values between 20 and 56% in the curd, depending on the cheese variety. Syneresis of the coagulum is controlled by combinations of time, temperature, pH, agitation, and pressure. At the end of manufacturing, all rennet cheeses are essentially very similar. The calcium paracaseinate matrix, with various levels of dispersed lipids and a moisture content in the range of 35 to 50%, has a rubbery texture and is essentially flavorless.[1,12]

During manufacturing, lactic acid is produced first by starter lactic acid bacteria, and most of the lactose in the cheese is fermented to lactate in the first 24 h of pressing. For Dutch-type cheeses, lactose levels are already undetectable after brining. The rate and extent of acidification have a major impact on cheese texture via demineralization of the casein micelles. About 98% of the lactose is removed in the whey during draining. The salt and moisture content of the curd determine the products of postmanufacture residual lactose fermentation. The complete and rapid metabolism of residual lactose by lactic acid bacteria is essential for the production of a good quality cheese.[5,13]

8.5.2 Cheese Ripening

The curd manufacturing process determines the basic composition and structure of cheeses. During ripening, cheeses develop their individuality and unique characteristics. Flavor and texture development is largely controlled by complex biochemical reactions, with glycolytic, proteolytic, and lipolytic activities being the primary events during cheese ripening.[13,14] The extent of protein and fat degradation is determined by the moisture, pH, and salinity in the cheese. Enzymes of various sources result in the production of peptides, amino acids, fatty acids, carbonyl components, and sulfur compounds. All biochemical reactions are usually limited to a certain degree because excessive lipolysis can lead to rancid flavors, and excessive proteolysis may produce bitter off-flavors or altered texture (softening) not characteristic for the specific product. The sources of cheese ripening enzymes are:

Indigenous milk enzymes (proteinases, lipases, phosphatases)
Psychrotrophic nonstarter bacteria
Added calf rennet, rennet substitutes, or recombinant chymosin
Starter lactic acid bacteria
Nonstarter lactic acid bacteria
Secondary microflora (molds, smear bacteria and yeasts, etc.)

With high levels of nonstarter lactic acid bacteria (e.g., *Lactobacillus* spp., *Pediococcus* spp.) present in the cheese milk, D-lactate may be formed in addition to L-lactate. During ripening, these bacteria may racemize L-lactate to D-lactate, which may not be significant for flavor characteristics but may have undesirable nutritional consequences, particularly in infants. In addition, calcium D-lactate is less soluble and may crystallize in cheese, causing undesirable white spots especially on cut surfaces. The nonstarter lactic acid bacteria have oxidative activities, oxidizing lactate to acetate. Acetate is usually present at high concentrations in Cheddar cheese and is considered to contribute to cheese flavor.[13]

The small amounts of citrate in milk can be metabolized by *L. lactis* ssp. *diacetylactis* and *Leuconostoc* species, which release diacetyl and CO_2. The CO_2 production is responsible for the characteristic eyes of some Dutch-type cheeses. Citrate metabolism is very important for aroma and flavor formation in cottage cheeses and quarg.[13]

Milk contains a very potent lipoprotein lipase. This enzyme probably has no significance for cheeses made from pasteurized milk, since it is completely inactivated by the pasteurization process. It may cause significant lipolysis in raw milk cheeses. Lactic acid bacteria (LAB) (*Lactococcus* and *Lactobacillus*) have low but measurable lipolytic and esterolytic activities. Lipolysis is generally considered undesirable because even a moderate level of free fatty acids makes cheeses smell rancid. The exceptions to this rule are certain Italian hard cheeses and blue-veined cheeses. The blue cheese flavor is dominated by methylketones and fatty acids.[13]

Phosphatases can play an important role in cheese maturation and cheese development because part of the casein is protected by phosphorylation and can only be degraded by the combined action of proteinases and phosphatases. The origin of acid phosphatases is the bovine milk and starter lactic acid bacteria. The enzymes are also found in *Penicillium roqueforti*.[13]

8.5.3 Proteolysis

Proteolysis is the most complex of the three primary events of cheese ripening and is probably the most important for development of flavor and texture. Proteolysis contributes directly to flavor via peptides, amino acids, and derivatives of amino acids (amines, acids, thiols, thioesters, etc.), an increased release of sapid compounds during mastication, a pH increase by formation of NH_3, and changes in texture from degradation of the protein matrix. Proteolysis is responsible for the generation of the smoother, softer texture of mature cheeses. Full flavor is probably caused by correct balance of a mixture of aromatic compounds (component balance theory).[15] The products of primary proteolysis, the water insoluble fraction consisting of proteins and large peptides, have no flavor or aroma. Secondary proteolysis includes the water soluble nonvolatile fraction consisting of small peptides, amino acids, and organic acids, which contains most of the compounds responsible for flavor, while the aroma is principally found in the volatile fraction.[13,16]

8.5.4 Milk Proteinases and Rennet

Plasmin and plasminogen are associated with the cheese micelles and accompany them into the cheese curd. In milk, the most susceptible plasmin substrate is β-casein. Plasmin activity in cheese and rennet curd increases with cooking temperature, apparently due to the activation of plasminogen. This suggests the importance of plasmin activities is higher for cooked Swiss-type cheeses than for cheddar or Dutch-type cheeses. Pasteurization increases plasmin activity in milk by inactivation of plasmin inhibitors or increasing the activation of plasminogen.[13] Between 3 and 6% of the rennet added to the cheese milk is retained in the curd. This is influenced by the pH at whey drainage; with increasing pH, a smaller amount of rennet is retained in the curd. Very little coagulant survives the high cooking temperatures used for Swiss-type cheeses. In the initial stages of maturation α_{S1}-casein is degraded by rennet. The amount of intact α_{S1}-casein is related to the firmness of the cheese. If the level of primary proteolysis is excessive, off-flavors due to bitter peptides can occur. β-casein and α_{S1}-casein are both hydrolyzed by rennet *in vitro*. In cheddar and Dutch-type cheeses, only α_{S1}-casein is completely degraded. In the cheese environment, the β-casein peptides liberated are different from the rennet specificity, suggesting that plasmin and/or bacterial proteinases are responsible for the hydrolysis.[13]

Proteolysis by rennet is believed to be responsible for the softening of cheese texture early during ripening via the hydrolysis of α_{S1}-casein to α_{S1}-I. Some rennet-produced peptides are bitter. Rennet-produced peptides can serve as substrate for microbial proteinases and peptidases that produce small peptides, which contribute to the background cheese flavor. All changes in cheese texture appear to influence the general release of aromatic compounds from proteolysis, lipolysis, and glycolysis. This may be the most significant contribution of this primary proteolysis to cheese flavor.[13]

8.5.5 Psychrotrophic Microorganisms

A very important source of potent lipases in milk is the psychrotrophic bacteria that dominate the microflora of refrigerated milk. Some of these lipases are heat stable, can adsorb on the surface of fat globules, and can act in the cheese environment. The negative effects of the lipases of psychrotrophic bacteria in cheesemaking are most significant in terms of their effects on proteinases.[5]

8.5.6 Starter and Nonstarter Bacteria

LAB have complex amino acid requirements and thus have a proteolytic system consisting of proteinases and peptidases that fulfills their need for a nitrogen supply.

Proteinases of starter lactococci are cell wall bound; all the known peptidases in *Lactococcus* are intracellularly located. A number of peptidases of various specificities were identified in *Lactococcus* and *Lactobacillus* species: aminopeptidases, dipeptidases, tripeptidases, and oligo-endopeptidases, as well as some peptidases with specificities for proline residues such as the prolinase and prolidase. Starter bacteria reach maximum numbers shortly after the end of manufacture, and viable cell counts usually decline quickly thereafter. It can be assumed that some of the bacterial cells lyse after cell death, liberating the intracellular enzymes into the cheese, which promotes the degradation of peptides and proteins.[8,17]

Thermophilic starters are used for a number of cheeses such as the Swiss-type cheeses and Parmesan and Romano; the species *S. thermophilus, Lb. delbrueckii* ssp. *bulgaricus, Lb. delbrueckii* ssp. *lactis,* and *Lb. helveticus* are used. Lactobacilli are generally used in combination with *S. thermophilus* because of mutual growth stimulation, e.g., *S. thermophilus* is stimulated by free amino acids produced by *Lb. bulgaricus,* while the production of formic acid and CO_2 by *S. thermophilus* stimulates the growth of *Lb. bulgaricus.* In Swiss-type cheeses, propionibacteria metabolize lactate to propionate, acetate, and CO_2, which is responsible for eye development. They contribute to proteolysis significantly and are responsible for the high concentration of proline in Emmental cheese, which contributes to the sweet taste of Emmental.[5,8,13]

Nonstarter lactic acid bacteria (NSLAB) reach populations of 10^8 CFU/g during the ripening of many cheeses. Lactobacilli usually dominate the flora. In pasteurized milk, NSLAB originate mainly as postpasteurization contaminants from the factory. They are a very heterogeneous group and include thermophilic and mesophilic, heterofermentative as well as homofermentative species. NSLAB have a significant impact on cheese quality. Due to the variability, the number, and the species composition of the NSLAB population, the contribution of these bacteria is difficult to characterize in detail. Sometimes pediococci are the predominant NSLAB species. They are weakly proteolytic and lipolytic, and their main contribution to cheese flavor may be their ability to oxidize lactate to acetate. They are also able to reduce acetaldehyde and propionaldehyde to alcohols and can produce diacetyl.[5,13]

8.5.7 Secondary Microorganisms

In surface ripened cheese varieties, yeasts and molds metabolize lactate to CO_2 and H_2O, causing an increase in pH. Amino acids are also metabolized, which liberates NH_3, further increasing the surface pH. The elevated pH stimulates the action of plasmin, which together with the coagulant is responsible for proteolysis in the cheese.[11,14,18]

Molds such as *Penicillium roqueforti* and *P. camemberti* possess a strong proteolytic system and are responsible for the liberation of amines resulting from decarboxylation, ammonia, keto acids, carbonyls, alcohols from deamination, other amino acids resulting from transaminations, and hydrogen sulfide, methanethiol, thioesters, and other sulfur compounds from conversion of cysteine and methionine.

Brevibacterium linens, Corynebacterium spp., *Arthrobacter* spp., and "food-grade" *Staphylococcus* species (formerly described as *Micrococcus* sp.) have active proteolytic systems as well as strong amino acid converting properties. The main contribution of the surface smear bacteria is the liberation of highly aromatic sulfur compounds released by the catabolism of amino acids.

8.6 CHEESE AS A CARRIER FOR PROBIOTIC MICROORGANISMS

8.6.1 General Considerations

Probiotic bacteria have to fulfill a number of basic technological requirements when used in commercial probiotic products. Most importantly, probiotic bacteria have to be present in sufficient numbers in the product at the date of consumption, and their properties essential for expressing health benefits after consumption have to be maintained up to that date. In addition, no adverse effects on taste and aroma of the product should be exerted by the probiotic organisms.

For exploitation of the functional properties of the probiotic bacteria, the processes of manufacture of cheese products may have to be modified and adapted to the requirements of the probiotics. Where this is not possible, either other probiotic strains may be applied or new products may have to be developed. In the following, some of the parameters necessary for or influencing the application of probiotic bacteria in cheese will be addressed.[19]

Dairy products containing living bacteria have to be cooled during storage. This applies in particular to products containing live probiotic bacteria. Cooling is necessary to guarantee high survival rates of the probiotics and to yield sufficient stability of the product.[20,21] In addition, oxygen content, redox potential, and water activity of the product have to be considered,[22] since the target of probiotic bacteria is the intestinal tract. This may be of considerable importance for prepackaged cheese. Cooling of probiotic cheese is also necessary to reduce or inhibit the interaction of the active microorganisms with the components of the food. The degree of interaction depends on the kind and amount of carbohydrates available, degree of hydrolysis of milk proteins and thus availability of essential amino acids, and composition and degree of hydrolysis of milk lipids, determining the availability of short chain fatty acids.[13,23] However, the proteolytic[24] and lipolytic properties of the probiotic bacteria may have considerable effects on taste and flavor of the product.[13]

Interactions between probiotics and starter organisms also have to be considered. The intensity of interaction very much depends on when the probiotics are added to the product; whether they are present during or added after fermentation. If they are added after fermentation, interactions may be kept to a minimum, since addition is possible immediately before or even after cooling below 8°C, and metabolic activities of starters and probiotics are drastically reduced at these temperatures. Nevertheless, during extended storage, even small interactions may yield measurable effects. These considerations, however, are not entirely new, since application in cheese manufacture of adjunct cultures[23] has already created some experience in the interaction of starter cultures with additional active bacteria.

Antagonism between bacteria is often based on the production of metabolites that inhibit or inactivate more or less specifically other related starter organisms or even unrelated bacteria. While antagonism caused by bacteriocins, peptides, or proteins exhibiting antibiotic properties[25] has been described as a limiting factor for combinations of starters and probiotics,[26] antagonism caused by other substances also has to be considered. Substances that may be of importance include hydrogen peroxide, benzoic acid, biogenic amines, and lactic acid.[27–31]

If probiotics are added to the cheese after fermentation, the physiological state of the probiotics may be of considerable importance for survival during ripening and/or storage.[32–34] This state depends on:

1. The nutritional composition of the growth medium of the probiotics in relation to the nutritional composition of the cheese to which they will be added
2. Harvesting of the culture (whether in logarithmic or stationary phase)
3. Conditions leading to transition to stationary phase
4. Treatment of the probiotics during and after harvesting

However, clues on how to handle probiotics may be drawn from current experience in production of commercial starter cultures.[34]

8.6.2 Fresh Cheese as a Carrier

Due to its manufacturing process, fresh cheese appears to be ideally suited to serve as a carrier for probiotic bacteria. One reason is that fresh cheese is an unripened cheese; thus, storage occurs at refrigeration temperatures, shelf life is rather limited, and no prolonged periods of ripening are necessary. As an example of fresh cheese, cottage cheese will be discussed in some detail. For this type of cheese, two options exist for the addition of probiotics:

1. Together with the starter culture
2. Together with cream and salt

For addition together with the starter culture, two problems can be seen:

1. The number of the probiotic bacteria in the final product may be difficult to control since a considerable number of bacterial cells are lost from the coagulant during drainage of the whey.
2. Survival of the probiotic bacteria in the product may be negatively affected by the rather high scalding temperature of up to 55°C.

An advantage for the producer would certainly be the fact that the size of the probiotic inoculum and thus the costs would be rather small. However, for cottage cheese, addition of the probiotics together with cream and salt appears to be a desirable alternative. The advantages are that the numbers of probiotics added can be exactly controlled, adverse effects of the high scalding temperature are avoided, and after addition and mixing, the product can be immediately cooled to below 8°C.

To date, two reports on the suitability of fresh cheese as a carrier for probiotic bacteria have been published. In one report,[36] probiotic bacteria (*Lactobacillus acidophilus, Lb. casei, Bifidobacterium bifidum, B. longum*) were added as adjuncts, and survival during refrigerated storage was analyzed. The data showed that although viable counts decreased in 16 days by about one log, final counts after this period were still acceptable. In contrast, Blanchette et al.[37] reported an increase of *B. infantis* within the first day after manufacture. However, large losses in viability were observed after 15 days at 4°C.

In Germany, the first cottage cheese called "probiotic" appeared on the market in 1998. The product contained *Lb. acidophilus* La5 and *B. animalis* BB12. However, while the probiotic properties of the bacterial strains added to the product have been demonstrated,[38] the product itself has never been tested for probiotic properties (health benefits).

8.6.3 Ripened Cheese as a Carrier

In order to apply probiotics in ripened cheese, the same considerations as for cottage cheese must be taken into account with regards to the time of addition of the probiotics and the impairment of survival by the scalding temperature. In ripened cheese, however, the long period of ripening causes an additional problem. It is by no means clear to what extent different probiotic strains will survive the ripening period and to what extent their functional properties will be conserved during this period. Several reports for ripened cheese have been published. In the reports, probiotic strains of the following species were added: *Lb. acidophilus, Lb. paracasei, B. infantis, B. lactis,* and *Enterococcus faecium.*

While several studies either showed unsatisfactory survival of the probiotic microorganisms during ripening or adverse effects on flavor,[39,40] the suitability of the approach of adding probiotics as adjuncts together with the starter culture appeared to be successful in others.[41–49] In one study, by Gardiner et al.,[44] it was demonstrated that cheddar cheese yielded significantly higher mean fecal probiotic counts than a yogurt produced with the same probiotic organism. In 1999, a patent for production of probiotic cheese was granted,[50] and in 2000 a probiotic cheese containing *Lactobacillus* GG was introduced into the Finnish market. *Lb.* GG is one of the best characterized probiotic bacterial strains with well-established probiotic properties.[51] However, so far no data exist on its probiotic properties when supplied in a cheese matrix.

8.7 RESULTS OF *IN VITRO* ANIMAL TESTS SHOWING BENEFICIAL EFFECTS OF CHEESE

Up to now there have been no clinical studies showing beneficial health effects of the consumption of cheese containing health-promoting bacteria, although this is, due to the strain and matrix specificity of probiotic effects, an essential prerequisite for the use of the claim as probiotic.[52,53] "Probiotic cheese" that contains bacteria of human origin or other bacterial strains has been reported with a wide range of possible beneficial properties (Table 8.5). Most of these published investigations are confined to providing proof of survival and sufficient numbers of probiotic bacteria in a cheese matrix or in ready-to-eat cheese. "Sufficient" is assumed when a regular daily serving contains ca. 10^8 probiotic bacteria. Accordingly, hard cheese (daily consumption 1 to 3 slices at 30 g per slice) should contain $\geq 3 \times 10^6$ CFU/g.

It has been postulated that the embedding of probiotic bacteria in the fat–protein matrix of cheese as well as the buffering capacity and the low acidity of ripened cheese may assist survival of probiotic bacteria during gastrointestinal passage. Vinderola et al.[36] have demonstrated the pH tolerance of strains of *B. longum, B. infantis, Lb. acidophilus* and *Lb. casei* in homogenates of Argentinian fresco cheese using an HCl solution of pH 3. *Propionibacterium freudenreichii* and *Propionibacterium acidopropionici* from Emmental-like cheeses showed survival and acid and bile tolerance *in vitro* when cells in Argentinian Emmental juice were exposed to artificial gastric and intestinal fluids.[54,55] In two feeding trials involving three or eight pigs per group, respectively, it was demonstrated that consuming Cheddar cheese containing *Lb. paracasei* NFBC 338[43,46] or *E. faecium* PR 68[44,45] led to significantly higher mean fecal counts of probiotic bacteria than consuming yogurts produced with the same bacteria. A positive serum IgG response in the probiotic group was observed, but no influence on fecal coliforms or on pig growth and food efficiency was found.

The high microbial β-galactosidase activity in Canestrato hard cheese[41] or cheddar-like cheese[42] supports lactose digestion in lactose-intolerant people but may be provided by any other lactic acid bacteria producing β-galactosidase and is not bound to the viability of the microorganisms. Angiotensin-1 converting enzyme (ACE)-inhibitory bioactive peptides are released in Festivo cheese during ripening by microbial proteolysis and were shown to decrease blood pressure when this cheese was fed to rats.[56] Again, the survival of the microorganisms during gastrointestinal transit is not required for the expression of this health effect.

Most of the proposed beneficial effects of bacteria that fulfill the criteria for the definition "probiotic" have been tested in clinical trials where probiotic bacteria were provided to subjects in fermented, yogurt-type milk products, in nonfermented milk, or as lyophilized pharmaceutical preparations. Table 8.5 lists those strains that have already been applied for the production of probiotic cheese. The data indicate that the strains applied show good survival during ripening of the cheese and during the gastrointestinal passage after consumption. Apparently, the cheese matrix is very well suited to protect the probiotic bacteria against oxygen, low pH, and bile salts. In the two studies carried out, fecal recovery of the probiotic bacteria was better when cheese instead of yogurt was used as carrier. However, only two studies addressing this topic have been performed so far.

8.8 CONCLUSIONS

Several reports have been published in which the production of probiotic cheese has been described. However, the designation as "probiotic" relies on the application of bacterial strains for which probiotic properties have been demonstrated. So far, no clinical studies have shown that cheese may in fact serve as a functional carrier for probiotics.

Compared to yogurt, the problem for cheese — especially semi-hard and hard cheese — acting as carrier for probiotics results from the high fat and salt content

TABLE 8.5
Probiotic Bacterial Strains Used in Cheesemaking and Postulated Health-Related Effects

Strain	Cheese	Survival in Cheese cfu/g cheese (time of storage)	Effects (tested in cheese)	Postulated Strain-Specific Beneficial Effects (on)[a]	Ref.
Lb. rhamnosus GG	Gefilus cheese: Emmentaler, Edam	3–4 slices of Edamer are equivalent to 150 ml yogurt		Survival of gastrointestinal passage, colonization of the gut;[59] rotavirus-induced diarrhea;[50,61] traveler's diarrhea;[62] antibiotic-induced diarrhea;[63–65] morbus Crohn;[66] obstipation;[67] premature infants;[68] immune modulation;[69–71] allergy, atopic diseases;[71,72] prophylaxis of respiratory and gastrointestinal infections;[73] decrease of cancer promoting enzymes;[67] reduction of caries risk[74]	Valio, Finland 2000 (in preparation)
Lb. acidophilus LA 5 + *B. animalis* BB12	Soft cheese			Survival of gastrointestinal passage;[75,76] rotaviral diarrhea;[77] traveler's diarrhea;[78] antibiotic-induced diarrhea;[79] infants;[76,80] modulation of the immune system;[81,82] allergy;[71] cancer;[83] serum cholesterol[84]	Hansen, Denmark
B. animalis BB12	Cheddar	6×10^7 (2 months)			57
B. animalis BB12	Cheddar	$\geq 10^8$ (6 months)			58
B. longum BB536	Cheddar	$\sim 10^5$ (6 months)		Antibiotic-induced diarrhea[85]	58

Lb. paracasei NFBC 338[b]	Cheddar	Survival and growth in cheese $>10^8$ (6 months)	Better fecal recovery in cheddar than in yogurt	Bile tolerance[46]	43
E. faecium PR 88	Cheddar	Survival and growth in cheese $>10^8$ (15 months)	Better survival of GI passage of *E. faecium* PR 88 in cheddar than in yogurt[c]	Bile and acid tolerance;[86] alleviation of symptoms of irritable bowel syndrome[d,86]	44, 45
B. lactis Bo plus *Lb. acidophilus* Ki	Gouda; goat cheese	Bo: $>10^8$ (9 weeks) Ki: $>10^6$ (9 weeks)		Bile tolerant, establishment in intestinal ecology;[87] cholesterol control;[88] bactericidal effects on *Salmonella typhimurium* or *Clostridium difficile*.[89]	48
B. longum, B. bifidus, Lb. acidophilus, Lb casei	Argent, fresco	$-1\ \log_{10}$/2 mo or less	3 h survival in a cheese-HCl homogenate at pH 3		36
Propionibacterium freudenreichii/acidi propionici isolated from cheese	Emmentaler-like		Survival in cheese juice of bacteria exposed to artificial gastric and intestinal fluid; bile and acid tolerant		54, 55

[a] Not tested in cheese.

[b] Feeding 10^9 or 10^{11} CFU/d NFBC 338 in cheddar or yogurt, respectively, to three pigs led to a recovery of 10^5 or $10^{4.5}$ CFU/ml small intestinal chyme.

[c] Feeding 1.3×10^{10} or 3.7×10^9 CFU/d PR 68 in cheddar or yogurt, respectively, to eight pigs led to a fecal recovery of 2×10^6 or 5.2×10^5 CFU/g feces.

[d] After an initial load by gastric intubation, 17 patients with otherwise incurable irritable bowel syndrome (IBS) received for 4–30 months lyophilised *E. faecium*. Weekly examination of fecal samples; assessment of condition scores before and after treatment.

and the relatively low recommended daily intake. It follows that the concentration of probiotics in cheese should be about four to five times higher than in yogurt. However, this does not apply to fresh cheese, such as cottage cheese, which can easily be adjusted to low fat and salt contents, and for which recommended daily intake is rather high. Fresh cheese may thus serve as a food with a high potential to be applied as a carrier for probiotics.

In countries in which yogurt or other fermented milks are less popular, and for lactose-intolerant persons who do not even tolerate yogurt, probiotic cheese may be a realistic alternative.

REFERENCES

1. Fox, P.F., Cheese: an overview, in *Cheese: Chemistry, Physics, and Microbiology; General Aspects,* Vol. 1, Fox, P.F., Ed., Chapman and Hall, London, 1993, pp. 1–36.
2. Scott, R., *Cheese Making Practice,* Elsevier Applied Science Publishers, London, 1986.
3. Dillon, J C. and Berthier, A.M., Nutritional characteristics of cheese, in *Cheesemaking — from Science to Quality Assurance,* Eck, A. and Gillis, J.C., Eds., Lavoisier Publishing, Paris, 2000, pp. 663–673.
4. Gesner, C., *Libellus de Lacte, et Operibus Lactariis, Philologus Pariter ac Medicus,* C. Froschauer, Zürich, 1541, p. 22.
5. Johnson, M. and Law, B.A., The origins, development and basic operations of cheese-making technology, in *Technology of Cheese Making,* Law, B.A., Ed., Sheffield Academic Press, Sheffield, 1999, pp. 1–32.
6. Banks, J.M., Cheese, in *Technology of Dairy Products,* 2nd ed., Early, R., Ed., Blackie Academic Professional Press, London, 1998, p. 81.
7. Ramet, J.P., Comparisons in the processing of different types of curd, in *Cheesemaking: from Science to Quality Assurance,* Eck, A. and Gillis, J.C., Eds., Lavoisier Publishing, Paris, 2000, pp. 312–340.
8. Høier, E., Janzen, T., Henriksen, C.M., Rattray, F., Brockmann, E., and Johansen, E., The production, application and action of lactic cheese starter cultures, in *Technology of Cheese Making,* Law, B.A., Ed., Sheffield Academic Press, Sheffield, 1999, pp. 99–131.
9. Harboe, M.K. and Budtz, P., The production, action and application of rennet and coagulants, in *Technology of Cheese Ripening,* Law, B.A., Ed., Sheffield Academic Press, Sheffield, 1999, pp. 33–65.
10. Lomholt, S.B. and Qvist, K.B., The formation of cheese curd, in *Technology of Cheese Making,* Law, B.A., Ed., Sheffield Academic Press, Sheffield, 1999, pp. 66–98.
11. Bockelmann, W., Secondary cheese cultures, in *Technology of Cheese Making,* Law, B.A., Ed., Sheffield Academic Press, Sheffield, 1999, p. 132.
12. Law, B.A., Cheese ripening and cheese flavour technology, in *Technology of Cheese Making,* Law, B.A., Ed., Sheffield Academic Press, Sheffield, 1999, pp. 163–192.
13. Fox, P.F., Law, J., McSweeney, P.H.L., and Wallace, J., Biochemistry of cheese ripening, in *Cheese: Chemistry, Physics, and Microbiology; General Aspects,* Vol. 1, Fox, P.F., Ed., Chapman and Hall, London, 1993, pp. 389–438.
14. Choisy, C., Desmazeaud, M., Gueguen, M., Lenoir, J., Schmidt, J.L., and Tourneur, C., Cheese ripening: Microbial phenomena, in *Cheesemaking: from Science to Quality Assurance*, Eck, A. and Gillis, J.C., Eds., Lavoisier Publishing, Paris, 2000, pp. 353–417.

15. Kosikowski, F.V. and Mocquot, G., *Advances in Cheese Technology,* FAO, Rome, 1958, p. 133.
16. Sousa, M.J., Ardo, Y., and McSweeney, P.L.H., Advances in the study of proteolysis during cheese ripening, *Int. Dairy J.,* 11, 327–345, 2001.
17. Law, J. and Haandrikman, A., Proteolytic enzymes of lactic acid bacteria, *Int. Dairy J.,* 7, 1–11, 1997.
18. Gripon, J.C., Mould-ripened cheeses, in *Cheese: Chemistry, Physics and Microbiology,* Vol. 2, Fox, P.F., Ed., Chapman and Hall, London, 1993, pp. 111–136.
19. Heller, K.J., Probiotic bacteria in fermented foods: product characteristics and starter organisms, *Am. J. Clin. Nutr.,* 73, 374S–379S, 2001.
20. Shah, N.P., Lankaputhra, W.E.V., Britz, M.L., and Kyle, W.S., Survival of *L. acidophilus* and *Bifidobacterium bifidum* in commercial yogurt during refrigerated storage, *Int. Dairy J.,* 5, 515–521, 1995.
21. Roy, D., Mainville, I., and Mondou F., Bifidobacteria and their role in yogurt-related products, *Microecol. Ther.,* 26, 167–180, 1997.
22. Dave, R.I. and Shah, N.P., Effect of cysteine on the viability of yogurt and probiotic bacteria in yogurts made with commercial starter cultures, *Int. Dairy J.,* 7, 537–545, 1997.
23. Fox, P.F., Wallace, J.M., Morgan, S., Lynch, C.M., Niland, E.J., and Tobin, J., Acceleration of cheese ripening, *Antonie van Leeuwenhoek,* 70, 175–201, 1996.
24. Kunji, E.R.S., Mierau, I., Hagting, A., Poolman, B., and Konings, W.N., The proteolytic systems of lactic acid bacteria, *Antonie van Leeuwenhoek,* 70, 187–221, 1996.
25. De Vuyst, L. and Vandamme, E.J., *Bacteriocins of Lactic Acid Bacteria,* Chapman and Hall, London, 1994.
26. Joseph, P.J., Dave, R.I., and Shah, N.P., Antagonism between yogurt bacteria and probiotic bacteria isolated from commercial starter cultures, commercial yogurts, and a probiotic capsule, *Food Aust.,* 50, 20–23, 1998.
27. Desmazeaud, M., Growth inhibitors of lactic acid bacteria, in *Dairy Starter Cultures,* Cogan, T.M. and Accolas, J.-P., Eds., VCH Publishers, New York, 1996, pp. 131–155.
28. Sieber, R., Buetikofer, U., and Bosset, J.O., Benzoic acid as a natural compound in cultured dairy products and cheese, *Int. Dairy J.,* 5, 227–246, 1995.
29. Weber H., Starter cultures in dairy industry, in *Mikrobiologie der Lebensmittel: Milch und Milchprodukte* (in German), Weber, H., Ed., Behr's Verlag, Hamburg, 1996, pp. 105–152.
30. Lankaputhra, W.E.V., Shah, N.P., and Britz, M.L., Survival of bifidobacteria during refrigerated storage in the presence of acid and hydrogen peroxide, *Milchwissenschaft,* 51, 65–70, 1996.
31. Leuschner, R.G., Heidel, M., and Hammes, W.P., Histamine and tyramine degradation by food fermenting microorganisms, *Int. J. Food Microbiol.,* 39, 1–10, 1998.
32. Kolter, R., Siegele, D.A., and Tormo, A., The stationary phase of the bacterial life cycle, *Annu. Rev. Microbiol.,* 47, 855–874, 1993.
33. Hartke, A., Bouche, S., Gansel, X., Boutibonnes, P., and Auffray, Y., Starvation-induced stress resistance in *Lactococcus lactis* subsp. *lactis* IL1403, *Appl. Environ. Microbiol.,* 60, 3474–3478, 1994.
34. Wetzel, K., Menzel, M., and Heller, K.J., Stress response in *Lactococcus lactis* and *Streptococcus thermophilus* induced by carbon starvation, *Kieler Milchwirtschaftl. Forsch. Ber.,* 51, 319–332, 1999.
35. Sandine, W.E., Commercial production of starter cultures, in *Dairy Starter Cultures,* Cogan, T.M. and Accolas, J.-P., Eds., VCH Publishers, New York, 1996, pp. 191–206.

36. Vinderola, C.G., Prosello, W., Ghiberto, T.D., and Reinheimer, J.A., Viability of probiotic (*Bifidobacterium, Lactobacillus acidophilus* and *Lactobacillus casei*) and non-probiotic microflora in Argentinian fresco cheese, *J. Dairy Sci.,* 83, 1905–1911, 2000.
37. Blanchette, L., Roy, D., Belanger, G., and Gauthier, S.F., Production of cottage cheese using dressings fermented by *Bifidobacteria, J. Dairy Sci.,* 79, 8–15, 1996.
38. Juntunen, M., Kirjavainen, P.V., Ouwehand, A.C., Salminen, S.J., and Isolauri, E., Adherence of probiotic bacteria to human intestinal mucus in healthy infants and during rotavirus infection, *Clin. Diagn. Lab. Immunol.,* 8, 293–296, 2001.
39. Ghoddusi, H.B. and Robinson, R.K., The test of time, *Dairy Ind. Int.,* 61, 25–28, 1996.
40. Gomez, A.M.P., Malcata, F.X., Klaver, F.A.M., and Grande, H.G., Incorporation and survival of *Bifidobacterium* sp. strain Bo and *Lactobacillus acidophilus* strain Ki in a cheese product, *Neth. Milk Dairy J.,* 49, 71–95, 1995.
41. Corbo, M.R., Albenzio, M., De Angelis, M., Sevi, A., and Gobbetti, M., Microbial and biochemical properties of Canestrato Pugliese hard cheese supplemented with bifidobacteria, *J. Dairy Sci.,* 84, 551–561, 2001.
42. Daigle, A., Roy, D., Belanger, G., and Vuillemard, J.C., Production of probiotic cheese (cheddar-like cheese) using enriched cream fermented by *Bifidobacterium infantis, J. Dairy Sci.,* 82, 1081–1091, 1999.
43. Stanton, C., Gardiner, G., Lynch, P.B., Collins, J.K., Fitzgerald, G., and Ross, R.P., Probiotic cheese, *Int. Dairy J.,* 8, 491–496, 1998.
44. Gardiner, G.E., Ross, R.P., Wallace, J.M., Scanlan, F.P., Jagers, P.P., Fitzgerald, G.F., Collins, J.K., and Stanton, C., Influence of the probiotic adjunct culture of *Enterococcus faecium* on the quality of Cheddar cheese, *J. Agric. Food Chem.,* 47, 4907–4917, 1999.
45. Gardiner, G.E., Stanton, C., Lynch, P.B., Collins, J.K., Fitzgerald, G., and Ross, R.P., Evaluation of Cheddar cheese as a food carrier for delivery of a probiotic strain to the gastrointestinal tract, *J. Dairy Sci.,* 82, 1379–1387, 1999.
46. Gardiner, G.E., Ross, R.P., Collins, J.K., Fitzgerald, G., and Stanton, C., Development of a probiotic Cheddar cheese containing human-derived *Lactobacillus paracasei* strains, *Appl. Environ. Microbiol.,* 64, 2192–2199, 1998.
47. Dinakar, P. and Mistry, V.V., Growth and viability of *Bifidobacterium bifidum* in Cheddar cheese, *J. Dairy Sci.,* 77, 2854–2864, 1994.
48. Gomez, A.M.P. and Malcata, F.X., Development of probiotic cheese manufactured from goat milk: response surface analyses via technological manipulation, *J. Dairy Sci.,* 81, 1492–1507, 1998.
49. Gobbetti, M., Corsetti, A., Smacchi, E., Zocchetti, A., and De Angelis, M., Production of Crescenza cheese by incorporation of bifidobacteria, *J. Dairy Sci.,* 81, 37–47, 1997.
50. Ross, R.P., Fitzgerald, G.F., Collins, J.K., O'Sullivan, G.C., and Stanton, C.G., Process for manufacture of probiotic cheese, PCT International Patent Application WO 99/62348 A1, 1999.
51. Reid, G. and Burton, J., Use of *Lactobacillus* to prevent infection by pathogenic bacteria, *Microbes Infect.,* 4, 319–324, 2002.
52. Schaafsma, G., State of the art concerning probiotic strains in milk products, *IDF Nutr. Newslett.,* 5, 23–24, 1996.
53. Schrezenmeir, J. and de Vrese, M., Probiotics, prebiotics and synbiotics — approaching a definition, *Am. J. Clin. Nutr.,* 73, 361S–364S, 2001.
54. Jan, G., Rouault, A., and Maubois, J.L., Acid stress susceptibility and acid adaptation of *Propionibacterium freudenreichii* ssp. *shermanii, Lait,* 80, 325–336, 2000.

55. Zarate, G., Chaia, A.P., Gonzalez, S., and Oliver, G., Viability and beta-galactosidase activity of dairy propionibacteria subjected to digestion by artificial gastric and intestinal fluids, *J. Food Prot.,* 63, 1214–1221, 2000.
56. Ryhanen, E.L., Pihlanto-Leppala, A., and Pahkala, E.B., A new type of ripened, low-fat cheese with bioactive properties, *Int. Dairy J.,* 11, 441–447, 2001.
57. Auty, M.A.E., Gardiner, G.E., McBrearty, S., O'Sullivan, E.O., Mulvihill, D.M., Collins, J.K., Fitzgerald, G.F., Stanton, C., and Ross, R.P., Direct *in situ* viability assessment of bacteria in probiotic dairy products using viability staining in conjunction with confocal scanning laser microscopy, *Appl. Environ. Microbiol.,* 67, 420–425, 2001.
58. McBrearty, S., Ross, R.P., Fitzgerald, G.F., Collins, J.K., Wallace, J.M., and Stanton, C., Influence of two commercially available bifidobacteria cultures on Cheddar cheese quality, *Int. Dairy. J.,* 11, 599–610, 2001.
59. Alander, M., Satokari, R., Korpela, R., Saxelin, M., Vilpponen-Salmela, T., Mattila-Sandholm, T., and von Wright, A., Persistence of colonisation of human colonic mucosa by a probiotic strain, *Lactobacillus rhamnosus* GG, after oral consumption, *Appl. Environ. Microbiol.,* 65, 351–354, 1999.
60. Guandalini, S., Probiotics in the treatment of diarrhoeal diseases in children, *Gastroenterol. Intern.,* 11, 87–90, 1998.
61. Isolauri, E., Juntunen, M., Rautanen, T., Sillanaukee, P., and Koivula, T., A human *Lactobacillus* strain (*Lactobacillus casei* sp. strain GG) promotes recovery from acute diarrhea in children, *Pediatrics*, 88, 90–97, 1991.
62. Hilton, E., Kolakowski, P., Smith, M., and Singer, C., Efficacy of *Lactobacillus* GG as a diarrhoea preventive, *J. Travel Med.,* 4, 3–7, 1997.
63. Arvola, T., Laiho, K., Torkkeli, S., Mykkanen, H., Salminen, S., Maunula, L., and Isolauri, E., Prophylactic *Lactobacillus* GG reduces antibiotic-associated diarrhoea in children with respiratory infections, *Pediatrics*, 104, 64–69, 1999.
64. Gorbach, S.L., Chang, T.W., and Goldin, B.R., Successful treatment of relapsing *Clostridium difficile* colitis with *Lactobacillus* GG, *Lancet*, 26, 1519, 1998.
65. Siitonen, S., Vapaatalo, H., Salminen, S., Gordin, A., Saxeli, M., Wikberg, R., and Kirkkola, A.-L., Effect of *Lactobacillus* GG yoghurt in prevention of antibiotic associated diarrhoea, *Ann. Med.,* 22, 57–59, 1990.
66. Ling, W.H., Hänninen, O., Mykkänen, H., Heikura, M., Salminen, S., and Von Wright, A., Colonization and fecal enzyme activities after oral *Lactobacillus* GG administration in elderly nursing home residents, *Ann. Nutr. Metab.,* 36, 162–166, 1992.
67. Malin, M., Suomalainen, H., Saxelin, M., and Isolauri, E., Promotion of IgA immune response in patients with Crohn's disease by oral bacteriotherapy with *Lactobacillus* GG, *Ann. Nutr. Metab.,* 40, 137–145, 1996.
68. Millar, M.R., Bacon, C., Smith, S.L., Walker, V., and Hall, M.A., Enteral feeding of premature infants with *Lactobacillus* GG, *Arch. Dis. Child.,* 69, 483–487, 1993.
69. Isolauri, E., Joensuu, J., Suomalainen, H., Luomala, M., and Vesikari, T., Improved immunogenicity of oral DxRRV reassortant rotavirus vaccine by *Lactobacillus casei* GG, *Vaccine*, 13, 310–312, 1995.
70. Miettinen, M., Matikainen, S., Vuopio-Varkila, J., Pirhonen, J., Varkila, K., Kurimoto, M., and Julkunen, I., Lactobacilli and streptococci induce interleukin-12 (IL-12), IL-18, and gamma interferon production in human peripheral blood mononuclear cells, *Infect. Immun.,* 66, 6058–6062, 1998.
71. Pelto, L., Isolauri, E., Lilius, E.-M., Nuutila, J., and Salminen, S., Probiotic bacteria down-regulate the milk-induced inflammatory response in milk-hypersensitive subjects but have an immunostimulatory effect in healthy subjects, *Clin. Exp. Allergy,* 28, 1471–1479, 1998.

72. Kalliomäki, M., Salminen, S., Arvilommi, H., Kero, P., Koskinen, P., and Isolauri, I., Probiotics in primary prevention of atopic disease: a randomised placebo-controlled trial, *Lancet,* 357, 1076–1079, 2001.
73. Hatakka, K., Savilahti, E., Pönkä, A., Meurman, J.H., Poussa, T., Näse, L., Saxelin, M., and Korpela, R., Effect of long term consumption of probiotic milk on infections in children attending day care centres: double blind, randomised trial, *Br. Med. J.,* 322, 1327–1329, 2001.
74. Näse, L., Hatakka, K., Savilahti, E., Saxelin, M., Pönka, A., Poussa, T., Korpela, R., and Meurman, J.H., Effect of long-term consumption of a probiotic bacterium, *Lactobacillus rhamnosus* GG, in milk on dental caries and caries risk in children, *Caries Res.*, 35, 412–420, 2001.
75. Hoier, E., Säure-und Gallentoleranz von *Lactobacillus acidophilus* und Bifidobakterien, *Deutsche Molkereizeitung,* 26, 770–772, 1992.
76. Langhendries, J.P., Detry, J., Van Hees, J., Lamboray, J.M., Darimont, J., Mozin, M.J., Secretin, M.C., and Senterre, J., Effect of a fermented infant formula containing viable *Bifidobacteria* on the fecal flora composition and pH of healthy full-term infants, *J. Pediatr. Gastroenterol. Nutr.,* 21, 177–181, 1995.
77. Saavedra, J.M., Baumann, N.A., Oung, I., Perman, J.A., and Yolken, R.H., Feeding of *Bifidobacterium bifidum* and *Streptococcus thermophilus* to infants in hospital for prevention of diarrhoea and shedding of rotavirus, *Lancet*, 344, 1046–1049, 1994.
78. Black, F.T., Andersen, P.L., Ørskov, J., Ørskov, F., Gaarslev, K., and Laulund, S., Prophylactic efficiacy of lactobacilli on traveler's diarrhea, in *Travel Medicine,* Steffen R., Ed., Springer, Berlin, 1989, pp. 333–335.
79. Black, F.T., Einarsson, K., Lidbeck, A., Orrhage, K., and Nord, C.E., Effect of lactic acid producing bacteria on the human intestinal microflora during ampicillin treatment. *Scand. J. Infect. Dis.,* 23, 247–254, 1991.
80. Saavedra, J., Abi-Hanna, A., Moore, N., and Yolken, R., Effect of long term consumption of infant formulas with *Bifidobacteria* (B) and *S. thermophilus* (St) on stool patterns and diaper rash in infants, *J. Pediatr. Gastroenterol. Nutr.,* 47, Abstract 82, 1998.
81. Link-Amster, H., Rochat, F., Saudan, K.Y., Mignot, O., and Aeschlimann, J.M., Modulation of a specific humoral immune response and changes in intestinal flora mediated through fermented milk intake, *FEMS Immunol. Med. Microbiol.,* 10, 55–64, 1994.
82. Schiffrin, E.J., Brassart, D., Servin, A.L., Rochat, F., and Donnet-Hughes, A., Immune modulation of blood leukocytes in humans by lactic acid bacteria: criteria for strain selection, *Am. J. Clin. Nutr.,* 66, 515S–520S, 1997.
83. Kim, H.-Y., Bae, H.-S., and Baek, Y.-J., *In vivo* antitumor effects of lactic acid bacteria on sarcoma 180 and mouse Lewis lung carcinoma, *J. Korean Cancer Assoc.,* 23, 188–195, 1991.
84. Klaver F.A.M., Klaasen, H.L.B.M, and Weerkamp, H.A., Selection of lactobacilli and bifidobacteria for use as human probiotic adjuncts (abstract), *FEMS Microbiol. Rev.,* 12(1–3), P186, 1993.
85. De Vrese, M., Fenselau, S., Feindt, F., Laue, C., Kristen, H., Lick, S., Heller, K.J., Schrezenmeir, J., Plock, J., Kleinbach-Sauter, H., Ishibashi, N., Hayasawa, H., and Tomita, M., Effects of Yogurt Containing *Bifidobacterium longum* BB536 on Diarrhea Induced by Antibiotic Administration for Eradication of *H. pylori,* International Conference of Intestinal Bacteriology (ICIB), Tokyo, 2001.
86. Allen, W.D., Linggood, M.A., and Porter, P., Probiotic Eur. Pat. No. 0 508 701, B1, 1992.

87. Klaver, F.A.M. and Van der Meer, R., The assumed assimilation of cholesterol by *Lactobacilli* and *Bifidobacterium bifidum* is due to their bile salt-deconjugating activity, *Appl. Environ. Microbiol.,* 59, 1120–1124, 1993.
88. Weerkamp, A.H., Klaver, F.A.M., and Van der Meer, R., Assimilation of cholesterol by lactobacilli is due to their bile salt deconjugation activity (abstract), *FEMS Microbiol. Rev.,* 12(1–3), P185, 1993.
89. Weerkamp, A.H., Bovee-Oudenhoven, I.M., and Van der Meer, R., Effect of Probiotics on Infection with Pathogenic Bacteria in Human Gut Flora-Associated Rats, Poster J36, presented at 5th Symposium on LAB, Veldhoven, The Netherlands, 1996.

9 Natto — A Food Made by Fermenting Cooked Soybeans with *Bacillus subtilis* (*natto*)

Tomohiro Hosoi and Kan Kiuchi

CONTENTS

0-8493-1372-4/03/$0.00+$1.50

9.1 FERMENTED SOYBEAN FOODS IN ASIA

Fermented soybean foods made with *Bacillus subtilis* cells are produced in China, where they are called dou chi (or dauchi). They include salted, sweet, and nonsalted types. The salted type of dou chi (xian-dou chi) contains 10 to 20% salt to inhibit putrefaction by contaminating bacteria. The most typical sweet dou chi is called tian-dou chi; it is used as a seasoning for Beijing duck. Nonsalted dou chi has been developed into various kinds of natto.[1] Foods of this type are called itohiki-natto (hereafter abbreviated to natto) in Japan, kinema in Nepal and Myanmar, tua nao in Thailand, and chungkuk-jang in Korea. Natto is produced only with *B. subtilis* (*natto*) (formerly called *B. natto*, see Section 9.2.1).

These fermented soybean foods are consumed in a variety of forms. For instance, tua nao is used as a raw ingredient in salads. Chungkuk-jang, which contains cayenne peppers and garlic, is used as an ingredient for a Korean soup called chige. In Japan, natto is mixed with soy sauce, sliced Welsh onion (similar to stone leeks), mustard, dried seaweed, and/or raw egg. Seasoned natto is usually eaten with rice in Japan. A looser natto (hikiwari-natto) and a dried type of natto are also used. Before the Second World War, the transportation infrastructure was not well developed in Japan. Hence, dried, hard natto, having a long shelf life, was usually manufactured and consumed. Of the three types of natto, itohiki-natto is the most popular at present in Japan. Hikiwari-natto is used for preparing sushi. Consumption of natto increased compared to other soybean foods during the 1990s, generating sales of 160 billion yen in 1996.[2] (See Chapter 1 for more details on the history of natto.)

9.2 INGREDIENTS OF NATTO

The raw materials required for natto production are *B. subtilis* (*natto*) spores called natto bacilli, soybeans, and water. *B. subtilis* (*natto*) strains are used to initiate fermentation of steamed whole soybeans. The quality of natto is affected by the quality of the soybeans (see following) and the *B. subtilis* (*natto*) strains but not the water. Normal tap water is adequate for the production of natto.

9.2.1 *BACILLUS SUBTILIS* (*NATTO*) SPORES

In 1913, Dr. S. Sawamura of Imperial University in Tokyo isolated the natto bacillus and named it *Bacillus natto*.[3] The first commercially available natto using a pure culture of *B. natto* was sold through the efforts of Professor J. Hanzawa of Hokkaido Imperial University in 1928.[4] In 1957, *B. natto* strains were included in and documented as *B. subtilis* strains in the 7th edition of *Bergey's Manual of Determinative Bacteriology*.[5] However, only *B. natto* and no other *B. subtilis* strains are used for natto production. This is because natto with desired characteristics can be produced only with *B. natto* strains.[6] Using *B. natto* strains yields natto with a distinctive aroma, a wrinkly bacterial layer on the surface of the soybeans, and a desirable degree of stickiness. When natto of high quality is picked up, it tends to elongate into strings of soybeans and not to lump. In addition, biotin greatly enhances the growth of *B. natto*.[7] Japanese scientists, therefore, usually refer to the natto bacillus as *B. subtilis* (*natto*) to distinguish it from other strains of *B. subtilis* which are not suited for natto production. Lyophilized spores of *B. subtilis* (*natto*) or spore suspensions in water are sold as a starter for natto by three companies in Japan. Some natto manufacturers have utilized their own *B. subtilis* (*natto*) strains in order to produce natto of characteristic quality. Contamination of natto products by other microorganisms must be avoided, as it affects the quality of the natto and may cause food poisoning.

9.2.2 Soybeans

The total volume of soybeans consumed in Japan has been approximately 5 million metric tonnes over the past ten years.[8] Total consumption of soybeans for food production was 1.01 million metric tonnes in 2000.[9] Japan imports the majority of the soybeans (more than 80%) required for domestic food production. There was a 13% increase in soybean consumption for natto products between 1991 and 2000, and consumption reached 122 thousand metric tonnes in 2000.[9] Domestic soybeans are considered to be superior to foreign ones for natto production by Japanese natto manufacturers, although the supply of Japanese-grown soybeans is insufficient to meet the demands for natto production. Because they are grown in limited domestic regions and production is small, Japanese soybeans command a high price. Considering the low rate of self-sufficiency in soybeans, and in order to increase quality using domestic soybeans, a national project has been initiated to increase the production of domestic cultivars of soybeans in Japan. To date, some new cultivars of soybeans have been bred in many districts of Japan and have been used in the production of natto.[10] The names of these soybean cultivars and the locations where they were bred have been compiled and published by the Japanese Ministry of Agriculture, Forestry, and Fisheries.[11]

Natto manufacturers prefer certain kinds of soybean cultivars, and their preferences can vary from region to region because of consumer preferences. Numerous processing tests have been conducted in an attempt to elucidate which cultivars of soybeans are most appropriate for making high-quality natto. Popular soybean varieties for natto production are "suzuhime" and "suzumaru," which are grown in Hokkaido; "kosuzu," in Iwate, Miyagi, and Akita Prefectures; and "natto-shoryu," in Ibaraki Prefecture.

Desirable qualities of soybeans for natto are generally as follows:

1. Extra small or small size
2. Easily washable

3. Yellow surfaces and hila
4. A suitable degree of stickiness when made into natto
5. Sweet taste
6. Slight changes in constituents during storage

9.2.2.1 Color

Soybeans with a brilliant light yellow or yellow colored surface and with a light yellow hilum are favored. Natto made from brown or black soybeans has traditionally not sold well in Japan due to its unappetizing appearance and lack of characteristic aroma. However, natto made from black soybeans has recently been sold, and its high content of polyphenols has been emphasized.

9.2.2.2 Size

Soybean size is classified by diameter into four groups in Japan: extra small refers to less than 5.5 mm in diameter, small ranges from 5.5 mm to no more than 7.3 mm, medium ranges from 7.3 mm to no more than 7.9 mm, and large refers to a diameter of 7.9 mm or more. Consumers in Kanto and more northern regions of Japan regard the extra small and small soybeans as the most suitable ones for natto. On the other hand, those in Kansai region prefer larger soybeans. Natto processed from small and extra small soybeans tends to have a distinctive natto flavor and a strong taste due to excess fermentation. It was reported that activities of alkaline protease (subtilisin) were higher in natto produced from small soybeans than that from large soybeans.[12,13] However, the activities of neutral proteases (metalloprotease) did not differ with the size of soybeans used.[12] This result suggests that degradation of proteins in natto produced from small soybeans progresses more strongly than that from large soybeans during the latter half of the fermentation period when the pH value of the product is increased by metabolites of the *B. subtilis* (*natto*) cells. This may cause the taste and smell of natto made from small soybeans to be stronger than that of natto made from large soybeans. Natto made from large soybeans has a weak smell of ammonia and shows a low degradation rate of proteins, although its natto-like taste is still perceivable.[12]

9.2.2.3 Protein Content

B. subtilis (*natto*) cells utilize the proteins, peptides, and amino acids in soybeans for their growth. The kinds and quantities of peptides and amino acids produced by the activities of *B. subtilis* (*natto*) during fermentation affect the flavor appeal of natto. Hence, soybeans with high protein content are preferred. By these criteria, "suzuhime" and "zizuka" cultivars, with high protein contents, bright color, and polished appearance, are highly regarded for natto production.

9.2.2.4 Sugar Content

In order to produce natto with good quality and flavor, it is important that available carbohydrates be supplied to *B. subtilis* (*natto*) and that the hydrolysis of proteins proceed appropriately during fermentation. Because extra small soybeans tend to have a higher sugar content, they are regarded as superior for the production of natto. Because the storage life of natto is determined by the ammonia flavor, the content of ammonia is considered to be the most important quality control characteristic in natto production.[14] However, it has been also reported that soybeans with high sugar content are not necessarily best; free sugar content is more important than total sugar content for natto processing.[15,16] This is because *B. subtilis* (*natto*) can utilize certain saccharides such as sucrose, raffinose, and stachyose but not starch for growth.[17]

Small soybeans tend to become softer than medium or large ones do when steamed under identical conditions.[14] A positive correlation between the firmness of steamed soybeans and the ammonia nitrogen levels of natto products has been reported.[14] In this report, it was suggested that in hard, steamed soybeans, hydrolysis of the constituents by *B. subtilis* (*natto*) cells probably occurs just under the surface of the soybeans. Degradation and usage of the constituents inside the soybeans by the bacteria does not occur quickly. This means that *B. subtilis* (*natto*) cells can only utilize sugars near the surface in a relatively short time, and then protein degradation begins in the early stages of fermentation. This induces a strong ammonia flavor. Hence, it is important to select soybeans with a high sugar content and to steam them until they become soft. A typical steam treatment is 1.5 kg/cm^2 steam pressure for 20 min.

9.2.2.5 Washing and Storage Methods

Natto is produced year round, and therefore harvested soybeans must be stored before they are processed. Soybeans dirty with soil are cleaned, washed, and then stored for months in a cool room. It is desirable that they be stored in a refrigerated room at a temperature below 15°C and a relative humidity of about 60%.[18] If soybeans are stored at temperatures between 25 and 35°C, the raffinose in the soybeans increases and the stachyose decreases.[18] Germination rates of soybeans are often examined to check their quality. A low germination rate suggests that the soybeans have been preserved under undesirable temperature and humidity conditions.[19] There was a negative correlation between germination rates and solid matter content of soybeans soaked in water.[19] The quality of natto made with such soybeans is generally not acceptable. Therefore, soybeans with a low germination rate are not suitable for natto production. However, germinated soybeans are not used to produce natto.

9.3 NATTO PROCESSING

Natto is processed as shown in Figure 9.1.

9.3.1 Washing and Soaking of Soybeans

Sieves are used to separate small or extra small soybeans at the beginning of the process. Contaminants and foreign substances such as plant stalks and leaves, soil, and sand are removed. Soybeans are weighed, washed in a screw wash press, rubbed in a maelstrom flow, polished with burhstone, and finally washed in clean tap water.

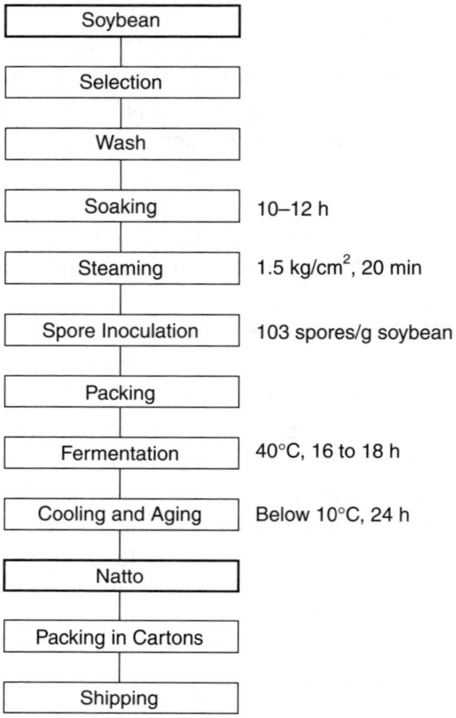

FIGURE 9.1 Procedure of natto processing.

Soybeans must be softened in water before natto processing because natto produced from unsoftened soybeans is too hard to eat. Soaking should be performed until the weight of soybeans increases approximately 2.2- to 2.7-fold.[14,20] To achieve this, soybeans are soaked in tap water at 10°C for about 20 h. It is important to clean the soaking tanks regularly because the tanks are prone to contamination with lactic acid-producing bacteria. Lactic acid may inhibit the growth of *B. subtilis* (*natto*) and impair fermentation. Some processors soak soybeans overnight, flowing tap water through the tanks for cooling. However, too much changing of water during soaking or soaking for too long a time will reduce the taste of the natto.

9.3.2 Steaming of Soybeans

After soaking, soybeans must be steamed (using steam at 1.5 kg/cm^2 pressure for 20 min) in order to soften the beans further and denature undesirable soybean proteins such as hematoglutenin and trypsin inhibitor.[21] At the same time, contaminating bacteria are killed. Steaming vats that can hold 60 to 120 kg of raw soybeans at a time are usually used.

9.3.3 Inoculation with *Bacillus subtilis* (*natto*) Spores

Bacillus subtilis (*natto*) exists both as vegetative cells and in spore form. The spores are more suitable for storage. Therefore, spore suspensions for natto production are sold. Immediately after steaming, while the soybeans are still hot (for example, 85°C), the soybeans are tipped from the vat and sprayed with a *B. subtilis* (*natto*) spore suspension. The concentration of *B. subtilis* (*natto*) spores in the soybean stock should be approximately 10^3 colony forming units (CFU)/g soybeans. Inoculation of *B. subtilis* (*natto*) spores at much higher concentrations inhibits the development of the desired stickiness and appearance.[12]

9.3.4 Packaging

Paper cups or polystyrene paper trays are commonly used to package 25 to 150 g of processed soybeans. The most common size holds 50 g of natto (Figures 9.2a and 9.2b). The historical changes in packaging methods are described later (Section 9.3.7). In recent years, almost all filling processes have been mechanized, and manual assistance is not required. In general, 60 g of steamed soybeans inoculated with *B. subtilis* (*natto*) spores are packed as a 50 g package by an automatic filler because moisture loss causes an approximate 20% reduction in weight during fermentation. After filling, soybeans inside the packages are covered with a perforated polyethylene film (Figure 9.2b). Plastic sachets of soy sauce seasoning and/or other condiments such as mustard and freeze-dried sliced Welsh onion can be put on the films. The lid of the package is fixed in place with a bond, using a heat sealer or a rubber ring.

9.3.5 Fermentation

Fermentation rooms are equipped with ventilators and air conditioners in order to control the humidity of the fermenting soybeans. Conditions in the fermentation room are controlled automatically by a computer for optimum fermentation. Packages are placed horizontally in a box and stacked in the fermentation room. In order to produce natto of high quality, a sufficient oxygen supply and fine adjustment of room temperature, humidity, and fermentation period are essential.[22] Typical fermentation conditions and product status are as follows. The room temperature is set at 40°C, and the initial humidity is controlled between 85 and 90%. The humidity is reduced to 75% between hours 6 and 16 and to 55% between hours 16 and 24. Under these conditions, after 8 h of fermentation, a viscous substance begins to be produced, creating the stickiness of natto. After 10 h of fermentation, the internal temperature of the product begins to increase, and it reaches 48 to 52°C by hour 14. If this optimum temperature of the product is not reached (i.e., if it is too low or too high), the quality of the final natto tends to be lower. The temperature largely affects the stickiness and taste of natto. Thereafter, the temperature of the product is gradually dropped down to 40°C. After hour 18 to 20 of fermentation, the temperature of the fermentation room is set so that the product temperature decreases to 10°C at hour 24. At hour 24, the product in the boxes is transferred from the fermentation room to a refrigerated room where the temperature is below 10°C and is stored for more than 1 day. It is important to control the condition of the fermentation rooms so that the temperature of the product changes as described above. The temperature near the ceiling of the fermentation room can easily increase more than that of the lower part of the room near the floor. This difference causes different

FIGURE 9.2 Photographs of natto. (a) Natto in polystyrene paper tray package. (b) Natto covered with polyethylene film with numerous small holes. (c) Natto stirred with chopsticks.

fermentation conditions and qualities of products. Therefore, it is necessary to adjust fermentation periods slightly among products at different positions in the room and/or to regularly change the position of the products in the room during fermentation. The optimum conditions for fermentation also vary depending upon the strains of *B. subtilis* (*natto*) used.

9.3.6 Packing for Shipment

Automated shrink packaging of two or three polystyrene paper (PSP) trays or cups is the conventional packing method, and an automated carton packer has recently been developed specifically for natto. Cartons are stored in a refrigerated room until they are transported to stores by trucks. Since the odor of natto easily intensifies due to the progressive production of ammonia under warm temperature conditions, natto is treated as a perishable product, and its shelf life is one week if refrigerated below 10°C.

9.3.7 Changes in Packages

The type of packaging affects the fermentation process, but in addition, the following factors are important for natto packaging:

1. Easy to manufacture
2. Easy to distribute
3. Easy to serve
4. Disposable and able to be incinerated

Before the Second World War, china jars and packages made of rice straw or wood shavings were often used to pack natto in Japan. Rice straw is an excellent packaging material for natto because it promotes growth of *B. subtilis* (*natto*) by maintaining a warm internal temperature and absorbing extra moisture emitted from the steamed soybeans during the initial stages of fermentation. Adequate moisture is required for good growth of *B. subtilis* (*natto*) cells. If moisture inside the packages is lost during fermentation, the resulting natto tends to be dry, with sticky and strong strings. However, these traditional packaging methods did not meet the requirement of being easy to store or the trend toward spoil-proof and mass-produced food after the 1960s. The development of fillers has also promoted the standardization of natto packaging. Paper tips replaced the rice straw first and then PSP packages were developed. At present, straw packages are used only for souvenirs or special local products. When straw packages are used, natto is packed in thin polyethylene films first and further packed in the steamed and dried straw.

The initial PSP packages containing 100 to 120 g natto were considered to be too large because most consumers want to eat an entire cup of natto at one time. Hence, PSP packages and easily printable paper cups containing 40 or 50 g natto are now mainly being used. Smaller cups containing 20 or 30 g natto are also used in school meals. However, since PSP packaging and paper cups do not regulate temperature and moisture automatically during fermentation, control of the conditions of the fermentation room becomes even more important. Moreover, steamed soybeans mixed with *B. subtilis* (*natto*) spores should be covered with a perforated inner packaging. The purpose of this inner packaging is to assist airflow and maintain the moisture inside the package. It also allows sachets of seasoning to be included in one package. The corners of the perforated lining are pressed down in order to prevent the steamed soybeans from drying out.

A functional plastic membrane that allows for automatic handling as described above was developed as a packaging material for natto.[23] This was called the "repiramy cup," and it did indeed meet many of the requirements for natto packaging. It functioned to maintain the temperature and allowed for the free movement of oxygen, carbon dioxide, ammonia gas, and other gases, but not of water. This membrane, however, was too expensive to use as a packaging material and is not used at the present time.

TABLE 9.1
Composition of Soybeans and Itohiki-natto (/100 g Dry)

	Dried and Raw Soybean			
	Japan	United States	China	Itohiki-natto
Energy (kcal)	476	490	482	493
Protein (g)	40.3	37.4	37.5	40.7
Lipid (g)	21.7	24.6	22.2	24.7
Carbohydrate (g)	32.2	32.6	35.2	29.9
Ash (g)	5.7	5.4	5.0	4.7
Vitamin A				
Retinol (μg)	0	0	0	0
Carotene (μg)	7	8	10	0
Retinol equivalents (μg)	1	1	1	0
Vitamin D (μg)	0	0	0	0
Vitamin E (mg)	4.1	3.9	4.9	3.0
Vitamin K (μg)	21	39	39	2148[a]
Vitamin B_1 (mg)	0.95	1.00	0.96	0.17
Vitamin B_2 (mg)	0.34	0.34	0.34	1.38
Niacin (mg)	2.5	2.4	2.5	2.7
Vitamin B_6 (mg)	0.61	0.52	0.67	0.59
Vitamin B_{12} (μg)	0	0	0	Tr
Folate (μg)	263	249	297	30
Pantothenic acid (mg)	1.74	1.69	1.87	8.89
Ascorbic acid (mg)	Tr	Tr	Tr	Tr

Note: Water contents of Japanese, American, and Chinese dried and raw soybeans and itohiki-natto are 12.5, 11.7, 12.5, and 59.5 (/100 g wet), respectively. Tr: Trace.

[a] Including menaquinone-7.

Source: Modified from Resources Council, Science and Technology Agency, Standard Tables of Food Composition in Japan, 5th revised ed., Ministry of Finance, Japan, 2000.

9.4 ASSESSMENT OF QUALITY

9.4.1 Chemical Composition

The chemical composition of natto does not differ greatly from that of soybeans except for the vitamin K content (Table 9.1).[24] *B. subtilis* (*natto*) produces vitamin K_2 (menaquinone-7).[25,26] Domestic soybeans, U.S. soybeans, Chinese soybeans, and natto contain 21, 39, 39, and 2148 μg/100 g dry weight of vitamin K, respectively. Natto is covered with sticky substances produced by *B. subtilis* (*natto*) during fermentation. These sticky substances are composed of polyglutamic acid and levan (fructan).[27] During the fermentation, *B. subtilis* (*natto*) also produces proteases and an amylase.[28] Peptides or amino acids produced during fermentation constitute a part of the natto taste. *B. subtilis* (*natto*) utilizes soybean saccharides and produces the characteristic flavor of natto.

9.4.2 Sensory Tests

The quality of natto is determined by the sensory tests described in the book *Methods of Natto Research.*[29] Natto of high quality has the following properties:

1. *Even surface layer* — The bacterial layer should be formed entirely on the surface of the soybeans.
2. *No lysis and no glistening* — Uneven spots or glistening of soybeans by bacterial cell lysis should not be observed.
3. *No damaged soybeans* — There should be few split, crushed, and/or peeled soybeans.
4. *Bright color* — The color of the soybean surfaces should be brown or light brown and should not be dark brown or blackish.
5. *Good aroma* — The product should have a sweet aroma without an ammonia-like, scorched, undesirable, or acidic odor.
6. *Proper firmness* — The soybeans should be properly soft and have a smooth texture.
7. *Good taste* — Relish, sweet, not bitter, and slightly astringent. The preferred taste is created by amino acids, peptides, and saccharides.
8. *Proper stickiness* — When stirred with a pair of chopsticks, the viscosity of the natto should increase to form strong strings (Figure 9.2c).

Taste panels can evaluate natto according to these criteria. If any foreign substances are found or tyrosine crystals are formed on the surface of the soybeans, they are noted and the natto is given a low quality score.

9.4.3 Changes in Consumers' Preferences

Traditionally, extra small and small soybeans have been used to make natto. However, in 1999, a natto made from medium-sized soybeans won the contest held by the Federation of Japan Natto Manufacturers Cooperative Society. Nattos made with large soybeans are also preferred now by Japanese consumers.[30] Present-day consumers are tending to buy a natto with markedly weaker odors and strings.[30] Natto with traditional characteristics (distinctive odor and strong strings) may no longer have mass appeal. Some manufacturers have used their own *B. subtilis* (*natto*) strains to produce less aromatic products.

9.5 HEALTH BENEFITS

Bacillus subtilis, Bacillus subtilis (*natto*), and natto are considered to have potential as probiotics. Ingestion of bacterial cells probably affects the intestinal microflora and the mucosal immune system. *B. subtilis* (*natto*) cells produce many enzymes and vitamin K_2. A serine protease, subtilisin, can degrade soybean allergens and shows fibrinolytic activity. Ingestion of vitamin K_2 (menaquinone-7) will help coagulant activity and prevent osteoporosis.[31,32] Natto contains the phytoestrogens (isoflavones) that originate in soybeans.[33,34] Isoflavones seem to have preventive effects on breast and prostate cancer, osteoporosis, menopausal symptoms, and heart disease.[33,34]

9.5.1 *Bacillus subtilis* (*natto*) Cells

9.5.1.1 Effects on Intestinal Microflora and Feed Efficiency

Lactobacillus spp. and *Bifidobacterium* spp. are mainly used as probiotics for humans and animals.[35] However, other bacteria and fungi can also be used as probiotics. For example, *Bacillus* spp., *Enterococcus* spp., *Streptococcus* spp., and *Saccharomyces cerevisiae* seem to have potential as probiotics.[35] In the screening and selection of certain microbial strains as probiotics, phenotype and genotype stability, carbohydrate and protein utilization patterns, safety, acid and bile stability, adhesion characterization, production of antimicrobial substances, antibiotic resistance patterns, immunogenicity, and viability and properties during processing and storage are considered to be important.[36] *B. subtilis* is an aerobic spore-forming bacterium. *B. subtilis* spores are relatively resistant to oxygen, active oxygen species, acid, drying, and heating compared to other bacteria.[5,27] *B. subtilis* can also grow under O_2-reduced conditions. These characteristics are desirable for potential probiotics. Unfortunately, however, *B. subtilis* is not strongly resistant to bile acid and is not a predominant bacterium in the human intestine.[37]

Several reports have demonstrated the effects of orally administered *B. subtilis* on the intestinal microflora, body weight gain, and increased feed efficiency of animals and birds.[38–42] These results indicate that ingestion of live *B. subtilis* cells can actually improve the intestinal microflora. When weanling piglets were fed a diet including spores of *B. subtilis* (*natto*), the changes in intestinal microflora varied depending upon the region of the intestine examined.[38] In the jejunum, the numbers of *Streptococcus* spp. and *Bifidobacterium* spp. increased, while no difference was observed in the colon when compared with the control diet group. When turkeys were fed *B. subtilis* culture, body weight gain and cumulative feed efficiency significantly increased, both by 2.5%.[39] When chickens were given *B. subtilis,* the detection rate of the intestinal pathogen *Campylobacter jejuni* decreased in the laboratory portion of the experiment.[40] The cell number of *Salmonella typhimurium* also decreased. In the field trial, feeding a *B. subtilis* strain decreased the cell number and/or detection rates of intestinal Enterobacteriaceae, *Clostridium perfringens,* and *Campylobacter* sp. When sows and gilts were fed an experimental diet containing *B. subtilis,* the number and/or detection rates of fecal *Bifidobacterium* spp. and *Lactobacillus* spp. increased, but *Streptococcus* spp., Enterobacteriaceae, *Clostridium perfringens,* and Bacteroidaceae decreased.[41] The diarrhea rate of piglets up to 10 days old and mortality rate up to 25 days old also decreased. When mice were intubated with intact and autoclaved *B. subtilis* (*natto*) spores for 8 days, only intact spores changed the fecal microflora, and the patterns of the changes differed depending upon the diets fed.[42] Feeding a diet including egg white decreased fecal *Lactobacillus* spp., although the administration of *B. subtilis* (*natto*) spores inhibited the decrease. On the other hand, feeding a diet including casein and administering *B. subtilis* (*natto*) spores increased only Bacteroidaceae but not lactobacilli.

Ingestion of natto (50 g/day) significantly affected the composition and metabolic activity of the human fecal microflora.[43] Ingestion of natto increased the number of *B. subtilis* (*natto*) and *Bifidobacterium* spp. (the latter increased from 15% of the total bacterial count before consumption to 39% after 14 days' consumption), although it decreased the number and detection rates of lecithinase-positive clostridia including *Clostridium perfringens*. The concentrations of fecal acetic acid, total organic acids, and succinic acid increased, while fecal concentrations of indole, ethylphenol, and skatol decreased. Fecal ammonia, cresol, and fecal pH values also decreased.

Mechanisms of the above effects have not been clarified. However, germination and/or some metabolites from *B. subtilis* cells seem to be necessary in order to explain their effects, because it was shown that in mice, administration of autoclaved spores did not affect the fecal microflora.[42] The possibility of germination of *Bacillus* spp. spores in the intestine has been examined. When *B. subtilis* (*natto*) spores were inoculated in the ligated loops of the ileum of dogs, some spores did germinate but died after germination.[44] It has been shown that *B. thuringiensis* spores germinate in the gut fluid of the tobacco horn worm.[45] *B. subtilis* spores also germinate in the mouse gut.[46] In this report, the number of spores excreted in the feces of the mice was, in some experiments, larger than the number of spores inoculated. However, this is inconsistent with the report that vegetative cells of *B. subtilis* could not be detected when spores were inoculated in mice that were left without food for 16 h.[37] Live *Bacillus* cells could be detected only in some organs after ingestion. In general, when foods are ingested, the pH value in the stomach sometimes increases to 3 to 4. Spores of *Bacillus* spp. appear to be resistant to such pH values. Some of the *B. subtilis* spores ingested together with other food may be able to sustain their viability and germinate in the upper intestine once the surrounding pH value is neutralized. This then allows them to produce probiotic activity.

Catalase and subtilisin have been proposed as the active molecules responsible for the effects of *B. subtilus* (*natto*) on intestinal microflora.[47] The growth of three strains of lactobacilli co-cultured aerobically with *B. subtilis* (*natto*) has been examined. Addition of *B. subtilis* (*natto*) to the culture medium *in vitro* resulted in an increase in the number of viable cells of all lactobacilli tested. Both catalase and *B. subtilis* (*natto*) enhanced the growth of *Lactobacillus reuteri*, whereas *B. subtilis* (*natto*), but not catalase, enhanced the growth of *Lactobacillus acidophilus*. In a medium containing 0.1 m*M* hydrogen peroxide, its toxic effect on *Lb. reuteri* was abolished by catalase or *B. subtilis* (*natto*). Catalase has been reported to exhibit a growth-promoting effect on lactobacilli.[48] The viability of lactobacilli readily decreases in the presence of active oxygen species. The decrease of viable cell number is partly attributable to the fact that lactobacilli do not generally produce a defense molecule against active oxygen species. However, aerobic bacteria, including *B. subtilis* (*natto*), can produce catalase. Vegetative cells of *B. subtilis* primarily produce catalase-1 in the logarithmic phase of growth, and additionally produce catalase-2 and -3 as growth progresses.[49–51] Intact *B. subtilis* spores contain only catalase-2 in the spore coat. Some other anaerobic bacteria in the intestine, such as

Escherichia coli, Bacteroides spp., and *Eubacterium* spp., also produce catalase.[52–55] It may be important for these bacteria to scavenge hydrogen peroxide in order to colonize the intestine where active oxygen species are produced. The addition of a serine protease, subtilisin, from *Bacillus licheniformis* to the culture medium improved the growth and viability of *Lb. reuteri* and *Lb. acidophilus* in the absence of hydrogen peroxide.[47] *B. subtilis* (*natto*) secretes two serine proteases, subtilisin NAT with an isoelectric point (pI) of 8.7 and a 90-kDa serine proteinase (pI 3.9).[56–59] Taken together, these results indicate that *B. subtilis* (*natto*) can enhance the growth and/or viability of lactobacilli possibly through production of catalase and subtilisin.

9.5.1.2 Effects on the Immune System

The mucosal immune system is a first line of defense against foreign antigens. The system consists of many kinds of cells, including epithelium, macrophage, dendritic cells, intraepithelial T lymphocytes, B lymphocytes, and neutrophils.[60] The effects of *B. subtilis* cells on some of these intestinal cells have been examined in order to determine whether *B. subtilis* cells possess immunostimulating effects. A recent study questioned whether human intestinal epithelium-like Caco-2 cells can produce cytokines to *B. subtilis* (*natto*) strains, in addition to other pathogenic and nonpathogenic bacteria.[61] It is not clear whether epithelial cells can respond to nonpathogenic strains of bacterial cells. Live cells of nonpathogenic *B. subtilis* or *B. subtilis* (*natto*) strains, as well as nonpathogenic *Escherichia coli*, pathogenic *Salmonella enteritidis,* and *Pseudomonas aeruginosa,* all induced secretion of interleukin 6 (IL-6) and/or IL-8 but not of IL-7 and IL-15. The amounts of cytokines induced by *B. subtilis* (*natto*) cells were dependent upon the strain used. Cytokine induction of epithelial cells may differ between bacterial species or strains regardless of their pathogenicity. Some nonpathogenic bacteria as well as pathogenic ones seem to be able to induce cytokine secretion from normal intestinal epithelial cells when they are orally ingested. Nitrite formation in the macrophage cell line J774.2 in the presence of heat-killed *B. subtilis* cells has been reported.[62] Peptidoglycan from *B. subtilis* induced nitrite formation in macrophages. Lipoteichoic acid from *Staphylococcus aureus* was more potent than lipoteichoic acid from *B. subtilis.*

Translocation of *Bacillus* spp. spores and/or cells has been examined.[37] When mice were fed spores intragastrically, both spores and vegetative cells were detected in the lymph nodes and spleens. In another report, low numbers of spores were detected in mesenteric lymph nodes, livers, and spleens. However, the effect of spore ingestion on the increase of bacterial numbers in these organs was not significant.[46] The authors indicated that spores do not appear to translocate substantially across mucosal surfaces. The effect of *B. subtilis* (*natto*) on T and B lymphocytes in the chicken spleen has also been examined.[63] When chickens were fed 10^7 CFU/g of *B. subtilis* (*natto*) spores, the percentages of T and B lymphocytes in the spleens increased compared to those of control groups. Although *B. subtilis* cells are not predominant in the intestine, these results indicate that ingested bacterial cells can affect the mucosal immune system.

Mechanisms of the effects of *B. subtilis* cells on the immune system remain unclear, although interaction between bacterial cellular components and Toll-like receptors seems to be essential to their effects.[64] Many bacterial components, including peptidoglycans, lipoproteins, lipoteichoic acid, flagellin, and unmethylated CpG dinucleotides in bacterial DNA are known to bind to Toll-like receptors and induce cytokine responses.[64] *B. subtilis* and *B. subtilis* (*natto*) cells would presumably also contain some of these active substances.

9.5.1.3 Anti-Allergy Effect of Subtilisin

About 15 soybean proteins have been shown to be recognized by sera of soybean-sensitive patients with atopic dermatitis.[65] Three major allergens were identified and designated as *Gly m* Bd 60K, *Gly m* Bd 30K, and *Gly m* Bd 28K, respectively. *Gly m* Bd 60K is an α-subunit of β-conglycinin.[66] *Gly m* Bd 30K is a soybean oil–body-associated glycoprotein homologous to Der p (or f) 1, a major allergen of house dust mites.[66] *Gly m* Bd 28K is a vicilin-like glycoprotein, which is a minor component fractionated into the 7S globulin fraction.[66] *Bacillus subtilis* (*natto*) produces a serine protease of subtilisin NAT during its growth.[56–58] Subtilisin NAT appears to degrade *Gly m* Bd 28K.[67] Various nonfermented soybean products, such as soybean protein isolate (SPI), tofu, kori-dofu, and yuba, contain *Gly m* Bd 28K at high concentrations, although fermented soybean products such as natto, soy sauce, and miso do not.[67]

Recent studies have implied that the intestinal microbial flora are important to the interrelationship between infection and allergy.[68] Endogenous flora of the gut stimulate the immune system. It is likely that *Lactobacillus* spp. administration may have preventive and therapeutic effects on allergic diseases.[69,70] It has also been indicated that allergic children are less colonized with *Lactobacillus* spp. compared to nonallergic children.[71] In addition, *Lactobacillus* spp. differently modulate expression of cytokines and maturation surface markers in murine dendritic cells dependent upon the strains.[72] Dendritic cells, present throughout the gastrointestinal tract, play a pivotal immunoregulatory role in the Th1 and Th2 cell balance.[68] Th2 cells produce IL-4, IL-13, and IL-5, which coordinately regulate the allergic response.[68] Unfortunately, it has not been examined whether similar effects can be obtained by the administration of *B. subtilis* strains.

9.5.1.4 Fibrinolytic Activity of Subtilisin

Circulating platelets and blood-derived proteins (fibrin) are essential for the formation of blood clots, which prevent bleeding long enough for healing to occur.[73] However, excess coagulation prevents normal physiologic blood flow, which causes thrombotic disorders. Thrombolytic therapy is the most direct means of restoring blood flow.[74] *Bacillus* spp. produce serine proteases called subtilisins, which are known to have fibrinolytic activity.[56–58,75–79] Subtilisin NAT produced by *B. subtilis* (*natto*) (also called nattokinase) is 99.5% homologous to subtilisin E.[58] Oral administration of subtilisin or natto induced mild and frequent enhancement of fibrinolytic activity in plasma and production of tissue plasminogen activator.[77] Euglobulin fibrinolytic activity, degradation products from fibrin/fibrinogen, and the amount of tissue plasminogen activator increased by administration of subtilisin. However, whole blood clot lysis time did not significantly decrease. In addition, ingestion of natto decreased euglobulin lysis time and increased euglobulin fibrinolytic activity.[77] The mechanisms of the fibrinolytic activity of subtilisin NAT are not fully understood, although subtilisin NAT appears to digest fibrin directly and cleave and inactivate plasminogen activator inhibitor-1 (PAI-1).[79] Elevation of PAI-1 in plasma is found in patients with thrombotic disease.[80] High PAI-1 activity is related to impaired fibrinolysis, and low activity is associated with bleeding disorders.[81–83]

9.5.1.5 Role of Vitamin K_2 (Menaquinone-7) in the Prevention of Osteoporosis

Vitamin K has important roles in blood coagulation and bone metabolism.[31,32] The most abundant forms of vitamin K are phylloquinone (vitamin K_1) and menaquinone (vitamin K_2). Menaquinone refers to a series of vitamin K homologues with polyunsaturated aliphatic side chains of varying lengths. These compounds are generally referred to as menaquinone-*n* (MK-*n*), where *n* is the number of isoprenoid residues of which the side chain is composed. A marked deficiency of menaquinone-7 and menaquinone-8 has been demonstrated in patients with osteoporotic fractures.[84] An effect of vitamin K_2 treatment on spinal bone mineral density (BMD) in postmenopausal women has been demonstrated.[85] In addition, low serum and bone vitamin K status in patients with longstanding Crohn's disease has been reported.[86] Some bacteria including *B. subtilis* (*natto*) produce vitamin K_2 (menaquinone-7) (see Section 9.4.1).[25,26] Ingestion of 100 g natto increases vitamin K levels in serum.[25] There were no significant changes in serum vitamin K_1 levels after intake of natto, although vitamin K_2 levels and total vitamin K levels after 24 h intake significantly increased from 0.90 to 6.21 and from 1.95 to 7.14 ng/ml. A large geographic difference in serum vitamin K_2 levels in postmenopausal women in Tokyo and Hiroshima in Japan and in England has been reported.[87] A trial using a bacterial strain with a high productivity of vitamin K_2 has been performed where vitamin K_2 production was compared to that of a commercial strain.[88] The results showed that the selected bacterial strain was capable of producing twice the concentration of vitamin K_2 produced by the original commercial strain.

9.5.2 Phytoestrogens — Effects on Cancer and Osteoporosis

It has been hypothesized that ingestion of phytoestrogens (isoflavones) contained in soybeans plays an important role in the prevention of cancer, osteoporosis, menopausal symptoms, and heart disease.[33,34] Many epidemiologic studies have examined the relationship between soybean consumption and cancer risk (reviewed in references 33 and 34), although no significant conclusion has been reached. Early exposure (during the neonatal or prepubertal periods of life) to soybean products is thought to be essential for cancer protection.[89–92] Some soybean products with a high isoflavone content have been developed and sold in Japan by using selected soybean varieties and specific strains of *B. subtilis* (*natto*).

Isoflavones are a subclass of the more ubiquitous flavonoids. The primary isoflavones in soybeans are genistein and daidzein and their respective β-glycosides, genistin and daidzin. Much smaller amounts of glycitein and its glycoside, glycitin, are also present in soybeans.[93,94] Isoflavones in nonfermented soybean foods appear mostly as the conjugate, whereas in fermented foods such as natto and miso, the aglycones dominate.[95] Hence, levels of genistein in fermented soybean products are higher than those in soybeans and nonfermented foods. The calculated daily intake levels of genistein and genistin ingested from soybeans and related soybean products by the Japanese are 1.5 to 4.1 and 6.3 to 8.3 mg/person, respectively.

In order to clarify the mechanisms of the effects of isoflavones, many studies have been performed. Intestinal microbial cells can convert daidzein into several different products, including the isoflavonoids equol, dihydrodaidzein, and *O*-desmethylangolensin.[96] It has been proposed that genistein was metabolized to dihydrogenistein and 6'-hydroxy-*O*-desmethylangolensin in humans.[96] Estrogenic activities of isoflavones are quite weak when compared to those of physiologic estrogens.[97] However, consumption of soybean foods increased blood isoflavone concentrations to several orders of magnitude higher than those of physiologic estrogens.[98] Isoflavones can also have antiestrogenic effects when placed in a high-estrogen environment.[99,100] Genistein inhibits several enzyme activities involved in signal transduction and DNA topoisomerases I and II.[101–107] In addition, genistein increases the *in vitro* concentrations of transforming growth factor β (TGF-β), an inhibitor of epithelial cell growth.[108] This effect is thought to be an important contributor to the anticancer effect. Isoflavones have been shown to bind the estrogen receptors (ERs). Recently, a new receptor, ERβ, has been discovered, to which genistein can bind, but the binding is weaker than for natural estrogen, 17β-estradiol.[109] It is possible that this finding, and the fact that ERβ is expressed in differing amounts depending on the types of cells, support the observation that isoflavones have antiosteoporotic but weakly antiuterotrophic effects.[110,111]

9.6 CONCLUSIONS

Natto is a simple, low-priced, popular soybean food made by fermenting cooked soybeans with *Bacillus subtilis* (*natto*) in Japan. Natto has characteristic aroma and stickiness. The consumption of natto products has increased during the 1990s in Japan. Natto contains many nutrients originating from soybeans and the metabolites of *B. subtilis* (*natto*) cells, which show many physiological functions. In 1999, the Food and Drug Administration (FDA) in the U.S. approved the use of health claims for soy protein related to the reduction of the risk of coronary heart disease by lowering blood cholesterol levels when 25 grams of soy protein per day are consumed. In addition, many physiological functions of *B. subtilis* cells have recently been reported. Although *B. subtilis* (*natto*) is not a predominant bacterium in the human intestine, it is thought to have potential as a probiotic. Some pharmaceutical products containing *B. subtilis* spores have been sold and utilized in Japan and European countries. Both scientific researchers and consumers are paying attention now to natto and *B. subtilis* cells. However, their effects and mechanisms are not yet clearly understood. Further careful studies are necessary to obtain data on their effects, safety, and efficacy. We hope that studies on the nutrition and physiological function of natto products, *B. subtilis* cells, and related products will progress steadily and that such products will become popular all over the world.

REFERENCES

1. Ito, H., Tong, J., and Li, Y., Chinese dauchi, from itohiki-natto to nonmashed miso, *Miso Sci. Technol.,* 44, 224–250, 1996.
2. Statistics Bureau, Ministry of Public Management, Home Affairs, Post and Telecommunications, Japan, *Annual Report on the Family Income and Expenditure Survey 2000,* Japan Statistical Association, Shinjuku-ku, 2001.
3. Sawamura, S., On *Bacillus natto*, Bulletin of Agricultural College, Tokyo, 5, 107–109, 1913.
4. Federation of Japan Natto Manufacturers Cooperative Society, in *History of Natto,* Taito-ku, 1971, pp. 154–170.
5. Smith, N.R. and Gordon, E., Genus *Bacillus subtilis* Cohn, 1872, in *Bergey's Manual of Determinative Bacteriology*, 7th ed., Breed, R.S., Murray, E.G.D., and Smith, N.R., Eds., Williams & Willkins, Baltimore, 1957, pp. 620–622.
6. Itaya, M. and Matsui, K., Conversion of *Bacillus subtilis* 168: *natto* producing *Bacillus subtilis* with mosaic genomes, *Biosci. Biotechnol. Biochem.,* 63, 2034–2037, 1999.
7. Ohsugi, M. and Imanishi, Y., Microbiological activity of biotin-vitamers, *J. Nutr. Sci. Vitaminol.,* 31, 563–572, 1985.
8. Ministry of Agriculture, Forestry, and Fisheries, Japan, Soybean Balance Sheet, http://www.maff.go.jp/soshiki/nousan/hatashin/daizu.siryo/jukyu.html.
9. Ministry of Agriculture, Forestry, and Fisheries, Japan, Changes in Consumption of Soybeans for Food Production, http://www.maff.go.jp/soshiki/nousan/hatashin/daizu.siryo/youtobetu.html.
10. Sohma, S. and Matsukawa, I., Soybean, in *Legume Encyclopedia*, Watanabe, T., Ed., Saiwai Shobo, Chiyoda-ku, 2000, pp. 47–59.
11. Ministry of Agriculture, Forestry, and Fisheries, Japan, Encyclopedia of Domestic Soybeans, http://www.maff.go.jp/soshiki/nousan/hatashin/jiten/sakuin.html.
12. Takahashi, J., Furuguchi, K., Miyama, K., and Kikuchi, K., Effect of inoculation size on quality of natto, *Rpt. Food Res. Inst. Tochigi Prefecture*, 10, 17–18, 1996.
13. Pero, J. and Sloma, A., Proteases, in Bacillus subtilis *and Other Gram-Positive Bacteria*, Sonenshein, A.L., Hoch, J.A., and Losick, R., Eds., American Society for Microbiology, Washington, D.C., 1993, pp. 939–952.
14. Taira, H., Suzuki, N., Tsukamoto, C., Kainuma, Y., Tanaka, H., and Saito, M., Quality of soybeans seeds grown in Japan. Part 15. Suitability for natto processing of small seed cultivars and quality of the natto, *Rpt. Natl. Food Res. Inst. (Japan),* 51, 48–58, 1987.
15. Taira, H., Quality and suitability of domestic soybeans for food processing, *Nogyo Gijutsu,* 44, 385–391, 1989.
16. Taira, H., Quality of soybeans for processed foods in Japan, *Jpn. Agric. Res. Quart.,* 24, 224–230, 1990.
17. Kanno, A., Takamatsu, H., Takano, N., and Akimoto, T., Change of saccharides in soybeans during manufacturing of natto, *Nippon Shokuhin Kogyo Gakkaishi,* 29, 105–110, 1982.
18. Hase, S., Yasui, T., Nagashima, S., and Ohta, T., Changes in quality of soybeans during storage. Changes in sugars and starch content, *Rpt. Natl. Food Res. Inst. (Japan),* 36, 7–13, 1980.

19. Taira, H., Quality of soybeans seeds grown in Japan. Part 3. Correlation between physical property, chemical component, and suitability for food processing pairs, *Rpt. Natl. Food Res. Inst. (Japan),* 42, 27–39, 1983.
20. Taira, H., Takahashi, H., Okano, H., and Nagashima, S., Quality of soybeans seeds grown in Japan. Part 4. Suitability of soybean varieties grown in Ibaraki Prefecture for natto production, *Rpt. Natl. Food Res. Inst. (Japan),* 43, 62–71, 1983.
21. Chang, C.J., Tanksley, T.D., Knabe, D.A., and Zebrowska, T., Effects of different heat treatments during processing on nutrient digestibility of soybean meal in growing swine, *J. Anim. Sci.,* 65, 1273–1282, 1987.
22. Muramatsu, K., Kanai, S., Kimura, N., Miura, N., Yoshida, K., and Kiuchi, K., Production of natto with high elastase activity, *Nippon Shokuhin Kogyo Gakkaishi,* 42, 575–582, 1995.
23. Ishikawa, T., Utility Model: S60–150035, 1985.
24. Resources Council, Science and Technology Agency, Japan, *Standard Tables of Food Composition in Japan*, 5th revised ed., Printing Bureau, Ministry of Finance, Japan, Minato-ku, 2000, pp. 60–63.
25. Kudo, T., Warfarin antagonism of natto and increase in serum vitamin K by intake of natto, *Artery,* 17, 189–201, 1990.
26. Sumi, H., Determination of the vitamin K (menaquinone-7) content in fermented soybean natto and in the plasma of natto-ingesting subjects, *J. Home Econ. Jpn.,* 50, 309–312, 1999.
27. Claus, D. and Berkeley, R.C.W., Genus *Bacillus subtilis* Cohn, 1872, 174 in *Bergey's Manual of Determinative Bacteriology,* Vol. 2, Sneath, P.H.A., Mair, N.S., Sharpe, M.E., and Holt, J.G., Eds., Williams & Willkins, Baltimore, 1986, pp. 1105–1139.
28. Ferrari, E., Jarnagin, A.S., and Schmidt, B.F., Commercial production of extracellular enzymes, in Bacillus subtilis *and Other Gram-positive bacteria,* Sonenshein, A.L., Hoch, J.A., and Losick, R., Eds., American Society for Microbiology, Washington, D.C., 1993, pp. 917–937.
29. Society for Study of Natto and National Food Research Institute, Ministry of Agriculture, Forestry and Fisheries, *Methods of Natto Research,* Kohrin, Taito-ku, 1989, chaps. 1 and 2.
30. Federation of Japan Natto Manufacturers Cooperative Society, Report on the Promoting Enterprise of Cooperative Domestic Soybean Utilization, 2000.
31. Vermeer, C., Jie, K.-S.G., and Knapen, M.H.J., Role of vitamin K in bone metabolism, *Annu. Rev. Nutr.,* 15, 1–22, 1995.
32. Shearer, M.J., Role of vitamin K and Gla proteins in the pathophysiology of osteoporosis and vascular calcification, *Curr. Opin. Clin. Nutr. Metab. Care,* 3, 433–438, 2000.
33. Kurzer, M.S. and Xu, X., Dietary phytoestrogens, *Annu. Rev. Nutr.,* 17, 353–381, 1997.
34. Messina, M.J., Legumes and soybeans: overview of their nutritional profiles and health effects, *Am. J. Clin. Nutr.,* 70, 439S–450S, 1999.
35. Goldin, B.R., Health benefits of probiotics, *Br. J. Nutr.,* 80, S203–S207, 1998.
36. Tuomola, E., Crittenden, R., Playne, M., Isolauri, E., and Salminen, S., Quality assurance criteria for probiotic bacteria, *Am. J. Clin. Nutr.,* 73, 393S–398S, 2001.
37. Spinosa, M.R., Braccini, T., Ricca, E., De Felice, M., Morelli, L., Pozzi, G., and Oggioni, M.R., On the fate of ingested *Bacillus* spores, *Res. Microbiol.,* 151, 361–368, 2000.
38. Ozawa, K., Yokota, H., Kimura, M., and Mitsuoka, T., Effects of administration of *Bacillus subtilis* strain BN on intestinal flora of weanling piglets, *Jpn. J. Vet. Sci.,* 43, 771–775, 1981.

39. Jiraphocakul, S., Sullivan, T.W., and Shahani, K.M., Influence of a dried *Bacillus subtilis* culture and antibiotics on performance and intestinal microflora in turkeys, *Poult. Sci.,* 69, 1966–1973, 1990.
40. Maruta, K., Miyazaki, H., Masuda, S., Takahashi, M., Marubashi, T., Tadano, Y., and Takahashi, H., Exclusion of intestinal pathogens by continuous feeding with *Bacillus subtilis* C-3102 and its influence on the intestinal microflora in broilers, *Anim. Sci. Technol.,* 67, 273–280, 1996.
41. Maruta, K., Miyazaki, H., Tadano, Y., Masuda, S., Suzuki, A., Takahashi, H., and Takahashi, M., Effects of *Bacillus subtilis* C-3102 intake on fecal flora of sows and on diarrhea and mortality rate of their piglets, *Anim. Sci. Technol.,* 67, 403–409, 1996.
42. Hosoi, T., Ametani, A., Kiuchi, K., and Kaminogawa, S., Changes in fecal microflora induced by intubation of mice with *Bacillus subtilis* (*natto*) spores are dependent upon dietary components, *Can. J. Microbiol.,* 45, 59–66, 1999.
43. Terada, A., Yamamoto, M., and Yoshimura, E., Effect of the fermented soybean product "natto" on the composition and metabolic activity of the human fecal flora, *Jpn. J. Food Microbiol.,* 16, 221–230, 1999.
44. Hisanaga, S., Mase, M., Mizuno, A., Hayakawa, Y., Ohkubo, T., and Hachisuka, Y., Germination of *Bacillus natto* spores in the canine intestine, *Nippon Saikingaku Zasshi,* 33, 689–696, 1978.
45. Wilson, G.R. and Benoit, T.G., Activation and germination of *Bacillus thuringiensis* spores in *Manduca sexta* larval gut fluid, *J. Invertebr. Pathol.,* 56, 233–236, 1990.
46. Hoa, T.T., Duc, L.H., Isticato, R., Baccigalupi, L., Ricca, E., Van, P.H., and Cutting, S.M., Fate and dissemination of *Bacillus subtilis* spores in a murine model, *Appl. Environ. Microbiol.,* 67, 3819–3823, 2001.
47. Hosoi, T., Ametani, A., Kiuchi, K., and Kaminogawa, S., Improved growth and viability of lactobacilli in the presence of *B. subtilis* (*natto*), catalase, or subtilisin, *Can. J. Microbiol.,* 46, 892–897, 2000.
48. Kono, Y. and Fridovich, I., Functional significance of manganese catalase in *Lactobacillus plantarum, J. Bacteriol.,* 155, 742–746, 1983.
49. Loewen, P.C. and Switala, J., Multiple catalases in *Bacillus subtilis, J. Bacteriol.,* 169, 3601–3607, 1987.
50. Loewen, P.C. and Switala, J., Purification and characterization of catalase-1 from *Bacillus subtilis, Biochem. Cell Biol.,* 65, 939–947, 1987.
51. Loewen, P.C. and Switala, J., Purification and characterization of spore-specific catalase-2 from *Bacillus subtilis, Biochem. Cell Biol.,* 66, 707–714, 1988.
52. Claiborne, A. and Fridovich, I., Purification of the *O*-dianisidine peroxidase from *Escherichia coli* B, *J. Biol. Chem.,* 254, 4245–4252, 1979.
53. Claiborne, A., Malinowski, D.P., and Fridovich, I., Purification and characterization of hydroperoxidase II of *Escherichia coli* B, *J. Biol. Chem.,* 254, 11664–11668, 1979.
54. Wilkins, T.D., Wagner, D.L., Vertri, B.J., Jr., and Gregory, E.M., Factors affecting production of catalase by *Bacteroides, J. Clin. Microbiol.,* 8, 553–557, 1978.
55. Takamine, F. and Imamura, T., Isolation and characterization of bile acid 7-dehydroxylating bacteria from human feces, *Microbiol. Immunol.,* 39, 11–18, 1995.
56. Ichishima, E., Takada, Y., Taira, K., and Takeuchi, M., Specificities of extracellular and ribosomal serine proteinases from *Bacillus natto*, a food microorganism, *Biochim. Biophys. Acta,* 869, 178–184, 1986.
57. Kamata, H., Yamagata, Y., Nakamura, T., Nakajima, T., Oda, K., Murano, S., and Ichishima, E., Characterization of the complex between α2-macroglobulin and a serine proteinase from *Bacillus natto, Agric. Biol. Chem.,* 53, 2695–2702, 1989.

58. Nakamura, T., Yamagata, Y., and Ichishima, E., Nucleotide sequence of the subtilisin NAT gene, *aprN,* of *Bacillus subtilis* (*natto*), *Biosci. Biotech. Biochem.,* 56, 1869–1871, 1992.
59. Kato, T., Yamagata, Y., Arai, T., and Ichishima, E., Purification of a new extracellular 90-kDa serine proteinase with isoelectric point of 3.9 from *Bacillus subtilis* (*natto*) and elucidation of its distinct mode of action, *Biosci. Biotech. Biochem.,* 56, 1166–1168, 1992.
60. Nagler-Anderson, C. Man the barrier! Strategic defences in the intestinal mucosa, *Nat. Rev. Immunol.,* 1, 59–67, 2001.
61. Hosoi, T., Hirose, R., Saegusa, S., Ametani, A., Kiuchi, K., and Kaminogawa, S., Cytokine responses of human intestinal epithelial-like Caco-2 cells to the non-pathogenic bacterium *Bacillus subtilis* (*natto*), *Int. J. Food Microbiol.* (in press)
62. Kengatharan, K.M., De Kimpe, S., Robson, C., Foster, S.J., and Thiemermann, C., Mechanism of Gram-positive shock: identification of peptidoglycan and lipoteichoic acid moieties essential in the induction of nitric oxide synthase, shock, and multiple organ failure, *J. Exp. Med.,* 188, 305–315, 1998.
63. Inooka, S., Uehara, S., and Kimura, M., The effect of *Bacillus natto* on the T and B lymphocytes from spleens of feeding chickens, *Poult. Sci.,* 65, 1217–1219, 1986.
64. Medzhitov, R., Toll-like receptors and innate immunity, *Nat. Rev. Immunol.,* 1, 135–145, 2001.
65. Ogawa, T., Bando, N., Tsuji, H., Okajima, H., Nishikawa, K., and Sasaoka, K., Investigation of the IgE-binding proteins in soybeans by immunoblotting with the sera of the soybean-sensitive patients with atopic dermatitis, *J. Nutr. Sci. Vitaminol.,* 37, 555–565, 1991.
66. Ogawa, A., Samoto, M., and Takahashi, K., Soybean allergen and hypoallergenic soybean products, *J. Nutr. Sci. Vitaminol.,* 46, 271–279, 2000.
67. Bando, N., Tsuji, H., Hiemori, M., Yoshizumi, K., Yamanishi, R., Kimoto, M., and Ogawa, T., Quantitative analysis of *Gly m* Bd 28K in soybean products by a sandwich enzyme-linked immunosorbent assay, *J. Nutr. Sci. Vitaminol.,* 44, 655–664, 1998.
68. Wills-Karp, M., Santeliz, J., and Karp, C.L., The germless theory of allergic disease: revisiting the hygiene hypothesis, *Nat. Rev. Immunol.,* 1, 69–75, 2001.
69. Pelto, L., Isolauri, E., Lilius, E.M., Nuutila, J., and Salminen, S., Probiotic bacteria down-regulate the milk-induced inflammatory response in milk-hypersensitive subjects but have an immunostimulatory effect in healthy subjects, *Clin. Exp. Allergy,* 28, 1474–1479, 1998.
70. Kalliomaki, M., Salminen, S., Arvilommi, H., Kero, P., Koskinen, P., and Isolauri, E., Probiotics in primary prevention of atopic disease: a randomized placebo controlled trial, *Lancet*, 357, 1076–1079, 2001.
71. Majamaa, H. and Isolauri, E., Probiotics: a novel approach in the management of food allergy, *J. Allerg. Clin. Immunol.,* 99, 179–185, 1997.
72. Christensen, H.R., Frøkiær, H., and Pestka, J.J., Lactobacilli differentially modulate expression of cytokines and maturation surface markers in murine dendritic cells, *J. Immunol.,* 168, 171–178, 2002.
73. Brass, S., Platelets and proteases, *Nature*, 413, 26–27, 2001.
74. Becker, R.C., Seminars in thrombosis, thrombolysis and vascular biology. 4. Fibrinolysis, *Cadiology,* 79, 188–210, 1991.
75. Fayek, K.I. and El-Sayed, S.T., Fibrinolytic activity of enzyme produced by *Bacillus subtilis, Z. Ernährungswiss,* 19, 21–23, 1980.

76. Sumi, H., Hamada, H., Tsushima, H., Mihara, H., and Muraki, H., A novel fibrinolytic enzyme (nattokinase) in the vegetable cheese natto; a typical and popular soybean food in the Japanese diet, *Experientia,* 43, 1110–1111, 1987.
77. Sumi, H., Hamada, H., Nakanishi, K., and Hiratani, H., Enhancement of the fibrinolytic activity in plasma by oral administration of nattokinase, *Acta. Haematol.,* 84, 139–143, 1990.
78. Fujita, M., Nomura, K., Hong, K., Ito, Y., Asada, A., and Nishimuro, S., Purification and characterization of a strong fibrinolytic enzyme (nattokinase) in the vegetable cheese natto, a popular soybean fermented food in Japan, *Biochem. Biophys. Res. Commun.,* 197, 1340–1347, 1993.
79. Urano, T., Ihara, H., Umemura, K., Suzuki, Y., Oike, M., Akita, S., Tsukamoto, Y., Suzuki, I., and Takada, A., The profibrinolytic enzyme subtilisin NAT purified from *Bacillus subtilis* cleaves and inactivates plasminogen activator inhibitor type 1, *J. Biol. Chem.,* 276, 24690–24696, 2001.
80. Wiman, B., Plasminogen activator inhibitor 1 (PAI-1) in plasma: its role in thrombotic disease, *Thromb. Haemost.,* 74, 71–76, 1995
81. Hamsten, A., de Faire, U., Walldius, G., Dahlen, G., Szamosi, A., Landou, C., Blomback, M., and Wiman, B., Plasminogen activator inhibitor in plasma: risk factor for recurrent myocardial infarction, *Lancet*, 2, 3–9, 1987.
82. Schleef, R.R., Higgins, D.L., Pillemer, E., and Levitt, L.J., Bleeding diathesis due to decreased functional activity of type 1 plasminogen activator inhibitor, *J. Clin. Invest.,* 83, 1747–1752, 1989.
83. Fay, W.P., Shapiro, A.D., Shih, J.L., Schleef, R.R., and Ginsburg, D., Brief report: complete deficiency of plasminogen-activator inhibitor type 1 due to frame-shift mutation, *N. Engl. J. Med.,* 327, 1729–1733, 1992.
84. Hodges, S.J., Pilkington, M.J., Stamp, T.C., Catterall, A., Shearer, M.J., Bitensky, L., and Chayen, J., Depressed levels of circulating menaquinones in patients with osteoporotic fractures of the spine and femoral neck, *Bone,* 12, 387–389, 1991.
85. Iwamoto, I., Kosha, S., Noguchi, S., Murakami, M., Fujino, T., Douchi, T., and Nagata, Y., A longitudinal study of the effect of vitamin K_2 on bone mineral density in postmenopausal women: a comparative study with vitamin D3 and estrogen–progestin therapy, *Maturitas*, 31, 161–164, 1999.
86. Schoon, E.J., Muller, M.C., Vermeer, C., Schurgers, L.J., Brummer, R.J., and Stockbrugger, R.W., Low serum and bone vitamin K status in patients with longstanding Crohn's disease: another pathogenetic factor of osteoporosis in Crohn's disease? *Gut,* 48, 473–477, 2001.
87. Kaneki, M., Hedges, S.J., Hosoi, T., Fujiwara, S., Lyons, A., Crean, S.J., Ishida, N., Nakagawa, M., Takechi, M., Sano, Y., Mizuno, Y., Hoshino, S., Miyao, M., Inoue, S., Horiki, K., Shiraki, M., Ouchi, Y., and Orimo, H., Japanese fermented soybean food as the major determinant of the large geographic difference in circulating levels of vitamin K_2: possible implications for hip-fracture risk, *Nutrition,* 17, 315–321, 2001.
88. Tsukamoto, Y., Kasai, M., and Kakuda, H., Construction of a *Bacillus subtilis* (*natto*) with high productivity of vitamin K_2 (menaquinone-7) by analog resistance, *Biosci. Biotechnol. Biochem.,* 65, 2007–2015, 2001.
89. Brown, N.M. and Lamartiniere, C.A., Xenoestrogens alter mammary gland differentiation and cell proliferation in the rat, *Environ. Health Perspect.,* 103, 708–713, 1995.
90. Lamartiniere, C.A., Moore, J.B., Brown, N.M., Thompson, R., Hardin, M.J., and Barnes, S., Genistein suppresses mammary cancer in rats, *Carcinogenesis,* 16, 2833–2840, 1995.

91. Murrill, W.B., Brown, N.M., Zhang, J.X., Manzolillo, P.A., Barnes, S., and Lamartiniere, C.A., Prepubertal genistein exposure suppresses mammary cancer and enhances gland differentiation in rats, *Carcinogenesis,* 17, 1451–1457, 1996.
92. Wu, A.H., Ziegler, R.G., Horn-Ross, P.L., Nomura, A.M., West, D.W., Kolonel, L.N., Rosenthal, J.F., and Hoover, R.N., Tofu and risk of breast cancer in Asian-Americans, *Cancer Epidemiol. Biomarkers Prev.,* 5, 901–906, 1996.
93. Wang, H.-J. and Murphy, P.A., Isoflavone composition of American and Japanese soybeans in Iowa: effects of variety, crop year, and location, *J. Agric. Food Chem.,* 42, 1674–1677, 1994.
94. Wang, H.-J. and Murphy, P.A., Isoflavone content in commercial soybean foods, *J. Agric. Food Chem.,* 42, 1666–1673, 1994.
95. Fukutake, M., Takahashi, M., Ishida, K., Kawamura, H., Sugimura, T., and Wakabayashi, K., Quantification of genistein and genistin in soybeans and soybean products, *Food Chem. Toxicol.,* 34, 457–461, 1996.
96. Joannou, G.E., Kelly, G.E., Reeder, A.Y., Waring, M., and Nelson, C., A urinary profile study of dietary phytoestrogens. The identification and mode of metabolism of new isoflavonoids, *J. Steroid Biochem. Mol. Biol.,* 54, 167–184, 1995.
97. Farmakalidis, E., Hathcock, J.N., and Murphy, P.A., Oestrogenic potency of genistin and daidzin in mice, *Food Chem. Toxicol.,* 23, 741–745, 1985.
98. Xu, X., Harris, K.S., Wang, H.-J., Murphy, P.A., and Hendrich, S., Bioavailability of soybean isoflavones depends upon gut microflora in women, *J. Nutr.,* 125, 2307–2315, 1995.
99. Folman, Y. and Pope, G.S., The interaction in the immature mouse of potent oestrogens with coumesrol, genistein and other utero-vaginotrophic compounds of low potency, *J. Endocrinol.,* 34, 215–225, 1966
100. Mäkela, S.I., Pylkkänen, L.H., Santti, R.S.S., and Adlercreutz, H., Dietary soybean may be antiestrogenic in male mice, *J. Nutr.,* 125, 437–445, 1995.
101. Akiyama, T., Ishida, J., Nakagawa, S., Ogaware, H., Watanabe, S., Itoh, N., Shibuya, M., and Fukami, Y., Genistein, a specific inhibitor of tyrosine-specific protein kinases, *J. Biol. Chem.,* 262, 5592–5595, 1987.
102. Thorburn, J. and Thorburn, T., The tyrosine kinase inhibitor, genistein, prevents α-adrenergic-induced cardiac muscle cell hypertrophy by inhibiting activation of the Ras-MAP kinase signaling pathway, *Biochem. Biophys. Res. Commun.,* 202, 1586–1591, 1994.
103. Linassier, C., Pierre, M., Le Pecq, J.B., and Pierre, J., Mechanisms of action in NIH-3T3 cells of genistein, an inhibitor of EGF receptor tyrosine kinase activity, *Biochem. Pharmacol.,* 39, 187–193, 1990.
104. Constantinou, A., Kiguchi, K., and Huberman, E., Induction of differentiation and DNA strand breakage in human HL-60 and K-562 leukemia cells by genistein, *Cancer Res.,* 50, 2618–2624, 1990.
105. Kiguchi, K., Constantinou, A., and Huberman, E., Genistein induced cell differentiation and protein-linked DNA strand breakage in human melanoma cells, *Cancer Commun.,* 2, 271–278, 1990.
106. Markovits, J., Linassier, C., Fosse, P., Couprie, J., Jacquemin-Sablon, A., Saucier, J.M., Le Pecq, J.B., and Larsen, A.K., Inhibitory effects of the tyrosine kinase inhibitor genistein on mammalian DNA topoisomerase II, *Cancer Res.,* 49, 5111–5117, 1989.
107. Okura, A., Arakawa, H., Oka, H., Yoshinari, T., and Monden, Y., Effect of genistein on topoisomerase activity and on the growth of [Val 12]Ha-ras-transformed NIH 3T3 cells, *Biochem. Biophys. Res. Commun.,* 157, 183–189, 1988.

108. Sathyamoorthy, N., Gilsdorf, J.S., and Wang, T.T., Differential effect of genistein on transforming growth factor beta 1 expression in normal and malignant mammary epithelial cells, *Anticancer Res.,* 18, 2449–2453, 1998.
109. Kuiper, G.G.J.M., Enmark, E., Pelto-Huikko, M., Nilsson, S., and Gustafsson, J.-Å., Cloning of a novel receptor expressed in rat prostate and ovary, *Proc. Natl. Acad. Sci. U.S.A.,* 93, 5925–5930, 1996.
110. Kuiper, G.G.J.M., Grandien, K., Enmark, E., Haggblad, J., Nilsson, S., and Gustafsson, J.–Å., Comparison of the ligand binding specificity and transcript tissue distribution of estrogen receptors alpha and beta, *Endocrinology,* 138, 863–870, 1997.
111. Barnes, S., Kim, H., Darley-Usmar, V., Patel, R., Xu, J., Boersma, B., and Luo, M., Beyond ERα and ERβ: estrogen receptor binding is only part of the isoflavone story, *J. Nutr.,* 130, 656S–657S, 2000.

10 Fermented Meat

Walter P. Hammes, Dirk Haller, and Michael G. Gänzle

CONTENTS

10.1 INTRODUCTION

A number of reviews and even monographs have dealt with the properties and characteristics of fermented meats.[1–5] In this chapter, the emphasis is placed on the nutritional and health-related aspects of the fermentation process or, more specifically, the products obtained with the aid of this process. Points to consider are the inherent properties of the nonfermented matrix (i.e., the nutritive value, hazards, and shelf life), as well as the effects of the fermentation process on these criteria.

0-8493-1372-4/03/$0.00+$1.50

TABLE 10.1
Meat Consumption in the 1980s and 1990s (g per person per day)

Country	1980s	1990s
United States	310	231
Australia	296	290
United Kingdom	201	204
France	290	305
Germany	269	261
Japan	100	123
China	26	108

Source: The values quoted are from the Organization for Economic Development and Cooperation (OECD) for the years 1982–1984 and 1992, respectively. The exception is China; these values are taken from *Asia Pacific Food Industry* 7, 14, 1995, using the years 1979 and 1994. The figures include bone weight.

10.1.1 Nutritional Role of Meat in the Human Diet

Meat has traditionally been considered an essential component of the human diet to ensure optimal growth and development. With a limited range of foods available in societies throughout history, meat was important as a concentrated source of a wide range of nutrients. Anthropological research shows that the length of the gut in primates and humans became shorter with the introduction of animal-derived food. Smaller quantities of food of high digestibility required relatively smaller guts, characterized by simple stomachs and proportionally longer small intestines, emphasizing absorption.[6,7] It is perhaps due to the fact that meat has been eaten as much for enjoyment as for its nutritional qualities that consumption of meat and meat products has increased with the affluence of the consumer (Table 10.1).

The meat consumption and production figures published by the U.S. Department of Agriculture and the European Union (EU) do not distinguish between fresh meat and processed or fermented meat products. Therefore, only estimates can be provided concerning the size of production of fermented meat products. Data concerning fermented meat production and consumption in the EU were compiled from EU statistics and various national sources by Fisher and Palmer.[8] The EU production of fermented meats amounted to 689,000 tonnes of fermented sausages and approximately the same amount of raw ham in 1988. Approximately 5% of the total meat production (carcass weight) is further processed by fermentation. The major producers of fermented meat products in the EU are Germany, Italy, Spain, and France. In these countries, 20 to 40% of processed meat products can be classified as fermented meat products. Fermented meat products include products listed in Table 10.2.

TABLE 10.2
Fermented Meat Products

Group	Water Activity	Examples
		Comminuted Fermented Meats[a]
Dry		
Mold ripened	<0.9	Classical Italian salami such as Tipo Milano, Felino, Narzi etc.; French saucisson sec; Hungarian salami (additionally smoked); Spanish chorizos
Smoked		German Katenrauch, Dauerwurst, etc.
Semi Dry		
Mold ripened	0.9–0.95	Various French and Spanish fermented sausages
Smoked		The majority of U.S. and Northern European fermented sausages (e.g., the Netherlands, Scandinavia, Germany)
Nondried		
Spreadable	0.94–0.96	German Mettwurst, Teewurst, Spanish Sobrasada

Group	Examples
	Whole Meat Products
Classical Ham	
Made from thigh of hog with or without bone	Prosciutto di Parma or San Daniele (Italy); Jambon de Beyonne (France); Jamón Serrano (Spain); Kraški pršut (Slovenia); Virginia ham (U.S.); Yunnan ho-twe and Tshingwa ho-twe (China)
Cuts of Meat	
Pork	Bacon (U.K., U.S.); Pancetta (Italy); Capocollo (Italy); Culatello di Zipello (Italy); Schwarzwälder, Westphälischer, Lachs-Schinken (Germany); Tirolean Speck (Austria)
Other Animal Sources	
Beef	Bresaola (also equine and venison, Italy); Bündner Fleisch (Switzerland); Pastirma (Near Orient); Biltong (also antelope, South Africa)
Mutton	Fenelar (Norway)

[a] Sausages produced from varying sizes of meat particles, obtained from various animals (most commonly pork), in various casings of varying diameter and containing additives and spices of varying compositions.

The fat content of meat as consumed is around 2 to 5%, even though total fat content varies with species, feeding regimes, and age. The principal fatty acids in meat are saturated fatty acids, including palmitic acid (C16:0) and stearic acid (C18:0). Around 40% of the fat in meat is monounsaturated, of which oleic acid (C18:1) is one of the main contributors.[9] Protein of high biological value and micronutrients such as iron, zinc, vitamin B_1, niacin equivalents, and vitamin B_{12} significantly contribute to the nutritional value of meat.[10] Hambraeus[11] reported that the requirement for iron is one of the most difficult nutritional requirements for humans to meet, since iron deficiency is caused not only by a low intake but is also the result of low bioavailability. Increased iron requirements may result from physiological variables or clinical problems. Red meat contains 50 to 60% of its iron in the heme form (from hemoglobin and myoglobin), which is absorbed in humans by a more efficient mechanism than is nonheme iron, the source of iron in plant foods.

The importance of meat and meat products in everyday food culture and consumer health may be questioned by the fact that populations of vegetarians living in rich countries are characterized by lower rates of cancer and cardiovascular disease.[12–14] The analysis of dietary patterns, as a possible approach to examining diet–disease relations, identified two major eating patterns defined by factor analysis using dietary data collected from food frequency questionnaires.[15] The first factor, the "prudent dietary pattern," was characterized by a high intake of vegetables, fruits, legumes, whole grains, and fish or other seafood, whereas the second factor, the "Western pattern," showed a high intake of processed meat, red meat, butter, high-fat dairy products, eggs, and refined grains. A study has been published involving Seventh-day Adventists, a well-characterized population, in which the effect of dietary intake of nutrients on biochemical parameters in blood and urine was compared in vegetarian and nonvegetarian subjects.[16,17] The dietary intake of cholesterol was higher in nonvegetarian subjects (560 to 710 mg/day) compared to vegetarians (<20 mg/day) and was associated with elevated serum cholesterol levels in the nonvegetarian population. These results demonstrated a correlation between dietary intake of certain food components (e.g., cholesterol) relevant for diseases (e.g., coronary heart disease) and their blood concentrations.

10.1.2 The Relationship between Meat in the Diet and Disease

Numerous studies have compared the health status and mortality of vegetarians to those of omnivores. The results show a strong correlation between *per capita* consumption of meat and the incidence of colon cancer among various countries.[18] In more detailed case-control and cohort studies, in which lifestyle factors were better controlled for, the consumption of red meat was associated with a high risk of colon cancer.[19–21] Results from a meta-analysis by Howe et al.,[22] including 13 of the case-control studies, indicated that total energy intake was positively associated with a higher risk of colon cancer. Surprisingly, the intakes of fat, protein, and carbohydrates were not related to cancer risk, independent of their contribution to total energy. Compared with Western vegetarians, nonvegetarians have a higher mean body mass index (BMI) by about 1 kg/cm^2, suggesting that higher total energy intake and meat consumption might be associated with the "Western diet pattern."[23]

The mechanisms that increase the risk of colon cancer are not yet clear. Several *in vitro* studies suggest that DNA damage in human cell lines can be caused by food ingredients or their metabolic products.[24,25] High meat consumption, for example, leads to higher levels of bile acids and *N*-nitroso compounds in the feces. Bile acid and *N*-nitroso compounds, as well as their metabolites, potentially promote colon cancer development.[26,27] Animal studies show that large intestinal *N*-nitrosation does not occur in germ-free rats, but it has been shown to occur in the presence of a conventional flora.[28] The effect of diet on the composition of the intestinal microflora was shown by Finegold et al.[29] Subjects eating a Western diet were compared with subjects eating a Japanese diet. The subjects eating the Japanese diet had a lower risk of colon cancer and had significantly higher numbers of *Enterococcus faecalis, Eubacterium lentum, E. contortum, Klebsiella pneumoniae,* and various *Lactobacillus* species in their feces. The Japanese diet has been associated with low incidence of large bowel cancer. Japanese people who migrate to the United States and adopt the Western diet develop this cancer with increased frequency, approaching that of native-born Americans.[30,31] The high risk group with Western dietary patterns had increased counts of species of the genera *Bacteroides, Bifidobacterium, Peptostreptococcus*, and *Clostridium* in their fecal flora. Goldin and Gorbach[32] reported that rats fed a high-fat (meat) diet had higher enzyme activity of the fecal bacterial enzymes β-glucuronidase, nitroreductase, and azoreductase than rats on a low-fat (no meat) diet. These results were confirmed and extended in subsequent animal and human studies demonstrating that a high-fat diet, independent of the meat content, elevated the activity of these bacterial enzymes, which are implicated in the generation of mutagens, carcinogens, and various tumor promoters.[33]

Regular consumption of meat is also associated with increased risk of death from coronary heart disease (CHD). The most compelling evidence comes from studies with Seventh-day Adventists. It was found that men and women who consumed red meat daily had around 60% greater chance of dying from CHD than those who consumed red meat less than once per week.[16,34] A review of studies of the association between blood homocysteine concentrations and atherosclerotic disease showed that 16 of 21 investigations reported significantly higher homocysteine concentrations in case subjects compared with control subjects.[35] Because red meat is a major source of methionine in the diet, and methionine is the direct metabolic precursor of homocysteine, a higher intake of red meat may be involved in cardiovascular disease initiation and progression.

In summary, high dietary intakes of energy, saturated fat, and red meat, all associated with the Western diet pattern, are likely to have adverse effects on chronic disease risks, particularly those of colon cancer and coronary heart disease. On the other hand, little evidence indicates that the consumption of moderate amounts of meat or meat products is harmful in regard to either cancer or cardiovascular disease.

10.2 THE HISTORY AND CULTURE RELATED TO FERMENTED MEAT

Meat is extremely susceptible to microbial spoilage. Virtually all ecological factors characterizing meat as a substrate are optimal for the growth of bacteria, which are the most efficient agents in remineralization of organic matter. For example, in meat, water activity and pH are 0.96 to 0.97 and 5.6 to 5.8, respectively, and nutrients and growth factors are abundantly available. Any storage of this nutritionally rich food and preservation of the nutrients contained therein requires the suppression of microbial growth or the elimination of microorganisms and prevention of recontamination.

The traditional methods employed for prevention of microbial spoilage are still in use, though with a different meaning in various products. These methods comprise reduction of water activity (drying, salting) and/or pH (fermentation, acidification), smoking, storage at refrigeration or freezing temperatures, and use of curing aids (nitrite and nitrate). Commonly, these methods act together in different combinations, building up hurdles against microbial growth. With regard to fermented sausages, these hurdles are low water activity (0.85 to 0.95) and pH (5.6 to 4.7), the use of nitrite (nitrate), and smoke. In addition, during fermentation and ripening, ecological factors, such as a reduced redox potential and low temperatures (10 to 12°C, at least for dry sausages), together with antagonistic compounds produced by the fermenting flora exert a selective effect against the growth of undesirable microorganisms. Basically, the same antimicrobial hurdles are effective in achieving the microbial stability of ham, except for the effect of a low pH. Since no lactic fermentation takes place, the reduced water activity is the most effective hurdle against microbial growth in ham. The understanding of these ecological factors and their control is not only a prerequisite in quality assurance, but also provides a basis for understanding to what extent these food matrices might be used to serve as probiotic foods.

The production of dried and cured meat (ham) can be traced back to prehistoric times. It cannot be excluded that among the sausages that are mentioned in historical literature (e.g., Homer's Iliad), fermented products were included, although we do not have sufficient knowledge of their production processes to permit a conclusion that a fermentation step was part of the technology. The origin of fermented sausages can be traced back with accuracy to ca. 1730, when salami was first mentioned in Italy.[36] From Italy, the art of producing fermented sausages spread to other European countries and was established, for example, in Germany in 1735 and Hungary in 1835. Today in various parts of the world, a large number of different types of fermented sausage exist. For example, 330 different types are produced in Germany.[37] A very large number of ham varieties are also produced.[38] This very high consumption of fermented meats is an indication that such products have a long tradition of being safe. However, some specific safety aspects deserve consideration.

The high fat content commonly found in fermented sausages (usually around 50% of dry matter) has been of nutritional concern, and leaner products are now available (some as low as 5%). The sensory quality of the traditionally high-fat sausages is, however, unique and a standard for the gourmet. The body fat content of pigs has been already drastically reduced by breeding, but with respect to ham, it is left to the consumer to cut off the fat layer before consumption.

Mold-ripened varieties of both sausages and ham (e.g., Tirolean speck and Bündner Fleisch) exist. The production of such products free of mycotoxin is a concern, because of the potential of fungi for contamination. In addition, a carryover from animal feed to the meat may be a source of mycotoxin contamination.[39] This hazard is the target of general meat inspection and control. In rare cases, mycotoxins have been detected in fermented sausages and ham. One way to overcome this hazard is the use of competitive mold strains that have a proven absence of a mycotoxigenic potential.[40,41] These starter cultures usually contain strains of *Penicillium nalgiovense* or *P. chrysogenum,* and are already widely in use in Europe. For ham production, the absence of mycotoxins is still a matter of rigorous quality control.

Meat may also contain bacterial food pathogens. Because fermented meat products usually do not undergo a physical treatment to eliminate pathogenic microorganisms, the meat has to be of high quality with regard to hygiene and microbial counts. The control of pathogens is achieved by appropriate fermentation technology, including the use of starter cultures.

10.3 THE FERMENTATION PROCESS

The traditional aim of the fermentation process is to transform the highly perishable substrate meat into a shelf stable and safe product ensuring an optimum nutritive value and sensory quality. The factors affecting the process are the nature of the raw materials and the activity of microorganisms, as well as endogenous enzymes and process technology. For all fermented meat products, the raw material is meat with a variable amount of fat that has not been subjected to a thermal or any other germ-reducing process. Meat is the flesh (muscle tissue) of warm-blooded animals, but fermented specialties from poultry (sausages as well as cured and smoked fermented poultry) are also available. Two groups of products can be differentiated on the basis of the microbial populations involved in the fermentation process — foods from a comminuted matrix and whole meat products.

10.3.1 Fermentation of a Comminuted Meat Matrix

10.3.1.1 Variables in Sausage Production

The comminution of muscle tissue to particles varying in size between 1 and 30 mm, together with the homogenous distribution of fermenting organisms, is the prerequisite for a fermentation process taking place throughout the matrix. Curing salt, nitrate, ascorbic acid, and, in some cases, sodium glutamate and glucono-δ-lactone are added to the particles together with spices and above all a carbohydrate source, which is commonly glucose. These compounds exert strong effects on the growth and performance of the fermentative flora.

The fatty tissue should be as fresh as possible, as any initiated oxidative process will strongly affect the shelf life by causing early rancidity. The whole comminution process of chopping or grinding together with a mixing procedure requires temperatures below 2 to 3°C. Thereafter, the temperature is raised usually to >20°C and <28°C to initiate the fermentation process. Semidry sausages of the U.S. summer sausage type are fermented at even higher temperatures (32 to 38°C). The many types of fermented sausages are the result of a great variety of process conditions. Variables include:

- The particle size of the comminuted meat and fatty tissue
- The selection of additives
- The temperature/humidity conditions prevailing in the course of fermentation until the final ripening
- The diameter of the sausages
- The nature of the casings
- Smoking
- Heating after fermentation

Supporting the development of mold growth on the surface or establishing a special tight surface film (e.g., coating with a titanium dioxide film)
Dipping in antifungal preparations (sorbic acid or pimaricin)[3,42]

A flow scheme of the process of production of common fermented sausage is depicted in Figure 10.1.

The fermentation process, together with the effects of the temperature/humidity conditions, ensures that the originally highly perishable raw materials turn into a spoilage-resistant, flavor-rich product with a defined texture and stable color. Great variation exists with regard to texture, which may range from spreadable to sliceable, from soft to very hard. With regard to the microbial effects, the first days are of great importance. During that time the organisms multiply, reduce the pH to values varying between 5.4 and 4.8, exhibit enzymatic activity, and interfere with undesired microorganisms, which constitute the indigenous flora of the meat. In the course of ripening, the pH usually rises again. This rise in pH does not constitute a safety

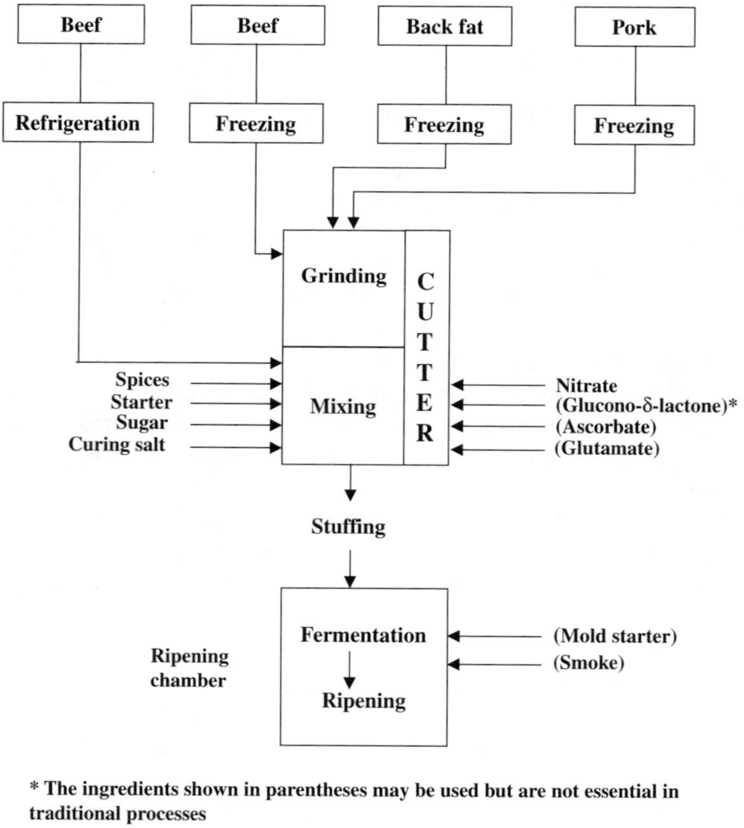

FIGURE 10.1 Flow scheme of the process of production of dry fermented sausage of a common German type.

TABLE 10.3
Current Status of Species Employed in Meat Starter Cultures

Bacteria

Lactic Acid Bacteria

Lactobacillus acidophilus,[a] *Lb. alimentarius,*[b] *Lb. paracasei,*[a] *Lb. rhamnosus,*[a] *Lb. curvatus, Lb. plantarum, Lb. pentosus, Lb. sakei, Lactococcus lactis, Pediococcus acidilactici, P. pentosaceus*

Actinobacteria

Kocuria varians,[c] *Streptomyces griseus, Bifidobacterium spp.*[a]

Staphylococci

Staphylococcus xylosus, S. carnosus ssp. *carnosus, S. carnosus* ssp. *utilis, S. equorum*[b]

Halomonadaceae

Halomonas elongata[b]

Fungi

Penicillium nalgiovense, P. chrysogenum, P. camemberti

Yeasts

Debaryomyces hansenii, Candida famata

[a] Used in probiotic cultures.
[b] Used in premarket studies at industrial scale (Laboratorium Wiesby, Niebüll and Rudolf Müller and Co. Gießen, personal communication).
[c] Formerly *Micrococcus varians.*

hazard, because at the same time the water activity is decreased to levels that discourage bacterial growth.

The study of the fermentation processes has revealed that lactobacilli and micrococci play a decisive role. These organisms develop under the specific prevailing ecological conditions and were sometimes inoculated by back slopping, i.e., adding chopped fermented sausage back to a new meat mixture. Especially between 1950 and 1960, microorganisms were isolated and turned into preparations of starter cultures, which are now commonly used, as they exhibit numerous advantages when compared with the classical indigenous fermentation. For example, the reproducibility with regard to process time and product quality is greatly enhanced and microbial risks are reduced in such a way that the application of starters plays a role in the Hazard Analysis and Critical Control Points (HACCP) concepts for quality assurance. In Table 10.3, microorganisms are listed that can be found as components in some starter cultures. Many of the bacteria used are lactic acid bacteria (LAB), which are of primary importance, but included in this table are also nonlactic acid bacteria, which are used mainly in combination with LAB and contribute to the fermentation process as they have unique properties. For example, *Kocuria* spp. and *Staphylococcus* spp. exhibit nitrite and nitrate reductase activity, respectively, which is important for the reddening of the sausages, i.e., the formation of the stable red color of nitrosomyoglobin. In addition, these organisms exhibit catalase activity, which counteracts the formation of hydrogen peroxide and thus helps to prevent color defects and rancidity. Yeast and fungi contribute mainly to the development of flavor and to a minor extent also to color stability.

10.3.1.2 Sausages as Possible Probiotics

Sausages have been developed with the claim that they contain probiotic bacterial strains. The use of fermented sausages to create a probiotic food that can fulfill the requirements necessary to justify a claim related to the health of the consumer has numerous obstacles. These have been discussed by Hammes and Haller.[43] It was argued that the sausages should contain the probiotic bacterial strain at counts that are needed to be effective in the intestines. They have to be in a state that supports survival and metabolic activity in the intestinal tract, and they should be consumed in adequately high numbers on a daily basis. Finally, adequate human studies with the probiotic food should have been performed to substantiate any health claim. Intestinal isolates as well as potential probiotic strains of lactobacilli and bifidobacteria had been used for sausage production, and it can be shown that these bacteria were present in finished sausage or model sausages; some even contribute to the fermentation process comparable to commonly used starters.[44–47]

In studies by Bunte et al.[48] and Jahreis et al.,[49] a food isolate (*Lactobacillus paracasei*) was used as a starter culture for production of a moist type of fermented sausage with a defined shelf life. Human studies were performed in which the microbiology of the sausages and occurrence of the test strain in the feces were investigated. Volunteers had consumed 50 g sausage containing ca. 10^8 colony forming units (CFU)/g of the test strain for a period of 4 weeks. The analyses of blood samples from these volunteers revealed that in those persons with high counts of the test strain in their feces, certain blood parameters were affected during the challenge period. In particular, the values of CD4 T helper cells were elevated and the phagocytosis index increased. Furthermore, the expression of CD54 (ICAM-I) decreased. This glycoprotein is otherwise up-regulated in response to a variety of inflammatory regulators. Finally, an increased titre of antibodies against oxidized LDL was measured. In this case, the starter strain brought about the "probiotic effect." In this example, the probiotic strain served as a starter culture because of its technological and sensory effects. Thus, in principle, it is possible to produce probiotic fermented sausages. However, the sausages have to be designed such that the probiotic bacterial strain can exhibit its beneficial effects. It can be assumed that any large reduction of pH (e.g., pH < 5.0), extended ripening (e.g., > 1 month), drying, or excessive heating (e.g., summer sausage) has the potential to make the probiotic effect questionable, as most strains of bacteria may be damaged or killed.

10.3.2 Fermentation of Whole Meat Products (Ham)

Immediately after slaughter, enzyme-catalyzed reactions start to act on the physical and chemical nature of muscle, turning it into meat. These reactions continue even when technological/processing measures, such as cool storage and lowering the water activity by drying or salting, are imposed. However, the reactions proceed in a predictable and controlled way. This process provides the foundation of ham production. With the exception of examples given below, microorganisms do not play a role in the fermentative processes taking place in ham.

By far the majority of fermented raw ham is made from pork, but in some regions beef (Bresaola, Bündner Fleisch, Pastirma) and even meat from game, reindeer, or bear is used to produce similar products.[1,38] The traditional ham in ancient Greek and Roman times as well as in China was made from the bone-containing ham of hogs. This type of ham is still considered the gold standard of quality and is produced in many countries, e.g., Prosciutto di Parma (Italy), Jambon de Bayonne (France), Jamón Serrano (Spain), Kraški pršut (Slovenia), Virginia ham (United States), and Yunnan ho-twe and Tshingwa ho-twe (China).

The process of ham production follows a rather simple principle.[38] It consists of curing by salting (with or without the use of nitrite and/or nitrate) to achieve a water activity of <0.96, which is equivalent to 4.5% sodium chloride. At low temperatures (5°C), the salt will diffuse to the deepest part of the meat, thus overcoming the hazard of food poisoning through *Clostridium botulinum* contamination. After a phase of equilibrating the salt concentration and flavor development, the temperature is raised to 15 to 25°C to ripen the ham. This phase lasts at least 6 to 9 months and may be extended even to 18 months to achieve the optimum flavor. At the end of the ripening step, the moisture has been reduced by about 25% and a salt concentration between 4.5 and 6% results. In some countries (e.g., Germany), in addition to a nitrate cure, smoking is used to obtain a characteristic flavor and to suppress surface growth of molds. When ham is produced from only one or a few muscles, numerous methods are applied to accelerate production time and to control flavor development and water content. For example, the curing is performed in brine or by injection of curing salt and, above all, the ripening period is drastically reduced. Microorganisms may cause spoilage by growing in the inner muscle parts before the water activity is reduced to safe levels. However, microorganisms also contribute favorably to the process as they are involved in nitrate curing in brine by the formation of the reactive nitrite, which affects color by reddening, flavor, and microbial safety. The microorganisms involved are Gram-negative bacteria such as vibrios and *Halomonas* spp. Unlike fermented sausages, ham has no apparent potential to be used as a probiotic food.

10.4 COMPOSITION AND CHANGES DURING FERMENTATION

The physical, biochemical, and microbial changes during sausage fermentation are summarized as follows: growth of LAB and concomitant acidification of the product, reduction of nitrates to nitrites and formation of nitrosomyoglobin, solubilization and gelification of myofibrillar and sarcoplasmic proteins, degradation of proteins and lipids, and dehydration.

10.4.1 Fermentation Microflora

The initial microbial population of sausage minces is highly variable and strongly depends on the microbial load of the raw materials. The ecological conditions of sausage minces favor the growth of Micrococcacea and lactobacilli. Lactobacilli generally grow to cell counts of 5×10^8 to 10^9 CFU/g, and these numbers remain stable throughout ripening. Micrococcacea (predominantly *Kocuria varians, Staphylococcus carnosus,* or *S. xylosus*) generally grow to cell counts of 10^6 to 10^7 CFU/g, when nitrate cure is applied. The growth of these organisms is inhibited by the application of nitrite cure as well as the decrease of pH. Higher cell counts are reached at the outer layer of the sausages, where a higher oxygen partial pressure occurs.

Because of their high salt tolerance, the predominant microorganisms found in dry cured ham fermentation belong to the classification that was formerly included in the family Micrococcacea. The species most often isolated are *Staphylococcus xylosus, S. equorum*, and *S. sciuri,* but *K. varians* is also found at appreciable cell counts.[38,50,51] Growth of staphylococci occurs primarily at the surface of hams.[52]

The growth of yeasts and fungi on mold-ripened products is restricted to the surface of the product, where cell counts reach 10^5 to 10^7 CFU/cm^2 within four weeks of ripening. Many of the traditional fermentations involving fungal ripening rely on inoculation by the "house flora" associated with the building or equipment used for fermentation and maturation. The occurrence of yeasts and other fungi on fermented meats was reviewed by Cook.[53] *Penicillium* constituted 96% of the microflora; the nontoxigenic species *Penicillium nalgiovense* was most frequently isolated. The halotolerant yeast *Debaryomyces hansenii* is the predominant yeast in meat fermentations.

10.4.2 Acidification, Dehydration, and Microbial Antagonism

Acidification to the isoelectric point of meat proteins (pH 5.3 to 5.4) and the increase of the ionic strength induces gel formation of the proteins and thus confers important structural changes. The high levels of sodium chloride and lactate in fermented sausages contribute to the development of the characteristic taste of the product. Rapid acidification and subsequent drying are of paramount importance for inhibition of the growth of pathogens and their subsequent inactivation during ripening.[54] As there exist nondried fermented sausages of a spreadable type (e.g., in Germany, known as Streichmettwurst, Teewurst, Rohpolnische), the highest hygienic standards of the raw materials and production facilities are key to product safety. In addition to low pH and water activity, specific microbial metabolites such as diacetyl or short chain fatty acids exert an inhibitory effect towards pathogens. Several meat starter cultures produce bacteriocins — small, heat stable peptides with antimicrobial activity.[55] The use of bacteriocinogenic starters has been shown to contribute to the elimination of *Listeria* during sausage fermentation.[56] However, because of the resistance of Gram-negative organisms, including *Salmonella* and *Escherichia coli* O157:H7 strains, the contribution of bacteriocins to the overall hygienic safety of fermented meats is limited.[57,58]

Growth and metabolism of LAB result in a fast drop of pH during the first days of sausage fermentation, followed by a slight increase during the ripening period. Lactic and acetic acids are the major fermentation products, and the molar ratio of lactate to acetate ranges between 7 and 20.[59,60] The product pH depends on the buffering capacity of the meat, the metabolic activities of the fermentation microflora, and the addition of fermentable carbohydrates. In Northern European and in U.S. summer sausage, the pH typically ranges from 4.8 to 5.2, corresponding to a content of 200 mmol lactate/kg dry weight. In Mediterranean-type products involving longer ripening periods of up to several months, the final pH typically ranges from 5.4 to 5.8. In mold-ripened products, the sausage pH may increase to levels close to 6.0 due to lactate consumption and the formation of ammonium. The dry matter content of fermented sausages ranges from 50 to 75% or more, corresponding to water activity (a_w) values ranging from 0.86 to 0.92 upon ripening.

10.4.3 Proteolytic and Lipolytic Degradation during Fermentation

The proteolytic events during fermentation of raw sausages and dry cured ham were recently reviewed by Toldra and Flores, and Ordónez et al.[61,62] In the course of ripening, peptides and amino acids accumulate to levels of about 1% dry matter.[63,64] Peptides and amino acids themselves may contribute to the characteristic taste of dry cured products and act as flavor enhancers and synergists. Excess proteolysis may result in bitter and metallic off-flavors because of the presence of bitter peptides. Furthermore, amino acids and peptides are utilized by microorganisms for the conversion to flavor volatiles. Several food proteins, mainly milk proteins, contain peptide sequences with opioid or angiotensin converting enzyme inhibitory activity and are released in their active form upon digestion. The release of these bioactive peptides is known to be influenced by lactic fermentation. However, to date no information is available on bioactive peptides from meat or fermented meat products.

The hydrolysis of muscular proteins to peptides is mainly achieved by the activity of endogenous enzymes. The endopeptidases — cathepsins B, B + L and H — were shown to remain active throughout the fermentation of dry cured ham and fermented sausages, whereas tissular calpains are inactivated during fermentation and do not contribute significantly to overall proteolysis. Furthermore, muscle exopeptidases contribute to peptide conversion to amino acids. The proteolytic system of lactobacilli consists mainly of cell wall-associated proteinases, which convert proteins to oligopeptides. Oligopeptide transport is the main route for nitrogen entry into the bacterial cells, and virtually all peptidases are located intracellularly.[65,66] The proteolytic activity of starter cultures is weak compared to that of tissue enzymes. Correspondingly, the inoculation of sausage minces with starter cultures leads to only a minor increase in amino acid levels of the sausages compared to aseptic control batches.[63] The proteolytic activity of *Kocuria varians* is inhibited by environmental conditions prevailing during sausage ripening, yet the peptidase activity of this organism may contribute to the formation of amino acids.[67] It was recently shown that *Lb. casei* utilizes peptides released from pork muscle sarcoplasmic and myofibrillar proteins under conditions of sausage ripening.[68,69]

The fat content of fermented sausages typically ranges from 40 to 60% of dry matter. During fermentation, long-chain fatty acids are released from triglycerides and phospholipids. Typically, an increase in the levels of free fatty acids up to approximately 5% of the total fatty acids has been found.[64] The fatty acid composition of fat varies considerably depending on the previous feeding regime of the animal. The specific release of polyunsaturated fatty acids is higher than that of monounsaturated or saturated fatty acids. This may reflect a preference of microbial and meat endogenous enzymes for the sn3 position of the tryglyceride most frequently occupied by unsaturated fatty acids or a preference for the polar lipid fraction.[62,64] Lysosomal muscle acid lipase and adipose tissue lipases remain active throughout several months of dry cured ham ripening.[61] Comparisons of aseptic fermented batches of sausages with batches inoculated with starter cultures has shown that lipolysis during fermentation is attributed mainly to meat endogenous enzymes.[70–72] Lactobacilli are considered to be weakly lipolytic. Strains of *K. varians* and *S. carnosus* or *S. xylosus* have been found to exhibit lipolytic activity, which is, however, inhibited at pH values below 6.[70,73] In mold-ripened products, lipolytic activities of the surface mold flora contributed to the generation of long-chain fatty acids.[74]

10.4.4 Generation of Flavor Volatiles

Flavor compounds are generated during sausage fermentation by the following general routes:

1. Flavor volatiles are generated by lipolysis and hydrolysis of phospholipids, followed by oxidation of free fatty acids.
2. Microorganisms produce organic acids; convert amino acids and peptides to flavor-active alcohols, aldehydes, and acids; and modify products of lipid oxidation, e.g., by esterification of acyl moieties or reduction of aldehydes.
3. Depending on the product formula and maturation conditions, the sausage aroma is determined by the addition of spices, smoking, or surface-ripening with yeasts or molds.

An overview of mechanisms for generation of flavor compounds during sausage fermentation is shown in Figure 10.2. Despite the differences in the process technology and the fermentation microflora, it may be assumed that the generation of flavor during fermentation of dry cured ham is governed by the same principles.[61,75]

Data are available from aroma extract dilution analysis showing the identification of those volatile compounds that have a significant impact on sausage flavor.[76] The most important odor compounds in French, Italian, and Spanish salami are those originating from added spices, i.e., sulfur compounds (e.g., diallylsulfide) originating from garlic, and eugenol from nutmeg. Products of lipid oxidation, fatty acids, as well as fermentation volatiles such as acetic acid, diacetyl, and phenylethanol, further contribute to overall flavor. The comparison of flavor volatiles and sensory attributes of sausages of various origins reveals that a high level of fatty acids negatively affects the sausage aroma.

Unsaturated fatty acids are prone to autoxidation in the presence of oxygen. Many of the products of lipid oxidation are highly volatile and have a low odor threshold. Hexanal, nonenal, 2(Z)octenal, 1-octen-3-ol, and 1-octen-3-on were identified as the most potent flavor volatiles resulting from autoxidation of linoleic acid, and these compounds are present in fermented sausages.[54,64,76,77] Aldehydes origi-

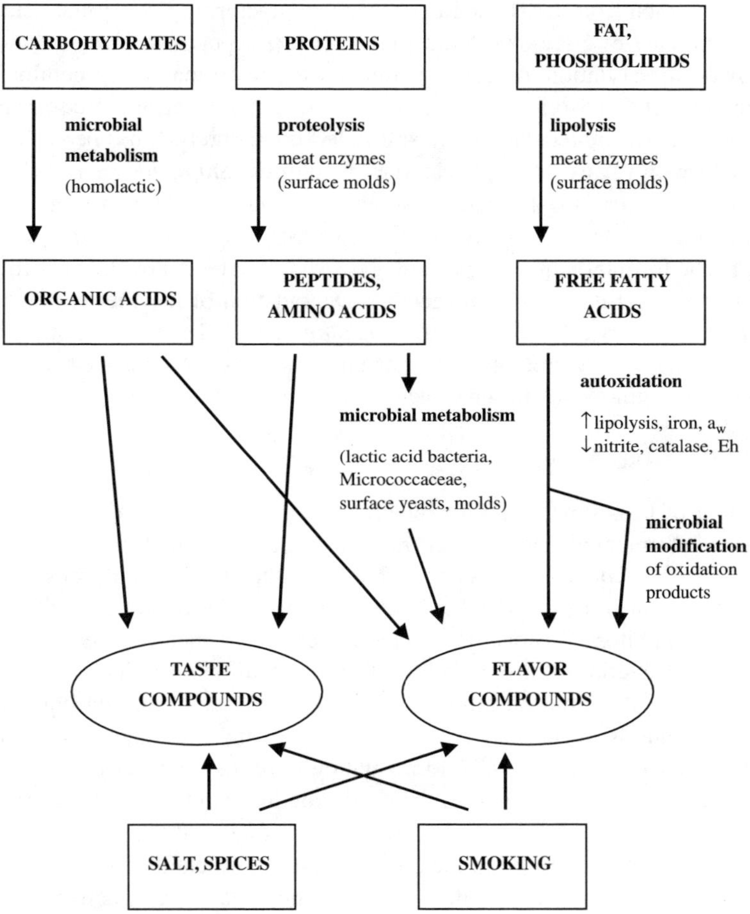

FIGURE 10.2 Schematic overview of mechanisms for generation of flavor compounds during sausage fermentation. The relative importance of the biochemical, chemical, and microbial transformations to overall taste and flavor remains to be determined and is expected to differ considerably depending on regional or product-related variations in the product formula and ripening conditions. Smoking of sausages or ripening with a surface mold flora is used for a significant proportion of fermented sausages depending on regional preferences.

nating from lipid oxidation are subjected to further modification by microbial reductases.[78,79] The rate of lipid oxidation is greatly enhanced by heme or nonheme iron. The increase in NaCl concentration during ripening also favors lipid oxidation. Nitrite present as part of the sausage formula or formed from nitrate by microbial nitrate reductase acts as an antioxidant. The removal of peroxides by catalase, pseudocatalase, or manganese-dependent superoxide dismutase activities of the Micrococcaceae and/or lactobacilli is crucial to limit the extent of fatty acid oxidation and to prevent off-colors.[60,78,80]

Amino acid catabolism by staphylococci and lactobacilli yields volatile products contributing to meat flavor. Insight into the relevant metabolic pathways has primarily been obtained from LAB used in cheese manufacturing.[81–83] Amino acid degradation is initiated by pyridoxal-5′-phosphate (PLP)-dependent transamination, followed by decarboxylation. A second pathway for transformation of methionine and cysteine involves β-γ-elimination and subsequent formation of methanethiol, methional, and low molecular weight sulfhydryl compounds. It has been shown that comparable pathways occur in meat starter cultures. *Staphylococcus xylosus* and *Staphylococcus carnosus* have been shown to produce a large variety of flavor volatiles originating from amino acid degradation during growth on sausage minces.[84] The formation of flavor volatiles is strongly affected by the growth parameters salt, nitrate, glucose, and oxygen.[85,86] Degradation of leucine was also shown for strains of *Lb. curvatus, Lb. sakei,* and *Lb. plantarum.*[87] The addition of proteolytic enzymes or the use of proteolytic starter cultures does not enhance the microbial conversion of amino acids to flavor-active derivatives.[73,88]

10.4.5 Biogenic Amines

The content of histamine, tyramine, phenylethylamine, tryptamine, putrescine, and cadaverine in fermented meat products ranges from not detectable to levels exceeding 100 mg/kg.[60,89] At concentrations above that level, the pressor amines histamine and tyramine may constitute a health hazard for patients under treatment with monoamine oxidase inhibitors.[90] Biogenic amines in fermented meat products are mainly derived from bacterial decarboxylation of amino acids. Putrescine and cadaverine are produced by the Gram-negative spoilage flora of raw meat, and high levels of these compounds indicate poor hygienic quality of the raw materials for sausage production. LAB used for sausage fermentation have the potential to decarboxylate tyrosine, histidine, and ornithine to the corresponding amines tyramine, histamine, and putrescine.[91] LAB are assumed to be the main source of tyramine in fermented sausages. The formation of tyramine during sausage fermentation occurs during the initial phase of the fermentation, in which LAB are metabolically active, rather than during the ripening phase. Proteolytic events during sausage fermentation do not appreciably affect levels of tyramine and histamine.[92] The use of starter cultures with low tyrosine decarboxylase activity to rapidly inhibit metabolism of Gram-negative bacteria and to suppress growth of the indigenous fermentation flora with potentially higher decarboxylase activities has been shown to effectively reduce tyramine levels in fermented sausages.[93] A reduction of tyramine was observed in sausages where both tyramine and high levels of tyramine oxidizing *K. varians* were added. Only a limited effect of microbial tyramine oxidases was found in experiments where the same strain was applied in combination with a tyramine producing strain of *Lb. curvatus.*[94,95]

10.5 POTENTIAL BENEFICIAL HEALTH EFFECTS OF BACTERIA INVOLVED IN MEAT FERMENTATION

It is characteristic of LAB that certain species are found not only as members of the human intestinal microflora but also as part of the man-made ecosystems present in fermented food.[96,97] Denaturating gradient gel electrophoresis (DGGE) of DNA fragments generated by polymerase chain reaction (PCR) with 16S rDNA targeted group-specific primers has been used to demonstrate that food-associated bacteria and especially meat fermenting bacteria such as *Lb. sakei, Lb. curvatus, Leuconostoc mesenteroides,* and *Pediococcus pentosaceus* are present but not culturable in human feces.[98] Bunte et al.[48] demonstrated that *Lb. paracasei* LTH 2579 ingested as a component of dry-fermented sausage can be recovered from the human feces, suggesting that food-associated microorganisms may contribute to the microbial ecosystem of the gastrointestinal tract.

Although many variables can determine the degree to which bacteria survive gastrointestinal transit, the survival in and temporary colonization of the human digestive tract by some LAB have been predicted with the use of *in vitro* experiments. Studies by Pochart et al.[99] and Marteau et al.[100] showed that by using intestinal perfusion, the recovery of viable bacterial cells in the small intestine correlated with their tolerance to *in vitro* acid/bile treatment. It has recently been shown that certain strains of bacteria important in meat fermentation, such as *Lb. plantarum, Lb. paracasei, Lb. sakei, Lb. curvatus,* and *Staphylococcus carnosus,* have a similar capability to survive low pH (1.5 to 2.5) and bile (10 m*M*), to hydrolyze bile salts, and to attach to enterocyte-like CaCO-2 cells when compared with bacteria of intestinal origin or probiotics.[101] These results may suggest that the properties of microorganisms that determine survival and function in the gastrointestinal tract may also be present in food-associated bacteria. Colonization studies in animals have shown that murine *Lactobacillus* spp. were permanently reestablished in mice that had been rendered free of lactobacilli by antibiotic treatment, and that germ-free mice were susceptible to permanent colonization with bifidobacteria.[102,103] However, in human trials, the recovery of probiotic or food-associated strains soon stopped after their oral administration was discontinued.[104] Consequently, the probiotic function of microorganisms may be determined by their ability to directly interact with the gut-associated tissue or to affect metabolic parameters of the indigenous microflora while passing through the gastrointestinal tract.

The intestinal epithelial cells that are the first cells to interact with intestinal bacteria are thought to participate in the initiation and regulation of the mucosal immune response. Experiments in germ-free animals have shown that lymphoid populations in the lamina propria are considerably reduced in the germ-free intestine but assume their normal appearance of physiological inflammation after bacterial colonization.[105–107] This is consistent with the concept that intestinal epithelial cells, upon activation with proinflammatory mediators or enteropathogens, express higher levels of molecules responsible for antigen presentation such as HLA class II molecules, classical class I and nonclassical HLA class Ib molecules, the adhesion molecule ICAM-1, complement factors, cytokines, and cytokine receptors.[108]

Epithelial cell transduction of information from the luminal environment to the mucosal immune system is not limited to pathogenic microorganisms. Commensal bacteria, including the meat starter strain *Lb. sakei* LTH 681 and nonpathogenic *E. coli,* were shown to elicit a characteristic cytokine response in leukocyte-sensitized intestinal epithelial cell lines (CaCO-2 and HT-29) *in vitro* and to deliver a discriminative signal to underlying immunocompetent cells.[109,110] These results suggest that bidirectional cross talk between intestinal epithelial cells and immunocompetent cells may constitute a crucial step in the recognition of nonpathogenic bacteria at the mucosal surface. Moreover, Rescigno et al.[111] showed that dendritic cells, an important population of antigen-presenting cells in the intestinal mucosa, specifically open tight junctions not only between the epithelium and sample pathogenic bacteria but also nonpathogenic bacteria from the gut. The epithelial compartment and the lamina propria are thus potential sites where bacteria of the intestinal microflora may directly encounter cells of the specific and nonspecific immune system. Several studies have shown that lactobacilli isolated from the human gastrointestinal tract activate human mononuclear cells and are potent inducers of monocyte derived cytokine IL-12.[112–114] On the other hand, *Lb. sakei* LTH 681 and the intestinal isolate *Lb. johnsonii* La1 showed a similar capability to induce natural killer (NK) cell activation, suggesting that the immunogenic activity of lactobacilli is also present in food fermenting microorganisms.[115,116]

There is evidence that the intestinal luminal microenvironment is responsible for the initiation and/or perpetuation of chronically relapsing inflammatory bowel disease (IBD). The role of enteric bacteria as a requirement for immune-mediated chronic intestinal inflammation is strongly indicated by experiments in rodent models.[117,118] Although studies of probiotics in IBD are at an early stage, encouraging data have already been obtained in experimental murine models, including IL-10 knock-out mice, where certain strains of *Lactobacillus* and *Bifidobacterium* were effective in reducing mucosal inflammation.[119] Although probiotic maintenance therapy may appear more rational in patients with Crohn's disease, human studies have been performed primarily in patients with ulcerative colitis or pouchitis.[120,121] Gionchetti et al.[121] showed that a mixture of lactobacilli, bifidobacteria, and *Streptococcus thermophilus* was highly effective in maintenance of remission in patients with chronic pouchitis. Although fecal concentration of all bacteria increased significantly from baseline levels in the treated group, bacterial translocation across the disturbed intestinal epithelial cell barrier of IBD patients or other adverse effects were not reported.

10.6 CONCLUSIONS

In summary, LAB as part of a traditional human diet or probiotic therapy may influence the homeostasis between the intestinal microflora and the host, but their efficacy in the prevention or even treatment of certain diseases remains to be clarified. Since mechanisms by which LAB confer therapeutic effects may be multiple, the use of a multitude of bacterial species in various food matrices, including those present in fermented sausages, should be of advantage when used as part of the normal diet. Furthermore, it may be possible to select bacterial strains that are capable of producing a fermented meat product with all the sensory qualities preferred by consumers and that at the same time provide beneficial probiotic effects.

REFERENCES

1. Campbell-Platt, G. and Cook, P.E., Eds. *Fermented Meats*, Blackie Academic & Professional, Glasgow, 1995.
2. Lücke, F.-K., Fermented sausages, in *Microbiology of Fermented Foods*, Vol. 2., 2nd ed., Wood, B.J.B., Ed., Blackie Academic & Professional, Glasgow, 1998, pp. 441–483.
3. Ricke, S.C., Diaz, I.Z., and Keeton, J.J.T., Fermented meat, poultry and fish products, in *Food Microbiology, Fundamentals and Frontiers,* 2nd ed., Doyle, M.P., Beuchat, L.R., and Montville, T.J., Eds., ASM Press, Washington, 2001, pp. 681–700.
4. Hammes, W.P., Bantleon, A., and Min, S., Lactic acid bacteria in meat fermentation, *FEMS Microbiol. Lett.,* 87, 165–174, 1990.
5. Hammes, W.P. and Hertel, C., New developments in meat starter cultures, *Meat Sci.,* 49, S125–S138, 1998.
6. Aiello, L.C. and Wheeler, P., The expensive tissue hypothesis. The brain and the digestive system in human and primate evolution, *Curr. Anthropol.,* 36, 199–221, 1995.
7. Williams, G., Fish and meat in man's diet: an historical perspective, *Nutr. Health,* 6, 99–104, 1988.
8. Fisher, S. and Palmer, M., Fermented meat production and consumption in the European Union, in *Fermented Meats,* Campbell-Platt, G. and Cook, P.E., Eds., Blackie Academic & Professional, Glasgow, 1995, pp. 217–233.
9. Enser, M., Hallett, K., Hewitt, B., Fursey, G., and Wood, J., Fatty acid content and composition of English beef, lamb, and pork at retail, *Meat Sci.,* 42, 443–456, 1996.
10. Mann, N., Dietary lean red meat and human evolution, *Eur. J. Nutr.,* 39, 71–79, 2000.
11. Hambraeus, L., Animal- and plant-food–based diets and iron status: benefits and costs, *Proc. Nutr. Soc.,* 58, 235–242, 1999.
12. Phillips, R.L., Kuzma, J.W., Beeson, W.L., and Lotz, T., Influence of selection versus lifestyle on risk of fatal cancer and cardiovascular disease among Seventh-day Adventists, *Am. J. Epidemiol.,* 112, 296–314, 1980.
13. Key, T.J., Thorogood, M., Appleby, P.N., and Burr, M.L., Dietary habits and mortality in 11,000 vegetarians and health conscious people: results of a 17 year follow up, *BMJ,* 313, 775–779, 1996.
14. Willett, W.C., Convergence of philosophy and science: the Third International Congress on Vegetarian Nutrition, *Am. J. Clin. Nutr.,* 70, 434S–438S, 1999.
15. Hu, F.B., Rimm, E., Smith-Warner, S.A., Feskanich, D., Stampfer, M.J., Ascherio, A., Sampson, L., and Willett, W.C., Reproducibility and validity of dietary patterns assessed with a food-frequency questionnaire, *Am. J. Clin. Nutr.,* 69, 243–249, 1999.
16. Phillips, R.L., Lemon, F.R., Beeson, W.L., and Kuzma, J.W., Coronary heart disease mortality among Seventh-day Adventists with differing dietary habits: a preliminary report, *Am. J. Clin. Nutr.,* 31, S191–S198, 1978.
17. Kritchevsky, D., Tepper, S.A., and Goodman, G., Diet, nutrition intake, and metabolism in populations at high and low risk for colon cancer. Relationship of diet to serum lipids, *Am. J. Clin. Nutr.,* 40, S921–S926, 1984.
18. Armstrong, B. and Doll, R., Environmental factors and cancer incidence and mortality in different countries, with special reference to dietary practices, *Int. J. Cancer,* 15, 617–631, 1975.
19. Giovannucci, E. and Willett, W.C., Dietary factors and risk of colon cancer, *Ann. Med.,* 26, 443–452, 1994.

20. Willett, W.C., Stampfer, M.J., Colditz, G.A., Rosner, B.A., and Speizer, F.E., Relation of meat, fat, and fiber intake to the risk of colon cancer in a prospective study among women, *N. Engl. J. Med.,* 323, 1664–1672, 1990.
21. Thun, M.J., Calle, E.E., Namboodiri, M.M., Flanders, W.D., Coates, R.J., Byers, T., Boffetta, P., Garfinkel, L., and Heath, C.W., Jr., Risk factors for fatal colon cancer in a large prospective study, *J. Natl. Cancer Inst.,* 84, 1491–1500, 1992.
22. Howe, G.R., Aronson, K.J., Benito, E., Castelleto, R., Cornee, J., Duffy, S., Gallagher, R.P., Iscovich, J.M., Deng-ao, J., Kaaks, R., Kune, G.A., Kune, S., Lee, H.P., Lee, M., Miller, A.B., Peters, R.K., Potter, J.D., Riboli, E., Slattery, M.L., Trichopoulos, D., Tuyns, A., Tzonou, A., Watson, L.F., Whittemore, A.S., Shu, Z., et al., The relationship between dietary fat intake and risk of colorectal cancer: evidence from the combined analysis of 13 case-control studies, *Cancer Causes Control,* 8, 215–228, 1997.
23. Key, T.J., Davey, G.K., and Appleby, P.N., Health benefits of a vegetarian diet, *Proc. Nutr. Soc.,* 58, 271–275, 1999.
24. Glinghammar, B., Venturi, M., Rowland, I.R., and Rafter, J.J., Shift from a dairy product–rich to a dairy product–free diet: influence on cytotoxicity and genotoxicity of fecal water — potential risk factors for colon cancer, *Am. J. Clin. Nutr.,* 66, 1277–1282, 1997.
25. Venturi, M., Hambly, R.J., Glinghammar, B., Rafter, J.J., and Rowland, I.R., Genotoxic activity in human faecal water and the role of bile acids: a study using the alkaline comet assay, *Carcinogenesis,* 18, 2353–2359, 1997.
26. Hughes, R., Cross, A.J., Pollock, J.R., and Bingham, S., Dose-dependent effect of dietary meat on endogenous colonic *N*-nitrosation, *Carcinogenesis,* 22, 199–202, 2001.
27. Rieger, M.A., Parlesak, A., Pool-Zobel, B.L., Rechkemmer, G., and Bode, C., A diet high in fat and meat but low in dietary fibre increases the genotoxic potential of "faecal water," *Carcinogenesis*, 20, 2311–2316, 1999.
28. Massey, R.C., Key, P.E., Mallett, A.K., and Rowland, I.R., An investigation of the endogenous formation of apparent total *N*-nitroso compounds in conventional microflora and germ-free rats, *Food Chem. Toxicol.,* 26, 595–600, 1988.
29. Finegold, S.M., Sutter, V.L., Sugihara, P.T., Elder, H.A., Lehmann, S.M., and Phillips, R.L., Fecal microbial flora in Seventh Day Adventist populations and control subjects, *Am. J. Clin. Nutr.,* 30, 1781–1792, 1977.
30. Berg, J.W., Howell, H.A., and Silverman, S.J., Dietary hypothesis and diet-related research in the etiology of colon cancer, *Health Serv. Rept.,* 88, 915–924, 1973.
31. Finegold, S.M., Attebery, H.R., and Sutter, V.L., Effect of diet on human fecal flora: a comparison of Japanese and American diets, *Am. J. Clin. Nutr.,* 27, 1456–1464, 1974.
32. Goldin, B.R. and Gorbach, S.L., The relationship between diet and rat fecal bacterial enzymes implicated in colon cancer, *J. Natl. Cancer Inst.,* 57, 371–375, 1976.
33. Mastromarino, A., Reddy, B.S., and Wynder, E.L., Metabolic epidemiology of colon cancer: enzymic activity of fecal flora, *Am. J. Clin. Nutr.,* 29, 1455–1460, 1976.
34. Snowdon, D.A., Phillips, R.L., and Fraser, G.E., Meat consumption and fatal ischemic heart disease, *Prev. Med.,* 13, 490–500, 1984.
35. Ueland, P.M., Refsum, H., and Brattstrom, L., Blood homocyst(e)ine and cardiovascular disease, in *Atherosclerotic Cardiovascular Disease, Homeostasis and Endothelial Function,* Francis, R.B., Jr, Ed., Marcel Dekker Inc., New York, 1992, pp. 183–236.
36. Leistner, L., Allgemeines über Rohwurst, *Fleischwirtschaft,* 66, 290–300, 1986.

37. Lerche, M., Die deutschen Wursterzeugnisse. Eine Systematik der Wurstarten und Wurstsorten in der Bundesrepublik und in Westberlin, Arbeiten der DLG., vol. 134, DLG Verlag, Frankfurt/Main, 1975.
38. Leistner, L., Allgemeines über Rohschinken, *Fleischwirtschaft,* 66, 496–510, 1986.
39. Leistner, L. and Eckardt, C., Schimmelpilze und Mykotoxine in Fleisch und Fleischerzeugnissen, in *Mykotoxine in Lebensmitteln,* Reiß, J., Ed., Gustav Fischer Verlag, Stuttgart, 1981, pp. 297–341.
40. Fink-Gremmels, J., Abd El-Banna, A., and Leistner, L., Entwicklung von Schimmelpilz-Starterkulturen für Fleischerzeugnisse, *Fleischwirtschaft,* 68, 24–26, 1988.
41. Hwang, H.-J., Vogel, R.F., and Hammes, W.P., Entwicklung von Schimmelpilzkulturen für die Rohwurstherstellung. Charakterisierung der Stämme und toxikologische Bewertung, *Fleischwirtschaft,* 73, 89–92, 1993.
42. U.S. Department of Agriculture, Food Safety Inspection Service, Prescribed Treatment for Pork and Products Containing Pork to Destroy Trichinae, in *Code of Federal Regulations,* Part 318.10, Title 9, Office of the Federal Register, Washington, D.C., 1995.
43. Hammes, W.P. and Haller, D., Wie sinnvoll ist die Anwendung von Probiotika in Fleischwaren? *Fleischwirtschaft,* 78, 301–306, 1998.
44. Erkkilä, S., Suihko, M.-L., Eerola, S., Petäjä, E., and Mattila-Sandholm, T., Dry sausage fermented by *Lactobacillus rhamnosus* strains, *Int. J. Food Microbiol.,* 64, 205–210, 2001.
45. Sameshima, T., Magome, C., Takeshita, K., Arihara, K., Itoh, M., and Kondo, Y., Effect of intestinal *Lactobacillus* starter cultures on the behaviour of *Staphylococcus aureus* in fermented sausage, *Int. J. Food Microbiol.,* 41, 1–7, 1998.
46. Andersen, L., Fermented dry sausages produced with the admixture of probiotic cultures, in *Proc. 44th Int. Congress Meat Sci. Technol.,* Diestre, A. and Monfort, J.M., Eds., Estrategias Alimentarius S.L. EUROCARNE, Madrid, 1998, pp. 826–827.
47. Arihara, K., Ota, H., Itoh, M., Kondo, Y., Sameshima, T., Yamanaka, H., Akimoto, M., Kanai, S., and Miki, T., *Lactobacillus acidophilus* group lactic acid bacteria applied to meat fermentation, *J. Food Sci.,* 63, 544–547, 1998.
48. Bunte, C., Hertel, C., and Hammes, W.P., Monitoring and survival of *Lactobacillus paracasei* LTH 2579 in food and the human intestinal tract, *Syst. Appl. Microbiol.,* 23, 260–266, 2000.
49. Jahreis, G., Vogelsang, H., Kiessling, G., Schubert, R., Bunte, C., and Hammes, W.P., Influence of probiotic sausage (*Lactobacillus paracasei*) on blood lipids and immunological parameters of healthy volunteers, *Food Res. Int.,* 35, 133–138, 2002.
50. Rodriguez, M., Nunez, F., Cordoba, J.J., Sanabria, C., Bermudez, E., and Asensio, M.A., Characterisation of *Staphylococcus* spp. and *Micrococcus* spp. isolated from Iberian ham throughout the ripening process, *Int. J. Food Microbiol.,* 24, 329–335, 1994.
51. Molina, I., Silla, H., Flores, J., and Monzo, J.L., Study of the microbial flora in dry-cured ham. II., Micrococcaceae, *Fleischwirtschaft,* 69, 1433–1434, 1989.
52. Cornejo, I., Carrascosa, A.V., Marin, M.E., and Martin, P.J., Raw ham. Studies on the origin of microorganisms multiplying during manufacture of dry cured ham, *Fleischwirtschaft,* 72, 1422–1425, 1992.
53. Cook, P.E., Fungal ripened meats and meat products, in *Fermented Meats,* Campbell-Platt, G. and Cook, P.E., Eds., Blackie Academic & Professional, Glasgow, 1995, pp. 110–129.
54. Leistner, L., Stable and safe fermented sausages world-wide, in *Fermented Meats,* Campbell-Platt, G. and Cook, P.E., Eds., Blackie Academic & Professional, Glasgow, 1995, pp. 160–175.

55. Hugas, M., Bacteriocinogenic lactic acid bacteria for the biopreservation of meat and meat products, *Meat Sci.,* 49, S139–S150, 1998.
56. Hugas, H., Neumeyer, B., Pagés, F., Garriga, M., and Hammes, W.P., Comparison of the antilisterial potential of bacteriocin-producing lactobacilli in fermented sausages, *Fleischwirtschaft,* 76, 649–652, 1996.
57. Gänzle, M.G., Hertel, C., and Hammes, W.P., Resistance of *Escherichia coli* and *Salmonella* against nisin and curvacin A, *Int. J. Food Microbiol.,* 48, 37–50, 1999.
58. Lahti, E., Johansson, T., Honkanen-Buzalski, T., Hill, P., and Nurmi, E., Survival and detection of *Escherichia coli* O157:H7 and *Listeria monocytogenes* during the manufacture of dry sausage using two different starter cultures, *Food Microbiol.,* 18, 75–85, 2000.
59. Kröckel, L., Bacterial fermentation of meats, in *Fermented Meats,* Campbell-Platt, G. and Cook, P.E., Eds., Blackie Academic & Professional, Glasgow, 1995, pp. 69–109.
60. Demeyer, D., Raemaekers, M., Rizzo, A., Holck, A., De Smedt, A., ted Brink, B., Hagen, B., Montel, C., Zanardi, E., Murbrekk, E., Leroy, F., Vandendriessche, F., Lorentsen, K., Venema, K., Sunesen, L., Stahnke, L., De Vuyst, L., Talon, R., Chizzolini, R., and Eerda, S., Control of bioflavour and safety in fermented sausages: first results of an European project, *Food Res. Int.,* 33, 171–180, 2000.
61. Toldra, F. and Flores, M., The role of muscle proteases and lipases in flavour development during the processing of dry-cured ham, *Crit. Rev. Food Sci. Nutr.,* 385, 331–352, 1998.
62. Ordónez, J.A., Hierro, E.M., Bruna, J.M., and de la Hoz, L., Changes in the components of dry-fermented sausages during ripening, *Crit. Rev. Food Sci. Nutr,* 39, 329–367, 1999.
63. Hierro, E., de la Hoz, L., and Ordónez, J.A., Contribution of the microbial and meat endogenous enzymes to the free amino acid and amine contents of dry fermented sausages, *J. Agric. Food Chem.,* 47, 1156–1161, 1999.
64. Dainty, R. and Blom, H., Flavour chemistry of fermented sausages, in *Fermented Meats,* Campbell-Platt, G. and Cook, P.E., Eds., Blackie Academic & Professional, Glasgow, 1995, pp. 176–193.
65. Kunji, E.R., Mierau, I., Hagting, A., Poolman, B., and Konings, W.N., The proteolytic systems of lactic acid bacteria, *Antonie van Leeuwenhoek,* 70, 187–221, 1996.
66. Siezen, R.J., Multi-domain, cell-envelope proteinases of lactic acid bacteria, *Antonie van Leeuwenhoek,* 76, 139–155, 1999.
67. Kenneally, P.M., Fransen, N.G., Grau, H., O'Neill, E.E., and Arendt, E.K., Effects of environmental conditions on microbial proteolysis in a pork myofibril model system, *J. Appl. Microbiol.,* 87, 794–803, 1999.
68. Sanz, Y., Fadda, S., Vignolo, G., Aristoy, M.C., Oliver, G., and Toldra, F., Hydrolytic action of *Lactobacillus casei* CRL 705 on pork muscle sarcoplasmic and myofibrillar proteins, *J. Agric. Food Chem.,* 47, 3441–3448, 1999.
69. Fadda, S., Vignolo, G., Aristoy, M.C., Oliver, G., and Toldra, F., Effect of curing conditions and *Lactobacillus casei* CRL705 on the hydrolysis of meat proteins, *J. Appl. Microbiol.,* 91, 478–487, 2001.
70. Hierro, E., de la Hoz, L., and Ordónez, J.A., Contribution of microbial and meat endogenous enzymes to the lipolysis of dry fermented sausages, *J. Agric. Food Chem.,* 45, 2989–2995, 1997.
71. Kenneally, P.M., Schwarz, G., Fransen, N.G., and Arendt, E.K., Lipolytic starter culture effects on production of free fatty acids in fermented sausages, *J. Food Sci.,* 63, 538–543, 1998.

72. Molly, K., Demeyer, D., Civera, T., and Verplaetle, A., Lipolysis in a Belgian sausage: relative importance of endogenous and bacterial enzymes, *Meat Sci.,* 43, 235–244, 1996.
73. Montel, M.-C., Reitz, J., Talon, R., Berdague, J.-L., and Rousset-Akrim, S., Biochemical activities of Micrococcaceae and their effects on the aromatic profiles and odours of a dry sausage model, *Food Microbiol.,* 13, 489–499, 1996.
74. Selgas, M.D., Casas, C., Toledo, V.M., and Garcia, M.L., Effect of selected mould strains on lipolysis in dry fermented sausages, *Eur. Food Res. Technol.,* 209, 360–365, 1999.
75. Hinrichsen, L.L. and Pedersen, S.B., Relationship among flavour, volatile compounds, chemical changes, and microflora in Italian-type dry-cured ham during processing, *J. Agric. Food Chem.,* 43, 2932–2940, 1995.
76. Schmidt, S. and Berger, R.G., Aroma compounds in fermented sausages of different origins, *Lebens. Wiss-u. Technol.,* 31, 559–567, 1998.
77. Ullrich, F. and Grosch, W., Identification of the most intense volatile flavour compounds formed during autooxidation of linoleic acid, *Z. Lebensm. Unters. Forsch.,* 184, 277–282, 1987.
78. Montel, M.C., Masson, F., and Talon, R., Bacterial role in flavour development, *Meat Sci.,* 49, S111–S124, 1998.
79. Heiler, C. and Schieberle, P., Bodel studies on the precursors and formation of the metallic smelling (*E,Z*)-2,6-nonadienol during the manufacture and storage of buttermilk, *Int. Dairy J.,* 7, 667–674, 1997.
80. Wolf, G. and Hammes, W.P., Effect of hematin on the activities of nitrite reductase and catalase in lactobacilli, *Arch. Microbiol.,* 149, 220–224, 1988.
81. Yvon, M., Thirouin, S., Rijnen, L., Fromentier, D., and Gripon, J.C., An aminotransferase from *Lactococcus lactis* initiates conversion of amino acids to cheese flavour compounds, *Appl. Environ. Microbiol.,* 63, 414–419, 1997.
82. Gao, S., Mooberry, E.S., and Steele, J.L., Use of ^{13}C nuclear magnetic resonance and gas chromatography to examine methionine catabolism by lactococci, *Appl. Environ. Microbiol.,* 64, 4670–4675, 1998.
83. Engels, W.J.M., Alting, A.C., Arntz, M.M.T.G., Gruppen, H., Voragen, A.G.J., Smit, G., and Visser, S., Partial purification and characterization of two aminotransferases from *Lactococcus lactis* subsp. *cremoris* B78 involved in the catabolism of methionine and branched-chain amino acids, *Int. Dairy J.,* 10, 443–452, 2000.
84. Stahnke, L.H., Volatiles produced by *Staphylococcus xylosus* and *Staphylococcus carnosus* during growth in sausage minces. Part I. Collection and identification, *Lebens. Wiss-u. Technol.,* 32, 357–364, 1999.
85. Stahnke, L.H., Volatiles produced by *Staphylococcus xylosus* and *Staphylococcus carnosus* during growth in sausage minces. Part II. The influence of growth parameters, *Lebens. Wiss-u. Technol.,* 32, 365–374, 1999.
86. Masson, F., Hinrichsen, L., Talon, R., and Montel, M.C., Factors influencing leucine catabolism by a strain of *Staphylococcus carnosus, Food Microbiol.,* 49, 173–178, 1999.
87. Larrouture, C., Ardaillon, V., Pépin, M., and Montel, M.C., Ability of meat starter cultures to catabolise leucine and evaluation of the degradation products by using an HPLC method, *Food Microbiol.,* 17, 563–570, 2000.
88. Ansorena, D., Astiasarán, I., and Bello, J., Influence of the simultaneous addition of the protease Flavourzyme and the lipase Novozym 677BG on dry fermented sausage compounds extracted by SDE and analyzed by GC-MS, *J. Agric. Food Chem.,* 48, 2395–2400, 2000.

89. Hernández-Jover, T., Izquierdo-Pulido, M., Vaciana-Nogués, T., Mariné-Font, A., and Vidal-Carou, M.C., Biogenic amine and polyamine contents in meat and meat products, *J. Agric. Food Chem.,* 45, 2098–2102, 1997.
90. McCabe, B.J., Dietary tyramine and other pressor amines in MAOI regimens: a review, *J. Am. Diet. Assoc.,* 86, 1059–1064, 1986.
91. Straub, B.W., Kicherer, M., Schilcher, S.M., and Hammes, W.P., The formation of biogenic amines by fermentation organisms, *Z. Lebensm. Unters. Forsch.,* 201, 79–82, 1995.
92. Bover-Cid, S., Izquierdo-Pulido, M., and Vidal-Carou, M.C., Effect of proteolytic starter cultures of *Staphylococcus* spp. on biogenic amine formation during the ripening of dry fermented sausages, *Int. J. Food Microbiol.,* 46, 95–104, 1999.
93. Bover-Cid, S., Izquierdo-Pulido, M., and Vidal-Carou, M.C., Effect of the interaction between a low tyramine-producing *Lactobacillus* and proteolytic staphylococci on biogenic amine production during ripening and storage of dry sausages, *Int. J. Food Microbiol.,* 65, 113–123, 2001.
94. Leuschner, R.G., Heidel, M., and Hammes, W.P., Histamine and tyramine degradation by food fermenting microorganisms, *Int. J. Food Microbiol.,* 39, 1–10, 1998.
95. Leuschner, R.G. and Hammes, W.P., Tyramine degradation by micrococci during ripening of fermented sausage, *Meat Sci.,* 49, 289–296, 1998.
96. Stiles, M.E., Biopreservation by lactic acid bacteria, *Antonie van Leeuwenhoek,* 70, 331–345, 1996.
97. Tannock, G.W., The normal microflora: new concepts in health promotion, *Microbiol. Sci.,* 5, 4–8, 1988.
98. Walter, J., Hertel, C., Tannock, G.W., Lis, C.M., Munro, K., and Hammes, W.P., Detection of *Lactobacillus, Pediococcus, Leuconostoc,* and *Weissella* species in human feces by using group-specific PCR primers and denaturing gradient gel electrophoresis, *Appl. Environ. Microbiol.,* 67, 2578–2585, 2001.
99. Pochart, P., Marteau, P., Bouhnik, Y., Goderel, I., Bourlioux, P., and Rambaud, J.C., Survival of bifidobacteria ingested via fermented milk during their passage through the human small intestine: an *in vivo* study using intestinal perfusion, *Am. J. Clin. Nutr.,* 55, 78–80, 1992.
100. Marteau, P., Vaerman, J.P., Dehennin, J.P., Bord, S., Brassart, D., Pochart, P., Desjeux, J.F., and Rambaud, J.C., Effects of intrajejunal perfusion and chronic ingestion of *Lactobacillus johnsonii* strain La1 on serum concentrations and jejunal secretions of immunoglobulins and serum proteins in healthy humans, *Gastroenterol. Clin. Biol.,* 21, 293–298, 1997.
101. Haller, D., Colbus, H., Gaenzle, M., Scherenbacher, H., Bode, C., and Hammes, W.P., Metabolic and functional properties of lactic acid bacteria in the gastro-intestinal ecosystem: A comparative *in vitro* study between bacteria of intestinal and fermented food origin, *Syst. Appl. Microbiol.,* 24, 218–226, 2001.
102. Tannock, G.W., Crichton, C., Welling, G.W., Koopman, J.P., and Midtvedt, T., Reconstitution of the gastrointestinal microflora of *Lactobacillus* free mice, *Appl. Environ. Microbiol.,* 54, 2971–2975, 1988.
103. Romond, M.B., Haddon, Z., Mialcareck, C., and Romond, C., Bifidobacteria and human health: regulatory effect of indigenous bifidobacteria on *Escherichia coli* intestinal colonization, *Anaerobe,* 3, 131–136, 1997.
104. Schiffrin, E.J., Rochat, F., Link-Amster, H., Aeschlimann, J.M., and Donnet-Hughes, A., Immunomodulation of human blood cells following the ingestion of lactic acid bacteria, *J. Dairy Sci.,* 78, 491–497, 1995.

105. Kenworthy, R., Observations on the reaction of the intestinal mucosa to bacterial challenge, *J. Clin. Pathol.,* 24, 138–142, 1971.
106. Berg, R.D. and Savage, D.C., Immunological responses and microorganisms indigenous to the gastrointestinal tract, *Am. J. Clin. Nutr.,* 25, 1364–1371, 1972.
107. Cebra, J.J., Influences of microbiota on intestinal immune system development, *Am. J. Clin. Nutr.,* 69, 1046S–1051S, 1999.
108. Kagnoff, M.F. and Eckmann, L., Epithelial cells as sensors for microbial infection, *J. Clin. Invest.,* 100, 6–10, 1997.
109. Haller, D., Bode, C., Hammes, W.P., Pfeifer, A.M., Schiffrin, E.J., and Blum, S., Nonpathogenic bacteria elicit a differential cytokine response by intestinal epithelial cell/leucocyte co-cultures, *Gut,* 47, 79–87, 2000.
110. Haller, D., Serrant, P., Peruisseau, G., Hammes, W.P., Bode, C., Schiffrin, E.J., and Blum, S., IL-10 producing $CD14^{low}$ monocytes inhibit lymphocyte-dependent activation of intestinal epithelial cells by commensal bacteria, *Microbiol. Immunol.,* 46, 195–205, 2002.
111. Rescigno, M., Urbano, M., Valzasina, B., Francolini, M., Rotta, G., Bonasio, R., Granucci, F., Kraehenbuhl, J.P., and Ricciardi-Castagnoli, P., Dendritic cells express tight junction proteins and penetrate gut epithelial monolayers to sample bacteria, *Nat. Immunol.,* 2, 361–367, 2001.
112. Hessle, C., Hanson, L.A., and Wold, A.E., Lactobacilli from human gastrointestinal mucosa are strong stimulators of IL-12 production, *Clin. Exp. Immunol,* 116, 276–282, 1999.
113. Herias, M.V., Hessle, C., Telemo, E., Midtvedt, T., Hanson, L.A., and Wold, A.E., Immunomodulatory effects of *Lactobacillus plantarum* colonizing the intestine of gnotobiotic rats, *Clin. Exp. Immunol.,* 116, 283–290, 1999.
114. Haller, D., Bode, C., and Hammes, W.P., Cytokine secretion by stimulated monocytes depends on the growth phase and heat treatment of bacteria: a comparative study between lactic acid bacteria and invasive pathogens, *Microbiol. Immunol.,* 43, 925–935, 1999.
115. Haller, D., Blum, S., Bode, C., Hammes, W.P., and Schiffrin, E.J., Activation of human peripheral blood mononuclear cells by nonpathogenic bacteria *in vitro:* evidence of NK cells as primary targets, *Infect. Immun.,* 68, 752–759, 2000.
116. Haller, D., Serrant, P., Granato, D., Schiffrin, E.J., and Blum, S., Activation of human NK cells by staphylococci and lactobacilli required cell contact–dependent costimulation by autologous macrophages, *Clin. Diag. Lab. Immunol.,* 9, 649–657, 2002.
117. Sartor, R.B., Pathogenesis and immune mechanisms of chronic inflammatory bowel diseases, *Am. J. Gastroenterol.,* 92, 5S–11S, 1997.
118. Fiocchi, C., Inflammatory bowel disease: etiology and pathogenesis, *Gastroenterology,* 115, 182–205, 1998.
119. Madsen, K.L., Doyle, J.S., Jewell, L.D., Tavernini, M.M., and Fedorak, R.N., *Lactobacillus* species prevents colitis in interleukin 10 gene-deficient mice, *Gastroenterology,* 116, 1107–1114, 1999.
120. Rembacken, B.J., Snelling, A.M., Hawkey, P.M., Chalmers, D.M., and Axon, A.T.R., Nonpathogenic *Escherichia coli* versus mesalazine for the treatment of ulcerative colitis: a randomized trial, *Lancet,* 354, 635–639, 1999.
121. Gionchetti, P., Rizzello, F., Venturi, A., Brigidi, P., Matteuzzi, D., Bazzocchi, G., Poggioli, G., Miglioli, M., and Campieri, M., Oral bacteriotherapy as maintenance treatment in patients with chronic pouchitis: a double-blind, placebo-controlled trial, *Gastroenterology,* 119, 305–309, 2000.

11 Miso: Production, Properties, and Benefits to Health

Yukiko Minamiyama and Shigeru Okada

CONTENTS

11.1 INTRODUCTION

The history of fermented foods and drinks dates back more than 4000 years. Wine already existed around 5000 B.C., and the original forms of soy sauce and fermented milk existed around 3000 to 2000 B.C. Microorganisms obtained from the environment were put to use for the fermentation and maturation of fermented foods. Regional differences in starting materials, climate, culture, and other environmental factors have led to the development of unique fermented products in various parts of the world. At the same time, regional and racial differences have had a big effect on whether some fermented products are considered good tasting or disgusting. This chapter describes the traditional Japanese fermented soybean product miso and its effects on human health and metabolism.

0-8493-1372-4/03/$0.00+$1.50

11.2 MISO

Miso is a fermented soybean paste and is one of the essential seasonings in Japanese cuisine. Its taste and aroma resemble those of soy sauce. Miso is made from steamed soybeans mixed with salt and koji. Koji is mold-treated rice, barley, or soybean that acts as a fermentation starter.

Miso soup is usually served for breakfast in Japan; it contains seasonal vegetables. Miso can also be used in marinades, dressings, stews, dips, and casseroles. There are regional differences in flavor and taste; shiro (white) miso paste has a mild taste and is low in salt, while aka (red) miso is very salty and has a different, stronger odor than that of shiro miso. The total commercial production of miso amounts to 600,000 tonnes/year in Japan. There are over 1500 miso factories in all parts of Japan today.

11.2.1 The History of Miso

Miso is one of the most traditional and characteristic fermented foods of Japan and an important soybean product. Soybeans are also used to produce soy sauce, soy milk, soy curd (tofu), sticky bean (natto), and bean sprouts and as a source of oil. (See Chapter 9 for more details on natto.) The soybean was probably introduced to Japan, most likely from Northern China through Korea, between 200 B.C. and 300 A.D. The first known book on miso is thought to be *Taiho-ritsuryo*, compiled in A.D. 702, but by that time, a type of miso was already known in Japan. The technique of miso production was introduced either from Korea or China. One theory proposed that miso developed from jan of Chinese origin, which is also a fermented food from rice or soybean. Jan is made in China as a seasoning. About the time when the Japanese came to know jan (about 550 to 600 A.D.), jan was a food made in Buddhist temples. It is believed that a Chinese Buddhist named Ganjin (688 to 763) came to Japan in 753, promoted Buddhism in Japan, and at the same time brought jan to Japan.

11.2.2 Manufacturing Miso

Various types of miso are produced that have differences in their soybean/salt/koji-starter ratio, aging periods, and other parameters. Figure 11.1 shows the simple manufacturing process for miso. Microorganisms called koji mold (fungi) in Japan are used for the fermentation process. Yellow koji mold, *Aspergillus oryzae,* is also used in the brewing process of Japanese sake. Enzymes (amylases) in the koji convert the rice starch to sugar. The koji used for soy sauce and miso is also high in proteases and peptidases, which convert proteins to amino acids.[1]

Miso can be classified based on its raw ingredients. Rice miso, the most common type of miso, is made with rice koji-starter. Barley miso and soybean miso are made with barley koji- and soybean koji-starters, respectively. *A. oryzae* is used in the fermentation for all three types of miso.

The main ingredients of miso are soybeans, rice, salt, and koji. Although many kinds of soybeans exist in the world, only a few cultivars of soybeans are of high enough quality for miso production in Japan. Salt made from seawater is recom-

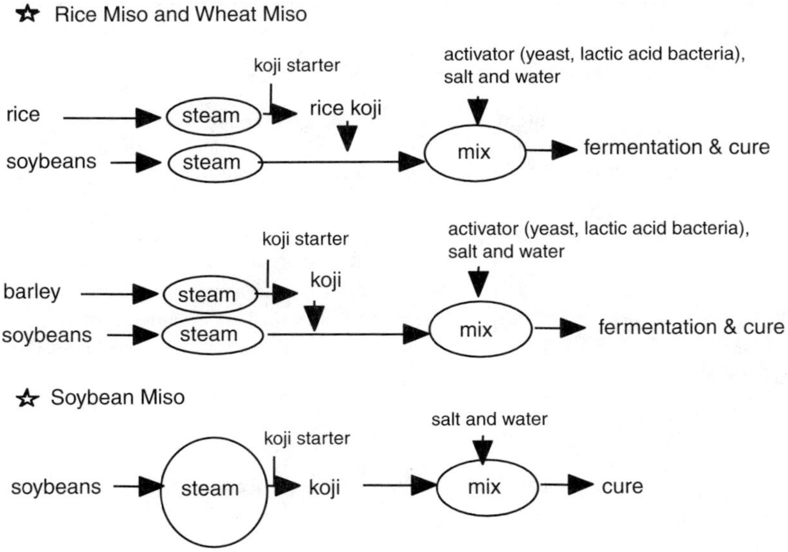

FIGURE 11.1 The miso manufacturing process.

mended for miso production because it contains metal elements such as magnesium and calcium that positively affect the fermentation. Traditionally in Japan, people used koji from previous batches as a starter culture. However, because there was no way to ensure the identity and quality of the fungus, batches of miso varied in quality (smell and taste). For many years, the supply of koji was controlled by miso manufacturers, and koji was not commercially available. Recently, public brewing institutes in many Japanese prefectures have started research on the selection of better strains of miso koji, and this has resulted in the production of several "standardized" miso products. Fungus strains are now available through the Internet or at miso shops in Japan.

11.2.3 Home Production of Miso

Japanese consumers have often made their own homemade miso. To make koji, rice (sometimes barley or soybean is substituted) is first steamed. After the temperature of the cooked rice drops to 25 to 30°C, a small amount of the koji fungus is added and mixed well with the rice. After about 10 h, the mixture produces its own heat, which is evidence that the fungi are growing. The depth of the rice pile is controlled to maintain the temperature between 35 and 40°C for about 40 h. Feather-like fungus fibers grow on the rice particles as the fermentation continues.

Soybeans are boiled and mashed while they are hot. The temperature of the mashed soybeans must drop to 35 to 40°C before mixing with the koji. The mashed soybean (40 to 60%), rice koji (25 to 40%), salt (12 to 20%), and warm water (usually the rest of soybean boiling water, ca. 15%) are mixed well until the mixture has the appearance and texture of soft dough. The mixture is then placed in a plastic container and sealed with a plastic sheet on the surface to prevent exposure to the

TABLE 11.1
Constituents of Miso

	Content	Metabolic Activity
Nonnutritional Substances (mg/kg soybean miso)		
p-Coumaric acid	1.3	Antioxidant
Ferulic acid	1.0	Antioxidant
Daidzein	14.5	Antioxidant (hydrolysis during the fermentation)
Genistein	5.3	Antioxidant (hydrolysis during the fermentation)
8-OH-Daidzein	1.2	Potent antioxidant (hydrolysis during the fermentation)
8-OH-Genistein	1.5	Potent antioxidant (hydrolysis during the fermentation)
Syringic acid	3.1	Potent antioxidant
Vanillic acid	8.5	
p-Hydroxybenzoic acid	1.0	
Nutritional Substances (mg/100 g soybean miso)		
Protein	17,000	
Fat	11,000	
Carbohydrate	15,000	
Ash	13,000	
Salt	12,000	
Iron	4.0	
Calcium	150	
Sodium	4,300	
Water	45,000	
Vitamins and Vitamin Precursors		
Ascorbic acid	0	
Tocopherols	2.4	
Retinol (IU)	0	
Carotene	0	
Vitamin B_6	0.13	

air. A lid is added and weighed down. The container is covered with wrapping paper, tied in place with strings, and then stored in a cool, dark place. After a month, the fermentation is checked. Miso can be stored and used for several years, and during this time it continues to smell good and retain its yellow-brown color. The composition, including nutrient content, of miso is shown in Table 11.1.[2]

11.2.4 Effects of Miso on Health

It is believed that the Japanese diet and methods of food preparation have contributed to the longevity of the Japanese people. The outstanding medicinal qualities of miso have been supported by scientific research. In 1981, Hirayama[3] of the Japan National Cancer Center conducted an epidemiologic study and reported that those who eat miso soup daily suffer significantly less from stomach cancer and heart disease. In 1972, Ito discovered an alkaloid in miso that removes heavy metals from the body,[4]

TABLE 11.2
The Major Active Ingredients of Miso and their Expected Health/Metabolism Effects

Nutrients (per 100 g)	Origin	Expected Effect
Protein (10–20 g)	Soybeans	Reduces blood cholesterol; maintains elasticity of blood vessels; prevents cerebral apoplexy
Vitamin B_2 (0.1 mg)	Aspergilli	Promotes oxidation reduction in the body
Vitamin B_{12} (0.1 mg)	Bacteria	Helps blood formation; reduces mental fatigue
Vitamin E (0.3–2.4 mg)	Soybeans	Inhibits lipid peroxidation; has antiaging effect
Enzymes	Koji, fungi, lactic acid bacteria	Help digestion
Saponin	Soybeans	Inhibits lipid peroxidation; reduces blood cholesterol; prevents hardening of the arteries; prevents hepatopathy
Trypsin inhibitor	Soybeans	Has anticancer effect; prevents diabetes
Isoflavones	Soybeans	Deoxidizes; alleviates stiff neck and shoulders; prevents breast cancer
Lecithin	Soybeans	Reduces blood cholesterol; prevents hardening of the arteries; prevents senile deterioration
Colin	Soybeans	Prevents fatty liver; has anti-aging effect
Prostaglandin E	Linoleic acid in soybeans	Prevents high blood pressure
Brown pigment	Soybeans	Inhibits lipid peroxidation; has anti-aging effect
Dietary fiber	Soybeans	Reduces blood cholesterol; prevents colon cancer

and recently, Yoshikoshi et al. at the Tohoku University of Japan isolated substances in miso that neutralize the effects of some carcinogens.[5] It has been known from antiquity that miso keeps the body in good condition, and it is said that miso is "a detoxicating drug in the morning" or "a doctor killer." Major ingredients of miso and their expected effects on health/metabolism are listed in Table 11.2.

11.2.4.1 Gastrointestinal Diseases

A study showed that people who eat miso soup regularly (daily) are less susceptible to stomach diseases, such as gastritis and gastric and duodenal ulcers, than those who seldom or never eat it.[6] A more detailed survey on the eating habits of people by age indicated that daily miso eaters in their sixties and older have a lowered risk of stomach diseases.[7] Recently, *Helicobacter pylori* has been identified as the causative organism for gastric inflammation and peptic ulcers, and it is associated with gastric cancer. Among the isoflavones of miso, genistein, which has an inhibitory activity of tyrosine kinase, especially showed a potent anti-*H. pylori* activity.[8–12] A large portion of the proteins contained in soybeans is degraded by enzymes and microbes in the fermentation process of miso. In addition, miso contains highly active enzymes, which help digestion and absorption of other nutrients. It is believed that plant fibers in miso help "clean the intestines." Microbes in miso antagonize putrefactive bacteria in the intestines and decompose harmful substances in the body.[13]

11.2.4.2 Cancer Prevention

Epidemiologists have known for years that vegetarians and other people who eat diets rich in plant products have relatively low rates of various cancers.[14–18] It is now widely known that anticancer effects are associated with an intake of miso on a regular basis.[19] Miso contains such ingredients as unsaturated fatty acids, isoflavones, yeasts, and lactic acid. Soybeans also contain trypsin inhibitor and genistein, which have antimutagenic properties.[20]

Japanese women at home and abroad have a very low incidence of breast cancer as long as they maintain their native diet, but the incidence becomes higher if they adopt a relatively soy-free diet.[21] A similar relationship holds for Japanese men and prostate cancer.[22] In both cases, the intake of soy makes the difference, not the intake of fat. According to the results of epidemiologic research conducted by Hoshiyama and Sosaba,[23] a negative correlation was found between miso soup consumption and the incidence of death from stomach cancer. People who do not eat miso soup at all are at 50% higher risk of dying of stomach cancer than those who eat it every day.

Researchers at the National Cancer Institute of Japan have identified certain types of phytochemicals in soybeans that have anticancer properties.[24] The group of chemicals called isoflavones exhibits the most powerful anticancer effects. Soybeans are the only commonly consumed food that provides isoflavones in the diet. Certain sugars, in particular oligosaccharides, in soybeans promote the growth of beneficial bacteria, called bifidobacteria, in the colon.[25] High levels of bifidobacteria in the intestines have been associated with a lower risk of colon cancer.[26] The level of oligosaccharides is higher in soybeans than in any other food.[25,26]

11.2.4.3 Elimination of Radioactive Materials

The export of miso to European countries increased after the accident at the Chernobyl nuclear power plant in 1986 because it was believed that miso consumption could reduce the effects of radiation exposure.[27] Researchers working on microbe activities have shown that the consumption of miso helps to eliminate radioactive substances from the body.[28] After the Hiroshima and Nagasaki atomic bombings at the end of World War II, it was observed that miso factory workers were less affected by radiation than others in the general population. The reason for this protective effect of miso is not known. However, some experimental evidence indicates that rats fed miso eliminate radioactive materials from the body more rapidly than animals not receiving miso do.[29]

11.2.4.4 Effect on Cholesterol Levels and Aging

Eating soy foods appears to markedly lower blood cholesterol, thereby reducing the risk of heart disease.[7] Tofu contains no cholesterol and is generally low in saturated fats, which have been linked to coronary disease. Miso contains several important substances, such as linoleic acid, plant sterols, and vitamin E, among others, that have been shown to be cardioprotective. Clinical and experimental studies have shown that substituting soy protein for animal protein or simply adding soy protein to the diet significantly reduces cholesterol levels, regardless of the type or amount of fat in the diet.[7,30–32] For example, 15 healthy nonvegetarian premenopausal women were studied over 9 months. A significant reduction in total cholesterol was found with the consumption of 50 g miso/day (45 mg conjugated isoflavones).[31] In cholesterol-fed rabbits, the level of cholesteryl ester hydroperoxide (ChE-OOH) induced by $CuSO_4$ in plasma in the high isoflavone group was significantly less than that in the control.[32] Miso also has strong antioxidant properties.[33–35]

The brown colored substance in aged miso has been identified as melanoidin. It strongly suppresses the production of peroxides derived from fatty acids in the body and prevents aging of the body.[36] Vitamin E, daidzein, saponin, and the brown pigment contained in miso all work as antioxidants, which prevent peroxides from accumulating in the body. Experiments using rats have shown that the increase of peroxides is remarkably inhibited by feeding saponin and the brown pigment.[5]

Scientists are also looking at soy products in connection with both osteoporosis[37–41] and kidney disease.[42,43] A study showed that rats excreted 50% less calcium in their urine when the animal products in their diets were replaced with soyfoods.[44] The mechanism responsible for the calcium retention is not known at this time.

11.2.4.5 Harmful Effects of Tobacco

"Miso soup is for smokers" is a Japanese saying from old times. The saying may have originated from the habit of using miso soup to clean pipes plugged with tars, which was a practice in the Edo era (1603 to 1867). Miso soup has a superior cleansing ability for nicotine when it is poured into a nicotine-stained pipe, compared to that of hot or cold water. Miso soup contains B vitamins, which are believed to protect the smoker's throat by neutralizing harmful substances from tobacco.[45]

11.3 CONCLUSIONS

A daily intake of miso soup is recommended by many researchers. The beneficial effects of miso have attracted worldwide attention, and studies are being conducted in various institutes to provide further experimental evidence on the health benefits of consuming miso. Many of the beneficial effects of miso appear to be due to bioactive compounds found in the soybeans themselves. However, the fermentation used in the production of miso imparts added benefits.

ACKNOWLEDGMENTS

The authors thank Toshihiko Osawa (Nagoya University) and Hajime Ohigashi (Kyoto University) for their valuable suggestions for this manuscript.

REFERENCES

1. Sugawara, E. and Sakurai, Y., Effect of media constituents on the formation by halophilic yeast of the 2 (or 5)-ethyl-5 (or 2)-methyl-4-hydroxy-3 (2H)-furanone aroma component specific to miso, *Biosci. Biotechnol. Biochem.,* 63, 749–752, 1999.
2. Nakamura, Y., Tsuji, S., and Tonogai, Y., Determination of the levels of isoflavonoids in soybeans and soy-derived foods and estimation of isoflavonoids in the Japanese daily intake, *J. AOAC Int.,* 83, 635–650, 2000.
3. Hirayama, T., Operational epidemiology of cancer, *J. Cancer Res. Clin. Oncol.,* 99, 15–28, 1981.
4. Ito, K., Achievements in the study of natural compounds in Japan, *Farmatsiia,* 21, 70–75, 1972.
5. Yoshikoshi, M., Yoshiki, Y., Okubo, K., Seto, J., and Sasaki, Y., Prevention of hydrogen peroxide damage by soybean saponins to mouse fibroblasts, *Planta Med.,* 62, 252–255, 1996.
6. Tsugane, S., Tei, Y., Takahashi, T., Watanabe, S., and Sugano, K., Salty food intake and risk of *Helicobacter pylori* infection, *Jpn. J. Cancer Res.,* 85, 474–478, 1994.
7. Yamaguchi, N., Yamamura, J., Takahashi, K., Nakamura, R., and Okubo, T., Familial history of cancer and dietary pattern, serum cholesterol, serum protein and blood hemoglobin, *Gan No. Rinsho. Spec. No.,* 377–381, 1990.
8. Shirataki, Y., Tani, S., Sakagami, H., Satoh, K., Nakashima, H., Gotoh, K., and Motohashi, N., Relationship between cytotoxic activity and radical intensity of isoflavones from Sophora species, *Anticancer Res.,* 21, 2643–2648, 2001.
9. Bae, E.A., Han, M.J., and Kim, D.H., *In vitro* anti-*Helicobacter pylori* activity of irisolidone isolated from the flowers and rhizomes of *Pueraria thunbergiana, Planta Med.,* 67, 161–163, 2001.
10. Odenbreit, S., Puls, J., Sedlmaier, B., Gerland, E., Fischer, W., and Haas, R., Translocation of *Helicobacter pylori* CagA into gastric epithelial cells by type IV secretion, *Science,* 287, 1497–1500, 2000.
11. Ohsaki, A., Takashima, J., Chiba, N., and Kawamura, M., Microanalysis of a selective potent anti-*Helicobacter pylori* compound in a Brazilian medicinal plant, *Myroxylon peruiferum,* and the activity of analogues, *Bioorg. Med. Chem. Lett.,* 9, 1109–1112, 1999.
12. Ding, S.Z., Cho, C.H., and Lam, S.K., *Helicobacter pylori* induces interleukin-8 expression in endothelial cells and the signal pathway is protein tyrosine kinase dependent, *Biochem. Biophys. Res. Commun.,* 240, 561–565, 1997.
13. Rajendran, R. and Ohta, Y., Binding of heterocyclic amines by lactic acid bacteria from miso, a fermented Japanese food, *Can. J. Microbiol.,* 44, 109–115, 1998.
14. Adlercreutz, C.H., Goldin, B.R., Gorbach, S.L., Hockerstedt, K.A., Watanabe, S., Hamalainen, E.K., Markkanen, M.H., Makela, T.H., Wahala, K.T., and Adlercreutz, T., Soybean phytoestrogen intake and cancer risk, *J. Nutr.,* 125, 757S–770S, 1995.
15. Kaplan, R.M. and Toshima, M.T., Does a reduced fat diet cause retardation in child growth?, *Prev. Med.,* 21, 33–52, 1992.
16. Denis, L., Morton, M.S., and Griffiths, K., Diet and its preventive role in prostatic disease, *Eur. Urol.,* 35, 377–387, 1999.
17. Thurnham, D.I., Zheng, S.F., Munoz, N., Crespi, M., Grassi, A., Hambidge, K.M., and Chai, T.F., Comparison of riboflavin, vitamin A, and zinc status of Chinese populations at high and low risk for esophageal cancer, *Nutr. Cancer,* 7, 131–143, 1985.

18. Reddy, B.S., Sharma, C., Darby, L., Laakso, K., and Wynder, E.L., Metabolic epidemiology of large bowel cancer. Fecal mutagens in high- and low-risk population for colon cancer. A preliminary report, *Mutat. Res.,* 72, 511–522, 1980.
19. Furukawa, K., Yamamoto, I., Tanida, N., Tsujiai, T., Nishikawa, M., Narisawa, T., and Shimoyama, T., The effects of dietary fiber from *Lagenaria scineraria* (yugao-melon) on colonic carcinogenesis in mice, *Cancer,* 75, 1508–1515, 1995.
20. Wang, H.L., Swain, E.W., Wallen, L.L., and Hesseltine, C.W., Free fatty acids identified as antitryptic factor in soybeans fermented by *Rhizopus oligosporus, J. Nutr.,* 105, 1351–1355, 1975.
21. Yan, S., A socio-medical study of adult diseases related to the life style of Chinese in Japan, *Nippon Eiseigaku Zasshi,* 44, 877–886, 1989.
22. Messina, M.J., Persky, V., Setchell, K.D., and Barnes, S., Soy intake and cancer risk: a review of the *in vitro* and *in vivo* data, *Nutr. Cancer,* 21, 113–131, 1994.
23. Hoshiyama, Y. and Sasaba, T., A case-control study of single and multiple stomach cancers in Saitama Prefecture, Japan, *Jpn. J. Cancer Res.,* 83, 937–943, 1992.
24. Ohta, T., Nakatsugi, S., Watanabe, K., Kawamori, T., Ishikawa, F., Morotomi, M., Sugie, S., Toda, T., Sugimura, T., and Wakabayashi, K., Inhibitory effects of *Bifidobacterium*-fermented soy milk on 2-amino-1-methyl-6-phenylimidazo[4,5-b]pyridine-induced rat mammary carcinogenesis, with a partial contribution of its component isoflavones, *Carcinogenesis,* 21, 937–941, 2000.
25. Roberfroid, M.B., Health benefits of nondigestible oligosaccharides, *Adv. Exp. Med. Biol.,* 427, 211–219, 1997.
26. Gallaher, D.D. and Khil, J., The effect of synbiotics on colon carcinogenesis in rats, *J. Nutr.,* 129, 1483S–1487S, 1999.
27. Swinbanks, D. and Johnston, K., Chernobyl takes macaroni off Japan's menu, *Nature,* 329, 278, 1987.
28. Key, T.J., Sharp, G.B., Appleby, P.N., Beral, V., Goodman, M.T., Soda, M., and Mabuchi, K., Soya foods and breast cancer risk: a prospective study in Hiroshima and Nagasaki, Japan, *Br. J. Cancer,* 81, 1248–1256, 1999.
29. Watanabe, H., Okamoto, T., Takahashi, T., Ogundigie, P.O., and Ito, A., The effects of sodium chloride, miso or ethanol on development of intestinal metaplasia after x-irradiation of the rat glandular stomach, *Jpn. J. Cancer Res.,* 83, 1267–1272, 1992.
30. Matsuo, M., Acceleration of sterol excretion, little meteorism, and less inhibition of iron absorption by Okara Koji, a new foodstuff, in rats, *Biosci. Biotechnol. Biochem.,* 60, 571–574, 1996.
31. Cassidy, A., Bingham, S., and Setchell, K., Biological effects of isoflavones in young women: importance of the chemical composition of soyabean products, *Br. J. Nutr.,* 74, 587–601, 1995.
32. Yamakoshi, J., Piskula, M.K., Izumi, T., Tobe, K., Saito, M., Kataoka, S., Obata, A., and Kikuchi, M., Isoflavone aglycone-rich extract without soy protein attenuates atherosclerosis development in cholesterol-fed rabbits, *J. Nutr.,* 130, 1887–1893, 2000.
33. Esaki, H., Kawakishi, S., Morimitsu, Y., and Osawa, T., New potent antioxidative *o*-dihydroxyisoflavones in fermented Japanese soybean products, *Biosci. Biotechnol. Biochem.,* 63, 1637–1639, 1999.
34. Schubert, S.Y., Lansky, E.P., and Neeman, I., Antioxidant and eicosanoid enzyme inhibition properties of pomegranate seed oil and fermented juice flavonoids, *J. Ethnopharmacol.,* 66, 11–17, 1999.
35. Matsuo, M., *In vivo* antioxidant activity of Okara Koji, a fermented okara, by *Aspergillus oryzae, Biosci. Biotechnol. Biochem.,* 61, 1968–1972, 1997.

36. Chuyen, N.V., Ijichi, K., Umetsu, H., and Moteki, K., Antioxidative properties of products from amino acids or peptides in the reaction with glucose, *Adv. Exp. Med. Biol.,* 434, 201–212, 1998.
37. Anderson, J.J., Anthony, M.S., Cline, J.M., Washburn, S.A., and Garner, S.C., Health potential of soy isoflavones for menopausal women, *Pub. Health Nutr.,* 2, 489–504, 1999.
38. Adlercreutz, H. and Mazur, W., Phyto-oestrogens and Western diseases, *Ann. Med.,* 29, 95–120, 1997.
39. Knight, D.C. and Eden, J.A., A review of the clinical effects of phytoestrogens, *Obstet. Gynecol.,* 87, 897–904, 1996.
40. Knight, D.C. and Eden, J. A., Phytoestrogens — a short review, *Maturitas*, 22, 167–175, 1995.
41. Messina, M. and Messina, V., Soyfoods, soybean isoflavones, and bone health: a brief overview, *J. Ren. Nutr.,* 10, 63–68, 2000.
42. Messina, M., Modern applications for an ancient bean: soybeans and the prevention and treatment of chronic disease, *J. Nutr.,*125, 567S–569S, 1995.
43. Messina, M.J., Legumes and soybeans: overview of their nutritional profiles and health effects, *Am. J. Clin. Nutr.,* 70, 439S–450S, 1999.
44. Kalu, D.N., Masoro, E.J., Yu, B.P., Hardin, R.R., and Hollis, B.W., Modulation of age-related hyperparathyroidism and senile bone loss in Fischer rats by soy protein and food restriction, *Endocrinology*, 122, 1847–1854, 1988.
45. Barnouin, J., Perez Cristia, R., Chassagne, M., Barrios, T.V., Arnaud, J., Fleites Mestre, P., and Favier, A., Vitamin and nutritional status in Cuban smokers and nonsmokers in the context of an emerging epidemic neuropathy, *Int. J. Vit. Nutr. Res.,* 70, 126–138, 2000.

12 Korean Fermented Foods: Kimchi and Doenjang

Hoonjeong Kwon and Young-Kyung Lee Kim

CONTENTS

12.1 INTRODUCTION

Koreans have been known for their taste for fermented foods and their skill in making them for more than 1500 years.[1] The fermentation products span a whole spectrum of raw materials and use various methods of preparation. Major fermented food items, excluding alcoholic beverages, consumed today in Korea may be divided into three categories (see Table 12.1). The first category is kimchi, which is consumed

0-8493-1372-4/03/$0.00+$1.50

TABLE 12.1
Fermented Food Consumption by the Korean Population (g/person/day

Food Items	Age (years)								
	Overall	**1–2**	**3–6**	**7–12**	**13–19**	**20–29**	**30–49**	**50–64**	**>65**
Kimchi[a]									
Scallion	1.3	0.0	0.0	0.0	0.5	1.0	2.1	2.0	0.9
Kodulbbagi	0.6	0.0	0.0	0.0	0.2	0.5	0.9	0.8	0.6
Mu (radish)	24.3	1.9	4.4	11.8	24.4	29.0	29.4	30.6	20.8
Mool (mostly liquid)	13.6	2.0	4.2	5.6	4.6	9.1	15.7	21.7	34.0
Baechu (napa cabbage)	83.8	9.9	21.7	50.5	78.6	89.7	105.3	94.5	85.6
Sobagi (cucumber)	0.0	0.0	0.0	0.0	0.0	0.0	0.0	0.1	0.1
Got (mustard leaf)	0.7	0.0	0.0	0.1	0.1	1.1	1.1	1.1	0.3
Total	124.3	13.8	30.3	68.0	108.4	130.4	154.5	150.8	142.3
Soy Based									
Ganjang (soy sauce)	6.6	1.8	3.7	5.8	6.4	6.7	8.0	6.8	5.3
Gochujang (hot pepper–soybean paste)	3.7	0.2	1.2	3.0	3.4	4.0	5.2	3.5	1.9
Doenjang (soybean paste)	5.6	0.9	1.6	3.5	3.0	5.0	7.1	8.0	7.2
Jajang (black bean paste)	1.1	0.4	1.2	1.8	0.9	1.2	1.3	0.7	0.3
Chongkukjang (soybean paste, quick fermented)	1.0	0.2	0.7	0.5	0.4	0.6	1.3	1.6	2.1
Mixed bean paste	1.0	0.0	0.2	0.3	0.7	1.3	1.5	1.0	0.4
Total	19.0	3.5	8.6	14.7	14.8	18.8	24.4	21.6	17.2
Jeotgal (salted fish and shellfish)	2.7	0.2	0.2	0.9	1.7	1.9	3.8	3.7	3.9
Vinegar	0.5	0.1	0.1	0.3	0.4	0.6	0.8	0.6	0.3

[a] Classified based on the main ingredient

Source: Adapted from Ministry of Health and Welfare, Report on 1998 National Health and Nutritional Survey (Dietary Intake Survey), Korea Health Industry Development Institute, Seoul, 1999.

the most. Kimchi has napa cabbage and/or radish as its main ingredient, although virtually any kind of vegetable can be used. Fermentation is achieved in a comparatively short time. Second in consumption are soy-based products, which include ganjang (soy sauce), doenjang (soybean paste), chongkukjang (quick fermented soybean paste), and gochujang (hot pepper–soybean paste). This category comprises major condiments in Korean cuisine. Traditionally, most households prepare these condiments once every year and keep and use them for many years. Fermented products based on fish and shellfish comprise the third category. These can be used as one of the ingredients in kimchi or eaten as they are. Although dozens of fermented foods are consumed in Korea, scientific research regarding their health effects is mainly concentrated on kimchi and doenjang, the two fermented food items consumed in largest quantities. Therefore, the focus in this chapter will be on these two food items.

12.2 KIMCHI

Kimchi is mentioned in the book of *Samkuksaki,* published in 1145 A.D., and is thought to be a simple fermentation product of a vegetable in brine prepared in a stone jar.[2] Since then, kimchi has developed into many different products (more than 50) that use different vegetables such as baechu (napa cabbage; *Brassica rapa* L. ssp. *pekinensis* [Lour.] Han), radish, cucumber, or scallion as the major ingredient, with various methods of preparation. The essence of kimchi-making is to brine the major raw vegetable; drain; mix with minor ingredients such as scallion, garlic, ginger, red pepper, fermented fish, salt, and sugar; and store the mixture for fermentation. Most people consider kimchi to be well ripened when it has been kept for 3 weeks at 4°C or for 4 days at 15°C, although it may be eaten at any stage of storage until it becomes too sour. Raw kimchi is eaten like salad, often mixed with sesame seed oil and sugar, while overripened kimchi is made into jigae by boiling with meat. The average daily consumption of kimchi for the Korean population is 124.3 g, and 154.5 g for the age group (30 to 49 years old) consuming the most (see Table 12.1). Many households still make kimchi at home, but commercially prepared kimchi has been gaining popularity. Typical ingredient composition of commercial baechu-kimchi is as follows: baechu, 85.9%; radish, 2.8%; garlic, 1.4%; red pepper powder, 2.9%; scallion, 1.5%; fermented fish, 1.9%; ginger, 0.7%; salt, 2.5%; sugar, 0.8%; flavor enhancers, 0.3%.[3]

12.2.1 CHANGES DURING FERMENTATION

The microoganisms mainly responsible for the fermentation process in the production of kimchi are lactic acid bacteria (LAB), although fermentation by aerobic bacteria, yeasts, and molds occurs simultaneously.[2] Numerous biochemical changes occur during fermentation, forming flavor compounds and changing the texture of the vegetables.

Kimchi is an important source of vitamins and their precursors, such as the vitamin B group, β-carotene, and ascorbic acid, which change in concentration during fermentation. For example, the ascorbic acid content of baechu-kimchi (3.5% salt) was 15.2

mg% in a freshly made sample, decreased gradually by 10% over a 12-day period, increased to 18 mg% on day 18, and then decreased again when kimchi was kept at 7°C.[4] The vitamin B_{12} content went through similar changes in baechu-kimchi (salt 3.25%) when kept at 2 to 7°C. The percentage change in the vitamin B_{12} concentration was much more pronounced (about 60% increase) than for ascorbic acid. Vitamins B_1 and B_2 and niacin increased in concentration without showing any reduction during the initial stage. The maximum concentrations of the B vitamins and vitamin C are observed after around 3 weeks of aging, when kimchi is considered well ripened by most people. Vitamin B_1 shows the greatest percent increase (137%), followed by B_2 (117%), B_{12} (60%), and niacin (53%). Unlike the B vitamins and vitamin C, the β-carotene concentration does not increase, but rather decreases to about 50% of the initial concentration when kimchi is well ripened, which suggests that there is no biosynthesis of β-carotene by microorganisms found in kimchi.[4] Changes in the concentration of these particular nutrients during fermentation may have important implications in terms of the antimutagenic and anticarcinogenic properties of kimchi, since several of these vitamins are believed to prevent certain cancers.[5–7]

Kimchi is also a good source of dietary fiber, which may prevent colorectal cancer, diabetes, obesity, atherosclerosis, high blood pressure, etc.[8–10] The concentration of various fiber fractions seems to change as kimchi ripens. When, for example, concentrations of the pectic substances were studied in relation to texture softening upon aging, water-soluble pectins and pectic acid increased by 127 and 31%, respectively, while the alcohol-insoluble solids and protopectin decreased by 45 and 61%.[11] Both the total fiber contents and the crude fiber contents, however, remained relatively constant. The total dietary fiber contents of raw kimchi (0 week) and fermented kimchi (3 weeks at 5°C) were 20.7 and 24.0%, respectively, on a dry weight basis, and crude fiber contents were 8.2 and 9.3%.[12]

12.2.2 Cancer

12.2.2.1 Epidemiology

In 1985, Lee et al.[13] reported that patients with stomach and other cancers consumed unusually large amounts of kimchi. It was suggested, however, that the positive correlation might not be due to kimchi *per se*, since green vegetables, which are used as the main ingredient, are believed to reduce rather than increase the incidence of cancer.[5,8,14] Recently, a case-control study was conducted, and kimchi was cited as one of the risk factors associated with stomach cancer together with doenjang and gochujang.[15] It was concluded, however, that the salt in those foods was the most important contributing factor. Park et al.[16] investigated the effect of several different fermented vegetables — baechu-kimchi, danmuji (pickled radish), and cucumber kimchi — on the incidence of stomach cancer. Of the three food items, only consumption of cucumber kimchi showed a positive correlation with cancer incidence. Although consumption of regular baechu-kimchi did not show any effect, those kimchis made with a high level of fermented fish had a positive correlation. This may also be explained on the basis of a salt effect, since fermented fish made in Korea is high in salt (15 to 18%). In Asian communities, high salt consumption seems to be an important contributing factor to the incidence of stomach cancer.[15,17–20] Kimchi normally contains about 2.5% salt, although the salt content can be as high as 9%. Traditionally, people in the southern part of Korea made saltier kimchi because of the warm weather. With the availability of refrigerators, kimchi now tends to be made with less salt. When future epidemiological studies involving kimchi are planned, the salt level must be considered carefully in order to minimize confounding and to properly reflect the effect of kimchi as it is currently manufactured and consumed.

12.2.2.2 Anticarcinogenic and Antimutagenic Activities in *in Vitro* and Animal Models

Vegetables used for kimchi contain high levels of vitamin C, β-carotene, dietary fiber, flavonoids, and chlorophylls, which are believed to be effective in preventing cancer.[5,8,14] Kimchi also contains LAB, which have been reported to show antitumor activity.[21,22]

Some of the raw vegetables used for kimchi, such as baechu, parsley, perilla leaf, green pepper leaf,[23] garlic,[24] and red pepper,[25] have been tested and shown to be antimutagenic with an *in vitro* assay system. A methanol extract of kimchi was found to be antimutagenic against aflatoxin B_1 and *N*-methyl-*N*′-nitro-*N*-nitrosoguanidine (MNNG) in the Ames assay and the SOS chromotest.[26] It was also found to be antimutagenic in the wing hair spot test on *Drosophila melanogaster.*[27] When several organic solvents were used for extraction and the activities compared, the dichloromethane extract showed the highest antimutagenicity.[28] These authors suggested that flavonoids, steroids, fatty acids, and terpenoids extracted into the dichloromethane might be responsible for the observed antimutagenicity.

Park et al. varied the ingredient compositions in kimchi and reported that kimchis containing high levels of red pepper powder and garlic showed a stronger antimutagenic activity in the bacterial systems studied.[29] They also determined the effect on the growth of AGS human gastric cancer cells and observed a higher inhibitory effect for the same type of kimchis. The antimutagenic and anticarcinogenic activities of capsaicin in red pepper and sulfur-containing compounds in garlic are well documented in the literature.[30,31] Kimchi made with organically grown vegetables showed greater effects also, perhaps because no residual pesticides that could offset the antimutagenic and anticarcinogenic activities were present.[32]

Several LAB isolated from kimchi suppress mutagenicity induced by 4-NQO, MeIQ, and Trp-P-2 in the Ames test and SOS chromotest.[33] *Leuconostoc mensenteroides* was especially effective, although other organisms such as *Lactobacillus brevis, Lb. fermentum, Lb. plantarum,* and *Pediococcos acidilactici* were also as effective as *Lb. acidophilus* from yogurt. Yogurt administered to mice has been shown to reduce Ehrlich cancer, and extracts of microorganisms from yogurt have been shown to inhibit growth of Ehrlich carcinoma and sarcoma 180. More recent works on the anticarcinogenicity of yogurt have been reviewed.[21,34] Kimchi also contains LAB, but it is not known at this time whether the bacteria found in kimchi are also anticarcinogenic.

The antimutagenicity of kimchi depends on the length of fermentation. Well-ripened kimchi (3 weeks old) was found to be more effective than raw (0-week-old) or 6-week-old over-ripened kimchi.[35] These findings suggest that the chemical components present in the raw materials, and compounds produced during the fermentation process and destroyed in the later stages of fermentation, may also be responsible for the antimutagenicity of kimchi.

Various mammalian cell cultures have been utilized to investigate the effect of kimchi on the growth of cancer cells. Kimchi extract was shown to inhibit growth of AGS, HT-29, HL-60, K-562, MG-63, and sarcoma-180 cells.[28,29,35–37] Growth inhibition involved the induction of apoptosis in HL-60 human leukemia cells as evidenced by the result of a DNA fragmentation assay.[28] Both aqueous and organic extracts were effective. As was noted for the antimutagenicity assays, the degree of inhibition depended on the length of fermentation. The supernatants of kimchi kept at 5°C for 0, 3, and 6 weeks were used to determine the effect on the growth of K-562 (human leukemia) and MG-63 (human osteosarcoma) cell lines. In both cases, the 3-week-old kimchi showed a higher inhibition rate than the 0 and 6 week samples.

When Cho et al.[36] fractionated kimchi extract by using various organic solvents, the dichloromethane fraction showed the highest inhibitory effect on the growth of AGS and HT-29 cancer cells, although all other fractions (hexane, methanol, ethyl acetate, butanol, and aqueous) had some inhibitory effects. The dichloromethane fraction was also the most effective in reducing the cytotoxicity incurred by MCA and DMBA on mouse embryonic C3H/10T1/2 fibroblast cells.[38] On the transformation of cells treated with methyl cholantherene (MCA) and 10-dimethyl-1,2-benzanthracene (DMBA), the same fraction that inhibited the formation of type 2 and 3 foci reduced the number of type 3 most effectively.

The anticarcinogenicity properties of kimchi may be due to an enhancement of the immune system of the host. Phagocytic activity of the peritoneal macrophages of mice was significantly augmented by kimchi extract when tested both *in vitro* and *in vivo*.[37] Alternatively, the anticarcinogenicity of kimchi may be due to its ability to affect the metabolism of mutagens. In guinea pigs treated with polycyclic aromatic hydrocarbons, kimchi increased the activities of liver enzymes responsible for the removal of foreign compounds.[24] When rats were given a single injection of diethylnitrosamine (DEN) and an oral administration of 2-acetylaminofluorene (2-AAF) with or without kimchi extract for 6 weeks, the numbers of glutathione *S*-transferase placental form-positive (GST-P^+) foci in the livers of the kimchi-treated group were decreased to 8.8/cm^2 from 13.8/cm^2 for the control group, indicating a protection against these two hepatocarcinogens.[39] To date, no human data are available.

12.2.3 Cardiovascular Disease

Several ingredients of kimchi are thought to reduce the incidence of cardiovascular disease. Baechu has considerable levels of sitosterol and *S*-methylcysteine sulfoxide, both of which have been reported to lower blood cholesterol in animals.[40–42] Red pepper contains capsaicin, which inhibits platelet aggregation and induces blood vessel expansion.[43] Garlic was reported to reduce plasma cholesterol and triglyceride levels, and ginger was observed to reduce serum cholesterol levels in

TABLE 12.2
Biochemical Parameters of Korean Adult Men with Respect to Daily Kimchi Consumption

	Quartile of Kimchi Intake (g/day)			
	1st (68.13 ± 24.2)	2nd (118.5 ± 11.4)	3rd (208.5 ± 19.0)	4th (382.3 ± 71)
HDL-C (mg/dl)	58.5 ± 21.9	60.2 ± 15.2	63.5 ± 13.5	66.5 ± 13.6
LDL-C (mg/dl)	116.0 ± 54.2	107.2 ± 30.1	99.4 ± 25.5	93.4 ± 35.3
TG (mg/dl)	143.6 ± 82.2	149.4 ± 92.9	158.3 ± 77.3	152.8 ± 62.9

Note: HDL-C, high-density-lipoprotein cholesterol; LDL-C, low-density-lipoprotein cholesterol; TG, triglyceride.

Source: Adapted from Kwon, M.-J., Chun, J.-H., Song, Y.-S., and Song, Y.-O., *J. Korean Soc. Food Sci. Nutr.,* 28, 1144, 1999. With permission.

animals.[44,45] LAB appear to lower blood pressure and prevent the increase of plasma cholesterol concentrations.[46]

Recently, a clinical study involving 102 healthy Korean men was carried out to determine if kimchi is hypolipidemic for humans.[47] The men were divided into four groups, with the groups consuming an average of 68 g, 118 g, 208 g, and 383 g of kimchi daily. Kimchi consumption was positively correlated with high-density lipoprotein (HDL) cholesterol and negatively correlated with low-density lipoprotein (LDL) cholesterol (see Table 12.2). Preference for hot taste was negatively correlated with systolic blood pressure (correlation coefficient; –0.15 with $P < 0.05$).

To study the effect in animal models, male Sprague–Dawley rats were fed diets containing 0%, 3%, 5%, and 10% freeze-dried kimchi for 6 weeks.[48,49] Plasma cholesterol was lowered in rats fed all three levels of kimchi, but triglyceride levels were decreased only in the highest dose group of animals. More specifically, kimchi intake reduced very-low-density lipoprotein (VLDL) cholesterol and LDL triglyceride at all levels of kimchi consumption. Kimchi consumption also increased serum HDL-cholesterol. Intake of 5 and 10% kimchi also lowered cholesterol, triglyceride, total lipid, and apolipoprotein B in the liver, whereas it increased fat, cholesterol, triglyceride, and apolipoprotein A-1 levels in feces. Kimchi intake elevated the T3 levels in rats but had no effect on the thyroxine (T4) level (see Table 12.3). These results suggest that intake of kimchi in rats decreases plasma and hepatic lipid levels by increasing fecal excretion of triglyceride and cholesterol, elevating T3 level, and altering lipoprotein metabolism.[48]

To assess the effect of kimchi on thrombosis, plasma fibrinolytic activity was measured in rats fed with kimchi.[49] Fibrinolytic activity of plasma from rats consuming diets with 10% kimchi was higher than that of controls, while the activities for animals consuming 3 and 5% kimchi were the same as those of controls. Fibrinolytic activities were also measured for water and methanol extracts of kimchi using the fibrin plate method. The fibrinolytic activity of the methanol extract of

TABLE 12.3
Effect of Kimchi Intake on Various Markers of Lipid Metabolism in the Rat

	Kimchi Intake			
	Control	3% Kimchi	5% Kimchi	10% Kimchi
Liver cholesterol (mg/g)	20.4 ± 4.2	16.1 ± 2.2	14.2 ± 3.1	14.9 ± 2.8
Plasma cholesterol (mg/dl)	88.4 ± 15.1	74.4 ± 2.6	66.9 ± 3.9	74.9 ± 9.0
Liver triglyceride (mg/g)	11.1 ± 2.2	9.5 ± 1.2	7.2 ± 1.9	7.0 ± 1.3
Plasma triglyceride (mg/dl)	72.0 ± 15.1	58.0 ± 10.5	65.4 ± 14.3	42.4 ± 15.7
Total lipid in liver (mg/g)	73.3 ± 20.3	66 ± 13.1	55.8 ± 10.9	54.5 ± 6.0
Apo B in liver (mg/g)	16.4 ± 2.9	14.6 ± 2.8	12.8 ± 2.5	12.5 ± 0.9
Apo A-1 in liver (mg/g)	1.6 ± 0.5	2.5 ± 0.2	3.7 ± 1.2	3.1 ± 0.6
Plasma triiodothyrosine (μg/dl)	0.26 ± 0.21	0.76 ± 0.23	0.83 ± 0.29	0.7 ± 0.35
Plasma thyroxine (μg/dl)	8.1 ± 1.4	9.0 ± 2.1	8.7 ± 2.1	9.0 ± 1.7

Source: Adapted from Kwon, M.-J., Song, Y.-O., and Song, Y.-S., *J. Korean Soc. Food Sci. Nutr.*, 26, 507, 1997. With permission.

kimchi was much higher than that of the water extract. The active chemical components in the methanol extract have not yet been determined.[49]

12.2.4 Nitrosamines and Ethyl Carbamate

Nitrosamines are known carcinogens for animals and suspected to be carcinogenic for humans as well.[50] There have been concerns about the possible formation of nitrosamines in kimchi, since baechu-kimchi contains high levels of both nitrate and a fermented fish product that contains amines. When kimchi was analyzed for nitrate, nitrite, and nitrosamines during fermentation, the results showed very little nitrite (undetected up to 0.5 ppm) and nitrosamines (undetected up to 0.05 ppb) throughout the process.[35] It was thought that the reaction of phenols and ascorbic acid with nitrite were at least partly responsible for these results. Recently, Kim et al.[51] reported that up to 91 ppb of *N*-nitrosodimethylamine (NDMA) could be formed at low pH, high temperature, and low salt concentration (pH 4, 16°C, 1.5%). However, it would be rather difficult to meet these conditions under the usual practice of kimchi-making.

Ethyl carbamate, which is mainly present in fermented foods and beverages, has long been associated with cancer. Ethyl carbamate has been measured in some Korean foods; values in kimchi ranged from 0 to 16.2 ppb.[52] Under the conditions of commercial kimchi production, formation of nitrosamines and ethyl carbamate should not be a problem. However, considering the large daily intake of kimchi by Koreans, it would be wise to identify the specific conditions under which ethyl carbamate and other undesirable compounds could be produced, in order to be able to make recommendations on how to avoid them.

12.3 DOENJANG

Doenjang is a soy-based food consumed daily in Korean households. It is a soup base used for a regular broth-type soup (kook) and a much thicker stew-type soup (jigae), both of which are eaten with rice. Traditionally, most soybean-based fermented foods are prepared once a year and kept up to 2 to 3 years. The preparation starts in the late fall with the natural fermentation of meju, which are solid blocks (size varies by household, but approximately 20 × 15 × 10 cm) made from steamed soybeans. The fermentation lasts for about 2 weeks. The product is then dried over the winter and subjected to further fermentation in brine during the spring for 6 to 8 weeks. At the end of the wet fermentation, doenjang (soybean paste, solid material) is separated from ganjang (soy sauce, supernatant). Doenjang can be consumed after 30 to 50 days of maturation; however, longer maturation is considered to produce a better flavor.[1]

12.3.1 Cancer

12.3.1.1 Epidemiology

Doenjang has had to endure a rather negative reputation because it has been linked to stomach cancer in Korea. A positive association between doenjang intake and stomach cancer was first reported by Crane et al.[53] in 1970. In this case-control study, the authors simply speculated that the causal factor might be contamination by aflatoxin. However, this mycotoxin is presently considered to be a liver but not a stomach carcinogen. The main reason for the speculation was that one of the main fungi involved in the fermentation of meju is *Aspergillus oryzae,* which is in the same genus as the aflatoxin-producing *A. flavus.* However, the production of the toxin during and after natural fermentation of meju was found to be insignificant, even though *A. flavus* could grow quite well.[54]

Several other groups have reported positive correlations between the intake of fermented soyfoods and stomach cancer.[15,55,56] In a case-control study conducted in Korea, three dishes that contain fermented ingredients were cited as risk factors for stomach cancer; doenjang (soybean paste) jagae, gochujang (hot pepper soybean paste) jigae, and kimchi. It was also reported that the frequent consumption of tofu made from soybeans decreased the risk.[15] This study pointed out the importance of cooking methods as well as ingredients and analyzed food consumption by prepared food type, rather than raw materials. The authors singled out salt as the most important contributing factor.[15] It is noteworthy that kook (broth-type soup) made with doenjang and gochujang did not show any positive correlation with the incidence of cancer. Later, the same group pooled and analyzed the epidemiological data available in Korea and again reached the conclusion that salt, rather than soy components or fermentation byproducts, was the risk factor for stomach cancer.[17]

In a recent meta-analysis of 14 pooled epidemiological studies, Wu et al. cautiously suggested that fermented soyfoods may increase the relative risk for stomach cancer to 1.26. In the same work, nonfermented soyfoods were thought to lower the risk to 0.72, based on analysis of the data of ten pooled studies. The authors, however, warned of possible confounders such as salt and fruits–vegetables intake.[55] Responding to the Wu et al.[55] study, Ji and colleagues reanalyzed 1124 original data from the 1998 case-control study,[19] which had shown a decreased cancer risk in men with increased soy intake, and a small, insignificant increase in risk in women. The reanalysis showed that adjustment for salt intake and salt preference lowered the odds ratio slightly, while nonfermented soyfoods lowered the cancer risk in men, but not in women. It was also shown that fermented bean curd was responsible for the increased risk among women.[56] A prospective study involving a Japanese immigrant population to Hawaii reported that miso (a Japanese fermented soybean product) was mildly associated with gastric cancer, while salt consumption showed a stronger association.

Considering the results of many biochemical and animal studies with various soy components, which indicate protection against cancer, it may indeed be the salt, which constitutes as much as 20% of doenjang, that increases stomach cancer risk. Other studies have also indicated salty foods as etiological factors apart from fermented foods.[18–20] A cohort study in Japan showed that self-reduction of salty food consumption decreased the risk of progression of precancerous lesions to gastric cancer to a relative risk of 0.56 (95% CI, 0.32 to 0.96).[57]

Ever since the differences in diets and in certain cancer rates were noted for Western and Asian populations, consumption of soyfoods has been suggested as a health-promoting factor.[58] In cancers other than stomach cancer, the role of soyfood is more hopeful, especially in breast cancer, although epidemiologic data have been mostly inconsistent and inconclusive.[59] Also, no conclusive epidemiological evidence links the consumption of legumes and the reduction of cancer.[60] It is interesting to note two recent population-based case-control studies conducted in Shanghai. Originally, breast cancer risk was shown to have a weak inverse relationship to overall soyfood intake, albeit not a statistically significant one.[61] The same Shanghai study, however, was analyzed for the soyfood intake during adolescence and the later development of breast cancer, and a strong inverse correlation was observed, in both pre- and postmenopausal women.[62] An age-specific protection mechanism against breast cancer appears to exist, which was not recognized in most of previous studies (see Section 12.3.3 for further discussion).

12.3.1.2 Anticarcinogenic and Antimutagenic Activities in *in Vitro* and Animal Models

In contrast to epidemiological studies, most laboratory studies have shown a positive effect of doenjang on the prevention of the development and growth of cancer. One possible explanation may be that studies have used extracts (usually of organic solvents), which are virtually devoid of salt. Many reports have documented the antimutagenic properties of doenjang extracts.[63–68] In classical bacterial antimutagenicity tests to see whether the test material inhibits mutagenesis induced by known compounds, doenjang showed the highest overall antimutagenic activity among methanolic extracts of four Korean soy fermented products — doenjang, ganjang, chongkukjang, and gochujang — even though the effect varied among the mutagens employed. One noteworthy observation in this study was that homemade samples of these foods showed higher antimutagenicity than their commercial counterparts or raw materials, in all four items tested.[68] Commercial fermentation typically utilizes a shortened process time by employing an inoculation of a defined mixture of microbes at an elevated but controlled temperature. Chemical conversions needed to produce the antimutagenic ingredients may have slow kinetics and/or multi-step processes that require a variety of enzymes from different microorganisms. These complex conversions may be more common in slower fermentations carried out in home production systems, which typically display a more complex microbial composition than commercial production does. Fractionation experiments showed that these antimutagenic component(s) seem to be heat stable and have hydrophobic chemical characteristics.[66,67] Doenjang showed a higher activity than Japanese miso to protect *Salmonella typhimurium* against the mutagenesis induced by aflatoxin B1 or MNNG in a comparative study.[35,63]

Mammalian cell culture and animal transplantation experiments have shown that in addition to providing protection against genotoxicity, doenjang also protects against the advancement of tumor transformation steps. Again, fermented foods showed much higher activities than raw soybeans or soy flour.[63] The relationship between fermentation and antimutagenic activity was clearly shown in an experiment with chongkukjang, by comparing soybeans at 0 and 48 hours of fermentation and 20 days of maturation.[65] It should be pointed out that these experiments were again performed with organic solvent extracts, effectively eliminating salt from the test sample and also perhaps concentrating active components. Therefore, conclusions may not be directly applicable to whole foods as they are commonly eaten.

In an animal model, both organic solvent and boiling water extracts of doenjang showed an inhibition of solid tumor formation in BALB/c mice transplanted with sarcoma-180. Results also showed the extension of survival time for those mice receiving the extract treatments compared to control mice.[66]

Modification of enzyme activities involved in the metabolism of xenobiotics in liver has been suggested as a biochemical mechanism for the beneficial effect of doenjang extracts. During a mouse tumor transformation assay, reduced activities of glutathione *S*-transferase and glutathione reductase and decreased glutathione content in the liver observed after the tumor cell implantation were restored in the animals treated with doenjang extracts. Lowered lipid peroxide levels indicated that antioxidation could be involved in the mechanism as well.[67] These authors identified the active components as linoleic acid and genistein, both of which were shown to block the passage of the cancer lines from the G2 to the M phase.

12.3.2 Cardiovascular Disease

A meta-analysis of 38 clinical studies showed that soy protein consumption reduced the serum levels of cholesterol, LDL cholesterol, and triglyceride.[69] The American Heart Association recently stated that the daily consumption of soyfoods containing phytoestrogen could benefit hypercholesterolemic subjects. The possible mechanisms for the beneficial effect suggested by various authors involve trypsin inhibitors and/or other peptides, phytic acids, saponins, isoflavones, or any combination of these constituents of soy.[70] The protein component of soy seems to play an integral part in this lipid altering effect, based on the fact that isoflavones alone failed to show any effect.[71] Elimination of the phytoestrogen from soy reduced the effect on serum cholesterol levels.[70]

12.3.2.1 Inhibition of Angiotensin Converting Enzyme

Angiotensin converting enzyme (ACE) produces angiotensin II, a vasoconstrictor, and inhibits the vasodilator bradykinin, leading to increased blood pressure. It has been shown that a number of food items including fermented foods have inhibitory activities on ACE.[72] Doenjang was tested both *in vitro* and *in vivo* and shown to inhibit ACE with IC_{50} of 2.2 to 310 μg/ml. The inhibition of ACE may ultimately result in lower blood pressure.[73–76] Two ACE-inhibiting peptides found in doenjang have been isolated and characterized as Arg-Pro and His-His-Leu by two independent groups in Korea.[75,76] The tripeptide, which showed the higher ACE inhibitory activity, has been synthesized and shown to lower systolic blood pressure when administered to spontaneously hypertensive rats by intravenous injection at a level of 5 mg/kg body weight. The ACE activity in the aortic tissue was significantly lower in the treated rats, while the activity in serum was unchanged.[76] Since small peptides can be directly absorbed through the intestinal tract, the tripeptide can probably reach the circulatory system and exert its protective effect. However, it remains to be seen how much of the peptides ingested in food survives the digestive system and gets into the blood circulation.

12.3.2.2 Antithrombotic Peptides

Another cause of cardiovascular disease is thrombosis, which is the abnormal aggregation of blood platelets leading to atherosclerosis or hypertension. In his pioneering work, Shon searched for antithrombotic peptides in doenjang extract and showed that fractions containing basic peptides exhibited a higher activity against ADP-induced platelet aggregation than other known antithrombic peptides.[77] Recently, ADP receptor antagonists have been successfully developed and used to treat thrombosis.[78] Applying this same concept, it may be possible in the future to prevent the disease by daily consumption of doenjang and other antithrombotic peptide containing foods.

12.3.3 Isoflavones

Isoflavones, especially genistein from soy, have been implicated in protection against a spectrum of chronic diseases including breast and prostate cancers, postmenopausal syndromes, osteoporosis, and cardiovascular disease. The protective role has been shown in many animal and *in vitro* studies, but only miniscule positive effects, if any, have been observed so far in human studies. For the details of the effects of isoflavones on various diseases, readers are referred to several excellent reviews.[79–82]

Isoflavones are known as weak estrogens; estrogen antagonists; antioxidants; inhibitors of topoisomerase II, angiogenesis, and platelet aggregation; and inducers of cell differentiation in animal and *in vitro* models. It is likely that these associations were not clearly observed in human studies because it requires long periods of time, perhaps decades, between the intake and the results. Furthermore, timing of isoflavone exposure seems important for the effect in humans. In one of the epidemiological studies, early exposure to soyfoods was necessary for protection against breast cancer.[62] One hypothesis to explain this observation, based on an animal model, is that genistein, when administered to young subjects, promotes early cell differentiation during adolescence. This early differentiation in turn lowers epidermal growth factor signaling, which is frequently associated with breast cancer, in later mature years.[79]

Another likely explanation for the inconclusive human evidence is that while the experimental data were largely drawn from isoflavone exposure, epidemiological observations were focused only on the soy intake of subjects. Most isoflavones in soy exist as glycosides, and only small amounts are present as free isoflavones (aglycon), which are in the form that can be absorbed into the body and exert physiological activities. Hydrolysis of the ingested glycosides is carried out by intestinal microbes, which are another variable in the human population. The ratio of free isoflavones to glycosides has been shown to be much higher in fermented soyfoods than in their unfermented counterparts.[83,84] The urinary excretion of isoflavones was also higher in humans after the consumption of tempeh (Indonesian fermented soyfood) than after the consumption of soybean pieces.[85] Doenjang is a promising source of isoflavones compared to other soyfoods because it has a higher proportion of free to glycoside-bound isoflavone[86] (see Table 12.4).

TABLE 12.4
Isoflavone Contents in Selected Soyfoods Consumed in Korea

	Free Isoflavones[a] (mg/kg)			Total Isoflavones[b] (mg/kg)		
Sample	**Deinzein**	**Genistein**	**Glycitein**	**Deinzein**	**Genistein**	**Glycitein**
Soybean						
Range	2–25	0–25	0–15	142–1737	357–2885	0–408
Mean	10	8	3	655	1152	161
Number of samples	19 varieties, 25 samples			19 varieties, 60 samples		
Doenjang						
Range	56–1172	26–808	12–567	99–3382	89–5864	14–394
Mean	462	304	126	1227	2013	116
Number of samples		14			14	
Chongkukjang						
Range	80–376	7–254	21–66	161–628	221–948	34–64
Mean	256	123	41	448	653	51
Number of samples		5			5	
Tofu						
Range	10–104	3–86	0–11	67–200	12–306	0–70
Mean	50	26	1	110	171	9
Number of samples	3 varieties, 21 samples			3 varieties, 21 samples		
Soymilk						
Range	2–12	0–3	0	178–655	35–908	0–84
Mean	6	1	0	421	455	33
Number of samples		12			12	

[a] Aglycon; before hydrolysis.
[b] After hydrolysis.

Source: Kim, M.-J., Quantification of Isoflavones and Coumestrol in Korean Soyfoods and Estimations of their Intake, thesis, Seoul National University, Korea, 2001. With permission.

12.4 CONCLUSIONS

At this time, there is no conclusive evidence that shows human health benefits from the consumption of Korean fermented foods such as kimchi and doenjang. However, results from *in vitro* and animal studies are promising for kimchi and fermented soyfoods. It appears that the fermentation process results in higher quality products. Well-ripened kimchi has a higher antimutagenicity than raw (unfermented) kimchi. Doenjang has greater amounts of peptides and free isoflavones than soybeans do. Both kimchi and doenjang have antihypertensive as well as anticarinogenic potential. One obstacle that needs to be overcome is the preference of Korean consumers for the salty taste of kimchi and doenjang, because the consumption of high salt levels can cause many health problems. It will be a challenge to investigate the effect of traditional fermented foods on human health and to develop improved versions of these foods that have optimal health effects.

REFERENCES

1. Han, B.-J., Han, B.-R., and Whang, H.-S., Kanjang and doenjang, in *One Hundred Korean Foods that Koreans Must Know,* Hyun Am Sa, Seoul, 1998, pp. 500–507.
2. Cheigh, H.-S. and Park, K.-Y., Biochemical, microbiological, and nutritional aspects of kimchi (Korean fermented vegetable products), *Crit. Rev. Food Sci. Nutr.,* 34, 175–203, 1994.
3. Park, W.-S., Koo, Y.C., Ahn, B.H., and Choi, S.Y., *Process Standardization of Kimchi Manufacturing,* Korea Food Research Institute, Seoul, 1994.
4. Cheigh, H.-S., Critical review on biochemical characteristics of kimchi, *J. East Asian Soc. Diet. Life,* 5, 89–101, 1995.
5. La Vecchia, C., Altieri, A., and Tavani, A., Vegetables, fruit, antioxidants and cancer: a review of Italian studies, *Eur. J. Nutr.,* 40, 261–267, 2001.
6. Chen, H., Tucker, K.L., Graubrad, B.L., Heineman, E.F., Markin, R.S., Potischman, N.A., Russell, R.M., Weisenburger, D.D., and Ward, M.H., Nutrient intakes and adenocarcinoma of the esophagus and distal stomach, *Nutr. Cancer,* 42, 33–40, 2002.
7. Huang, H.Y., Appel, L.J., Croft, K.D., Miller, E.R., Mori, T.A., and Puddey, I.B., Effects of vitamin C and vitamin E on *in vivo* lipid peroxidation: results of a randomized controlled trial, *Am. J. Clin. Nutr.,* 76, 549–555, 2002.
8. Mason, J.B., Nutritional chemoprevention of colon cancer, *Semin. Gastrointest. Dis.,* 13, 143–153, 2002.
9. Fung, T.T., Hu, F.B., Pereira, M.A., Liu, S., Stampfer, M.J., Colditz, G.A., and Willett, W.C., Whole-grain intake and the risk of type 2 diabetes: a prospective study in men, *Am. J. Clin. Nutr.,* 76, 535–540, 2002.
10. Jenkins, D.J., Kendall, C.W., Augustin, L.S., Franceschi, S., Hamidi, M., Marchie, A., Jenkins, A.L., and Axelsen, M., Glycemic index: overview of implications in health and disease, *Am. J. Clin. Nutr.,* 76, 266S–273S, 2002.
11. Kim, S.-D., Lee, S.-H., Kim, M.-J., and Oh, Y.-A., Changes in pectic substance of lower salted Chinese cabbage kimchi with pH adjusted during fermentation, *J. Korean Soc. Food Sci. Nutr.,* 17, 255–261, 1988.
12. Park, K.-Y., Ha, J.-O., and Rhee, S.-H., A study on the contents of dietary fibers and crude fiber in kimchi ingredients and kimchi, *J. Korean Soc. Food Sci. Nutr.,* 25, 69–75, 1996.

13. Lee, K.-Y., Lee, Y.-C., Park, Y.-S., Yoon, K.-H., and Kim, B.-S., A study of relation between dietary vitamin A intake and serum vitamin A levels and cancer risk in Korea, *Korean J. Nutr.,* 18, 301–311, 1985.
14. Shikany, J.M. and White, G.L.J., Dietary guidelines for chronic disease prevention, *South Med. J.,* 93, 1135–1151, 2000.
15. Lee, J.-K., Park, B.-J., Yoo, K.-Y., and Ahn, Y.-O., Dietary factors and stomach cancer: a case-control study in Korea, *Int. J. Epidemiol.,* 24, 33–41, 1995.
16. Park, H.S., Kim, H.S., Choi, S.Y., and Chung, C.K., Effect of dietary factors in the etiology of stomach cancer, *Korean J. Epidemiol.,* 20, 82–101, 1998.
17. Ahn, Y.-O., Diet and stomach cancer in Korea, *Int. J. Cancer,* S10, 7–9, 1997.
18. You, W.-C., Blot, W.J., Chang, Y.-S., Ershow, A.G., Yang, Z.-T., An, Q., Henderson, B., Xu, G.-W., Fraumeni, J.F., Jr., and Wang, T.-G., Diet and high risk of stomach cancer in Shandong, China, *Cancer Res.,* 48, 3518–3523, 1988.
19. Ji, B.-T., Chow, W.-H., Yang, G., McLaughlin, J.K., Zheng, W., Shu, X.-O., Jin, F., Gao, R.-N., Gao, Y.-T., and Fraumeni, J.F., Jr., Dietary habits and stomach cancer in Shanghai, China, *Int. J. Cancer,* 76, 659–664, 1998.
20. Hirohata, T. and Kono, S., Diet/nutrition and stomach cancer in Japan, *Int. J. Cancer,* S10, 34–36, 1997.
21. Reddy, G.V., Shahani, K.M., and Banerjee, M.R., Inhibitory effect of yogurt on Ehrlich ascites tumor-cell proliferation, *J. Natl. Cancer Inst.,* 50, 815–817, 1973.
22. Bogdanov, I.G., Dalev, P.G., Gurevich, A.E., Kosolov, M.N., Mal'Kova, V.P., Plemyannikoa, L.A., and Sorokina, I.B., Antitumor glycopeptides from *Lactovacillus bulgaricus* cell wall, *FEBS Letters,* 57, 259, 1975.
23. Park, K.-Y., Lee, K.-I., and Rhee, S.-H., Inhibitory effect of green-yellow vegetables on the mutagenicity in *Salmonella* assay system and on the growth of AZ-521 human gastric cancer cells, *J. Korean Soc. Food Sci. Nutr.,* 21, 149–153, 1992.
24. Park, K.-Y., Kim, S.-H., Suh, M.-J., and Chung, H.-Y., Inhibitory effects of garlic on the mutagenicity in *Salmonella* assay system and on the growth of HT-29 human colon carcinoma cells, *J. Korean Soc. Food Sci. Nutr.,* 23, 370–374, 1991.
25. Kim, S.-H., Park, K.-Y., and Suh, M.-J., Inhibitory effect of aflatoxin B1 mediated mutagenicity by red pepper powder in the *Salmonella* assay system, *J. Korean Soc. Food Sci. Nutr.,* 20, 156–161, 1991.
26. Paik, K.-A., Antimutagenic Effect of Kimchi Extracts, thesis, Pusan National University, Korea, 1994.
27. Hwang, S.-Y., Hur, Y.-M., Choi, Y.-H., Rhee, S.-H., Park, K.-Y., and Lee, W.-H., Inhibitory effect of kimchi extracts on mutagenesis of aflatoxin B1, *Environ. Mut. Carcin.,* 17, 133–137, 1997.
28. Cho, E.-J., Rhee, S.-H., Lee, S.-M., and Park, K.-Y., *In vitro* antimutagenic and anticancer effects of kimchi fractions, *J. Korean Assoc. Cancer Prev.,* 2, 113–121, 1997.
29. Park, K.-Y., Cho, E.-J., and Rhee, S.-H., Increased antimutagenic and anticancer activities of Chinese cabbage kimchi by changing kinds and levels of subingredients, *J. Korean Soc. Food Sci. Nutr.,* 27, 169–182, 1998.
30. Surh, Y.-J., Lee, C.-J., Park, K.-K., Mayne, T., Liem, A., and Miller, J.A., Chemopreventive effects of capsaicin and diallyl sulfide against mutagenesis or tumorigenesis by vinyl carbamate and *N*-nitrosodiethylamine, *Carcinogenesis,* 16, 2647–2471, 1995.
31. Hong, J.-Y., Lin, M.C., Wang, Z.Y., Wang, E.-J., and Yang, C.S., Inhibition of chemical toxicity and carcinogenesis by diallyl sulfide and diallyl sulfone, in *Food Phytochemicals for Cancer Prevention 1: Fruits and Vegetables,* Huang, M.-T., Osawa, T., Ho, C.-T., and Rosen, R.T., Eds., American Cancer Society, Washington, 1994, pp. 97–101.

32. Ren, H., Endo, H., and Hayashi, T., The superiority of organically cultivated vegetables to general ones regarding antimutagenic activities, *Mutation Res.,* 496, 83–88, 2001.
33. Son, T.-J., Kim, S.-H., and Park, K.-Y., Antimutagenic activities of lactic acid bacteria isolated from kimchi, *J. Korean Assoc. Cancer Prev.,* 3, 65–74, 1998.
34. Teitelbaum, J.E. and Walker, W.A., Nutritional impact of pre- and probiotics as protective gastrointestinal organisms, *Annu. Rev. Nutr.,* 22, 107–138, 2002.
35. Park, K.-Y., The nutritional evaluation and antimutagenic and anticancer effect of kimchi, *J. Korean Soc. Food Sci. Nutr.,* 24, 169–182, 1995.
36. Cho, E.-J., Rhee, S.-H., Kang, K.-S., and Park, K.-Y., *In vitro* anticancer effect of Chinese cabbage kimchi fractions, *J. Korean Soc. Food Sci. Nutr.,* 28, 1326–1331, 1999.
37. Choi, M.-W., Kim, K.-H., and Park, K.-Y., Effects of kimchi extracts on the growth of sarcoma-180 cells and phagocytic activity of mice, *J. Korean Soc. Food Sci. Nutr.,* 26, 254–260, 1997.
38. Cho, E.-J., Rhee, S.-H., and Park, K.-Y., *In vitro* anticarcinogenic effect of kimchi fractions, *J. Korean Assoc. Cancer Prev.,* 4, 79–85, 1999.
39. Park, K.-Y., Kim, H.-C., and Jung, K.-O., Anticarcinogenic effect of kimchi extract on 2-AAF–induced rat hepatocarcinogenesis system, *J. Korean Assoc. Cancer Prev.,* 5, 81–86, 2000.
40. Fujiwara, M., Antihypercholesterolemic effect of sulfur containing amino acid, *S*-methyl-L-cysteine sulfoxide isolated from cabbage, *Experimenta,* 28, 254, 1972.
41. Plat, J. and Mensink, R.P., Effects of plant sterols and stanols on lipid metabolism and cardiovascular disease, *Nutr. Metab. Cardiovasc. Dis.,* 11, 31–40, 2001.
42. Liu, L. and Yeh, Y.Y., Water-soluble organosulfur compounds of garlic inhibit fatty acid and triglyceride syntheses in cultured rat hepatocytes, *Lipids,* 36, 395–400, 2001.
43. Oh, Y.-J., Hwang, I.-J., and Leitzmann, C., Physiological and nutritional evaluation of kimchi, in *Science of Kimchi,* Korean Society of Food Science and Technology, Seoul, 1994, pp. 226–245.
44. Kang, J.A. and Kang, J.S., Effect of garlic and onion on plasma and liver cholesterol and triacylglycerol and platelet aggregation in rats fed basal or cholesterol supplemented diets, *Korean J. Nutr.,* 30, 132–138, 1997.
45. Orekhov, A.N., Tertov, V.V., Sobenin, I.A., and Pivovarova, E.M., Direct antiatherosclerosis-related effects of garlic, *Ann. Med.,* 27, 63–65, 1995.
46. Gilliland, S.E., Nelson, C.R., and Maxwell, C., Assimilation of cholesterol by *Lactobacillus acidophilus, Appl. Environ. Microbiol.,* 49, 377–381, 1985.
47. Kwon, M.-J., Chun, J.-H., Song, Y.-S., and Song, Y.-O., Daily kimchi consumption and its hypolipidemic effect in middle-aged men, *J. Korean Soc. Food Sci. Nutr.,* 28, 1144–1150, 1999.
48. Kwon, M.-J., Song, Y.-O., and Song, Y.-S., Effects of kimchi on tissue and fecal lipid composition and apolipoprotein and thyroxine levels in rats, *J. Korean Soc. Food Sci. Nutr.,* 26, 507–513, 1997.
49. Kim, M.-J., The fibrinolytic activity of kimchi and its ingredients *in vivo* and *in vitro, J. Korean Soc. Food Sci. Nutr.,* 27, 633–638, 1998.
50. Concon, J.M., *Food Toxicology,* Marcel Dekker, New York, 1988.
51. Kim, J.-H., Chang, Y.-S., and Shin, H.-S., Effects of temperature, pH and salt concentration on formation of *N*-nitrosamines during kimchi fermentation, *J. Food Hyg. Safety,* 13, 332–336, 1998.
52. Kim, Y.-K.L., Koh, E., Chung, H.-J., and Kwon, H., Determination of ethyl carbamate in some fermented Korean foods and beverages, *Food Additives Contam.,* 17, 469–475, 2000.

53. Crane, P.S., Rhee, S.U., and Seel, D.J., Experience with 1079 cases of cancer of the stomach seen in Korea from 1962 to 1968, *Am. J. Surg.*, 120, 747–751, 1970.
54. Park, K.-Y., Destruction of aflatoxins during the manufacture of doenjang by traditional method and cancer preventive effects of doenjang, *J. Korean Assoc. Cancer Prev.*, 2, 27–37, 1997.
55. Wu, A.H., Yang, D., and Pike, M.C., A meta-analysis of soyfoods and risk of stomach cancer: the problem of potential confounders, *Cancer Epidemiol. Biomarkers Prev.*, 9, 1051–1058, 2000.
56. Ji, B.-T., Chow, W.-H., Yang, G., Jin, F., Gao, Y.-T., and Fraumeni, J.F., Jr., Correspondence re: A.H. Wu et al., A meta-analysis of soyfoods and risk of stomach cancer: the problem of potential confounders, *Cancer Epidemiol. Biomarkers Prev.*, 9, 1051–1058, 2000, *Cancer Epidemiol. Biomarkers Prev.*, 10, 570–571, 2001.
57. Inoue, M., Tajima, K., Kobayashi, S., Suzuki, T., Matsuura, A., Nakamura, T., Shirai, M., Nakamura, S., Inuzuka, K., and Tominaga, S., Protective factor against progression from atrophic gastritis to gastric cancer: data from a cohort study in Japan, *Int. J. Cancer,* 66, 309–314, 1996.
58. Messina, M.J., Persky, V., Setchell, K.D.R., and Barnes, S., Soy intake and cancer risk: a review of the *in vitro* and *in vivo* data, *Nutr. Cancer,* 21, 113–131, 1994.
59. Wu, A.H., Ziegler, R.G., Nomura, A.M.Y., West, D.W., Kolonel, L.N., Horn-Ross, P.L., Hoover, R.N., and Pike, M.C., Soy intake and risk of breast cancer in Asians and Asian Americans, *Am. J. Clin. Nutr.,* 68, 1437S–1443S, 1998.
60. Kushi, L.H., Meyer, K.A., and Jacobs, D.R.J., Cereals, legumes, and chronic disease risk reduction: evidence from epidemiologic studies, *Am. J. Clin. Nutr.,* 70, 451S–458S, 1999.
61. Dai, Q., Shu, X.-O., Jin, F., Potter, J.D., Kushi, L.H., Gao, Y.-T., and Zheng, W., Population-based case-control study of soyfood intake and breast cancer risk in Shanghai, *Br. J. Cancer,* 85, 372–378, 2001.
62. Shu, X.-O., Jin, F., Dai, Q., Wen, W., Potter, J.D., Kushi, L.H., Ruan, Z., Gao, Y.-T., and Zheng, W., Soyfood intake during adolescence and subsequent risk of breast cancer among Chinese women, *Cancer Epidemiol. Biomarkers Prev.,* 10, 483–488, 2001.
63. Park, K.-Y., Lim, S.-Y., and Rhee, S.-H., Antimutagenic and anticarcinogenic effects of doenjang, *J. Korean Assoc. Cancer Prev.,* 1, 99–107, 1997.
64. Lee, J.-M., Moon, S.-H., and Park, K.-Y., *In vitro* antimutagenic and anticarcinogenic effect of linoleic acid identified from doenjang, *J. Korean Assoc. Cancer Prev.,* 3, 74–82, 1998.
65. Ko, H.-S., Ju, H.-K., Jung, K.-O., and Park, K.-Y., Antimutagenic effect of chungkookjangs prepared with the different kinds of soybeans, *J. Korean Assoc. Cancer Prev.,* 4, 204–212, 1999.
66. Park, K.-Y., Son, M.-H., Moon, S.-H., and Kim, K.-H., Cancer preventive effects of doenjang *in vitro* and *in vivo:* antimutagenic and *in vivo* antitumor effect of deonjang, *J. Korean Assoc. Cancer Prev.,* 4, 68–78, 1999.
67. Son, M.-H., Moon, S.-H., Choi, J.-W., and Park, K.-Y., Cancer preventive effects of doenjang *in vitro* and *in vivo:* effect of doenjang extracts on the changes of serum and liver enzyme activities in sarcoma-180 transplanted mice, *J. Korean Assoc. Cancer Prev.,* 4, 143–154, 1999.
68. Hong, S.-S., Chung, K.-S., Yoon, K.-D., and Cho, T.-J., Antimutagenic effect of solvent extracts of Korean fermented soybean products, *Foods Biotechnol.,* 5, 263–267, 1996.
69. Anderson, J.W., Johnstone, B.M., and Cook-Newell, M.E., Meta-analysis of the effects of soy protein intake on serum lipids, *N. Engl. J. Med.,* 333, 276–282, 1995.

70. Erdman, J.W.J., Soy protein and cardiovascular disease: a statement for healthcare professionals from the nutrition committee of the AHA, *Circulation,* 102, 2555–2559, 2000.
71. Hodgson, J. M., Puddey, I.B., Beilin, L.J., Mori, T.A., and Croft, K.D., Supplementation with isoflavonoid phytoestrogens does not alter serum lipid concentrations: a randomized controlled trial in humans, *J. Nutr.,* 128, 728–732, 1998.
72. Okamoto, A., Hanagata, H., Matsumoto, E., Kawamura, Y., Koizumi, Y., and Yanagida, Y., Angiotensin-I converting enzyme inhibitory activities of various fermented foods, *Biosci. Biotechnol. Biochem.,* 59, 1147–1149, 1995.
73. Shin, J.-I., Ahn, C.-W., Nam, H.-S., Lee, H.-J., Lee, H.J., and Moon, T.W., Fractionation of angiotensin converting enzyme inhibitory peptide from soybean paste, *Korean J. Food Sci. Technol.,* 27, 230–234, 1995.
74. Hwang, J.-H., Angiotensin I converting enzyme inhibitory effect of doenjang fermented by *B. subtilis* SCB-3 isolated from meju, Korean traditional food, *J. Korean Soc. Food Sci. Nutr.,* 26, 775–783, 1997.
75. Kim, S.-H., Lee, Y.-J., and Kwon, D.Y., Isolation of angiotensin converting enzyme inhibitor from doenjang, *Korean J. Food Sci. Technol.,* 31, 848–854, 1999.
76. Shin, J.-I., Yu, R., Park, S.-A., Chung, D.K., Ahn, C.-W., Nam, H.-S., Kim, K.-S., and Lee, H.J., His-His-Leu, an angiotensin I converting enzyme inhibitory peptide derived from Korean soybean paste, exerts antihypertensive activity *in vivo, J. Agric. Food Chem.,* 49, 3004–3009, 2001.
77. Shon, D.-H., Lee, K.-A., Kim, S.-H., Ahn, C.-W., Nam, H.-S., Lee, H.-J., and Shin, J.-I., Screening of antithrombotic peptides from soybean paste by the microplate method, *Korean J. Food Sci. Technol.,* 28, 684–688, 1996.
78. Cusack, N.J. and Hourani, S.M.O., Platelet P2 receptors: from curiosity to clinical targets, *J. Auton. Nerv. Syst.,* 81, 37–43, 2000.
79. Lamartiniere, C.A., Protection against breast cancer with genistein, a component of soy, *Am. J. Clin. Nutr.,* 71, 1705S–1707S, 2000.
80. Adlercreutz, H., Mazur, W., Bartels, P., Elomaa, V.-V., Watanabe, S., Wahala, K., Landstrom, M., Lundin, E., Bergh, A., Damber, J.-E., Aman, P., Widmark, A., Johansson, A., Zhang, J.-X., and Hallmans, G., Phytoestrogens and prostate disease, *J. Nutr.,* 130, 658S–659S, 2000.
81. Kurzer, M.S., Hormonal effects of soy isoflavones: studies in premenopausal and postmenopausal women, *J. Nutr.,* 130, 660S–661S, 2000.
82. Kim, H., Xia, H., Li, L., and Gewin, J., Attenuation of neurodegeneration-relevant modifications of brain proteins by dietary soy, *Biofactors,* 12, 243–250, 2000.
83. Wang, H.-J. and Murphy, P., Isoflavone content in commercial soybean foods, *J. Agric. Food Chem.,* 42, 1666–1673, 1994.
84. Nakamura, Y., Tsuji, S., and Tonogai, Y., Determination of the levels of isoflavonoids in soybeans and soy-derived foods and estimation of isoflavonoids in the Japanese daily intake, *J. AOAC Int.,* 83, 635–650, 2000.
85. Hutchins, A.M., Slavin, J.L., and Lampe, J.W., Urinary isoflavonoid phytoestrogen and lignan excretion after consumption of fermented and unfermented soy products, *J. Am. Diet. Assoc.* 95, 545–551, 1995.
86. Kim, M.-J., Quantification of Isoflavones and Coumestrol in Korean Soyfoods and Estimations of Their Intake, thesis, Seoul National University, Korea, 2001.

13 The Role of *Lactobacillus plantarum* in Foods and in Human Health

Göran Molin

CONTENTS

0-8493-1372-4/03/$0.00+$1.50

13.1 HISTORY AND CULTURE

Consumption of live lactic acid bacteria (LAB) included in lactic acid fermented foods has been a regular part of the food intake of humans for a long time. In fact, archaeological evidence indicates that mankind has used this technique since prehistoric times.[1,2] This technique was presumably invented 1.5 million years ago by early humanoids (Figure 13.1). Humans have in this way consumed large numbers of live LAB, and presumably those associated with plant material were consumed before those associated with milk-based foods. Lactic acid fermentation is the simplest and often the safest way of preserving food, and before the Industrial Revolution, lactic acid fermentation was applied just as much in Europe as it still is in Africa. Thus, it could very well be that the human gastrointestinal (GI) tract evolved to adapt to a more or less daily supply of live LAB. This supply ceased in industrialized countries during the twentieth century, which might have led to GI problems, and even to immunologically dependent ones. (See Chapter 1 for more details on the history of fermented foods.)

Lactobacillus plantarum frequently occurs spontaneously in high numbers in most lactic acid fermented foods, especially when the food is based on plant material; for example, in brined olives,[6] sauerkraut,[7] salted gherkins,[8] sourdough,[9] Nigerian ogi (made from maize or sorghum),[10] Ethiopian Kocho (made from starch from *Ensete ventricosum*),[11,12] Ethiopian sourdough made from tef (*Eragrostis tef*),[12,13] and cassava.[14,15] Thus, it is obvious that individuals consuming lactic acid fermented products of plant origin also consume large numbers of *Lb. plantarum.*

One example of the importance of lactic acid fermented plant material for indigenous living humans involves the Tschuktscer people living in Siberia on the Tschuktsch peninsula along the shore of the North Polar Sea. They were described by the explorer A.E. Nordenskiöld during his expedition around Asia in his voyages to discover the North-East Passage (1878 to 1880).[16] At that time, this population was a primordial society of hunters and fishermen, and a major component of their diet was lactic acid fermented plant material. During the summer months, the Tschuktscer collected different kinds of plant material, such as the leaves from willow (*Salix*) and *Rhodiola*. After picking, the plant material was pressed into sealskin bags that were sealed and left to spontaneously ferment during the summer months. During the autumn, the contents froze in the shape of the outstretched bags. The frozen mass was cut in pieces and eaten as it was, or together with meat or fish,

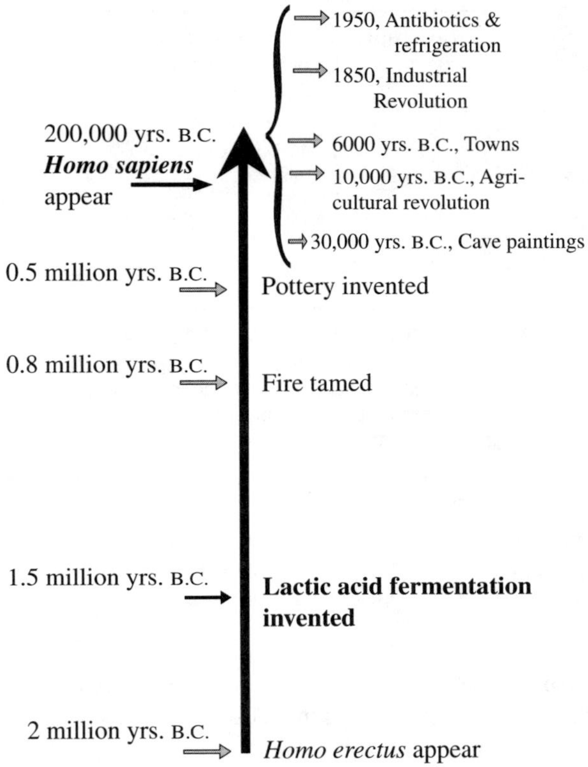

FIGURE 13.1 A suggested time scale for human development, showing how early the technique of lactic acid fermentation probably came into use. (Adapted from Leakey, R., *På Spaning Efter Människans Ursprung,* Natur och Kultur, Stockholm, 1993; Leakey R., *Hur Människan Blev Till,* Natur och Kultur, Stockholm, 1995; Arrhenius, B., Grötfrukost på stenåldern, *Forskning och Framsteg,* 4–8, 1984; Lagerqvist, L.O. and Åberg, N., *Mat och Dryck i Forntid och Medeltid,* Vincent Förlag and Statens Historiska Museum, Stockholm, 1994; Larsson, L., *Ett Fångstsamhälle för 7000 År Sedan,* Signum, Lund, Sweden, 1988. With permission.)

or it could also be used in hot soups. Nordenskiöld speculated that the observed consumption of lactic acid fermented plant material could mimic the way hunters ate during the Stone Age.[16]

In Sweden and in the rest of Northern Europe, lactic acid fermented vegetables have been widely consumed up to modern times. In a Swedish cookbook from 1755, the procedures for making sauerkraut, fermented spinach, and sorrel are described.[17]

Sauerkraut has had a reputation of being healthful for a long time. Captain James Cook, during his sailing trips around the world (1768 to 1780), forced his crew to eat sauerkraut. James Cook became famous not only for his geographic discoveries, but also for the extraordinary record of survival of the seamen on board his ships.[18] *Lb. plantarum* are often spontaneously the dominating bacteria in sauerkraut, and may have been responsible for the good health of Cook's crews.[7] (See Chapter 14 for more details on sauerkraut.)

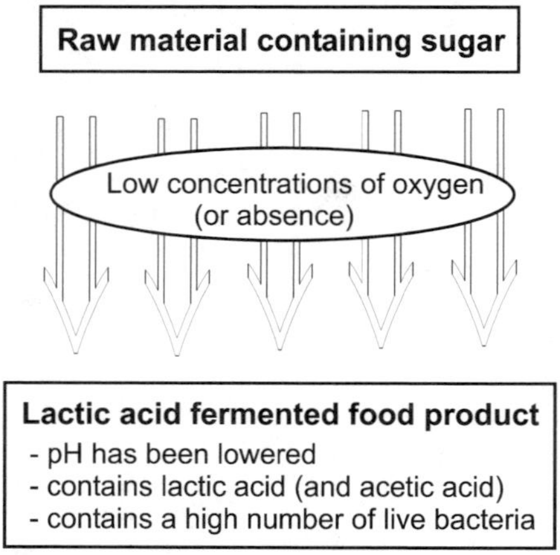

FIGURE 13.2 A schematic representation of lactic acid fermentation of food.

13.2 FOODS FERMENTED WITH *LACTOBACILLUS PLANTARUM*

13.2.1 FACTORS AFFECTING FERMENTATION

Lactic acid fermentation spontaneously occurs as soon as organic matter is enclosed in a limited space where access to oxygen is restricted. Thus, as the microorganisms grow, oxygen is consumed and carbon dioxide is produced. This change in gas atmosphere is the first environmental factor to control the microflora in favor of LAB. Production of organic acids and decreasing pH add to the altered gas atmosphere and become critical for microbial control. In addition to these major environmental controlling mechanisms, antimicrobial compounds besides organic acids, such as hydrogen peroxide,[19,20] nitrogen oxide,[21,22] or antimicrobial proteins or peptides,[23] can be produced by the LAB. The principle of a spontaneous lactic acid fermentation is shown in Figure 13.2.

In order to enhance the selection pressure of a spontaneous lactic acid fermentation, salt can be added or the water activity (a_w) can be decreased ($a_w = p/p_0$, where p = steam pressure over the product; p_0 = vapor pressure over pure water). The water activity over pure water is 1.0, and the relative humidity (RH) in percent, in the gas phase, is the same as $a_w \times 100$.

A more sophisticated way to improve the control of the lactic acid fermentation is to add a starter culture, either as a pure culture, which is the modern industrial method, or by adding some material from a previously produced product (back slopping), which is the more traditional method. The original purpose of performing a lactic acid fermentation was to increase the shelf life of the product. Empirically, humanity has learned that this is a safe way of preserving food, and also that the nutritional quality of the product persists for a considerable time and may even improve in some aspects. Lactic acid fermentation sometimes also has other advantages such as improving the taste and consistency of the product. In addition, it has also been recognized that beneficial health effects may result from consumption of the live LAB.

The method of pressing down plant material in an airtight bag of sealskin for spontaneous lactic acid fermentation as was done by the Tschuktscer people (see previous discussion) is a simple and efficient technique with the prime purpose of preserving the material for later consumption. It has a modern equivalent in the silage technique using large plastic bags that is much favored by Swedish farmers. *Lb. plantarum* frequently occurs in large numbers in silage;[23–25] presumably the fermented product of the Tschuktscer people also contained high numbers of *Lb. plantarum.*

An even simpler technique than tight bags is to make a hole in the ground into which the fermentable material is pressed. Such an application was used in the ancient times in the north of Sweden for the preservation of salmon — the so called "Gravad lax" (buried salmon). During the summer, fresh-caught salmon were salted and buried on the sandy riverbanks at the mouths of rivers and left to spontaneously ferment. The sour salmon could then be retrieved for consumption during the winter season. An example of the same technique of burying material for fermentation is still seen in Ethiopia today, where people make a lactic acid fermented product, called Kocho, containing large numbers of live *Lb. plantarum.* A slightly more advanced technique is traditionally used for the preservation of cucumbers. Instead of a hole in the ground, large open containers standing outdoors are used. The lactic acid fermentation is controlled with salt, and the result is salted gherkins. However, in a product such as brined olives, and even more so in products such as sourdough and ogi, the main purpose of the fermentation is not only to improve the shelf life, but to improve the eating and nutritional qualities of the food. In recent times, a new purpose has been added and that is to make use of the health beneficial effects of live LAB, as is done in the product ProViva.®

13.2.2 Ethiopian Kocho

Ensete ventricosum is a perennial, banana-like, starchy root crop that grows in Ethiopia at altitudes of 1500 to 3000 m above sea level. The height of the plant can reach 6 to 8 m when the plant is harvested, 6 to 8 years after planting. It is a leading staple crop for about 10 million inhabitants.[12] The plant is processed as follows. The pseudostem and corm are pulverized with a long wooden pestle and pounded into a pulp from which the fibers are removed. The remaining scrapings, the pulp, and the inner corm are kneaded together, rolled into balls, and wrapped in fresh enset leaves. The leaf packages with fresh enset mash are packed into a pit in the ground that has been completely lined with leaves, and left for prefermentation for 2 to 5 days.[11,12] After this prefermentation, the packages are opened, and the mash is once again mixed and thoroughly kneaded, rolled into balls, and wrapped in new fresh enset leaves. The packages are then pressed by hands and feet into the pit. Some of the waste mash and cellulosic material from the production are put on top to create a cover, and heavy stones are put on top of this cover. The aim is to limit the access of air into the pit. The major fermentation takes about 2 weeks at a temperature of 15 to 18°C (the same temperature range used for the fermentation of sauerkraut[26]), but the Kocho can be left in the closed pit for periods of time from 6 months up to years (in the colder regions).

In the colder regions of Ethiopia, the general belief is that the quality of Kocho improves with storage time. In the warmer regions, the product becomes excessively

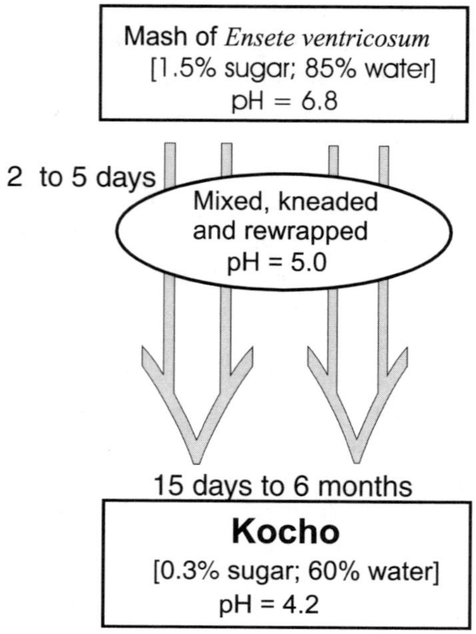

FIGURE 13.3 Flow scheme of the production of Ethiopian Kocho.

sour and discolored if it is left too long.[11] Kocho is mainly baked into bread or cooked and eaten alone or in combination with various indigenous foods. High quality Kocho or variants of Kocho are also eaten unheated and are considered by the public to possess beneficial health effects. The production of Kocho is schematically shown in Figure 13.3.

In the beginning of the fermentation (first days), the fermentation is dominated by *Leuconostoc mesenteroides,* but after approximately a week, *Lactobacillus* spp. reach similar numbers.[11] After 2 weeks, both *Lactobacillus* and *Leuconostoc* are present in large numbers (about 5×10^9 colony forming units (CFU)/g). The viable count of *Lactobacillus* remains high for months, while the count of *Leuconostoc* rapidly declines. The same sequential pattern for *Leuconostoc* and *Lactobacillus* has been described for both sauerkraut and salted gherkins.[26] The pH may be a controlling factor. *Leuconostoc* is less resistant than *Lactobacillus* to low pH, and the species *Lb. plantarum* is especially hardy at low pH.[23,27]

The dominating LAB in Kocho bought in the market were *Lb. plantarum, Weissella minor,* and *Pediococcus pentosaceus.*[12] Gashe[11] also found that *Lb. plantarum* was one of the dominating species of lactobacilli.

13.2.3 Salted Gherkins

Cucumbers contain (on a percentage basis) a higher proportion of water than the Mediterranean Sea and hence are susceptible to microbial spoilage. They are harvested during a short season in the autumn and are used mainly for the production of different mixtures of pickles. Only a relatively small proportion is sold as salted gherkins, the traditional lactic acid fermented product. In Sweden, the production of salted gherkins to be used in the food industry as raw material for pickles is around 2000 tons per year, but it can be up to 7000 tons depending on the harvest. Production in the United States is at least 100 times higher.

The cucumbers (*Cucumis sativis*) used for the lactic acid fermentation become heavily contaminated by microorganisms from the soil in the field during growth. A typical aerobic viable bacterial count can be 5×10^6 CFU/g; the count of Enterobacteriaceae can be 1×10^6 CFU/g, while the count of lactobacilli is only 5×10^3 CFU/g. Thus, the odds of successful preservation by spontaneous lactic acid fermentation apparently seem low. But, by the use of salt (NaCl), it is feasible to control the fermentation and obtain a product with a long shelf life. The level of salt used is higher than for the fermentation of cabbage, but otherwise the process is similar. The steps of a traditional process are depicted in Figure 13.4. In Sweden, the cucumber is put into a brine to achieve a final concentration of 5% NaCl. Traditionally, this occurs in open containers of wood holding 25 tons of cucumber each, situated outdoors. After fermentation, the container is covered by a tarpaulin. In warmer countries, the salt concentration is usually higher during fermentation (up to 8% NaCl), and after the completed fermentation, the salt concentration is increased from 8% to 16% NaCl to ensure a long shelf life of the product. In this way, salted gherkins can be stored for at least one year without problems.[26] In commercial processing plants, the containers have been modernized, and the cucumbers are purged with nitrogen (or air, which is less expensive) to displace accumulating carbon dioxide. Carbon dioxide can form pockets inside the cucumbers (so-called bloater formation). Control over the fermentation can also be achieved by acidifying the cucumbers with acetic acid (pH = 2.8) for a few days before starting the fermentation

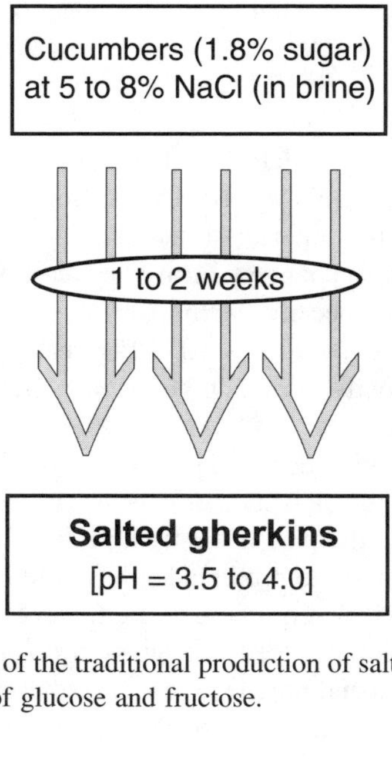

FIGURE 13.4 Flow scheme of the traditional production of salted gherkins. The sugar in the cucumbers consists mainly of glucose and fructose.

by increasing the pH to 4.6. In such a process, it can also be favorable to use a starter culture.

Under conditions where the salt concentration in the brine is not too high, the first LAB to increase to a dominating position are leuconostocs, and as the fermentation proceeds they will be succeeded by lactobacilli and pediococci; this is generally the same succession as has been seen in Ethiopian Kocho[11] and in sauerkraut.[26] A typical bacterial species found in high numbers at the end of fermentation is *Lb. plantarum,* both in salted gherkins and in sauerkraut.[26] However, when used as a starter culture, *Lb. plantarum* can cause gas pockets of carbon dioxide to form in the cucumbers.[28] Another problem is that lactic acid can be produced by malic acid fermentation.[29] A mutant strain of *Lb. plantarum* lacking the ability to ferment malic acid has been developed[30,31] and has been made commercially available for cucumber fermentation.[23]

13.2.4 Green Olives in Brine

The lactic acid fermentation of green olives (Spanish-style green olives) has major similarities to cucumber fermentation. After pretreatment with lye (1.3 to 2.6% [w/w] of NaOH) for 5 to 7 h to hydrolyze and remove some of the bitter-tasting phenolic compounds (mostly ortho-diphenols and their glucosides as oleuropein),[26] the olives are put in a brine solution and subjected to a spontaneous lactic acid fermentation. The production procedure is schematically shown in Figure 13.5. After the fermentation is finished, the salt in the brine is increased to 8% (w/w) to ensure the keeping qualities of the olives for extended storage periods.

Lb. plantarum is normally found in high numbers at the end of the fermentation.[32] These bacteria are thought to be coexisting with a yeast flora.[9] The yeasts are believed to release B vitamins that are utilized by the lactobacilli.[33,34]

13.2.5 Sourdough

Sourdough contains a mixed population of LAB and yeasts. Originally, the use of sourdough was a way of providing yeast for breadmaking. Baker's yeast was not generally available before the nineteenth century. In the early nineteenth century, the beermaking process was changed from top yeast to bottom yeast, and the production of baker's yeast became significant. A sourdough normally contains around 10^9 CFU LAB/g and 10^6 to 5×10^7 CFU yeasts/g,[9,35] where *Lactobacillus* normally is the most frequently occurring genus of LAB[36] and *Lb. plantarum* often are isolated in large numbers.[9,37]

The lactic and acetic acid added to the bread by the lactobacilli improve flavor and can also provide beneficial health effects to the consumer by reducing the glycemic index by reducing the gastric emptying rate[38,39] or by reducing the rate of starch digestion.[40] Furthermore, in rye breads, the acidification by the sourdough is essential for the technological requirements of the bread (comprising >20% rye flour) because the water-soluble proteins of rye flour do not form gluten. The dough structure relies on pentosans and mucilage, the contributions of which are enhanced in an acid environment.[9] Thus, the sourdough improves the water holding capacity

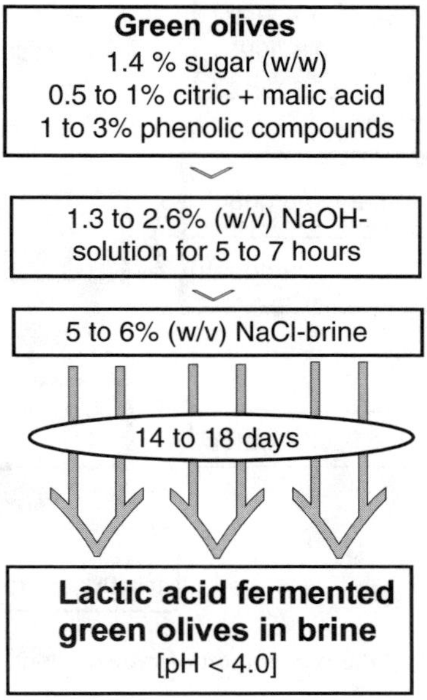

FIGURE 13.5 Flow scheme of the traditional production of green olives in brine.

of the starch and of the pentose containing polymers of the bread. It can also improve the shelf life[9] and increase the availability of minerals by enhancing the breakdown of phytic acid during the breadmaking process.[9,37] Phytic acids reduce the bioavailability of minerals by forming strong complexes.[41] The phytic acids can be broken down both by the microflora (especially by yeasts) and by indigenous phytases of the cereals, which can be activated by pH lowering.[42]

A typical procedure for the production of rye sourdough is illustrated in Figure 13.6. Different options exist for inoculating a sourdough to be made for mixing into bread dough (Figure 13.6). A new sourdough can be up-started by mixing rye flour with water; after 2 days at 30°C a primary sourdough will have formed (Figure 13.6). However, the quality of this sourdough will not be optimal due to high numbers of Enterobacteriaceae that are already present in the flour.[37] To improve the quality of the sourdough, it has to be reused; part of the sourdough is mixed with new flour and water. As this procedure is repeated, the quality gradually improves. Not all types of cereals can be used for starting a sourdough. Wheat flour will not have enough selective power to control the lactic acid fermentation; it will allow bacterial groups other than LAB to flourish, which will disturb the sourdough development.[9] It can be speculated that the high content of phytic acid in rye provides a selection pressure. Phytic acid binds iron, but this is not a problem for organisms such as *Lb. plantarum* that have no requirement for iron.[43,44] Thus, when sourdough is used in the making of wheat bread, the rye sourdough must gradually be exchanged with wheat flour by recycling the sourdough.

An alternative to inoculating with old sourdough is to add a starter culture containing, for example, a suitable *Lb. plantarum* strain, directly to the mixture of flour and water (Figure 13.6).

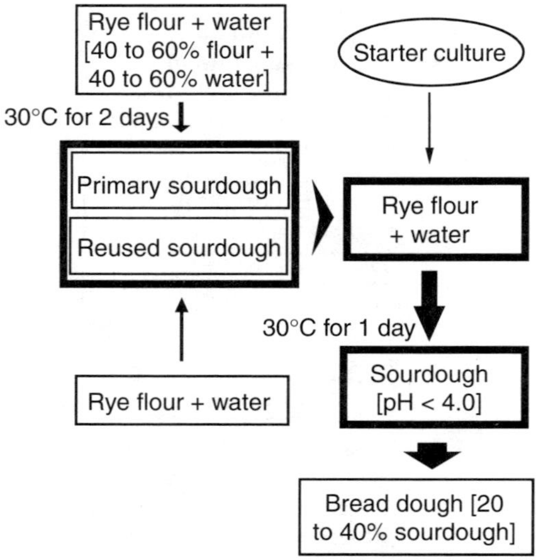

FIGURE 13.6 Flow scheme of the preparation of rye sourdough and rye bread dough.

The sourdough technique is used in many parts of the world. One example is the tef-based Injera-bread from Ethiopia. Injera, one of the major foods in Ethiopia, is normally prepared from cereals such as tef, wheat, barley, sorghum, maize, millet, or combinations.[12] However, injera from tef (*Eragrositis tef* of the grass family *Poaceae*) is the most popular. Tef flour is mixed with water (1 kg flour per 1.8 l water), inoculated with a part of an old sourdough, and left in a jar at 18 to 21°C for 2 to 4 days, where the pH drops from 6.7 to below 4.0.[12,13] It has been shown that antinutrients such as phytate and tannins are reduced during the fermentation period.[45] From the microbial point of view, the lactic acid fermentations of rye and tef sourdoughs have much in common; *Lb. plantarum* frequently occurs in high numbers in tef sourdough.[12]

13.2.6 Nigerian Ogi

Ogi is a traditional lactic acid fermented cereal-based product from Nigeria. It is used as a weaning food for children (gruel) but is also widely consumed by adults as porridge at, for example, breakfast, or as a cooked stiff gel (agidi) eaten together with stews, soups, or fried bean cakes. Ogi can be made from maize, sorghum, or millet, but the most popular type is made from maize (*Zea maize*) and sorghum (*Sorghum bicolor* or *Sorghum dabar*). Maize is frequently used for the ogi-gel (agidi), while the red sorghum (*Sorghum bicolor*) based food often is preferred as a weaning food. Red sorghum generally gives a gruel of lower viscosity than maize does, which is advantageous when the consumers are young children, as the bulking effect of maize will reduce intake and increase the risk of malnutrition. Paradoxically, the content of tannins is high in red sorghum, and tannins can react with proteins to make them indigestible in the gut.[46–48] However, the protein digestibility of high-tannin sorghum was significantly improved by lactic acid fermentation.[42]

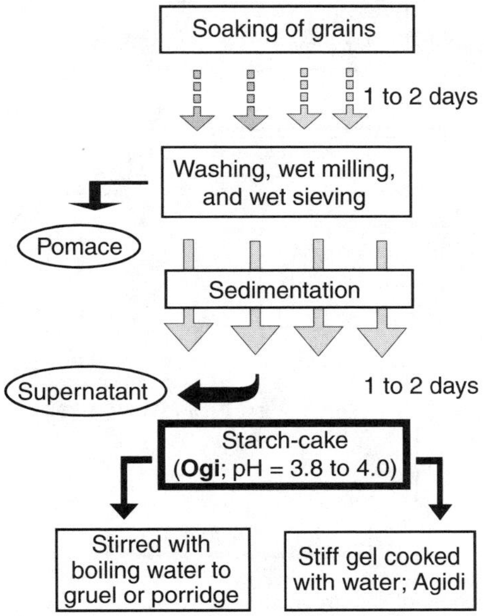

FIGURE 13.7 Flow scheme of the traditional wet-milling procedure for ogi production.

Interestingly, *Lb. plantarum* is able to degrade tannins.[49] *Lactobacillus plantarum* is frequently a dominant part of the bacterial flora in ogi.[50] Furthermore, it has been shown that *Lb. plantarum* could be used as a single-strain starter for producing high quality ogi.[10,51]

Ogi is traditionally produced by a labor-intensive procedure (Figure 13.7). The cereal grains are cleaned and steeped in water for 1 to 3 days, and the first spontaneous fermentation occurs. After the water is poured off, the grains are wet-milled and wet-sieved through a muslin cloth or a fine wire mesh.[10,51] The pomace, mostly consisting of hulls, is discarded and usually used for animal feed. The remaining flour suspension is left for sedimentation for 1 to 3 days. During this step, a spontaneous lactic acid fermentation occurs. When the ogi is sour enough, the supernatant is decanted, and the flour cake is stirred with boiling water or with the decanted supernatant to form a gruel or a porridge. The ogi can also be cooked in water into a thick gel (agidi) that is put in leaf packages.

13.2.7 Tanzanian Togwa

Togwa is a cereal-based, lactic acid fermented beverage, frequently consumed in Tanzania by all age groups as a refreshment and by infants as a weaning food.[52,53] As part of the daily diet, children can be fed togwa two to three times a day, and sometimes children as young as three months old are given togwa. The consumption of togwa as a beverage depends on the age of the consumer. It can be taken by adults 2 to 3 times a day (0.3 to 0.7 l per occasion), especially during the dry and hot season. Children are given 0.2 to 0.5 l of togwa per feeding.[53] The togwa is mostly made from flour of maize (*Zea mays*), sorghum (*Sorghum bicolour*) or finger millet

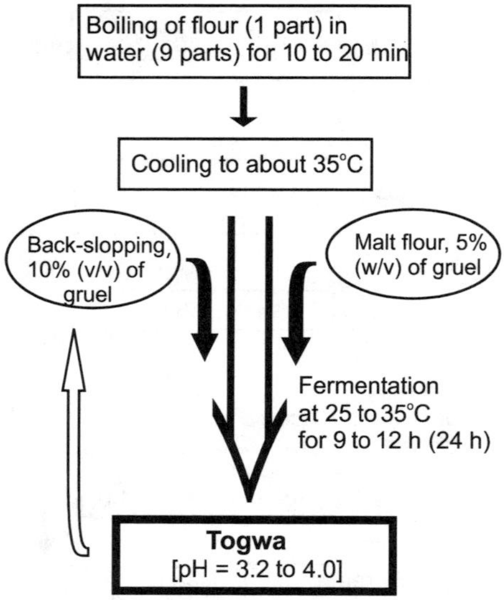

FIGURE 13.8 Flow scheme of the production of Tanzanian togwa.

(*Eleusine coracana*). Rice (*Oryza sativa*) and cassava (*Manihot esculenta*) flour or mixtures of cassava and cereals are also used in some areas.[53] It seems that sorghum-based togwa is preferred by many consumers. The different steps in the preparation of togwa are summarized in Figure 13.8. The flour is mixed with water (1 part flour and 9 parts of water) and cooked for 10 to 20 minutes. The gruel is then cooled down to about 35°C, and 10% (v/v) old togwa (back slopping) is mixed into the gruel together with 5% (w/v) malt flour.[52,53]

Malt flour is prepared from sorghum or millet. The malt flour is prepared by soaking the cereal grains of choice for 12 h; then they are drain-dried, spread on broad leaves, and covered with new leaves.[52] A more modern alternative is to use winnowing trays, jute mats, or aluminum trays and to cover the grains with wet cloths. The grains are allowed to germinate at about 30°C for 3 to 6 days until the roots and plumule become pinkish. After sun drying and milling of the germinated grains, the malt flour is ready for use.[52] The malting procedure reduces up to 99% of the tannins in the sorghum.[54] Presumably, the supplementation of malt flour also leads to a decrease in tannins during the fermentation (Figure 13.8).

After mixing the heat-treated gruel with malt flour and old togwa (as a starter culture), the mixture is left for 9 to 24 h (usually 10 to 12 h) for lactic acid fermentation to produce togwa (Figure 13.8). The fermentation can be performed in gourd bottles, covered earthenware pots, aluminum pots, or plastic containers.[53] Togwa has a pH of 3.2 to 4.0, mainly due to lactic acid formed during the fermentation, and contains high levels of *Lb. plantarum*.[53] It has been shown that togwa made using *Lb. plantarum* as a single-strain starter culture equals spontaneously fermented togwa in quality.[53]

In spite of the fact that togwa is often produced under poor hygienic conditions, where inferior water quality and the malting procedure expose the product to considerable hygienic obstacles, there have been no documented outbreaks of foodborne disease connected to togwa.[53] Furthermore, Mugula,[53] after interviewing mothers, did not find any indication of togwa being incriminated in food poisoning. It has also been demonstrated that enteropathogens (*Bacillus cereus, Campylobacter jejuni*, enterotoxigenic *Escherichia coli, Salmonella typhimurium*, and *Shigella flexneri*) inoculated before fermentation disappear after 24 h in the fermenting gruel, provided that the pH during this 24 h period has fallen to ≤4.0, which is the normal pH fall in togwa.[55,56] The fermenting gruel also inhibited enterotoxin production by *C. jejuni* and *E. coli* and even inactivated pure cholera toxin when they were added to the fermenting gruel.[56] It is thus obvious that the traditional lactic acid fermentation is a very safe process from a food hygiene point of view.

Togwa can lower the incidence of enteropathogens in feces of children,[57] but it has also been shown to improve the condition of the intestinal mucosa in children with acute diarrhea as shown by measurements of intestinal permeability to lactulose and mannitol.[58] Under pathological conditions, intestinal permeability to larger sugars increases while that to smaller ones stays the same or decreases.

13.2.8 Swedish ProViva

The probiotic strain *Lb. plantarum* 299v (DSM 9843) is included in a European "functional food" product with the brand name ProViva.[59–61] This *Lb. plantarum* strain was isolated from healthy human intestinal mucosa[62] and is patented. ProViva is a fruit beverage that today is marketed in Sweden, Finland, Denmark, and the United Kingdom. In Germany, it is marketed under the brand name PrimaVita,® and in Belgium, ProVie.® ProViva is produced and marketed in Scandinavia and the United Kingdom by Skåne Dairies (Malmö, Sweden), while the holder of the rights to the strain, *Lb. plantarum* 299v, is Probi AB (Lund, Sweden). Probi AB licenses the rights to use *Lb. plantarum* 299v to Skåne Dairies. This strain is also included in a low-fat ice cream, God Hälsa,® that is marketed by the SIA Ice Cream AB (Slöinge, Sweden).

The lactic acid fermented component in the drink ProViva and in the ice cream God Hälsa is an oatmeal gruel that has been fermented with *Lb. plantarum* 299v. The lactic acid fermentation produces about 1×10^9 CFU of *Lb. plantarum* 299v/ml of oatmeal gruel. This fermented oatmeal formula was originally developed as a new concept for enteral feeding (nasogastric feeding).[62] The lactic acid fermented oatmeal gruel is an integral part of ProViva, where 5% fermented oatmeal gruel has been mixed with different types of fruit drinks, including rose hip, strawberry, blueberry, black currant, and tropical fruits. In the final product (ProViva), about 5×10^7 CFU of *Lb. plantarum* 299v are present in each milliliter of fruit drink.

The ice cream God Hälsa contains $>5 \times 10^7$ CFU of *Lb. plantarum* 299v per gram of ice cream. Similar to ProViva, the carrier of the bacteria is the fermented oatmeal gruel. However, the water content of the oatmeal gruel used in God Hälsa is only about half of that of the oatmeal gruel used in ProViva. The ice cream God Hälsa contains 20% (w/w) oatmeal gruel.

The patented process to produce the lactic acid fermented oatmeal gruel is schematically shown in Figure 13.9. It is interesting to note that the general proce-

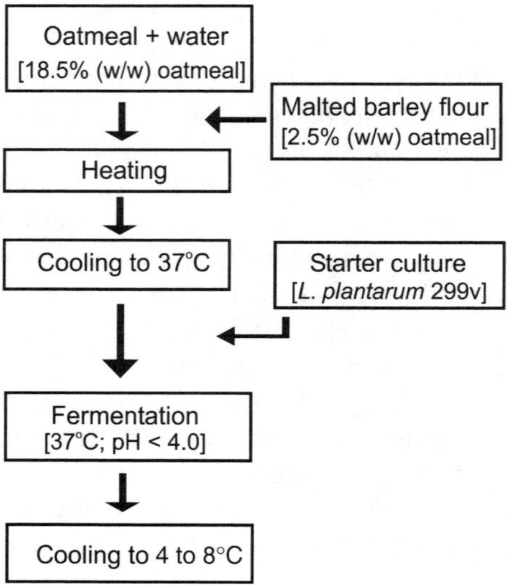

FIGURE 13.9 Flow scheme of the production of Swedish lactic acid fermented oatmeal gruel to be used in probiotic formulas.

dure of the industrially adopted Swedish process of producing lactic acid fermented oatmeal gruel has some important features in common with the traditionally produced Tanzanian togwa. Both are cereal based, lactic acid fermented gruels of relatively low viscosity, where the consumer is eating high levels of live *Lb. plantarum* (Figure 13.8). The viscosity of the products is lowered by supplemented malt flour (malt flour of barley in Sweden and malt of sorghum or millet in Tanzania) in combination with a heat treatment, followed by the decreased pH in the lactic acid fermentation. In the Swedish product that originally was intended as a base for a nutritional formula for enteral feeding, the low viscosity and high energy content of the liquid were prerequisites.[63] Without added malt flour, oatmeal gruel of the stated concentration of flour (18.5%; w/w) will form a thick porridge impossible to administer through a thin tube.[63–65] The decrease in viscosity is presumably in large part due to degradation of starch. Malt is rich in amylases. There is also an increased solubility of β-glucans, and if larger amounts of malt are used or extra malt flour is added after the heat treatment, the total amount of β-glucans is substantially reduced.[64,65] However, the β-glucans are considered valuable, as they are believed to delay intestinal absorption and beneficially affect cholesterol and glucose metabolism. The process shown in Figure 13.9 causes a relatively small, if any, reduction of the total content of β-glucans, even if the viscosity is significantly affected.

The lactic acid fermented oatmeal gruel (Figure 13.9) provides about 76% of the energy and 70 and 99% of the protein and carbohydrate content, respectively, compared to the average nutrient content in commercial nutritive solutions intended for enteral feeding.[64] The gruel is also relatively rich in β-glucans, thiamin, phosphorus, iron, copper, and manganese.[64] The amount of free amino acids in the lactic acid oatmeal gruel is affected by the malt and by the activity of *Lb. plantarum* 299v. It has been shown for *Lb. plantarum* strains genetically closely related to *Lb.*

plantarum 299v that the concentrations of aspartic acid, asparagine, and alanine decrease during fermentation, while the concentrations of glutamic acid, proline, glycine, and arginine increase.[65]

A drawback of oats from a nutrition point of view is that oats contain large amounts of phytate (myoinositol hexaphosphate), which is one of the main inhibitors of absorption of iron and zinc in humans.[66] Even small amounts of phytate in foods have a strong negative effect on the absorption of iron.[67] However, degradation of phytate during food processing can be accomplished by activation of intrinsic phytases in cereals. In fact, phytate can be completely degraded in wheat and rye, but to a lesser extent in oats by soaking at low pH (pH 5) and increased temperature (55°C).[68] In the oatmeal gruel fermented with *Lb. plantarum* 299v, the phytate concentration is somewhat reduced with the help of the supplemented phytases from the malt flour and the decrease in pH during fermentation at a temperature considered near optimum for the oat intrinsic phytases.[65,69] Due to the high stability of the oat phytate, substantial amounts remain in the fermented gruel. The level of inositol hexaphosphate is around 12 μmol/g.[65] There are no data to indicate that *Lb. plantarum* 299v possesses phytase activity.

13.3 THE SPECIES *LACTOBACILLUS PLANTARUM*

13.3.1 Systematics

13.3.1.1 Lactic Acid Bacteria

The organisms performing the conversion of carbohydrates to carboxylic acids, mainly lactic acid, are by tradition called LAB. The term was used early by food microbiologists, and by 1919 the Danish bacteriologist Orla Jensen tried to define key features of LAB as follows: "The true lactic acid bacteria form a large natural group of nonmotile, non-spore-formers, Gram-positive cocci and rods that at fermentation of sugar mainly produce lactic acid." Based on definitions such as this, different systematically defined taxa have been included in the group LAB. However, LAB is not a systematically defined group based on evolutionary relationships. It is a functional classification that food microbiologists apply to bacteria, harmless to both food quality and human health, that occur spontaneously in traditional lactic acid fermented foods. From the systematic point of view, this means a relatively wide variety of taxa. How many genera and species should be included depends much on how many different types of foods are included and how strict the quality definitions are set for these food products. For example, the higher the eating quality of a lactic acid fermented food product is, the fewer types of bacteria can generally be involved in the final fermentation. In a product of poorer quality, all types of unwanted organisms can be present in high numbers in the final product. The only absolute condition for the organisms involved in lactic acid fermentation must be that they produce lactic acid and that they are harmless to consume in high numbers, even for consumers with underlying illnesses that weaken their immunological defense. The taxa frequently occurring in high numbers in traditional and spontaneously fermenting lactic acid fermented foods are *Lactobacillus, Pediococcus, Weis-*

sella, Leuconostoc, Oenococcus, Lactococcus, and *Streptococcus (thermophilus).* The genera *Lactobacillus, Pediococcus, Leuconostoc, Weissella,* and *Oenococcus* have a relatively close phylogenetic relationship and might all be included in the trivial expression "lactobacilli." However, *Lactococcus* and *S. thermophilus* have, from a phylogenic point of view, nothing in common with the lactobacilli other than being members of the same general branch of evolution, i.e., the phylum (or division) of Gram-positive bacteria having a low ratio of guanine and cytosine in their genomes.

13.3.1.2 Phylogenetic Relationships

Lactobacillus plantarum is a bacterial species in the huge and relatively diverse genus *Lactobacillus,* which comprises more than 70 validly named species. The DNA guanine plus cytosine (G+C) content of the different species ranges from 32 to 54 mol%, which is about twice as large a range as that normally accepted for a well defined genus.[70,71,72] By tradition, the *Lactobacillus* spp. have been divided into three groups depending on their fermentation abilities; the obligate homofermentatives (group I), the facultative heterofermentatives (group II), and the obligate heterofermentatives (group III).[73] Group I species ferment hexoses exclusively to lactic acid by the fructose-1,6-biphosphate pathway but cannot ferment gluconate or pentoses, while group II species can ferment pentoses and/or gluconate. Group III species ferment hexoses to lactic acid, acetic acid and/or ethanol, and carbon dioxide. *Lactobacillus plantarum* is facultatively heterofermentative, while the type species of *Lactobacillus, Lb. delbrueckii,* is obligately homofermentative. The mol% G+C of *Lb. plantarum* is 44 to 46%, while it is 49 to 51 for *Lb. delbrueckii.*[73] Paradoxically, the mol% G+C for some of the other well known obligately homofermentative species, such as *Lb. acidophilus*, *Lb. crispatus*, *Lb. jensenii,* and *Lb. gasseri,* are 32 to 37, 35 to 38, 35 to 37, and 33 to 35%, respectively.[73] It is obvious that major genomic differences exist between *Lb. plantarum* and many of the obligately homofermentative species. *Lb. plantarum* has also a significantly larger genome than, for example, *Lb. acidophilus.*

More recent taxonomic efforts to understand the phylogenetic relationships have been directed towards comparative analysis of 16S (and/or 23S) ribosomal ribonucleic acid (rRNA) gene sequences. *Lactobacillus* is a genus included in the so-called *Clostridium* phylum; Gram-positives with a mol% G+C less than 54%. These analyses divide *Lactobacillus* into several subgroups but also point out that several of the facultatively heterofermentative *Lactobacillus* spp. are related to *Pediococcus* spp.[72,74–76] The *Lactobacillus* spp. were divided phylogenetically into three groups that were not altogether in agreement with the traditionally, phenotypically based subgroups (fermentation groups). Thus, many of the obligate homofoermentatives (for example, *Lb. delbrueckii, Lb. acidophilus, Lb. crispatus, Lb. jensenii,* and *Lb. gasseri*) formed one subgroup. This rRNA-group was called the "*Lb. delbrueckii* group." The second group was formed by more than 30 *Lactobacillus* spp., including *Lactobacillus* spp. of all three fermentation groups, and also some *Pediococcus* spp., and called the "*Lb. casei* group." The third group, the so-called "*Leuconostoc* group," included *Leuconostoc* spp., some obligately heterofermentative *Lactobacillus* spp. and *Weissella* spp.[72,74–76] *Lactobacillus plantarum* was included in the *Lb. casei* group.

The genus *Lactobacillus* was further subdivided[77] and is now in five phylogenetic rRNA groups:

1. The *Lb. acidophilus* group (*Lb. delbrueckii* was considered atypical for the obligate homofermentatives due to its high mol% G+C)
2. The *Lb. salivarius* group
3. The *Lb. reuteri* group
4. The *Lb. buchneri* group
5. The *Lb. plantarum* group

Surprisingly, the *Lb. plantarum* group included the obligately homofermentative *Lb. farciminis,* the facultatively heterofermentative *Lb. alimentarius,* and the obligately heterofermentative *Lb. collinoides.*[77] None of these species has in the past been regarded as *Lb. plantarum*-like. However, more closely related to *Lb. plantarum*, and definitely most *Lb. plantarum*-like, are *Lb. pentosus*[74,78] and *Lb. paraplantarum.*[79] These two species and *Lb. plantarum* not only have high similarities in the 16S rRNA gene, they have also phenotypical similarities. All also have a cell wall peptidoglycan of the *m*-diaminopalemic-direct type[79] which is not the most common type among *Lactobacillus. Lactobacillus agilis* also has this cell wall type and shows phenotypic similarities to *Lb. plantarum.*[80] The majority of *Lactobacillus* spp. have the peptidoglycan type, L-Lys-D-Asp.

13.3.1.3 Diagnostic Features

Key features of *Lb. plantarum,* according to *Bergey's Manual of Determinative Bacteriology,*[73] are rod-shaped cells, growth at 15 but not at 45°C, cell walls containing teichoic acid, cell wall peptidoglycan of the *m*-diaminopalemic-direct type, production of both isomers of lactic acid (DL lactic acid), inability to produce NH_3 from arginine, utilization of pentoses by the induction of phosphoketolase, and mol% G+C = 44 to 46%. The type strain of *Lb. plantarum* is ATCC 14917.[73] The ability of different *Lb. plantarum* strains to ferment different carbohydrates at 37°C in the API 50CH test kit is shown in Table 13.1. *Lb. plantarum* possess a striking ability to ferment many different carbohydrates. In view of its use for fermentation of cereals, it is also interesting to note that some strains are able to ferment starch.[50]

The phenotype of *Lb. plantarum* can be extremely heterogeneous, which may hamper a phenotypic identification.[50,81] However, ribotyping of the 16S rRNA genes, i.e., restriction fragment length polymorphism (RFLP) of the 16S rRNA genes, is extremely homogeneous within the species *Lb. plantarum*, irrespective of the phenotype and source of isolation.[82] Thus, a genomic identification is much more precise and must be strongly recommended.

Subgrouping of strains/isolates of *Lb. plantarum* below the hierarchical level of the species can conveniently be performed with the polymerase chain reaction (PCR)-based technique of randomly amplified polymorphic DNA (RAPD).[83] However, if individual strains are to be identified, restriction endonuclease analysis (REA) of total chromosomal DNA is recommended. By use of relatively frequently cutting restriction enzymes, such as *Eco*RI and *Cla*I, and traditional agarose gel electro-

TABLE 13.1
Percentage of *Lb. plantarum* Strains Able to Ferment Different Carbohydrates in the API 50CH Test Kit[a] at 37°C

Carbohydrate	% Positive *Lb. plantarum* Strains[b]
Ribose, mannose, galactose, glucose, fructose, mannitol, *N*-acetyl-glucosamine, amygdalin, arbutin, aesculin, salicin, cellobiose, maltose, lactose, melibiose, sucrose, gentiobiose	100
L-Arabinose	75
D-Xylose	15
Rhamnose	5
Inositol	5
Sorbitol	80
α-Methyl-D-mannoside	80
α-Methyl-D-glucoside	5
Trehalose	95
Melezitose	70
Raffinose	50
Starch	30
Glycogen	15
D-Turanose	75
Gluconate	85
Glycerol, erythriol, D-arabinose, L-xylose, adonitol, β-methyl-D-xyloside, sorbose, dulcitol, xylitol, D-lyxose, D-tagatose, D-fucose, L-fucose, L-arabitol, 2-keto-gluconate, 5-keto-gluconate	0

[a] API systems, S.A., Montalieu Versieu, France.

[b] Twenty tested *Lb. plantarum* strains isolated from human intestinal sites and traditional lactic acid fermented foods.

Source: Data from Johansson, M.-L., Molin, G., Pettersson, B., Uhlén, M., and Ahrné, S., *J. Appl. Bacteriol.*, 79, 536–541, 1995. With permission.

phoresis, a reproducible identification with high resolution capacity can be obtained.[84] This method was successfully used for reisolation of *Lb. plantarum* strains 299 and 299v from mucosal biopsies obtained in administration studies in humans.[85]

13.3.2 Physiology and Ecology

13.3.2.1 Ecological Niches

Lactobacillus plantarum is often the dominant *Lactobacillus* sp. in traditional lactic acid fermented foods based on plant material, as has been described previously in this chapter. The fermented foods can be inoculated with *Lb. plantarum* directly from the plant, as *Lb. plantarum* can be present in low numbers (less than 10 CFU/g of plant material[86]) on living plants. However, the plant material can also be inoculated with lactobacilli originating from animals and from the humans preparing the product for fermentation. *Lb. plantarum* is often present on the human mucosa, from the mouth to the rectum,[62,87] and is also present in the GI tracts of several domestic animals, such as dogs, pigs, and horses. *Lb. plantarum* may even be present in insects, spiders, and snails. Thus, *Lb. plantarum* may cyclically change its environment from the human or animal intestinal tract, via plants and lactic acid fermented foods, back to the mouth and the intestinal tract of humans and animals. This behavior places high demands on the adaptability and competitiveness of the organism and may be one reason for the relatively large genome of *Lb. plantarum* and its ability to utilize different carbohydrates.

One factor of importance contributing to the ability of *Lb. plantarum* to go from food to the GI tract is that the organisms have the ability to survive in the GI environment and to adhere to the mucosa in order to avoid immediate washout. Short duration establishment in high numbers of two *Lb. plantarum* strains on the intestinal mucosa after oral administration in lactic acid fermented oatmeal gruel (freeze dried) has been shown for two genetically closely related strains of *Lb. plantarum* (strain 299v [DSM 9843] and strain 299 [DSM 6595]).[85] The ability of *Lb. plantarum* 299v, when administrated in the probiotic fruit drink ProViva, to survive the passage through the human GI tract and to establish itself for a short period of time in the intestine after consumption has been confirmed.[88,89]

13.3.2.2 Adhesion

The *Lb. plantarum* strains 299 and 299v are included in a genetic subgroup within the species *Lb. plantarum*[84] where the members mostly originate from the intestinal mucosa but also can be found in traditional lactic acid fermented foods.[62,87] The strains of this subgroup have been shown to have a pronounced ability to attach to human mucosa cells *in vitro*; the adhesion is dependent on a mannose-binding adherence mechanism.[87,90] Moreover, *Lb. plantarum* strains of this particular genomic subtype frequently dominate the total *Lactobacillus* flora of healthy individuals, both on rectal and on oral mucosa.[87,90]

Yeasts also have mannose receptors on their surface, and *Lb. plantarum* strains with mannose-sensitive capacity for adhesion can bind to the surface of the yeast cell.[90] *Lb. plantarum* often occurs together with yeasts in traditional lactic acid fermented foods. Presumably, *Lb. plantarum* benefits from attaching to the surface of eucaryotic cells, perhaps by utilizing growth factors leaking out from the eucaryotes. *Lb. plantarum* requires calcium, pantothenate, and niacin.[73] The yeasts may also have a protective effect against the noxious effects of oxygen.

13.3.2.3 Oxidative Reactions

Lactobacillus plantarum possesses enzymatic systems of its own to handle oxygen radicals. However, this bacterium should be regarded as a microaerobe; it has less efficient systems for handling oxygen radicals than fully aerobic organisms do. Oxidative reactions that can occur with *Lb. plantarum* are:

1. Use of pyruvate oxidase to produce H_2O_2, CO_2, and acetyl phosphate from pyruvate, O_2, and phosphate

2. Use of L-lactate oxidase or NAD-independent D-lactate dehydrogenase to produce pyruvate and H_2O_2 from lactate and O_2
3. Use of NADH oxidase to produce NAD^+ and H_2O_2 from NADH, H^+, and O_2
4. Use of nonenzymatic superoxide reduction by manganese to produce H_2O_2 from O_{2-} and hydrogen[23]

Lactobacillus plantarum has a high growth requirement for manganese and can also accumulate high intracellular levels of manganese.[91] Interestingly, plants are rich in manganese, and manganese provides a defense for *Lb. plantarum* against oxygen toxicity by the reduction of oxygen radicals to H_2O_2.[92] The produced H_2O_2 can be converted to O_2 and water by manganese cofactored pseudocatalase in *Lb. plantarum*.[93,94]

13.3.2.4 Carbohydrate Fermentation

While lactic acid always is the major endproduct from glucose under anaerobic conditions (two moles lactic acid per mole hexose), considerable amounts of acetic acid have been shown to be produced by *Lb. plantarum* under aerobic conditions.[95] About one third acetic acid and two thirds lactic acid were produced by *Lb. plantarum* ATCC 8014 under aerobic conditions.[96]

Lactobacillus plantarum is not only able to ferment hexoses and pentoses (producing one mole each of lactate, acetate, and CO_2 per mole of pentose), but can also utilize many organic acids such as malic, tartaric, and citric acids to produce CO_2 and lactic or acetic acid and other byproducts. The breakdown of malic acid to lactic acid and CO_2 (malolactic fermentation) is important in winemaking.[80] The metabolic options of *Lb. plantarum* have been reviewed by Vescovo et al.[80] and Daeschel and Nes.[23] *Lb. plantarum* can also produce smaller amounts of diacetyl or acetoin, which have antimicrobial properties and may also affect the taste of foods.[23,80,96]

13.3.2.5 Resistance to Low pH

The facts that *Lb. plantarum* frequently predominates in spontaneously lactic acid fermented foods, where the pH usually is below 4.0 (see previous discussion), and also survives the passage through the acid conditions of the human stomach[85] point to its high resistance to acid conditions. This organism also has a high tolerance to low pH compared to other lactic acid bacteria.[23] For example, a comparison between *Lb. plantarum* and *Leuconostoc mesenteroides* showed that the growth of *Lc. mesenteroides* ceased when internal cellular pH reached 5.4 to 5.7 and growth of *Lb. plantarum* stopped when the internal pH dropped to 4.6 to 4.8.[97] *Lb. plantarum* maintained its pH gradient down to an external pH of 3.0.

13.3.2.6 Breakdown of Tannin

Besides pH, a controlling factor in the fermentation of plant material may be the presence of tannins. Tannins are naturally occurring water soluble polyphenols of varying molecular weight, which differ from most other natural phenolic compounds because of their ability to precipitate proteins from solutions.[98] Tannins inhibit the growth of a number of microorganisms and are resistant to microbial attack.[47,99] So-called condensed (nonhydrolyzable) tannins are more resistant to microbial degradation than hydrolyzable tannins. Tannins are commonly found in fruits and seeds such as grapes, apples, olives, beans, grains of sorghum and finger millets, coca, tea, and coffee. Fungi and yeasts and some aerobic bacteria are usually able to degrade tannins, but anaerobic degradation also occurs, for example in the intestinal tract.[47,100] Strains of *Lb. plantarum, Lb. pentosus,* and *Lb. paraplantarum* can posses tannase activity[49] and are also able to metabolize phenolic acids.[101,102] *Lb. plantarum,* which grows in environments where high concentrations of tannins are often present, has the relatively unusual ability to break up tannins and to metabolize the phenolic acids. This species may modify the tannins and produce breakdown products from them, for example, substituted phenyl-propionic acids.[102] It has also been shown that *Lb. plantarum* is able to produce small amounts of phenyl-acids such as benzoic acid[103] and phenyl-lactic acid,[104] with strong antifungal properties.

The degradation of tannins by *Lb. plantarum* will positively affect the nutritional value of tannin-rich fermented food products, and this may have physiological effects in the gastrointestinal (GI) tract of the host. It also seems likely that a tannin-rich environment will give *Lb. plantarum* a selective advantage compared with other microorganisms that are unable to degrade tannins and that even may be inhibited by them.

13.4 HEALTH EFFECTS

13.4.1 The Intestinal Microflora

13.4.1.1 Probiotics and the Bacterial Balance

It is well established that high numbers of lactobacilli counteract many pathogenic and potentially pathogenic bacteria, regardless of whether the system is a lactic acid fermented food or the human intestine.[105,106] The original concept of probiotics implies that the balance between beneficial and harmful bacteria in the microflora of the GI tract can be positively affected by eating the right type of living microorganisms.[107,108] Even if the term "probiotics" is used today in a broader sense — to refer to live microorganisms with beneficial health effects when administrated to animals and humans — the original concept of counteracting deleterious bacteria in the GI tract still remains interesting. In any case, the key question is: what components of the intestinal flora should be suppressed? That the probiotics should inhibit pathogens is self-evident, but the normal intestinal flora is much more than pathogens. Unfortunately, the human intestinal flora (and animal flora as well) is poorly defined, and many components have not been systematically described, not even at the hierarchical level of genus.[109] Examples of frequently occurring components of the human intestinal flora that presumably can have negative health implications and therefore should be counteracted are *Bacteroides fragilis* and species of the family Enterobacteriacieae (for example, *Escherichia coli* and *Klebsiella pneumoniae*). These groups found in the normal flora frequently are involved in abdominal infections and sepsis.

Lactobacillus spp. are usually present in varying numbers in the human GI tract, but are usually present in lower numbers than many other components of the normal flora such as *Bacteroides,* clostridia/eubacteria, and *Ruminococcus.*[110,111] An ingested probiotic will not only work in the colon, but will come in contact with the mucosa of the mouth and then the intestinal mucosa and its microbial inhabitants all along the small intestine. This means the probiotic has exposure to a huge interface that is harboring a smaller population of resident bacteria than that found in the colon. The effects and actions in the small intestine will probably also have an influence on the colonic environment.

Lactobacillus plantarum frequently occurs on the human GI mucosa.[62,87] The two strains of *Lb. plantarum,* 299 (DSM 6595) and 299v (DSM 9843), that have been shown to survive the passage through the human GI tract,[85] have also been shown *in vitro* to possess antimicrobial activity against potentially pathogenic species such as *Listeria monocytogenes, Bacillus cereus, E. coli, Yersinia enterocolitica, Citrobacter freundii, Enterobacter cloacae,* and *Enterococcus faecalis.*[112] Furthermore, when healthy volunteers consumed a mixture of lactobacillus strains, including *Lb. plantarum* 299/299v, the level of lactobacilli in the intestine increased and levels of Gram-negative anaerobes, Enterobacteriaceae, and sulfite-reducing clostridia decreased.[85] The inhibitory effect of *Lb. plantarum* 299v against Enterobacteriaceae and Gram-negative anaerobes has also been demonstrated in rat models simulating severe clinical conditions.[113,114]

Gram-negative anaerobes are often involved in secondary infections after abdominal surgery.[115–117] Furthermore, Gram-negative bacteria always contain endotoxins and they initiate, even when present in small numbers, violent inflammatory reactions. Gram-negative anaerobes are also suggested to be producers of carcinogenic substances in the intestine.[118,119] Rats pretreated with the Gram-negative anaerobe *Bacteroides fragilis* before the onset of an acute liver injury developed a significantly poorer liver status than control rats with the liver injury but without bacterial pretreatment.[120] The group of sulfite-reducing clostridia can contain subgroups that produce toxins. Sulfite-reducing clostridia also produce hydrogen sulfide, which has a general toxicity. Furthermore, clostridia can produce carcinogenic substances in the intestine.[118] Enterobacteriaceae is a genetically close family including many pathogenic taxa, and even normally nonpathogenic taxa have a pathogenic potential in situations where the immunological defense of the host is failing. *Lactobacillus plantarum* 299v has been shown to inhibit enteropathogenic *E. coli* adherence *in vitro* to HT-29 intestinal epithelial cells by inducing intestinal mucin gene expression. When this gene is expressed, epithelial cells produce more mucin, and the slime protects the cells from enteropathogenic *E. coli.*[121] It has also been shown that the colonization of *Lb. plantarum* 299v competes with that of *E. coli* in gnotobiotic rats.[122]

In a study in Tanzania, *Lb. plantarum* 299v was used as a starter culture for producing the cereal based lactic acid fermented beverage togwa. *Lb. plantarum* 299v was used for producing 50% of the test togwa, while the other 50% was made by traditional back slopping.[57] Spontaneously fermented togwa is frequently dominated by *Lb. plantarum.*[53] The product was given to children (<5 years) once a day for 13 consecutive days, and the presence of fecal enteropathogens such as *Campy-*

lobacter, enteropathogenic *E. coli, Salmonella,* and *Shigella* was evaluated. The proportion of children with isolated fecal enteropathogens decreased significantly ($P < 0.001$) during the study period.[57]

The ingestion of probiotics can positively alter the GI microflora, as has been seen by the decreased plate counts of Enterobacteriaceae and sulfite-reducing clostridia after ingestion of lactobacilli.[85] In a randomized, placebo controlled, double blind study in healthy volunteers who consumed *Lb. plantarum* 299v in a fruit drink (2×10^{10} CFU/d for 3 weeks), the total level of carboxylic acids in feces increased;[88] this increase was due to increases in the concentrations of acetic acid and propionic acid.[88] The carboxylic acids are produced by the GI microflora, and this change in acid composition points to significant changes in the flora. *Lb. plantarum* 299v is not known to be able to produce propionic acid. The increased concentration of acetic acid and propionic acid must be regarded as beneficial from a health perspective. Both types of short-chain fatty acids are utilized as energy sources by the mucosal cells of the intestine. Short-chain fatty acids are in fact the major energy source of the colonic mucosa cells. An increased level of short-chain fatty acids in the lumen is therefore beneficial for the condition of the mucosa. Moreover, absorbed propionic acid comes via the portal blood to the liver, and there it can have positive effects on both lipid metabolism and inflammatory responses in the liver. Healthy subjects receiving *Lb. plantarum* 299v also experienced a decrease in flatulence during the treatment period,[88] which might indicate that the concentration of gas-producing microorganisms in the GI tract decreased.

13.4.1.2 Intestinal Mucosal Status and Reduced Translocation

The effect of *Lb. plantarum* on the mucosal status and barrier function has been extensively studied in rat models. When the status of the intestinal mucosa was evaluated using the content of protein or content of rRNA and DNA as an indicator, an improvement in status was shown in rats with acute liver injury that had been pretreated with *Lb. plantarum* 299v.[123,124] An improved mucosal status was also seen in rats with enterocolitis that had been treated with *Lb. plantarum* 299v.[113] In this study, the permeability of ethylenediaminetetra acetic acid (EDTA) through the mucosa was measured and found to decrease in animals receiving *Lb. plantarum* 299v.[113]

Translocation (the passage of viable bacteria through the epithelial mucosa into the *lamina propria* and then to the mesenteric lymph nodes and possibly other tissues[125]) can be reduced due to the improved status of the intestinal mucosa. Translocation can be studied in rats with an acute liver injury induced by an injection with D-galactose-amine, which causes a severe liver inflammation.[126,127] Twenty-four hours after the onset of the liver injury, translocating bacteria can be found in organs such as the liver and spleen and in the portal and arterial blood. The liver injury does not directly affect the intestinal mucosa, but the immunological defense of the animal is severely weakened, which allows the translocating bacteria to travel beyond the mesenteric lymph nodes and the liver. However, by pretreating the animals with *Lb. plantarum* 299v, translocation can be significantly decreased.[114,120,123,128] Another strain of *Lb. plantarum* (DSM 6595) has been shown to have an effect in the liver failure model,[120] as have some strains of *Lactobacillus* spp. other than *Lb. plan-*

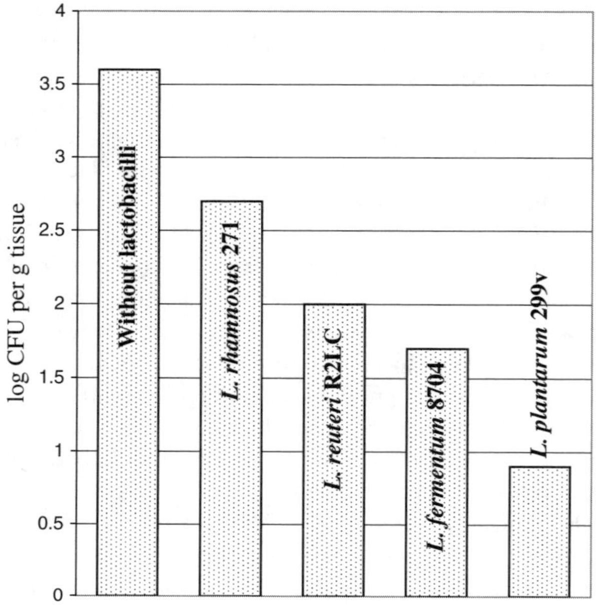

FIGURE 13.10 Bacterial translocation to the liver 24 h after onset of the liver injury of rats pretreated with different *Lactobacillus* strains. (From Adawi, D., Molin, G., Ahrné, S., and Jeppsson, B., *Microb. Ecol. Health Dis.*, 11, 47–54, 1999. With permission.)

tarum.[114] However, *Lb. plantarum* 299v seems to be an especially effective strain in this respect (see Figure 13.10).

It is interesting to identify what type of bacteria are translocating in the rats with liver failure. In rats that had not received any lactobacilli treatment, the majority of the bacteria found in the liver originated from the dominating population of the intestinal mucosal flora, i.e., *Lb. animalis, Lb. reuteri,* and *Lb. acidophilus* (*Lactobacillus* species are much more dominant in rats than in humans), but *Proteus vulgaris, Bacteroides distasonis, Enterococcus faecalis* and *Staphylococcus aureus* were also found in the liver. *Proteus vulgaris* and *S. aureus* were also found in the arterial blood.[128] However, pretreatment for 8 days with *Lb. plantarum* 299v before the liver injury not only decreased the rate of translocation to the liver, but no bacteria translocated to the blood and only *Lb. animalis, Lb. reuteri,* and *Lb. acidophilus* were found in the liver.[128] The *Lb. plantarum* treatment not only decreased the rate of translocation, it obviously had a controlling impact on the intestinal microflora and enhanced the domination of *Lactobacillus.* It can also be noted that *Lb. plantarum* was never found in extraintestinal sites in spite of the large pretreatment dose.[128] It has also been shown that pretreatment of rats with *Lb. plantarum* 299v in drinking water for a week inhibited *E. coli*-induced permeability of the intestine.[129] This was shown in intestinal segments mounted in Using chambers, where the permeability of mannitol was measured. Exposure to *E. coli* in the Using chamber normally increases the permeability, but the pretreatment of the living rats with *Lb. plantarum* abolished this increase in permeability.[129]

Many of the intestinal bacteria that translocate in the rats with liver failure end up in the liver, which enhances inflammation of the liver, causing the condition of the liver to worsen. This deterioration can be measured by the concentration of liver enzymes in the blood. In the liver failure model, it was shown that pretreatment with *Lb. plantarum* 299v decreased the concentration of the liver enzymes aspartate transaminase and alanine transaminase in the blood, indicating that the liver status was improved by the treatment.[114,123,124] This was also true for another, genetically similar *Lb. plantarum* strain, DSM 6595.[120]

The preventive effect of *Lb. plantarum* 299v on translocation has also been seen in other experimental models in rats. *Lb. plantarum* 299v significantly reduced translocation in rats with enterocolitis induced by methotrexate.[113] In this model, the mucosa is inflamed and damaged, in contrast to the liver failure model, where the mucosa is unaffected. The administration of lactobacilli to the rats with enterocolitis mitigated the mucosal injuries induced by the chemotherapy.[113] Also, another *Lb. plantarum* strain (genetically close to 299v; strain DSM 6595) possessed this effect, and in a comparison with the type strain of *Lb. acidophilus* (strain DSM 6595), the effect was shown to be more pronounced for the *Lb. plantarum* strain than for *Lb. acidophilus*.[130] Decreased translocation has also been observed by treatment with *Lb. plantarum* 299v in an experimental rat model with pancreatitis.[131]

There can be several explanations as to how *Lb. plantarum* can improve the mucous status and decrease the translocation rate. One is the traditional probiotic effect, i.e., that the administrated probiotic strain counteracts adverse bacteria. These aggressive adverse bacteria can induce and maintain an inflammation, and they may be especially suited for translocation and are capable of fighting off the host's immunological defense. It is also possible that the probiotic strain not only counteracts adverse components of the flora, it might also stimulate beneficial components that are part of the resident flora. In fact, the domination of resident intestinal lactobacilli of rats increased after treatment with *Lb. plantarum* 299v.[128] This was also indicated in humans, when the amount of propionic acid in feces increased after consumption of *Lb. plantarum* 299v, since propioinic acid is not produced by 299v.[88] However, the improved barrier effect of the mucosa could also be due to an immunomodulation (see following sections) and to stimulation of the mucin production of the human mucosa cells.

13.4.2 Risk Factors for Coronary Artery Disease

Lactobacillus plantarum 299v has surprisingly been shown to be able to decrease different risk factors for coronary artery diseases in individuals at risk. In a small randomized, placebo controlled and double blind study on men with slightly elevated cholesterol levels, it was shown that the concentrations of total cholesterol and low-density-lipoprotein (LDL) cholesterol were decreased after consumption of *Lb. plantarum* 299v in a fruit drink.[132] The study included 30 individuals divided into two groups, where the treatment group consumed 200 ml fruit drink (rose hip), containing 5×10^7 CFU/ml, for 6 weeks and the placebo group consumed fruit drink without lactobacilli. The fall in cholesterol level was small but statistically significant.[132] However, even more surprising, it was shown in the same study that the fibrinogen level of the serum also was decreased significantly ($P < 0.001$), representing a reduction of 13.5%.[132] Fibrinogen is an acute phase protein that reflects the inflammatory status of the individual and also is an independent risk factor for coronary artery disease.[133] In a subsequent placebo controlled randomized double blind study with 38 healthy smokers, it was shown that the consumption of 400 ml ProViva daily for 6 weeks not only significantly decreased the level of fibrinogen, but also F_2-isoprostans and interleukin (IL)-6, which are other inflammatory markers.[134] Moreover, *Lb. plantarum* 299v in the fruit drink also positively affected the systolic blood pressure and the insulin and leptin responses.[134]

13.4.3 Irritable Bowel Syndrome (IBS)

Irritable bowel syndrome (IBS) is common, but its cause is unknown. It is not a single condition, but rather a collection of disorders causing similar symptoms of abdominal pain, diarrhea, constipation, or variability of bowel habit. The absence of strict pathogenic features has made IBS a disease without a proper diagnosis. Attempts have been made to develop criteria for a positive diagnosis of IBS.[135,136] Among patients coming to gastroenterology clinics, 20 to 50% are suffering from IBS, even if most patients with IBS do not seek medical care.[137] IBS is a chronic relapsing condition that perhaps occurs in most adults at some point in their lives. Symptoms begin before age 35 in 50% of patients, and 40% of patients are aged 35 to 50.[137] IBS was found in 18% of the adult population in the Bristol area in the United Kingdom.[138]

Lb. plantarum 299v in the fruit drink ProViva (rose hip) was administrated to patients with IBS in two double blind, placebo controlled studies, one in Poland[139] and one in Sweden.[89] In both studies, the patients were divided into two groups: one was given *Lb. plantarum* 299v and the other a similar rose hip drink without *Lb. plantarum* 299v (placebo). In the Polish study, the magnitude of several of the IBS symptoms decreased in the *Lb. plantarum* group, and a higher proportion of the patients were free from symptoms in the treatment group than in the placebo group.[139] In the Swedish study, *Lb. plantarum* 299v significantly decreased the subjective bloating experienced during the treatment period.[89] Pain was also significantly reduced in both the treatment group and in the placebo group, but the decrease was more rapid and more pronounced in the *Lb. plantarum* group. Twelve months after the treatment, the patients given *Lb. plantarum* 299v in the study still experienced better overall gastrointestinal function than the patients who had received the placebo.[89]

The bloating and pain experienced by IBS patient might be due to abnormal colonic fermentation giving rise to an excess of gas production, especially of hydrogen.[140] Presumably, *Lb. plantarum* 299v suppresses the components of the intestinal microflora that are responsible for this gas production.

13.4.4 Inflammatory Bowel Disease (IBD)

Inflammatory bowel disease (IBD) is a chronic inflammation along the GI tract. It can be limited to the large bowel (ulcerative colitis) or it can be situated anywhere along the GI tract (Crohn's disease). Ulcerative colitis is a relatively superficial ulcerative inflammation, while Crohn's disease is a transmural, granulomatous inflammation. IBD is thought to be due to an abnormal and uncontrolled immune response to normally occurring constituents of the intestine. The etiology of IBD is unknown. Microbial agents appear to be involved in the pathogenesis of IBD, and intestinal bacteria seem to be an important factor in its development and chronicity.[141–143] In these conditions, the bacteria, mucosa, and immune system interact in complex ways, and this interaction is far from clear.[142]

Inflammation and the potential of *Lb. plantarum* 299v to counteract the inflammation have been studied in different animal models. In rats with enterocolitis induced by treatment with methotrexate, administration of *Lb. plantarum* 299v mitigated the mucosal injuries induced by the chemotherapy.[113] Furthermore, inflammation in the intestinal mucosa of rats after radiation was decreased by administration of *Lb. plantarum* 299v in fermented oatmeal gruel.[144]

In a study using interleukin-10–deficient mice in germ-free and specific pathogen-free (SPF) environments, *Lb. plantarum* 299v was able to attenuate the established colitis when the bacterium had colonized the gastrointestinal tract of the mouse before the mouse was transferred to the SPF environment.[143,145] It was also demonstrated that a mono-association with *Lb. plantarum* 299v (i.e., *Lb. plantarum* 299v was the only bacterium in the animal) did not induce colitis but only initiated a very mild immune response.

There are today no clinical data on humans with IBD or attempts to treat the disease with *Lb. plantarum.* However, oral administration of a mixture of probiotic strains (including one strain of *Lb. plantarum*) was shown to be effective in preventing flare-ups of chronic pouchitis.[146] Pouchitis is a nonspecific inflammation of the ileal reservoir made after surgery for ulcerative colitis.

13.4.5 Immune Modulation

13.4.5.1 Expression of Cytokines in Cells *in Vitro*

The cytokine response of human peripheral blood mononuclear cells differs between different *Lactobacillus* spp. It has been shown that different strains of *Lb. plantarum* of intestinal origin are able to induce the production of the cytokines IL-12 and IL-10 from blood mononuclear cells.[147] Compared to *E. coli,* less IL-10 was produced but considerably more IL-12 was produced. In the same study, *Lb. paracasei* induced the production of a higher proportion of IL-12, and *Lb. rhamnosus* induced a higher proportion of IL-10. The response of the mononuclear cells was more balanced with respect to IL-10 and IL-12 production when they were exposed to *Lb. plantarum* than to the other two *Lactobacillus* spp.[147]

The cytokine response of bone marrow-derived murine dendritic cells, when exposed to different probiotic strains of lactobacillus, has also been shown to vary.[148] Substantial differences could be seen between strains in their capacity to induce IL-12 and TNF-α production in dendritic cells. The ranking among the tested strains was as follows: *Lb. casei* ssp. *alactus* CHCC3137 >> *Lb. plantarum* Lb1 > *Lb. fermentum* Lb20 > *Lb. johnsonii* La1 > *Lb. plantarum* 299v >> *Lb. reuteri* DSM 12246.[148] Similar but less pronounced differences were observed among the test strains in the induction of IL-6 and IL-10.

13.4.5.2 Experimental Models in Rats

The subnormal levels of secretory IgA antibodies in the intestines of rats with enterocolitis were increased, and approached a more normal level, after the administration of *Lb. plantarum* 299v. Also the levels of CD4 and CD8 lymphocytes in the intestinal lamina propria increased to more normal levels after treatment with *Lb. plantarum* 299v.[149]

The levels of total serum IgA antibodies increased and the IgA and IgM antibody levels against *Escherichia coli* were marginally higher in gnotobiotic rats colonized with *E. coli* together with *Lb. plantarum* 299v, compared with rats that only were colonized with *E. coli*.[122] The group treated with *Lb. plantarum* 299v also showed a significantly increased density of CD25-positive cells in *lamina propria* and displayed a decreased proliferative spleen cell response after stimulation with ConA one week after colonization. The results indicated that *Lb. plantarum* 299v can modulate a response to antigens presented via the gut.

13.4.5.3 Immune Response in HIV Positive Children

Children congenitally exposed to human immunodeficiency virus (HIV) have received *Lb. plantarum* 299v in a fermented oatmeal gruel (freeze dried) in a pilot study. The results suggested that *Lb. plantarum* 299v elicits specific systemic immune responses after oral supplementation.[150,151]

13.5 SAFETY ASPECTS

The safety of consuming large numbers of live bacteria has been questioned, and there are reports that *Lactobacillus* spp., including *Lb. plantarum* strains, have been isolated from diseased sites in patients.[152] However, the potential of *Lactobacillus* spp. to cause serious infections has been assessed by studying the prevalence of bacteremia due to *Lactobacillus* spp. during a 4-year period; this study indicated that the pathogenic potential of *Lactobacillus* spp. is low.[153]

The fact that many traditional lactic acid fermented foods spontaneously contain high numbers of *Lb. plantarum*[6–9,11,13–15,50] and that these products, all over the world, have a reputation of being safe and wholesome, strongly indicates that live *Lb. plantarum* can safely be consumed. This becomes especially obvious if the long historical tradition of lactic acid fermented foods is taken into account (Figure 13.1). However, in the case of *Lb. plantarum* strain 299v, safety has been more directly confirmed in a series of studies.

Lactobacillus plantarum 299v has been given in high doses to immune-compromised children with HIV for extended time periods without any adverse effects.[150,151] It has also been given to patients receiving autologous stem cell transplantation.[151]

The risk of endocarditis has been tested in an experimental rat model.[155] A catheter was passed down the right common carotid artery into the lumen of the left ventricle. The catheter was tied in place, and the neck incision was closed. After 48 h, 10^8 CFU of *Lb. plantarum* 299v were injected (0.5 ml of bacterial suspension) through the tail vein. Four days after the injection of the *Lb. plantarum* strain, the rats were sacrificed and the blood, heart tissue, and catheter were sampled for bacteria. No *Lb. plantarum* 299v could be found at any of the sample sites.[155] Thus, even with this animal model, using a very unusual and challenging situation where a high dose of the bacteria is injected directly into the bloodstream of an animal with an implant of artificial material in the artery and heart, *Lb. plantarum* was removed from the system before causing any damage. This strain of *Lb. plantarum* appears to be perfectly safe, and so presumably are other strains of the species *Lb. plantarum.*

REFERENCES

1. Leakey, R., *På Spaning Efter Människans Ursprung,* Natur och Kultur, Stockholm, 1993 [Swedish translation of *Origins Reconsidered: In Search of What Makes Us Human,* Bantam Doubleday Dell Publishing Group, Inc., New York, 1992].
2. Leakey R., *Hur Människan Blev Till,* Natur och Kultur, Stockholm, 1995 [Swedish translation of *The Origin of Humankind,* HarperCollins Publishers, Inc., New York, 1994].
3. Arrhenius, B., Grötfrukost på stenåldern [Gruel breakfast in the stone age], *Forskning och Framsteg,* 4–8, 1984.
4. Lagerqvist, L.O. and Åberg, N., *Mat och Dryck i Forntid och Medeltid* [Food and drink in prehistoric time and in the Middle Ages], Vincent Förlag and Statens Historiska Museum, Stockholm, 1994.
5. Larsson, L., *Ett Fångstsamhälle för 7000 År Sedan* [A hunting society 7000 years ago], Signum, Lund, Sweden, 1988.
6. Fernández Gonzalez, M.J., García, P.G., Fernández, A.G., and Quintana, M.C.D., Microflora of the aerobic preservation of directly brined green olives from Hojiblanca cultivar, *J. Appl. Bacteriol.,* 75, 226–233, 1993.
7. Dedicatoria, R.F., Aspiras, R.B., and Sanchez, P.C., The fermentation inoculation with lactic acid bacteria to increase the nutritive value of sauerkraut, *Kalikasan,* 10, 214–219, 1981.
8. McDonald, L.C., Shieh, D.H., Fleming, H.P., McFeeters, R.F., and Thompson, R.L., Evaluation of malolactic-deficient strains of *Lactobacillus plantarum* for use in cucumber fermentations, *Food Microbiol.,* 10, 489–499, 1993.
9. Lönner, C. and Ahrné, S., *Lactobacillus:* Baking, in *Food Biotechnology Microorganisms,* Hui, Y.H. and Khachatourians, G.G., Eds., VCH Publishers Inc., Eureka, CA, 1995, pp. 797–844.
10. Johansson, M.L., Systematics and Starter Culture Selection of *Lactobacillus* for Human Intestine and Nigerian Ogi, with Special Reference to *Lactobacillus plantarum,* Ph.D. thesis, Lund University, Lund, Sweden, 1995.
11. Gashe, B.A., Kocho fermentation, *J. Appl. Bacteriol.,* 62, 473–474, 1987.
12. Nigatu, A., Systematics of *Lactobacillus* and *Pediococcus* Isolated from Fermented Tef (*Eragrostis tef*) and Kocho (*Ensete ventricosum*) and Microbiological Status of Baked Products, Ph.D. thesis, Addis Ababa University, Addis Ababa, Ethiopia, 1998.
13. Gashe, B.A., Involvement of lactic acid bacteria in the fermentation of tef (*Eragrostis tef*), an Ethiopian fermented food, *J. Food Sci.,* 50, 800–801, 1985.
14. Oyewole, O.B. and Odunfa, S.A., Characterization and distribution of lactic acid bacteria in cassava fermentation during fufu production, *J. Appl. Bacteriol.,* 68, 145–152, 1990.

15. Moorthy, S.N. and Mathew, G., Cassava fermentation and associated changes in physicochemical and functional properties, *Crit. Rev. Food Sci. Nutr.,* 38, 73–121, 1998.
16. Nordenskiöld, A.E., *Vegas Färd kring Asien och Europa* [The Voyage of Vega around Asia and Europe], F. & G. Beiers Förlag, Stockholm, 1881, part II, pp. 114–115.
17. Warg, C., *Cajsa Wargs Kokbok* [The Cookery Book of Cajsa Warg], P.A. Norstedt & Söner AB, Stockholm, 1999.
18. Lloyd, C., *Kapten James Cooks Resor* [Original title: The Voyages of Captain James Cook Round the World], Bohusläns Grafiska AB, Uddevalla, Sweden, 1991.
19. Price, R.J. and Lee, J.S., Inhibition of *Pseudomonas* species by hydrogen peroxide producing lactobacilli, *J. Milk Food Technol.,* 33, 13–17, 1970.
20. Gilliland, S.E. and Speck, M.L., Antagonistic action of *Lactobacillus acidophilus* towards intestinal and food-borne pathogens in associative cultures, *J. Food Prot.,* 40, 820–823, 1977.
21. Kelley, E., Morris, S.M. Jr., and Billiar, T.R., Nitric oxide, sepsis, and arginine metabolism, *JPEN,* 19, 234–238, 1995.
22. Malawista, S.E., Montgomery, R.R., and van Blaricom, G., Evidence for reactive nitrogen intermediates in killing staphylococci by human neutrophil cytoplasts, *J. Clin. Invest.,* 90, 631–636, 1992.
23. Daeschel, M.A. and Nes, I.F., *Lactobacillus plantarum:* physiology, genetics and applications in foods, in *Food Biotechnology Microorganisms,* Hui, Y.H. and Khachatourians, G.G., Eds., VCH Publishers, Inc., New York, 1995, chap. 21, pp. 721–743.
24. Langston, C.W. and Bouma, C., Types sequence change of bacteria in orchardgrass and alfalfa silages, *J. Dairy Sci.,* 43, 1575–1584, 1960.
25. Dellaglio, F., Vescovo, M., Morelli, L., and Torriani, S., Lactic acid bacteria in ensiled high moisture corn grain: physiological and genetic characterization, *System. Appl. Microbiol.,* 5, 534–544, 1984.
26. Nouts, M.J. and Rombouts, F.M., Fermented and acidified plant foods, in *The Microbiological Safety and Quality of Food,* Lund, B.M., Baird-Parker, T.C., and Gould G.W., Eds., Aspen Publishers, Inc., Gaithersburg, MD, 2000, chap 26, pp. 685–737.
27. Lund, B. and Eklund, T., Control of pH and use of organic acids, in *The Microbiological Safety and Quality of Food,* Lund, B.M., Baird-Parker, T.C., and Gould, G.W., Eds., Aspen Publishers, Inc., Gaithersburg, MD, 2000, chap. 8, pp. 175–199.
28. Flemming, H.P., Thompson, R.L., Etchells, J.L., Kelling, R.E., and Bell, T.A., Bloater formation in brined cucumbers fermented by *Lactobacillus plantarum, J. Food Sci.,* 38, 499–503, 1973.
29. McFeeters, R.F., Fleming, H.P., and Daeschel, M.A., Malic acid degradation and brined cucumber bloating, *J. Food Sci.,* 49, 999–1002, 1984.
30. Daeschel, M.A., McFeeters, R.F., Fleming, H.P., Klaenhammer, T.R., and Sanozky, R.B., Mutation and selection of *Lactobacillus plantarum* strains that do not produce carbon dioxide from malate, *Appl. Environ. Microbiol.,* 47, 419–420, 1984.
31. Daeschel, M.A., McFeeters, R.F., Fleming, H.P., Klaenhammer, T.R., and Sanozky, R.B., Method for Differentiating Between Lactic Acid Bacteria Which Do and Do Not Decarboxylate Malic Acid, U.S. Patent 4,666,849, 1987.
32. Garcia, G., Quintana, M.C.D., Balbuena, M.B., and Fernández, A.G., Lactic fermentation during the storage of "Alorena" cultivar untreated green table olives, *J. Appl. Bacteriol.,* 73, 324–330, 1992.
33. Ruiz-Barba, J.L. and Jimenez-Diaz, R., Vitamin and amino acid requirements of *Lactobacillus plantarum* strains isolated from green olive fermentation, *J. Appl. Bacteriol.,* 76, 350–355, 1994.

34. Ruiz-Barba, J.L. and Jimenez-Diaz, R., Availability of essential B-group vitamins to *Lactobacillus plantarum* in green olive fermentation brine, *Appl. Environ. Microbiol.,* 61, 1294–1297, 1995.
35. Lönner, C., Welander, T., Molin, N., Dostálek, M., and Blickstad, E., The microflora in a sour dough started spontaneously on typical Swedish rye meal, *Food Microbiol.,* 3, 3–12, 1986.
36. Legan, J.D., Cereals and cereal products, in *The Microbiological Safety and Quality of Food,* Lund, B.M., Baird-Parker, T.C., and Gould, G.W., Eds., Aspen Publishers, Inc., Gaithersburg, MD, 2000, chap. 28, pp. 759–783.
37. Lönner, C., Starter Cultures for Rye Sourdoughs, Ph.D. thesis, Lund University, Lund, Sweden, 1988.
38. Liljeberg, H.G.M. and Björck, I.M.E., Delayed gastric emptying rate as a potential mechanism for lowered glycaemia after eating sourdough bread: studies in humans and rats using test products with added organic acids or an organic salt, *Am. J. Clin. Nutr.,* 64, 883–893, 1996.
39. Liljeberg, H.G.M. and Björck, I., Delayed gastric emptying rate may explain improved glycaemia in healthy subjects to a starchy meal with added vinegar, *Eur. J. Clin. Nutr.,* 52, 368–371, 1998.
40. Liljeberg, H.G.M., Lönner, C.H., and Björck, I.M.E., Sourdough fermentation or addition of organic acids or corresponding salts to bread improves nutritional properties of starch in healthy humans, *J. Nutr.,* 125, 1503–1511, 1995.
41. Hallberg, L., Rossander, L., and Skånberg, A.B., Phytates and the inhibitory effect of bran on iron absorption in man, *Am. J. Clin. Nutr.,* 49, 140–144, 1987.
42. Lorri, W.S.M., Nutritional and Microbiological Evaluation of Fermented Cereal Weaning Foods, Ph.D. thesis, Chalmers University of Technology, Göteborg, Sweden, 1993.
43. Archibald, F., *Lactobacillus plantarum,* an organism not requiring iron, *FEMS Microbiol. Lett.,* 19, 29–32, 1983.
44. Elli, M., Zink, R., Rytz, A., Reniero, R., and Morelli, L., Iron requirement of *Lactobacillus* spp. in completely chemically defined growth media, *J. Appl. Microbiol.,* 88, 695–703, 2000.
45. Urga, K., Fite, A., and Biratu, E., Effect of natural fermentation on nutritional and antinutritional factors of tef (*Eragrostis tef*), *Ethiop. J. Health Dev.,* 11, 61–66, 1997.
46. Graham, G.G., Glover, D.V., Lopez de Romana, G., Morales, E., and MacLean, W.C., Nutritional value of normal opaque and sugary-2 opaque maize hybrids for infants and children. 1. Digestibility and utilization, *J. Nutr.,* 110, 1061–1069, 1980.
47. Chung, K.T., Wong, T.Y., Wei, C.I., Huang, Y.W., and Lin, Y., Tannins and human health: a review, *Crit. Rev. Food Sci. Nutr.,* 38, 421–464, 1998.
48. Parr, A.J. and Bolwell, G.P., Phenols in the plant and in man. The potential for possible nutritional enhancement of the diet by modifying the phenols content profile, *J. Sci. Food Agric.,* 80, 985–1012, 2000.
49. Osawa, R., Kuroiso, K., Goto, S., and Shimzu, A., Isolation of tannin-degrading lactobacilli from humans and fermented foods, *Appl. Environ. Microbiol.,* 66, 3093–3097, 2000.
50. Johansson, M-L., Sanni, A., Lönner, C., and Molin, G., Phenotypically based taxonomy using API 50CH of lactobacilli from Nigerian ogi, and the occurrence of starch fermenting strains, *Int. J. Food Microbiol.,* 25, 159–168, 1995.
51. Sanni, A.I., Lönner, C., Marklinder, I., Johansson, M.-L., and Molin, G., Starter cultures for the production of ogi, a fermented infant food from maize and sorghum, *Chemie Microbiol. Technol. der Lebensmittel,* 16, 29–33, 1994.

52. Kingamkono, R., Influence of Lactic Acid Fermentation on Enteropathogenic Bacteria: Growth Inhibition in Cereal Gruels and in Human Intestine, Ph.D. thesis, Chalmers University of Technology, Göteborg, Sweden, 1997.
53. Mugula, K.K., Microbiology, Fermentation and Shelf-Life Extension of *Togwa,* a Tanzanian Indigenous Food, Ph.D. thesis, Agricultural University of Norway, Ås, 2001.
54. Elmaki, H.B., Babiker, E.E., and El Tinay, A.H., Changes in chemical composition, grain malting, starch, and tannin contents and protein digestibility during germination of sorghum cultivars, *Food Chem.,* 64, 331–336, 1999.
55. Kingamkono, R., Sjögren, E., Svanberg, U., and Kaijser, B., pH and acidity in lactic-fermenting cereal gruels: effects on viability of enteropathogenic microorganisms, *World J. Microbiol. Biotech.,* 10, 664–669, 1994.
56. Kingamkono, R., Sjögren, E., Svanberg, U., and Kaijser, B., Inhibition of different strains of enteropathogens in a lactic-fermenting cereal gruel, *World J. Microbiol. Biotech.,* 11, 299–303, 1995.
57. Kingamkono, R., Sjögren, E., and Svanberg, U., Enteropathogenic bacteria in faecal swabs of young children fed on lactic acid-fermented cereals, *Epidemiol. Infect.,* 122, 23–32, 1999.
58. Willumsen, J.F., Darling, J.C., Kitundu, J.A., Kingamkono, R.R., Msengi, A.E., Mduma, B., Sullivan, K.R., and Tomkins, A.M., Dietary management of acute diarrhea in children: effect of fermented and amylase-digested weaning foods on intestinal permeability, *J. Pediatr. Gastroenterol. Nutr.,* 24, 235–241, 1997.
59. Molin, G., *Lactobacillus* strains for fermented oatmeal soup in PRO VIVA™, in Proc. Symp. Les Bactéries Lactiques: Quelles souches? Pur quels marchés? [Lactic acid bacteria: Which strain? For which markets?], Lactic 94, Caen, 1994, Centre de Publications de l'Université de Caen, Caen Cedex, France, 1995, pp. 213–223.
60. Molin, G. and Ahrné, S., Probiotic effects of *Lactobacillus plantarum* 299v, in *Food Microbiology and Safety into the Next Millennium,* Tuijtelaars, A.C.J., Samson, R.A., Rombouts, F.M., and Notermans, S., Eds., Seventeenth International Conference of the International Committee on Food Microbiology and Hygiene, Veldhoven, Ponsen & Looyen, Wageningen, 1999, pp. 850–853.
61. Molin, G., Probiotics in foods not containing milk or milk constituents, with special reference to *Lactobacillus plantarum* 299v, *Am. J. Clin. Nutr.,* 73, 380S–385S, 2001.
62. Molin, G., Jeppsson, B., Ahrné, S., Johansson, M.-L., Nobaek, S., Ståhl, M., and Bengmark, S., Numerical taxonomy of *Lactobacillus* spp. associated with healthy and diseased mucosa of the human intestines, *J. Appl. Bacteriol.,* 74, 314–323, 1993.
63. Molin, N., Albertsson, K.E., Bengmark, S., and Larsson, K., Nutrient Composition and Method for the Preparation Thereof, Swedish patent 463796, European patent 0415941, U.S. patent, 5.190755, 1991.
64. Marklinder, I. and Lönner, C., Fermented oatmeal soup: influence of additives on the properties of a nutrient solution for enteral feeding, *Food Microbiol.,* 11, 505–513, 1994.
65. Marklinder, I., Lactic Acid-Fermented Oats and Barley for Human Dietary Use, Ph.D. thesis, Acta Universitatis Upsaliensis, Uppsala, Sweden, 1996.
66. Rossander, L., Sandberg, A.S., and Sandström, B., The effect of dietary fibre on mineral absorption and utilization, in *Dietary Fibre: A Component in Food. Nutritional Function in Health and Disease,* Schweizer, T.S. and Edwards, C.A., Eds., Springer Verlag, London, 1992.
67. Brune, M., Rossander-Hultén, L., Hallberg, L., Gleerup, A., and Sandberg, A.S., Iron absorption from bread in humans: inhibiting effects of cereal fibre, phytate and inositol phosphates with different numbers of phosphate groups, *J. Nutr.,* 122, 442–449, 1992.

68. Sandberg, A.S. and Svanberg, U., Phytate hydrolysis by phytase in cereals; effects on *in vitro* estimation of iron availability, *J. Food Sci.,* 56, 1330–1333, 1991.
69. Marklinder, I.M., Larsson, M., Fredlund, K., and Sandberg, A.S., Degradation of phytate by using varied sources of phytases in an oat-based nutrient solution fermented by *Lactobacillus plantarum* 299v, *Food Microbiol.,* 12, 487–495, 1995.
70. Sueoka, N., Variations and heterogeneity of base composition of nucleic acids: a compilation of old and new data, *J. Mol. Biol.,* 3, 31–40, 1961.
71. Schleifer, K.H. and Stackebrandt, E., Molecular systematics of prokaryotes, *Annu. Rev. Microbiol.,* 37, 143–187, 1983.
72. Vandamme, P., Pot, B., Gillis, M., de Vos, P., Kersters, K., and Swings, J., Polyphasic taxonomy, a consensus approach to bacterial systematics, *Microbiol. Rev.,* 60, 407–438, 1996.
73. Kandler, O. and Weiss, N., Regular, nonsporing Gram-positive rods, in *Bergey's Manual of Determinative Bacteriology,* Sneath, H.A., Mair, N.S., Sharpe, M.E., and Holt, J., Eds., Williams & Wilkins, Baltimore, 1986, vol. 2, pp. 1208–1234.
74. Collins, M.D., Rodrigues, U., Ash, C., Aguirre, M., Farrow, J.A.E., Martinez-Murcia, A., Phillips, B.A., Williams, A.M., and Wallbanks, S., Phylogenetic analysis of the genus *Lactobacillus* and related lactic acid bacteria as determined by reverse transcriptase sequencing of 16S rRNA, *FEMS Microbiol. Lett.,* 77, 5–12, 1991.
75. Olsen, G.J., Larsen, N., and Woese, C.R., The ribosomal RNA database project. *Nucleic Acids Res.,* 19 (suppl.), 2017–2021, 1991.
76. Schleifer, K.H., Ludwig, W., Amann, R., Hertel, C., Ehrmann, M., Köhler, W., and Krause, A., Phylogenetic relationships of lactic acid bacteria and their identification with nucleic acid probes, in *Actes du Colloque Lactic 91,* Novel, G. and Le Querler, J.-F., Eds., Centre de Publications de l'Université de Caen, Caen, France, 1992, pp. 23–32.
77. Schleifer, K.H. and Ludwig, W., Phylogeny of the genus *Lactobacillus* and related genera, *System. Appl. Microbiol.,* 18, 461–467, 1995.
78. Zanoni, P., Farrow, J.A.E., Phillips, B.A., and Collins, M.D., *Lactobacillus pentosus* (Fred, Peterson and Anderson) sp. nov., nom. rev., *Int. J. Syst. Bacteriol.,* 37, 339–341, 1987.
79. Curk, M.-C., Hubert, J.-C., and Bringel, F., *Lactobacillus paraplantarum* sp. nov., a new species related to *Lactobacillus plantarum, Int. J. Syst. Bacteriol.,* 46, 595–598, 1996.
80. Vescovo, M., Torriani, S., Dellaglio, F., and Bottazzi, V., Basic characteristics, ecology and application of *Lactobacillus plantarum*: a review, *Ann. Microbiol. Enzymol.,* 43, 261–284, 1993.
81. Pot, B., Ludwig, W., Kersters, K., and Schleifer, K.-H., Taxonomy of lactic acid bacteria, in *Bacteriocins of Lactic Acid Bacteria*, De Vuyst, L. and Vandamme, E.J., Eds., Blackie Academic & Professional, Glasgow, 1994, pp. 13–90.
82. Johansson, M.-L., Molin, G., Pettersson, B., Uhlén, M., and Ahrné, S., Characterisation and species recognition of *Lactobacillus plantarum* strains by fragment length polymorphism (RFLP) of the 16S rRNA gene, *J. Appl. Bacteriol.*, 79, 536–541, 1995.
83. Johansson, M.-L., Quednau, M., Molin, G., and Ahrné, S., Randomly amplified polymorphic DNA (RAPD) for rapid typing of *Lactobacillus plantarum* strains, *Lett. Appl. Microbiol.,* 21, 155–159, 1995.
84. Johansson, M.-L., Quednau, M., Ahrné, S., and Molin, G., Classification of *Lactobacillus plantarum* by restriction endonuclease analysis of total chromosomal DNA using conventional agarose gel electrophoresis, *Int. J. Syst. Bacteriol.,* 45, 670–675, 1995.

85. Johansson, M.-L., Molin, G., Jeppsson, B., Nobaek, S., Ahrné, S., and Bengmark, S., Administration of different *Lactobacillus* strains in fermented oatmeal soup: *in vivo* colonization of human intestinal mucosa and effect on the indigenous flora, *Appl. Environ. Microbiol.,* 59, 15–20, 1993.
86. Mundt, J.O. and Hammer, J.L., Lactobacilli on plants, *Appl. Microbiol.,* 16, 1326–1330, 1968.
87. Ahrné, S., Nobaek, S., Jeppsson, B., Adlerberth, I., Wold, A., and Molin, G., The normal *Lactobacillus* flora of healthy human rectal and oral mucosa, *J. Appl. Microbiol.,* 85, 88–94, 1998.
88. Johansson, M.-L., Nobaek, S., Berggren, A., Nyman, M., Björck, I., Ahrné, S., Jeppsson, B., and Molin, G., Survival of *Lactobacillus plantarum* DSM 9843 (299v), and effect on the short-chain fatty acid content in faeces after ingestion of a rose-hip drink with fermented oats, *Int. J. Food Microbiol.,* 42, 29–38, 1998.
89. Nobaek, S., Johansson, M.-L., Molin, G., Ahrné, S., and Jeppsson, B., Alteration of intestinal microflora is associated with reduction in abdominal bloating and pain in patients with irritable bowel syndrome, *Am. J. Gastroenterol.,* 95, 1231–1238, 2000.
90. Adlerberth, I., Ahrné, S., Johansson, M.-L., Molin, G., Hanson, L.-Å., and Wold, A.E., A mannose-specific adherence mechanism in *Lactobacillus plantarum* conferring binding to the human colonic cell line HT-29, *Appl. Environ. Microbiol.,* 62, 2244–2251, 1996.
91. Archibald, F. and Fridovich, I., Manganese, superoxide dismutase, and oxygen tolerance in some lactic acid bacteria, *J. Bacteriol.,* 146, 928–936, 1981.
92. Archibald, F. and Fridovich, I., Manganese and defence against oxygen toxicity in *Lactobacillus plantarum, J. Bacteriol.,* 145, 442–451, 1981.
93. Kono, Y. and Fridovich, I., Isolation and characterization of the pseudocatalase of *Lactobacillus plantarum:* a new manganese containing enzyme, *J. Biol. Chem.,* 258, 6015–6019, 1983.
94. Kono, Y. and Fridovich, I., Functional significance of manganese catalase in *Lactobacillus plantarum, J. Bacteriol.,* 155, 742–746, 1983.
95. Sedewitz, B., Schleifer, K.H., and Götz, F., Physiological role of pyruvate oxidase in aerobic metabolism of *Lactobacillus plantarum, J. Bacteriol.,* 160, 462–465, 1984.
96. Tseng, C.-P. and Montville, T.J., Enzyme activities affecting end product distribution by *Lactobacillus plantarum* in response to changes in pH and O_2, *Appl. Environ. Microbiol.,* 56, 2761–2763, 1990.
97. McDonald, L.C., Fleming, H.P., and Hassan, H.M., Acid tolerance of *Leuconostoc mesenteroides* and *Lactobacillus plantarum, Appl. Environ. Microbiol.,* 56, 2120–2124, 1990.
98. Spencer, C.M., Cai, Y., Martin, R., Gaffney, S.H., Goulding, P.N., Magnolato, D., Lilley, T.H., and Haslam, E., Polyphenol complexation — some thoughts and observations, *Phytochemistry,* 27, 2397–2409, 1988.
99. Field, J.A. and Lettinga, G., Toxicity of tannic compounds to microorganisms, in *Plant Polyphenols: Synthesis, Properties, Significance,* Hemingway, R.W. and Laks, E., Eds., Plenum Press, New York, 1992, pp. 673–692.
100. Bhat, T.K., Singh, B., and Sharma, O.P., Microbial degradation of tannins: a current perspective, *Biodegradation*, 9, 343–357, 1998.
101. Barthelmebs, L., Divies, C., and Cavin, J.-F., Knockout of the *p*-coumarate decarboxylase gene from *Lactobacillus plantarum* reveals the existence of two other inducible enzymatic activities involved in phenolic acid metabolism, *Appl. Environ. Microbiol.,* 66, 3368–3375, 2000.

102. Barthelmebs, L., Divies, C., and Cavin, J.-F., Molecular characterization of the phenolic acid metabolism in the lactic acid bacteria *Lactobacillus plantarum, Lait,* 81, 161–171, 2001.
103. Niku-Paavola, M.-L., Laitila, A., Mattila-Sandholm, T., and Haikara, A., New types of antimicrobial compounds produced by *Lactobacillus plantarum, J. Appl. Microbiol.,* 86, 29–35, 1999.
104. Lavermicocca, P., Valerio, F., Evidente, A., Lazzaroni, S., Corsetti, A., and Gobbetti, M., Purification and characterization of novel antifungal compounds from sourdough *Lactobacillus plantarum* strain 21B, *Appl. Environ. Microbiol.,* 66, 4084–4090, 2000.
105. De Vuyst, L. and Vandamme, E.J., Antimicrobial potential of lactic acid bacteria, in *Bacteriocins of Lactic Acid Bacteria: Microbiology, Genetics and Applications,* De Vuyst, L. and Vandamme, E.J., Eds., Blackie Academic & Professional, London, 1994, pp. 91–142.
106. De Vuyst, L. and Vandamme, E.J., Lactic acid bacteria and bacteriocins: Their practical importance, in *Bacteriocins of Lactic Acid Bacteria: Microbiology, Genetics and Applications,* De Vuyst, L. and Vandamme, E.J., Eds., Blackie Academic & Professional, London, 1994, pp. 1–11.
107. Parker, R.B., Probiotics, the other half of the antibiotic story, *Anim. Nutr. Health*, 29, 4–8, 1974.
108. Fuller, R., Probiotics in man and animals, *J. Appl. Bacteriol.,* 66, 365–368, 1989.
109. Langendijk, P.S., Schut, F., Jansen, G.J., Raangs, G.C., Kamphuis, M.H., and Welling, G.W., Quantitative fluorescence *in situ* hybridization of *Bifidobacterium* spp. with genus-specific 16S rRNA-targeted probes and its application in fecal samples, *Appl. Environ. Microbiol.,* 61, 3069–3075, 1995.
110. Moore, W.E. and Holdeman, L.V., Human fecal flora: the normal flora of 20 Japanese-Hawaiians, *Appl. Microbiol.,* 27, 961–979, 1974.
111. Finegold, S.M., Sutter, V.L., and Mathisen, G.E., Normal indigenous flora, in *Human Intestinal Microflora in Health and Disease,* Hentges, D.J., Ed., Academic Press, New York, 1983, pp. 3–31.
112. Jacobsen, C.N., Rosenfeldt Nielsen, V., Hayford, A.E., Møller, P.L., Michaelsen, K.F., Pœrregaard, A., Sandström, B., Tvede, M., and Jakobsen, M., Screening of probiotic activities of forty-seven strains of *Lactobacillus* spp. by *in vitro* techniques and evaluation of the colonization ability of five selected strains in humans, *Appl. Environ. Microbiol.,* 65, 4949–4956, 1999.
113. Mao, Y., Nobaek, S., Kasravi, B., Adawi, D., Stenram, U., Molin, G., and Jeppsson, B., The effect of *Lactobacillus* strains and oat fiber on methotrexate-induced enterocolitis in rats, *Gastroenterology,* 111, 334–344, 1996.
114. Adawi, D., Kasravi, F.B., Molin, G., and Jeppsson, B., Effect of *Lactobacillus* supplementation with and without arginine on liver damage and bacterial translocation in an acute liver injury model, *Hepatology,* 25, 642–647, 1997.
115. Nichols, R.L., Infections following gastrointestinal surgery. Intraabdominal abscesses, *Surg. Clin. North Am.,* 60, 197–212, 1980.
116. Offenbartl, K. and Bengmark, S., Intraabdominal infection and gut origin sepsis, *World J. Surg.,* 14, 191–195, 1990.
117. Wittman, D.H., *Intraabdominal Infections. Pathophysiology and Treatment,* Marcel Dekker, New York, 1991, pp. 20–29.
118. Rowland, I.R., Metabolic interactions in the gut, in *Probiotics: The Scientific Basis,* Fuller, R., Ed., Chapman & Hall, London, 1992, pp. 29–53.

119. Roberfroid, M.B. and Gibbson, G.R., Workshop on Colonic Microflora: Nutrition and Health, ILSI Europe, Barcelona, 1994 [Summary published in *Nutr. Rev.*, 53, 127–130, 1995].
120. Adawi, D., Molin, G., Ahrné, S., and Jeppsson, B., Modulation of the colonic bacterial flora affects differently bacterial translocation and liver injury in an acute liver injury model, *Microb. Ecol. Health Dis.*, 11, 47–54, 1999.
121. Mack, D.R., Michail, S., Wei, S., McDougall, L., and Hollingsworth, M.A., Probiotics inhibit enteropathogenic *E. coli* adherence *in vitro* by inducing intestinal mucin gene expression, *Am. J. Physiol.*, 276, G941–G950, 1999.
122. Herías, M.V., Hessle, C., Telemo, E., Midtvedt, T., Hansson, L.Å., and Wold, A.E., Immunomodulatory effects of *Lactobacillus plantarum* colonizing the intestine of gnotobiotic rats, *Clin. Exp. Immunol.*, 116, 283–290, 1999.
123. Kasravi, F.B., Adawi, D., Molin, G., Bengmark, S., and Jeppsson, B., Effect of oral supplementation of lactobacilli on bacterial translocation in acute liver injury induced by D-galactosamine, *J. Hepatol.*, 26, 417–424, 1997.
124. Adawi, D., Molin, G., and Jeppsson, B., The role of nitric oxide inhibition on the effects of arginine and *Lactobacillus* administration in acute liver injury, *Ann. Surg.*, 228, 748–755, 1999.
125. Berg, R.D. and Garlington, A.W., Translocation of certain indigenous bacteria from the gastrointestinal tract to the mesenteric lymph nodes and other organs in a gnotobotic mouse model, *Infect. Immunol.*, 23, 403–411, 1979.
126. Kasravi, F.B., Wang, L., Wang, X.-D., Molin, G., Bengmark, S., and Jeppsson, B., Bacterial translocation in acute liver injury induced by D-galactosamine, *Hepatology*, 23, 97–103, 1996.
127. Kasravi, F.B., Adawi, D., Molin, G., Bengmark, S., and Jeppsson, B., Dynamics of bacterial translocation in acute liver injury induced by D-galactosamine in rat, *APMIS*, 104, 135–140, 1996.
128. Wang, M., Adawi, D., Molin, G., Pettersson, B., Jeppsson, B., and Ahrné, S., Identification of the translocating bacteria in rats with acute liver injury and their relation to the bacterial flora of the intestinal mucosa, *APMIS*, 109, 551–558, 2001.
129. Mangell, P., Nejdfors, P., Wang, M., Ahrné, S., Weström, B., Thorlacius, H., and Jeppsson, B., *Lactobacillus plantarum* 299v inhibits *Escherichia coli*-induced intestinal permeability, *Digest. Dis. Sci.*, 47, 511–516, 2002.
130. Mao, Y., Nobaek, S., Adawi, D., Molin, G., and Jeppsson, B., Comparison of the effects of different *Lactobacillus* strains in reducing bacterial translocation on methotrexate induced enterocolitis in rats, *Digest. Surg.*, 14, 284–291, 1997.
131. Mangiante, G., Colucci, G., Canepari, P., Bassi, C., Nicoli, N., Casaril, A., Marinello, P., and Bengmark, S., *Lactobacillus plantarum* reduces infection of pancreatic necrosis in experimental acute pancreatitis, *Digest. Surg.*, 18, 47–50, 2001.
132. Bukowska, H., Pieczul-Mróz, J., Jastrzebsk, K., Chelstowski, K., and Naruszewicz, M., Significant decrease in fibrinogen and LDL-cholesterol levels upon supplementation of the diet with *Lactobacillus plantarum* (ProViva) in subjects with moderately elevated cholesterol concentrations, *Atherosclerosis*, 137, 437–438, 1998.
133. Kannel, W.B., Wolf, P.A., Castelli, W.P., and D'Agnostino, R.B., Fibrinogen as a cardiovascular risk of cardiovascular disease: the Framingham study, *J. Am. Med. Assoc.*, 258, 1183, 1987.
134. Naruszewicz, M., Johansson, M.-L., Zapolska-Downar, D., and Bukowska, H., Effect of *Lactobacillus plantarum* 299v on cardiovascular risk factors in smokers, *Am. J. Clin. Nutr.*, 2002 (in press).

135. Manning, A.P., Thompson, W.G., Heaton, K.W., and Morris, A.F., Towards positive diagnosis of the irritable bowel, *BMJ,* 2, 653–654, 1978.
136. Thompson, W.G., Creed, F., Drossman, D.A., Heaton, K.W., and Mazzacca, G., Functional bowel disease and functional abdominal pain, *Gastroenterol. Int.,* 5, 75–91, 1992.
137. Maxvell, P.R., Mendall, M.A., and Kumar, D., Irritable bowel syndrome, *Lancet*, 350, 1691–1695, 1997.
138. Heaton, K.W., Creed, F., Drossman, D.A., Heaton, K.W., and Mazzacca, G., Symptoms of irritable bowel syndrome in a British urban community: consulters and nonconsulters, *Gastroenterol.*, 102, 1962–1967, 1992.
139. Niedzielin, K., Kordecki, H., and Birkenfeld, B., A controlled, double-blind, randomized study on the efficacy of *Lactobacillus plantarum* 299v in patients with irritable bowel syndrome, *Eur. J. Gastroenterol. Hepatol.,* 13, 1143–1147, 2001.
140. King, T.S., Elia, M., and Hunter, J.O., Abnormal colonic fermentation in irritable bowel syndrome, *Lancet*, 352, 1187–1189, 1998.
141. Ardizzone, S., Bollani, S., Manzionna, G., and Porro, G.B., Inflammatory bowel disease approaching the 3rd millennium: pathogenesis and therapeutic implications, *Eur. J. Gastroenterol. Hepatol.,* 11, 17–32, 1999.
142. Campieri, M. and Gionchetti, P., Bacteria as the cause of ulcerative colitis, *Gut*, 48, 132–135, 2001.
143. Schultz, M. and Sartor, R.B., Probiotics and inflammatory bowel diseases, *Am. J. Gastroenterol.,* 95, S19–S21, 2000.
144. Liu, Q., Nobaek, S., Adawi, D., Mao, Y., Wang, M., Molin, G., Ekelund, M., and Jeppsson, B., Administration of *Lactobacillus plantarum* 299v reduces side effects of external radiation on colonic anastomotic healing in an experimental model, *Colorect. Dis.,* 3, 245–252, 2001.
145. Schultz, M., Veltkamp, C., Dieleman, L.A., Wyrick, P.B., Tonkonogy, S.L., and Sartor, R.B., Continuous feeding of *Lactobacillus plantarum* attenuates established colitis in interleukin-10 deficient mice, *Gastroenterology* 114, A4426, 1998.
146. Gionchetti, P., Rizzello, F., Venturi, A., Brigidi, P., Matteuzzi, D., Bazzocchi, G., Poggioli, G., and Miglioli, M., Oral bacteriotherapy as maintenance treatment in patients with chronic pouchitis: a double-blind, placebo-controlled trial, *Gastroenterology* 119, 305–309, 2000.
147. Hessle, C., Hansson, L.Å., and Wold, A.E., Lactobacilli from human gastrointestinal mucosa are strong stimulators of IL-12 production, *Clin. Exp. Immunol.,* 116, 276–282, 1999.
148. Christensen, H.R., Frokiaer, H., and Pestka, J.J., Lactobacilli differentially modulate expression of cytokines and maturation surface markers in murine dendritic cells, *J. Immunol.,* 168, 171–178, 2002.
149. Mao, Y., Yu, J., Ljungh, Å., Molin, G., and Jeppsson, B., Intestinal immune response to oral administration of *Lactobacillus reuteri* R2LC, *Lactobacillus plantarum* DSM 9843, pectin and oatbase on methotrexate-induced enterocolitis in rats, *Microb. Ecol. Health Dis.,* 9, 261–270, 1996.
150. Cunningham-Rundles, S., Ahrné, S., and Bengmark, S., Probiotics and immune response, *Am. J. Gastroenterol.,* 95, S22–S25, 2000.
151. Cunningham-Rundles, S., Ahrné, S., Abuav-Nissbaum, B.S., and Dnistrian, A., Development of immunocompetence: role of micronutrients and microorganisms, *Nutr. Rev.,* 60, S68–S72, 2002.
152. Aguirre, M. and Collins, M.D., Lactic acid bacteria and human clinical infection, *J. Appl. Bacteriol.,* 75, 95–107, 1993.

153. Saxelin, M., Chuang, N.H., Chassy, B., Rautelin, H., Makela, P.H., Salminen, S., and Gorbach, S.L., Lactobacilli and bacteremia in southern Finland, 1989–1992, *Clin. Infect. Dis.,* 22, 564–566, 1996.
154. Sallerfors, B., Jeppsson, B., Johansson, M.-L., Lööf, L., and Molin, G., High doses of *Lactobacillus plantarum* DSM 9843 to patients receiving autologous stemcell transplantation: a pilot study, 2002 (in manuscript).
155. Adawi, D., Molin, G., Ahrné, S., and Jeppsson, B., Safety of the probiotic strain *Lactobacillus plantarum* DSM 9843 (= strain 299v) in an endocarditis animal model, *Microb. Ecol. Health Dis.,* 14, 50–53, 2002.

14 Sauerkraut

Wilhelm Holzapfel, Ulrich Schillinger, and Herbert J. Buckenhüskes

CONTENTS

14.1 HISTORY AND CULTURE RELATED TO SAUERKRAUT

Present-day sauerkraut is a product resulting from lactic acid fermentation of shredded, salted white cabbage (*Brassica oleracea* var. *capitata* for. *alba* L.). There is no doubt that the preservation of plant material by fermentation dates back to prehistoric times. Plinius the Elder, in the first century A.D., is said to have been the first to describe the production of sauerkraut by preservation of so-called salt cabbage in earthen vessels. It can be assumed that under the conditions described, the cabbage was fermented by microorganisms, some of which were typically associated with the plant phylloplane, but most of which were located in the pores of the fermentation vessels and/or originated from a former fermentation. Heads of white cabbage, the raw material of today's sauerkraut, seem to have been known as early as the eighth century A.D.[1] Lind[2] was the first to describe sauerkraut manufacture comparable to

0-8493-1372-4/03/$0.00+$1.50

contemporary processing. An important date in the history of sauerkraut was the year 1775, when Captain James Cook was awarded the Great Copley Medal for his observations and conclusions about sauerkraut as an effective food for the prevention of scurvy which, up to that time, was feared as the plague of the sea.[3] (See Chapter 1 for more details on the history of sauerkraut.)

14.2 PRODUCTION AND REGIONS OF IMPORTANCE

The results of the European Research Program "COST 91 bis,"[4] which were published in 1990, indicated that in Europe, 21 different vegetables are commercially preserved by lactic acid fermentation. The economic importance of the different products is not quite clear, since fermented vegetables are not separately listed in the national statistics. However, based on the European production figures of 1985, fermented olives represent the most important fermented product, with a total amount of more than 510,000 tons per annum. Second place was held by sauerkraut with a total of 220,000 tons, followed by 45,000 tons of fermented cucumbers. Of this total, 122,000 tons of sauerkraut were produced in Germany.[4] During the 1990s, the consumption of sauerkraut decreased in Germany from 1.7 kg per capita/year in 1989 to 1.2 kg in 1999. As a result of this development, production also decreased to approximately 103,000 tons in 1999.[5]

Outside Europe two other regions have significant production of sauerkraut: Korea, China, and other Far Eastern countries on the one side, and the United States on the other. Whereas American sauerkraut is very similar to the European products, the Asian "kimchi" is produced from Chinese cabbage. (See Chapter 12 for more details on the production and health properties of kimchi.)

14.3 FERMENTATION

14.3.1 The Manufacturing Process

At the present time, sauerkraut is manufactured in all European countries by small, medium, and large sized companies. Consequently, the production procedures differ within a wide range. Nevertheless, the basic principles are rather similar in all cases, and therefore it is possible to summarize them as is shown in the simplified flow sheet given in Figure 14.1.

Following quality control of the delivered cabbage (estimation of external quality criteria such as color, spoiled leaves, damages including insect damage, and internal tip burn, as well as the determination of the dry matter, sugar, vitamin C, and nitrate content and residues of plant protection agents), the outer green and dirty leaves are removed, and the core of each head is bored or partly removed. Subsequently, the cabbage is shredded into 0.7- to 2-mm wide strips, which in most cases are salted on a conveyor belt between the shredding station and the fermentor with 0.7 to 2.5% sodium chloride. Under commercial conditions, the shredded cabbage is placed into fermentation containers of volumes with capacity between 100 kg and 40 tons of cabbage.

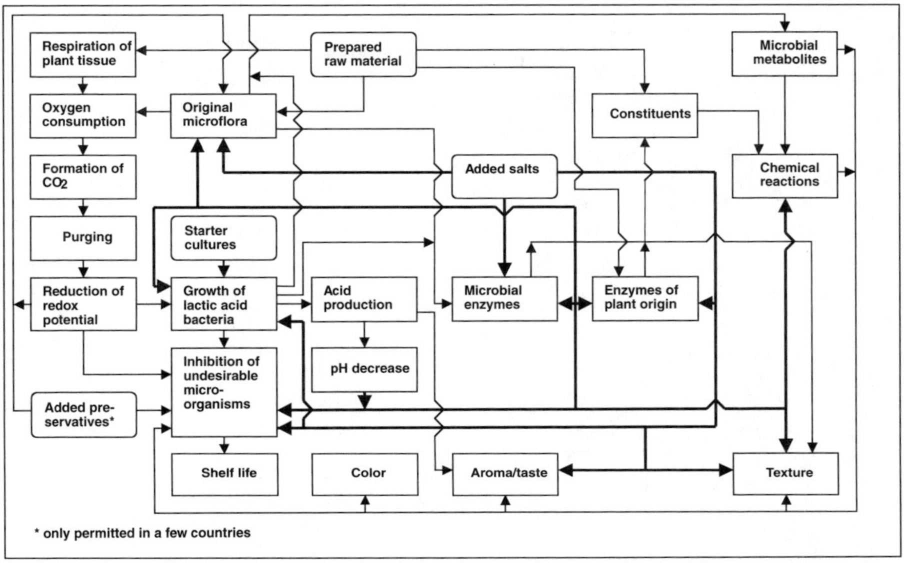

FIGURE 14.1 Interactions of various factors and processing steps involved in sauerkraut fermentation.

Apart from salting, filling is one of the critical control points in sauerkraut manufacture. It is of major importance that the air between the particles be removed as much as possible. Filling and pressing the cabbage into the fermentors together with added salt leads to an osmotic withdrawal of water out of tissue cells; the water replaces the air between the cabbage shreds. It is of utmost importance to exclude air in order to support the subsequent lactic acid fermentation and to prevent mold and yeast growth. Finally, the surface of the filled containers must be covered carefully in order to exclude oxygen and microbial contamination. Depending on the temperature, a "spontaneous" fermentation will start within a few hours to 1 to 2 days and will continue for between 7 days and several weeks.

If the sauerkraut is to be distributed as a fresh and unheated product, the fermentation has to be continued until all fermentable carbohydrates have been metabolized. Otherwise, a secondary fermentation by yeasts may occur, resulting in an alcoholic fermentation. If the sauerkraut is to be distributed as a pasteurized product, the fermentation can be stopped any time after the pH falls below 4.1. After the kraut is taken out of the fermentor, it is blanched, canned, pasteurized, and finally cooled in order to prevent heat-induced quality losses.

14.3.2 Factors Affecting Sauerkraut Fermentation

The fermentation of cabbage to sauerkraut represents a very complex system of microbial, biochemical, enzymatic, chemical, and physical processes. The complexity of the system is schematically illustrated in Figures 14.2a and b. Only the few major reactions and interrelations of the sauerkraut fermentation that influence the sensory properties and the shelf life of the final product are depicted. To complicate

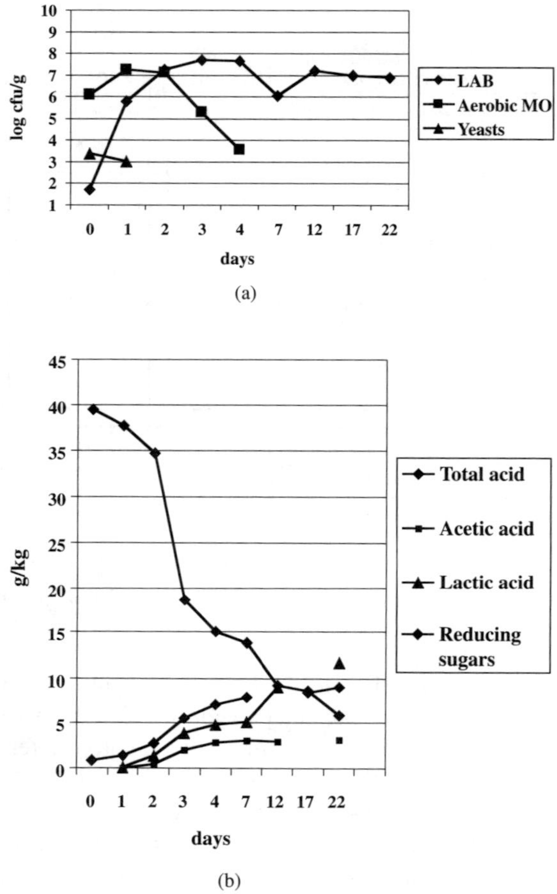

FIGURE 14.2 Sauerkraut fermentation from early harvested cabbage, conducted at 19°C in laminated plastic pouches. (a) Changes in major microbial groups. (b) Changes in some analytical parameters. (From Schneider, M., Microbiology of Sauerkraut Fermented in Small Ready-to-Sell Containers (in German), dissertation, Hohenheim University, Stuttgart, Germany, 1988. With permission.)

matters, fermentation can be influenced by a multitude of technological, microbiological, and raw material inherent factors, which are summarized in Table 14.1. The most important factors are discussed in more detail below.

14.3.2.1 Addition of Sodium Chloride

The addition of salt (sodium chloride) as well as its even distribution throughout the cabbage fibers, is one of the critical control points in sauerkraut production.[6] Not only the development of anaerobic conditions and the type and extent of microbial growth, but also the sensory properties of the final product are affected by the amount of salt used.

TABLE 14.1
Important Factors Influencing the Sauerkraut Fermentation Process

Raw Material Inherent Factors
Variety of cabbage
General quality (fresh harvested or stored heads)
Dry matter content
Content of fermentable carbohydrates
Vitamin C content
pH value
Buffering capacity
Microbial Factors
Spontaneous microorganisms or starter cultures
Home flora
Bacteriophages
Residues of plant protective substances
Technological Factors
Amount of added sodium chloride
Degree of shredding
Size and material of fermentor
Exclusion of oxygen
Temperature of the raw material

Directly after its addition as well as during compression of the shredded cabbage in the fermentation vessels, the sodium chloride fulfills its first function, namely causing the osmotic withdrawal of water from the cabbage cells. The emerging liquid fills up the space between the pieces of shredded cabbage and thereby supports the development of anaerobic conditions, which comprise the selective basis for the lactic acid fermentation.

From the onset of fermentation, the amount of salt added affects the microbial population, since it selectively favors growth of desired groups of bacteria. An increased salt content limits growth of undesirable microorganisms such as pseudomonads, flavobacteria, *Achromobacter,* or fungi, whereas growth of particular lactic acid bacteria (LAB) is promoted.[7] Heterofermentative LAB are more sensitive to high salt concentrations than homofermentative lactobacilli, with some species tolerating levels >3%. Therefore, high salt levels favor the growth of homofermentative LAB, resulting in an accelerated production of lactic acid. However, fermentations are more than acidification, and it has been shown that the flavor of sauerkraut produced under these conditions is unbalanced because of the lack of acetic acid and other metabolic products normally produced by the heterofermentative species. On the other hand, too low salt concentrations (<0.8%) may result in undesirable fermentation as well as in a poor-quality product such as soft sauerkraut.

Sodium chloride is a major flavor and modifying ingredient. Sauerkraut usually contains between 0.6 and 2% sodium chloride (Table 14.2). The amount used

TABLE 14.2
Initial Ecological Conditions within the Substrate

The substrate consists of solids in a liquid environment; in sauerkraut production it is not common to circulate the brine in order to distribute the microorganisms as well as released nutrients, sodium chloride, or metabolites throughout the fermentor	
Microorganisms originating from the raw material or proceeding from the equipment (home flora) are distributed throughout the fermentation stock during shredding and filling	
Salmonella, clostridia, listeriae, and other undesirable microorganisms are present in all probability	
The currently used cabbage varieties are rich in nutrients, growth factors, and minerals	
Water activity	a_w = 0.95–0.99
pH	5.9–6.5
Temperature	5–20°C
Sugar content	20–50 g/kg
Vitamin C content	300–700 mg/kg
Buffering capacity	0.45–0.65 g lactic acid/100 g cabbage
Sodium chloride	0.6–2.0% (sometimes up to 2.5%)

Source: Buckenhüskes, H.J., in *Food Microbiology — Fundamentals and Frontiers,* 2nd ed., Doyle, M.P., Beuchat L.R., and Montville, T.J., Eds., ASM Press, Washington, 2001, pp. 665–679. With permission.

depends on consumer demands and on traditional considerations in the producing countries. For instance, the salt content of canned sauerkraut produced in Germany tends to be lower (average content: 11.3 g/kg) than in the United States (average salt content: 16.7 g/kg). Following the general trend in industrialized countries to reduce the salt content of such products, the average amount in German sauerkraut decreased between the 1960s and 1980s from 12.9 to 11.3 g/kg.[8] The vegetable fermentation industry is also interested in low-salt fermentations as a means to reduce the chloride waste from sauerkraut fermentations.[9] Finally, it should be mentioned that the amount of salt used, to some extent, depends on the desired degree of acidity, since practical experience has shown that the total acid and salt contents should be in an certain ratio to ensure a balanced taste of the resulting sauerkraut.

14.3.2.2 Carbohydrates

The amount of fermentable sugars available is a major factor affecting the development of the LAB that convert the carbohydrates into lactic and acetic acids. The resulting pH decrease depends on the amount and kind of produced acids as well as on the buffering capacity of the fermenting cabbage, from which it is possible to estimate the minimum amount of fermentable carbohydrates required to reduce the pH below 4.1.[10]

14.3.2.3 Temperature

As with other biological processes, sauerkraut fermentation is influenced by temperature. From a sensory point of view, the best results are obtained at temperatures between 15 and 20°C.[11] Higher temperatures will cause an accelerated acid production, which leads to products with a so-called green and immature flavor.[7] Temperatures below 10°C hamper the start of fermentation and favor spoilage of the cabbage. Although some experiments have been conducted to control the temperature of the fermenting cabbage, this has not been very successful. The only method industrially used to affect the temperature is the warming up of cold cabbage by steam injection, resulting in a temperature increase of approximately 5°C.

14.3.2.4 Microbiology

As shown in Figure 14.2, sauerkraut fermentation is a complex microbiological process resulting from the metabolic activity of a definite sequence of different microorganisms, predominantly heterofermentative and homofermentative LAB. This succession is a consequence of the changing environmental conditions within the fermenting substrate. The initial ecological conditions are listed in Table 14.2. The growth sequence of spontaneously fermenting cabbage is invariably initiated by *Leuconostoc mesenteroides,* which comprises >90% of the LAB population at this stage, followed by heterofermentative lactobacilli and finally by homofermentative lactobacilli.[13,14]

Fresh plant material harbors numerous microorganisms of various types. The initial microflora is dominated by aerobic bacteria such as pseudomonads, Enterobacteriaceae, and coryneforms.[14] On cabbage leaves, LAB are present only in extremely small numbers, representing 0.15 to 1.5% of the total bacterial population. The identification of the LAB initially present on 30 different batches of cabbage has shown that nearly all belong to *Leuconostoc mesenteroides* ssp. *mesenteroides*, while only in a few cases some "streptococci" (probably enterococci and/or lactococci) were found. According to older literature, yeasts should play an important role in flavor formation; however, their share in the total microbial count is normally less than 0.1%.[11]

The traditional process of the spontaneous fermentation of cabbage can be divided into four steps:[12,15]

1. Fermentation starts as soon as the cabbage is filled into the containers. When the cabbage fibers are tightly packed, the number of strictly aerobic bacteria such as *Pseudomonas, Flavobacterium,* and *Acinetobacter* species initially present decreases immediately. Anaerobiosis is rapidly attained as a result of the respiration of the plant material and the consumption of oxygen by facultative anaerobic enterobacteria, which multiply for the first 2 or 3 days. Oxygen deprivation is accompanied by a change in pH resulting from the organic acids (lactic, acetic, formic, and succinic acids) formed. It is not clear whether or not the microorganisms present at this stage of fermentation have a significant influence on flavor development.
2. The more anaerobic atmosphere/lower redox potential, the added salt, and the reduced pH favor the facultatively anaerobic LAB, which soon become the predominant organisms. Although *Leuconostoc mesenteroides* is not as acid tolerant as other LAB species, it generally initiates the fermentation, as it is present at sufficiently high numbers and is well adapted to the substrate. Under ideal conditions, e.g., in cabbage juice, this species may achieve a maximum cell population of more than 10^8 colony forming

units (CFU)/ml after 12 to 14 h incubation.[16] *Lc. mesenteroides* produces lactic and acetic acids that quickly lower the pH. Since it is a heterofermentative LAB, it produces carbon dioxide, supporting the replacement of air and providing an anaerobic atmosphere favorable for the stabilization of vitamin C (ascorbic and dehydro-ascorbic acid) and the natural color of the cabbage. It has recently been shown that not only *Lc. mesenteroides* but also strains of *Lc. fallax* are involved in this stage of fermentation.[17] Growth and fermentation patterns of these strains have been found highly similar to those of *Lc. mesenteroides.*

Growth of *Lc. mesenteroides* and other leuconostocs is followed by growth of the heterofermentative species *Lactobacillus brevis,* which is more acid- and salt-tolerant than *Leuconostoc*. Depending on the temperature, the first two stages of sauerkraut fermentation are completed after 3 to 6 days. During that time, the concentration of lactic acid will increase up to approximately 1%.[15]

4. The third stage of fermentation starts with another shift in the lactic population: homofermentative lactobacilli become the predominant organisms, mainly due to the combined effect of anaerobiosis, lowered pH, and elevated levels of salt. Among these, streptococci (most probably enterococci) and pediococci represent a minor component, usually less than 10% of the total LAB population. The dominating homofermentative component consists of members of the former "subgenus" *Streptobacterium.* In the older literature, these organisms are usually ascribed to a single species, namely *Lactobacillus plantarum* (formerly referred to as *Lactobacillus cucumeris.*)[18] Investigations in the 1980s revealed that *Lb. plantarum* comprises only 30 to 80% of the "streptobacteria," and that *Lb. sakei* and *Lb. curvatus* are also present in large numbers during this stage of sauerkraut fermentation.[19] The LAB convert the largest part of the available carbohydrates (glucose, fructose, and sucrose) to organic acids, predominantly lactic acid. During the third stage of fermentation, the total acid content (calculated as lactic acid) will increase to 1.5 to 2.0%. Presently, most of the sauerkraut in Europe is unpacked and pasteurized when it reaches a pH of 3.8 to 4.1, since consumers increasingly prefer mild products, in terms of both acid and in salt content.
5. Only fresh distributed, unpasteurized sauerkraut will undergo the final stage of fermentation, when *Lb. brevis* and some other heterofermentative species able to metabolize pentoses such as arabinose and xylose become dominant. Living plant material normally does not possess free pentoses. However, they are liberated after harvest as a result of acid-induced hydrolysis of hemicellulose.[20] Due to the increase of the total acid content up to 2.5%, the pH of the sauerkraut will decrease to as low as 3.4.

14.3.3 Starter Cultures

Up to now, the majority of the sauerkraut produced in Europe and North America has still been prepared by spontaneous fermentation. Due to current thinking about

TABLE 14.3
Traits Considered Relevant to Starter Cultures for Sauerkraut Fermentation[12,24]

Criteria	Relevance
Technologically Relevant Criteria	
Rapid and predominant growth	Important
Salt tolerance	Advantageous
Acid production and tolerance	Important
Inability to metabolize organic acids	Important
Growth at low temperatures	Important
Formation of dextrans	Detrimental
Pectinolytic activities	Unacceptable
Formation of bacteriocins	Useful
Bacteriophage resistance	Not relevant
Sensorially Relevant Criteria	
Heterofermentative metabolism	Important
Formation of flavor precursors	Important
Nutritionally Relevant Criteria	
Reduction of nitrate and nitrite	Useful
Formation of L-(+) lactic acid	Advantageous
Formation of biogenic amines	Unacceptable

Sources: Buckenhüskes, H.J., in *Food Microbiology — Fundamentals and Frontiers,* 2nd ed., Doyle, M.P., Beuchat L.R., and Montville, T.J., Eds., ASM Press, Washington, 2001, pp. 665–679; Lücke, F.-K., Brümmer, J.-M., Buckenhüskes, H., Garrido Fernandez, A., Rodrigo, M., and Smith, J.E., in *Processing and Quality of Foods, Vol. 2. Food Biotechnology: Avenues to Healthy and Nutritious Products,* Zeuthen, P., Cheftel, J.C., Eriksson, C., Gormley, T.R., Linko, P., and Paulus, K., Eds., Elsevier Applied Science, London, 1990, p. 11. With permission.

food security and product quality, modern approaches include the use of defined starter cultures for sauerkraut fermentation. Since it is not possible to eliminate the natural microflora of the shredded cabbage by appropriate methods, suitable strains should be highly competitive and should lead to a product of consistent quality. Traits considered relevant in starter cultures for sauerkraut fermentation are listed in Table 14.3. Experiments using strains of *Lb. plantarum* as a starter culture showed that this starter caused a rapid decrease of the pH; however, the resulting products had a poor, unbalanced, and less complex flavor due to the lack of different metabolites normally produced by other microorganisms. On the other hand, sauerkraut with excellent flavor characteristics was obtained using selected strains of *Lc. mesenteroides.*[21]

Use of selected starter cultures, and especially strains of *Lb. plantarum,* resulted in a significant reduction in the formation of biogenic amines, particularly tyramine, putrescine, and cadaverine.[22,23] In addition, Kalač et al.[23] also observed a significant reduction in the formation of acetic acid, ammonia, and α-amino groups during the fermentation.

14.4 COMPOSITION AND CHANGES AS A RESULT OF FERMENTATION

Cabbage consists of crude fibers, carbohydrates, proteins, lipids, and ash in relatively high proportions (see Table 14.4). The major change that takes place during fermentation is the conversion of the carbohydrates to lactic and acetic acids, ethyl alcohol, carbon dioxide, mannitol, and dextrans. The proteins, lipids, glucosides, and other constituents of cabbage are also affected by the fermentation process, resulting in alterations of the chemical and physical properties of the product.

The major changes during fermentation are indicated in Table 14.5, where some analytical data from the salted cabbage and from the product after 23 days of fermentation at 19°C are listed.

A completely fermented sauerkraut contains from 1.8 to 2.25% acid; occasionally, total acidities of about 2.5% are attained. Lactic and acetic acids are the predominant acids and are normally formed in a ratio of 4:1. Other organic acids such as succinic, malic, and propionic acids may also be formed in much smaller quantities.[26] Ethanol and CO_2 are produced in variable amounts as a result of the metabolism of heterofermentative LAB.

Marked changes also occur in the lipid components of the cabbage including waxes, fats, and phospholipids, although these lipids are present in minor amounts.[27,28] For instance, a high proportion of the acetone-soluble true fats are hydrolyzed to glycerol and free fatty acids, and phospholipids may be fermented to yield glycerol, free fatty acids, phosphates, and free choline.[29]

Volatile sulfur compounds such as hydrogen sulfide, methanethiol (methyl mercaptan), dimethyl sulfide, and allyl isothiocyanate can be detected in the headspace of sauerkraut and sauerkraut juice and have a great impact on the flavor of the sauerkraut.[30] A variety of additional volatile compounds, including carbonyls such as diacetyl and acetaldehyde, are produced by the bacteria, by auto-chemical reactions, or by the intrinsic enzymes of the fermenting cabbage itself.[31] Acetal, isoamyl alcohol, *n*-hexanol, ethyl lactate, and *cis*-hex-3-ene-1-ol were among the substances identified as major volatiles in sauerkraut.[32]

White cabbage contains variable amounts of different glucosinolates.[33] After wounding of the vegetable tissue, the glucosinolates are hydrolyzed by the enzyme myrosinase (E.C. number 3.2.3.1), resulting, after some chemical reactions, in the formation of thiocyanates, which are responsible for the typical cabbage-like taste. During sauerkraut fermentation, the goal is to remove this taste, otherwise the sauerkraut is said to be green or immature. Gail-Eller and Gierschner[30] have found that the hydrolysis of the glucosinolates is temperature dependent. At a fermentation temperature of 19°C, hydrolysis is completed within 3 days, while at 5°C it takes up to 8 days.

TABLE 14.4
Constituents of White Cabbage and Sauerkraut (Average Values per 100 g)

Constituents	White Cabbage	Sauerkraut
Principal Constituents		
Water (g)	90.4	90.7
Total nitrogen (g)	0.22	0.24
Protein (N × 6.25) (g)	1.37	1.52
Fat (g)	0.20	0.31
Available carbohydrates (g)	4.18	0.77
Total dietary fiber (g)	2.96	2.14
Available organic acids (g)	0.23	1.60
Minerals (g)	0.66	2.35
Minerals and Trace Elements		
Sodium (mg)	12	355
Potassium (mg)	255	288
Magnesium (mg)	14	14
Calcium (mg)	45	48
Manganese (μg)	200	140
Iron (μg)	412	600
Copper (μg)	33	130
Zinc (μg)	224	320
Phosphorus (mg)	36	43
Vitamins and Related Substances		
Retinol equivalent (μg)	12	3.0
Total carotenoids (μg)	69	18
β-carotene (μg)	69	18
Vitamin K (μg)	70	62
Vitamin B_1 (μg)	43	27
Vitamin B_2 (μg)	45	50
Nicotinamide	320	170
Pantothenic acid	260	230
Vitamin B_6	190	210
Folic acid	31	31
Vitamin C	48	20
Specific Carbohydrates		
Glucose (mg)	2039	420
Fructose (mg)	1762	210
Sucrose (mg)	347	140

Source: Souci, S.W., Fachmann, W., and Kraut, H., *Food Composition and Nutrition Tables,* Medpharm. Scient. Publ., Stuttgart, 2000. With permission.

TABLE 14.5
Selected Analytical Data from Fermenting Sauerkraut on Days 1 and 23 of Fermentation at 19°C[25]

Attribute	Day 1	Day 23
pH	6.18	3.41
Total acidity (titrated to pH 8.4; calculated as lactic acid)	0.38 g/kg	12.45 g/kg
Lactic acid	0.09 g/kg	11.67 g/kg
D-(−) Lactic acid	0.02 g/kg	8.77 g/kg
L-(+) Lactic acid	0.07 g/kg	2.90 g/kg
Acetic acid	0.03 g/kg	3.07 g/kg
Reducing sugars	38.0 g/kg	5.8 g/kg
Dry matter	9.1%	8.3%
Sodium chloride	17.4 g/kg	17.5 g/kg

Source: Gail-Eller, R., Beitrag zur Kenntnis der Inhaltsstoffe von Sauerkraut in Kleinbehältern, Hergestellt Nach Einem Neuen Verfahren, dissertation, Hohenheim University, Stuttgart, Germany, 1984. With permission.

During the last centuries, sauerkraut was one of the major sources of vitamin C in the diet. From a technological point of view, the vitamin C content is a typical indicator for good manufacturing practice of sauerkraut, and it is important to stabilize the taste and the color of the final product as well.[34] Bohrer[34] investigated the influence of the different steps in sauerkraut processing on the vitamin C content. It could be demonstrated that all steps cause different losses in total vitamin C (sum of ascorbic and dehydro-ascorbic acid). Depending on the quality of the raw material and the processing parameters, the losses of vitamin C from the fresh cabbage to canned and pasteurized sauerkraut ranged from 13.2 to 52.7%.

Boskov Hansen et al.[35] investigated the effect of lactic acid fermentation of cabbage on the content and solubility of dietary fibers. In the fermented product, the content of total dietary fiber was higher than in the original cabbage, which can be explained by the concentration increase due to the removal of cell liquid. The percentage composition of soluble and insoluble fibers was only slightly changed. However, it could be shown that changes in the fibers occur during fermentation. The solubility of the dietary fiber decreased within 7 days of fermentation. During further fermentation for 3 weeks, the solubility increased to the same percentage as in raw vegetables.

14.5 PRODUCT QUALITY AND FORMATION OF METABOLITES

14.5.1 Biogenic Amines

Some microorganisms associated with food fermentations may cause the formation of biogenic amines (BA) from free amino acids by the activity of amino acid

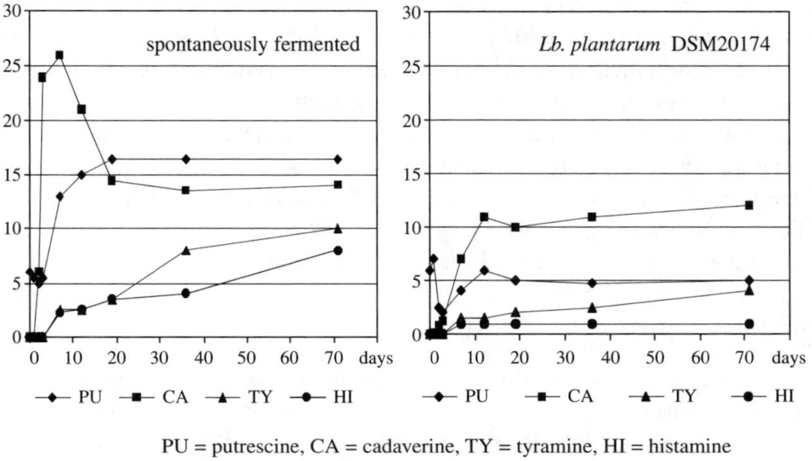

FIGURE 14.3 Biogenic amine formation (mg/100 g dry matter) during sauerkraut fermentation without starter culture and with *Lactobacillus plantarum* DSM20174. (From Halász, A., Baráth, Ă., and Holzapfel, W.H., The influence of starter culture selection on sauerkraut fermentation, *Z. Lebensm. Unters. Forsch.*, 208, 434–438, 1999. With permission.)

decarboxylase enzymes. BA are low-molecular-weight organic bases; they are biologically active and may cause intoxications. The "classic" BA intoxication is caused by histamine, which at levels of ≥50 mg/100 g may cause toxic effects or allergy-like symptoms such as sneezing, headache/migraine, shortness of breath, etc. Therefore, the presence of BA as products of microbial metabolism in fermented or spoiled foods is important. Histamine production is more typically associated with Gram-negative bacteria such as the Enterobacteriaceae. However, these bacteria only occur in significant numbers during the early stages of sauerkraut fermentation and are soon inhibited by the competing LAB and the concomitant reduction of pH (see Figure 14.3). Taylor et al.[36] suggested that histamine production mainly occurs during the first fermentation phase. This explains why histamine is not generally considered a typical major BA in sauerkraut. Still, its presence, together with that of putrescine has been noted. The histamine concentration in sauerkraut has been reported to fall within the range of 9 to 200 mg/kg,[37,38] although typical amounts seem to range between 12 and 78 mg/kg.[39] Average values of 174, 146, and 50 mg/kg have been reported for tyramine, putrescine, and cadaverine, respectively, in household and commercial sauerkraut from the Czech Republic and Austria, with the lowest concentrations in the household product.[40] Generally, BA concentrations are much higher in the brine than in the fermented cabbage,[22] although Kalač et al.[41] observed no differences. They also reported the highest levels for tyramine during sauerkraut storage, followed by putrescine and cadaverine.

It has clearly been shown that the concentration and type of BA in sauerkraut are influenced by fermentation conditions. Well-controlled fermentations followed by storage at 4°C resulted in reduced levels of cadaverine, putrescine, and spermidine,[22] whereas an increase in histamine appeared to be associated with reduction of the pH below 3.8 to 3.6.[39] Moreover, it is known that only a small proportion of the food-associated bacteria may be able to produce toxicologically significant amounts of BA, and even within a species, strains may have different levels of decarboxylase activities. In an attempt to reduce BA production during sauerkraut fermentation, selected strains of *Lb. plantarum* with low amino acid decarboxylase activities have been applied as starter cultures, and the generation of BA during fermentation was compared to that in spontaneously fermented cabbage.[22] In this way, the formation of all BA, and especially of cadaverine and putrescine, could be significantly reduced (see Figure 14.3). Starter culture concentrations necessary for this effect appear to amount to at least 5×10^6 CFU/100 g of cabbage.[22]

14.5.2 Bacteriophages

Bacteriophages do not appear to cause problems in sauerkraut fermentation. Pure culture fermentations are not common, and if one or several LAB strains are infected by phages, other naturally occurring strains will become dominant and carry out the fermentation. Dissemination of bacteriophages within a fermentor is very limited, since the brine is normally not circulated in sauerkraut fermentation. However, dissemination may occur during the early stages of fermentation as long as carbon dioxide is produced by tissue respiration or microbial metabolism. The gas bubbles force their way to the surface and therefore enable transport through the fermenting substrate.

In a recent study by Yoon et al.,[9] nine different phages were isolated from commercial sauerkraut fermentations. They were characterized as members of the Siphoviridae and Myoviridae families and were found to be active against *Lc. mesenteroides* and *Lb. plantarum* strains isolated from the same product. Phages were also detected against a *Lc. mesenteroides* used as a starter culture in brine samples from the first day of inoculation. *Leuconostoc* phages reported in the literature are of the Siphoviridae family and were mainly of dairy origin. The diversity of phages indicates that they could play a significant role in the ecology of "natural" sauerkraut fermentation.[9]

14.6 HEALTH PROPERTIES OF SAUERKRAUT

Sauerkraut is considered to be a healthful product because it is an important source of vitamins (especially vitamin C), mineral salts, and dietary fiber (see Table 14.5). In our modern Western diet, however, sauerkraut no longer plays an essential role as source of vitamin C.

The lactic acid produced during sauerkraut fermentation normally consists of both isomers, the L-(+) and D-(–) forms. There is still discussion about the importance of L-(+)- versus D-(–)-lactic acid. Since the majority of the lactate produced by the metabolism of mammals is of the L-(+) form, this isomer is called the physiological form. Whereas L-(+)-lactic acid is metabolized by a specific lactate hydrogenase, the D-(–) isomer can only be used by a nonspecific D-2-hydroxy carbonic acid dehydrogenase. The oxidation rate of the nonspecific enzyme is much lower than that of the lactate hydrogenase, and it can be inhibited by L-(+)-lactate. This inhibition functions at all steps, which means that intake to the liver and transport within the kidneys, as well as oxidation by D-2-hydroxy carbonic acid dehydrogenase, will be inhibited. For a long time, it was assumed that large amounts of D-(–)-lactic acid ingested with foods would lead to a lactate acidosis. Today, it is agreed that intake of D-(–)-lactic acid is not a problem for healthy adults. Only for babies in their first year is it recommended to exclude foods containing the D-(–) isomer from the diet.[42]

Despite these scientific opinions, there is still a need for products containing predominantly L-(+)-lactic acid, and therefore such products are offered by the industry. The investigation of facultatively heterofermentative lactobacilli in sauerkraut fermentation has led to the isolation of strains characterized by the exclusive formation of L-(+)-lactic acid. Originally, such strains were designated as a new species named *Lactobacillus bavaricus*.[43] However, this "species" was renamed *Lb. sakei* after a high DNA similarity of these strains was found with *Lb. sakei*.[44] The so-called L-(+) sauerkraut that is distributed in health food stores is produced by the application of a racemase defective strain of *Lb. sakei*. This starter culture is highly competitive and well adapted, and is able to suppress *Lc. mesenteroides,* which otherwise initiates the fermentation. *Lb. plantarum* strains are also outnumbered. Therefore, in sauerkraut freshly fermented by such L-(+)-lactate producing strains, the L-(+) isomer represents more than 90% of the total lactic acid. However, the flavor is poorly developed because of the suppression of the heterofermentative LAB.

A health-promoting effect of sauerkraut may be linked to the high content of glucosinolates (up to 1% of dry weight) of white cabbage.[45] Glucosinolates undergo hydrolysis during fermentation[46] by the enzyme myrosinase. Some of the resulting metabolic products, including indoles and isothiocyanates, are highly reactive compounds and were shown to be powerful inhibitors of carcinogenesis in laboratory animals.[47] Isothiocyanates are able to inhibit mitosis and stimulate apoptosis in human tumor cells[48] and influence phase I and phase II biotransformation enzyme activities, thereby possibly influencing several processes related to chemical carcinogenesis, e.g., the metabolism, DNA-binding, and mutagenic activity of promutagens.[49] One of the major underlying mechanisms appears to be the selective inhibition of cytochrome P450 enzymes involved in carcinogenic metabolic activation.[47]

Another health-promoting property of sauerkraut may result from the LAB involved in sauerkraut fermentation. As with other fermented products, unpasteurized sauerkraut contains high numbers of viable LAB, which may include organisms showing a beneficial effect on the intestinal ecosystem of the consumer. Up to now, however, there are no reports on the probiotic efficacy of typical sauerkraut LAB, and the health effects of these organisms still have to be demonstrated.

REFERENCES

1. Fritz, D. and Stolz, W., *Erwerbsgemüsebau,* 8th ed., Eugen Ulmer, Stuttgart, 1980, p. 77.
2. Crowder, S., Wilson, D., Nicholls, G., and Cadell, T., Eds., *A Treatise on Scurvy,* 3rd ed., Bechet and Co, London, 1772.
3. Eichholtz, F., *Die Biologische Milchsäure und Ihre Entstehung in Vegetabilischem Materal,* 2nd ed., Eden-Stiftung, Bad Soden, 1975.

4. Buckenhüskes, H., Aabye Jensen, H., Andersson, R., Garrido Fernández, A., and Rodrigo, M., Fermented vegetables, in *Processing and Quality of Foods. Vol. 2. Food Biotechnology: Avenues to Healthy and Nutritious Products*, Zeuthen, P., Cheftel, J.C., Eriksson, C., Gormley, T.R., Linko, P., and Paulus, K., Eds., Elsevier Applied Science, London, 1990, p. 162.
5. VdS: Verband der deutschen Sauerkonserven-Industrie e.V., Geschäftsbericht, Bonn, 1992/93 and 1999/2000.
6. Holzapfel, W.H., Haltbarmachung durch Säuren, in *HACCP in der Praxis: Lebensmittelsicherheit*, Sinell, H.-J. and Meyer, H., Eds., Behr's Verlag, Hamburg, 1996, pp. 111–132.
7. Pederson, C.S., Sauerkraut, in *Advances in Food Research,* Vol. 10, Chichester, C.O., Mrak, E.M., and Stewart, G.F., Eds. Academic Press, New York, 1960, pp. 233–291.
8. Buckenhüskes, H., Gessler, A., and Gierschner, K., Analytische Charakterisierung von pasteurisiertem Sauerkraut in Dosen, *Ind. e Obst- und Gemüseverwertung*, 73, 454–463, 1988.
9. Yoon, S.S., Barrangou-Poueys, R., Breidt, F., Jr., Klaenhammer, T.R., and Fleming, H.P., Isolation and characterization of bacteriophages from fermenting sauerkraut, *Appl. Environ. Microbiol.,* 68, 973–976, 2002.
10. Buckenhüskes, H. and Gierschner, K., Determination of the buffering capacity of some vegetables (in German), *Alimenta,* 24, 83–88, 1985.
11. Buckenhüskes, H., Schneider, M., and Hammes, W.P., Die milchsaure Vergärung pflanzlicher Rohware unter besonderer Berücksichtigung der Herstellung von Sauerkraut, *Chem. Mikrobiol. Technol. Lebensm.*, 10, 42–53, 1986.
12. Buckenhüskes, H.J., Fermented vegetables, in *Food Microbiology — Fundamentals and Frontiers,* 2nd ed., Doyle, M.P., Beuchat L.R., and Montville, T.J., Eds., ASM Press, Washington, 2001, pp. 665–679.
13. Pederson, C.S., Floral changes in the fermentation of sauerkraut, *N.Y. State Agr. Exp. Sta. Tech. Bull.,* 168, 1930.
14. Schneider, M., Microbiology of Sauerkraut Fermented in Small Ready-to-Sell Containers (in German), Dissertation, Hohenheim University, Stuttgart, Germany, 1988.
15. Müller, G., *Mikrobiologie pflanzlicher Lebensmittel.,* 4. Aufl., VEB Fachbuchverlag, Leipzig, 1988.
16. Stamer, J.R., Stoyla, B.O., and Dunckel, B.E., Growth rates and fermentation patterns of lactic acid bacteria associated with the sauerkraut fermentation, *J. Milk Food Technol.,* 34, 521–525, 1971.
17. Barrangou, R., Yoon, S.-S., Breidt, F., Jr., Fleming, H.P., and Klaenhammer, T.R., Identification and characterization of *Leuconostoc fallax* strains isolated from an industrial sauerkraut fermentation, *Appl. Environ. Microbiol.,* 68, 2877–2884, 2002.
18. Pederson, C.S., A study of the species *Lactobacillus plantarum* (Orla-Jensen) Bergey et al., *J. Bacteriol.,* 31, 217–224, 1936.
19. Kandler, O., Hammes, W.P., Schneider, M., and Stetter, K.O., Microbial Interaction in Sauerkraut Fermentation, in Proc. 4th Int. Symp. Microbial Ecology, 1987, pp. 302–308.
20. Dewar, W.A., McDonald, P., and Whittenbury, R., Hydrolysis of grass hemicelluloses during ensilage, *J. Sci. Food Agric.,* 14, 411–420, 1963.
21. Buckenhüskes, H.J. and Hammes, W.P., Starter cultures in fruit and vegetable processing (in German), *BioEngineering,* 6, 34–42, 1990.
22. Halász, A., Baráth, Ă., and Holzapfel, W.H., The influence of starter culture selection on sauerkraut fermentation, *Z. Lebensm. Unters. Forsch.,* 208, 434–438, 1999.
23. Kalač, P., Spička, J., Křižek, M., and Pelikánová, T., The effects of lactic acid bacteria on biogenic amines formation in sauerkraut, *Food Chem.,* 70, 355–359, 2000.

24. Lücke, F.-K., Brümmer, J.-M., Buckenhüskes, H., Garrido Fernandez, A., Rodrigo, M., and Smith, J.E., Starter culture development, in *Processing and Quality of Foods, Vol. 2. Food Biotechnology: Avenues to Healthy and Nutritious Products,* Zeuthen, P., Cheftel, J.C., Eriksson, C., Gormley, T.R., Linko, P., and Paulus, K. Eds., Elsevier Applied Science, London, 1990, p. 11.
25. Gail-Eller, R., Beitrag zur Kenntnis der Inhaltsstoffe von Sauerkraut in Kleinbehältern, Hergestellt Nach Einem Neuen Verfahren, Dissertation, Hohenheim University, Stuttgart, Germany, 1984.
26. Trail, A.C., Fleming, H.P., Young, C.T., and McFeeters, R.F., Chemical and sensory characterization of commercial sauerkraut, *J. Food Qual.,* 19, 15–30, 1996.
27. Vorbeck, M.L., Mattick, L.R., Lee, F.A., and Pederson, C.S., Volatile flavor of sauerkraut. Gas chromatographic identification of a volatile acidic off-flavor, *J. Food Sci.,* 26, 569–572, 1961.
28. Vorbeck, M.L., Albury, M.N., Mattick, L.R., Lee, F.A., and Pederson, C.S., Lipid alterations during the fermentation of vegetables by lactic acid bacteria, *J. Food Sci.,* 28, 495–552, 1963.
29. Steinkraus, K.H., Acid-fermented vegetables, in *Handbook of Indigenous Fermented Foods,* Steinkraus, K.H., Ed., Marcel Dekker, Inc., New York, 1983, pp. 99–131.
30. Gail-Eller, R., and Gierschner, K., Zum Gehalt und Verhalten der Glucosinolate in Weißkohl und Sauerkraut, *Deutsch. Lebenm,-Runds.,* 80, 341–346, 1984.
31. Hrdlicka, J., Curda, D., and Pavelka, J., Volatile carbonyl compounds during fermentation of cabbage, *Sb. vys. Sk. Chem.-technol.*, Praze, Potraviny, E15–51, 1967.
32. Lee, C.Y., Acree, T.E., Butts, R.M., and Stamer, J.R., Flavor constituents of fermented cabbage, *Proc. IV Int. Congress Food Sci. and Technol.,* 1, 175–178, 1976.
33. Watzl, B., Glucosinolate, *Ernährungs-Umschau,* 48, 330–333, 2001.
34. Bohrer, B., Untersuchungen über den Einfluß der Technologie auf den Vitamin C-Gehalt von pasteurisiertem Sauerkraut, Dissertation, Hohenheim University, Stuttgart, Germany, 1984.
35. Boskov Hansen, H., Bognar, A., Buckenhüskes, H., and Gierschner, K., The effect of lactic acid fermentation on dietary fibre content of vegetables, in *Processing and Quality of Foods. Vol.2. Food Biotechnology: Avenues to Healthy and Nutritious Products,* Zeuthen, P., Cheftel, J.C., Eriksson, C., Gormley, T.R., Linko, P., and Paulus, K. Eds., Elsevier Applied Science, London, 1990, p. 162.
36. Taylor, S.L., Lieber, E.R., and Leatherwood, M., Histamine in sauerkraut, *J. Food Sci.,* 43, 1030–1032, 1978.
37. Mayer, K. and Pause, G., Biogenic amine in sauerkraut, *Lebensm. Wiss. u Technol.,* 5, 108–109, 1972.
38. Mayer, K., Pause, G., and Vetsch, U., Formation of biogenic amines during sauerkraut fermentation, *Ind. Obst. Gemüseverwert.,* 58, 307–309, 1973.
39. Buckenhüskes, H.J., Sabatke, I., and Gierschner, K., Zur Frage des Vorkommens Biogener Amine in milchsauer fermentiertem Gemüse, *Industrielle Obst- und Gemüseverwertung,* 77, 255–263, 1992.
40. Kalac, P., Spicka, J., Krizek, M., Stewidlova, S., and Pelikanova, T., Concentration of seven biogenic amines in sauerkraut, *Food Chem.,* 67, 275–280, 1999.
41. Kalac, P., Spicka, J., Krizek, M., and Pelikanova, T., Changes in biogenic amine concentrations during sauerkraut storage, *Food Chem.,* 69, 309–314, 2000.
42. Bässler, K.H., Die physiologische Rolle von Laktat im Licht neuerer Erkenntnisse, Ernährungs-Umschau, 35, 71–74, 1988.
43. Stetter, H. and Stetter, K.O., *Lactobacillus bavaricus* sp. nov., a new species of the subgenus *Streptobacterium, Zbl. Bakt., I. Abt. Orig.,* C1, 70–74, 1980.

44. Kagermeier-Callaway, A.S. and Lauer, E., *Lactobacillus sakei* Katagiri, Kitahara, and Fukami 1934 is the senior synonym for *Lactobacillus bavaricus* Stetter and Stetter 1980, *Int. J. Syst. Bacteriol.,* 45, 398–399, 1955.
45. Kushad, M.M., Brown, A.F., Kurilich, A.C., Juvik, J.A., Klein, B.P., Wallig, M.A., and Jeffery, E.H., Variation of glucosinolates in vegetable crops of *Brassica oleracea, J. Agric. Food Chem.,* 47, 1541–1548, 1999.
46. Fenwick, G.R., Haeney, R.K., and Mullin, W.J., Glucosinolates and their breakdown products in food and food plants, *CRC Crit. Rev. Food Sci. Nutr.,* 18, 135–201, 1982.
47. Hecht, S.S., Inhibition of carcinogenesis by isothiocyanates, *Drug Metab. Rev.,* 32, 395–411, 2000.
48. Johnson, I.T., Glucosinolates: bioavailability and importance to health, *Int. J. Vitam. Nutr. Res.,* 72, 26–31, 2002.
49. Verhoeven, D.T., Verhagen, H., Goldblom, R.A., van den Brandt, P.A., and van Poppel, G., A review of mechanisms underlying anticarcinogenicity by brassica vegetables, *Chem. Biol. Interact.,* 28, 79–129, 1997.
50. Souci, S.W., Fachmann, W., and Kraut, H., *Food Composition and Nutrition Tables*, Medpharm Scientific, Stuttgart, 2000.

15 The Future for Fermented Foods

Edward R. Farnworth

CONTENTS

15.1 INTRODUCTION

Fermented foods are consumed in every country of the world and, as the chapters in this book show, there is growing scientific evidence that many fermented foods are good for health or contain ingredients that are good for health. Foods that improve or change the intestinal microflora are of particular interest because of our increased knowledge of the role the intestinal microflora plays in health and disease resistance. In the future, more fermented foods with health-promoting properties will become available on the market, with many directed towards consumers with very specific health and metabolic needs.

The desire to buy and consume foods that impact favorably on the gastrointestinal (GI) tract will be driven by consumer demands and by the advancement of knowledge

0-8493-1372-4/03/$0.00+$1.50

of both microbiology in general and the human intestinal microflora in particular. It is becoming evident that each individual has a unique microflora. As more sophisticated and rapid methods are developed that can characterize this population, the nutrition and health implications of maintaining or altering this population will become more evident.

In the future, fermented foods will become even more important in our diet and the maintenance of health, as we identify different microorganisms that can be used in the production of probiotic foods. Probiotic foods will be made that target specific age groups who have specific metabolic requirements (newborns, adolescents, seniors), people in specific disease states (irritable bowel syndrome, Crohn's disease, intestinal cancers) or those who have had their microflora compromised (irradiation patients, intestinal surgery patients, people who have received antibiotics). These advances will occur as we understand more about the role the intestinal bacteria play in human health and as we are able to identify the mechanisms involved in the interaction between food bacteria passing through the GI tract and the host intestinal bacteria.

New processes in food technology will allow for the incorporation of microorganisms in a wider range of foods and beverages. At this time, it is still not clear whether all microorganisms must be alive when they are consumed to exert their beneficial effects; however, it is apparent that many conditions during food processing (extremes of temperature, humidity, pH) challenge the viability of most microorganisms.[1] New ways will be found to add and protect microorganisms during manufacturing, packaging, and storage. Alternatively, new strains of microorganisms may be identified with characteristics that make them better ingredients in food products. In addition, new technologies and processes such as controlled release (by pH, by time, by enzymatic action) encapsulation will be developed that enable live bacteria to be delivered to very specific locations along the GI tract. In the future, novel foods termed synbiotics that contain both probiotic bacteria and prebiotics (substrates required by targeted bacteria) will be available to the consumer.[2]

Foods that are produced by or contain microorganisms when they are consumed can impact on many disease states that plague humans today. As we start exploring beyond Earth, other problems will arise. Health and metabolic problems have already been identified during our first ventures into space. In the future, astronauts may consume probiotic foods to protect them during space travel and aid in their adaptation when they return to earth.

15.2 ADVANCES IN MICROBIOLOGY

Our knowledge of what endogenous bacteria are contributing to the host will advance only when they are fully identified and enumerated. It has been estimated that even with our current knowledge of the growth requirements of bacteria, only about 60% of the bacteria observed in the human GI tract can be isolated and then grown and identified outside the body.[3] Welling et al.[4] quoted data from an experiment where total *Bifidobacterium* spp. were counted using plate culturing (3.87×10^{10}/g) compared to nonculturing techniques (2.71×10^{11}/g). Suau et al.[5] cited *Fusobacterium*

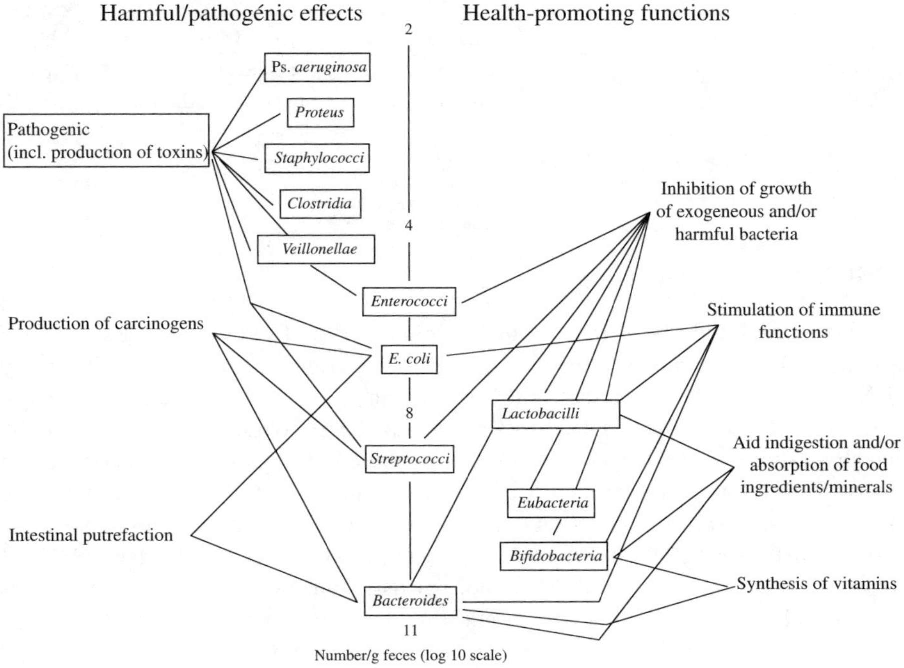

FIGURE 15.1 Bacterial population in the human gastrointestinal tract. (From Gibson, G.R. and Roberfroid, M.B., Dietary modulation of the human colonic microbiota: introducing the concept of probiotics, *J. Nutr.,* 125, 1401–1412, 1995. With permission.)

prausnitzii as an example of a bacterium that is believed to be one of the prominent species in the human gut but that is often reported as not detected, possibly due to problems with traditional enumeration techniques. Often more than one bacterial species is capable of growing on a specific medium, again adding to an inaccurate count of the microbial population. These technical shortcomings lead to an incomplete picture of the gut microbiota.

Various reports have been made as to the population of the microflora of the human intestine. Figure 15.1 lists what are believed to be the most common species. The figure also shows one estimate of the sizes of the populations of particular bacterial species, genera, or families. However, the population estimates appear to vary depending on the methodology used and can vary among individuals. Advances in molecular biology have meant that a large number of bacterial groups in the gut ecosystem can now be identified and counted unequivocally — even many bacteria that are not culturable using classical techniques.

15.2.1 New Methods of Identification and Enumeration of Bacteria

Ribosomal 16S rRNA has been used in molecular phylogeny, where the different degrees of variability in the nucleotide makeup allow for the construction of phy-

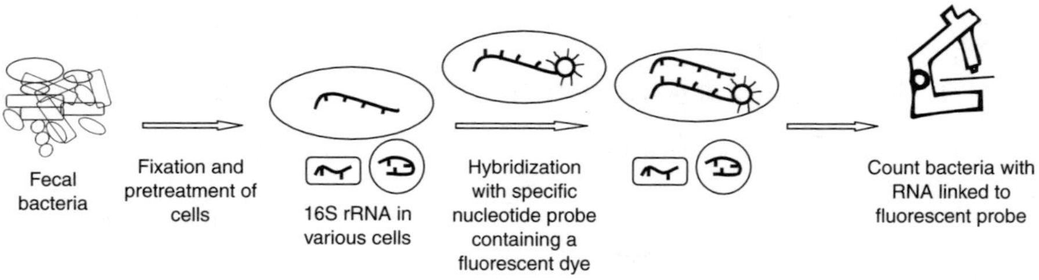

FIGURE 15.2 The FISH method.

logenetic trees and the establishment of evolutionary links between species.[6] Using the many bacterial nucleotide sequences that have been elucidated, complimentary DNA-probes can be made that will hybridize uniquely with a specific 16S rRNA sequence.[7] Probes generally need only be around 20 nucleotides in length. Depending on the uniqueness of the 16S rRNA being studied, probes can be customized to identify bacteria at different phylogenic levels (domain, family, genus, species). The fluorescence *in situ* hybridization (FISH) method uses synthesized oligonucleotide probes that have been labeled with fluorescent dyes that target the 16S rRNA of various bacteria (see Figure 15.2). Table 15.1 lists some of the nucleotide sequences that have been used to study human intestinal bacterial populations.

The FISH method is not without its own problems. Some counting of nontarget organisms can occur, and sensitivity may limit applications.[8,9] It has been also pointed out that in some cases, such as with Gram-positive bacteria (e.g., lactobacilli), penetration of the cell wall may be difficult and may require pretreatment of the sample with enzymes.[4,8] In spite of these problems, bacterial species such as bifidobacteria can be enumerated as accurately as with traditional counting techniques,[10] and as has been shown recently, with replicate sampling and counting, the variability of data using FISH can be much less than that obtained by plate counting methods.[11]

A DNA–RNA hybridization technique has been used by Sghir et al.[12] to estimate the relative proportions of the various intestinal bacterial populations in 27 human fecal samples. Individual bacterial groups were enumerated using a group-specific probe, and this count was compared to that obtained with a universal probe (Bact 338) that hybridizes to a conserved rRNA sequence that is found in the majority of bacterial cells. Seventy percent of the 16S rRNA hybridized by the universal probe was hybridized by the sum of six oligonucleotide probes (Bacto1080, Clept1240, Erecc482, Bif412, Lacto722, Enter1432). Total *Bacteroides* made up 37% of the total fecal bacteria probed, while *Lactobacillus* + *Streptococcus* + *Enterococcus* were found to contribute 1% of the total. The *Bifidobacterium* group accounted for less than 1% of the total. Sghir et al. emphasized the simplicity and specificity of this method and suggested it would be useful in nutrition studies where changes in the major microbial species could be monitored.

The FISH method is particularly suited for monitoring the effects of dietary changes on larger ($>10^6$ cells/g) populations of intestinal bacterial species.[11,13,14] Publications are now appearing in the literature in which this method has been

TABLE 15.1
Nucleotide Sequences Used in FISH Analyses of Fecal Bacteria

Nucelotide Sequence	Type of Bacteria
HIS 150 5'-TTA TGC GGT ATT AAT CT(CT) CCT TT-3'	Clostridia (*Clostridium perfringens/histolyticum* subgroup)
Bif 164 5'-CAT CCG GCA TTA CCA CCC-3'	Bifidobacteria
Bac 303 5'-CCA ATG TGG GGG ACC TT-3'	Bacteriodes
Lab 158 5'-GGT ATT AGC A(CT)C TGT TTC CA-3'	Lactobacilli–enterococci
COR653 5'-CCC TCC C(A/C)T ACC GGA CCC-3' (Harmsen)	*Coriobacterium* and *Collinsella*
ATO291 5'-GGT CGG TCT CTC AAC CC-3' (Harmsen)	*Atopobium* cluster (includes coriobacterium group)
E.had579 5'-GCA TCC ACC ATA CCG TTC AG-3' (Schwiertz)	*Eubacterium ventiosum*
S-*-Fprau-0645-a-A-23 5'-CCT CTG CAC TAC TCA AGA AAA CA-3' (Suau)	*Fusobacterium prausnitzii*

Sources: Schwiertz, A., le Blay, G., and Blaut, M., *Appl. Environ. Micrbiol.,* 66, 375–382, 2000; Franks, A.H., Harmsen, H.J.M., Raangs, G.C., Jansen, G.J., Schut, F., and Welling, G.W., *Appl. Environ. Microbiol.,* 64, 3336–3345, 1998; Langendijk, P.S., Schut, F., Jansen, G.J., Raangs, G.C., Kamphuis, G.R., Wilkinson, M.H.F., and Welling, G.W., *Appl. Environ. Microbiol.,* 61, 3069–3075, 1995; Manz, W., Amann, R., Ludwig, W., Vancanneyt, M., and Schleifer, K.-H., *Microbiology,* 142, 1097–1106, 1996; Harmsen, H.J.M., Wildeboer-Veloo, A.C.M., Grijpstra, J., Knol, J., Degener, J.E., and Welling, G.W., *Appl. Environ. Microbiol.,* 6, 4523–4527, 2000. With permission.

used.[15–17] As more probes become available in the future, the application of this technique will grow.

Denaturing gradient gel electrophoresis (DGGE), which is based on the principles of temperature gradient gel electrophoresis (TGGE), is the most recent technique to be used to study the intestinal microbial population.[18,19] Figure 15.3 is a schematic of this procedure. The bacterial mixture is first digested to obtain 16S rRNA fragments. The 16S rRNAs are amplified using polymerase chain reactions (PCR) and then separated by DGGE. The DGGE uses a polyacrylamide gel that contains a

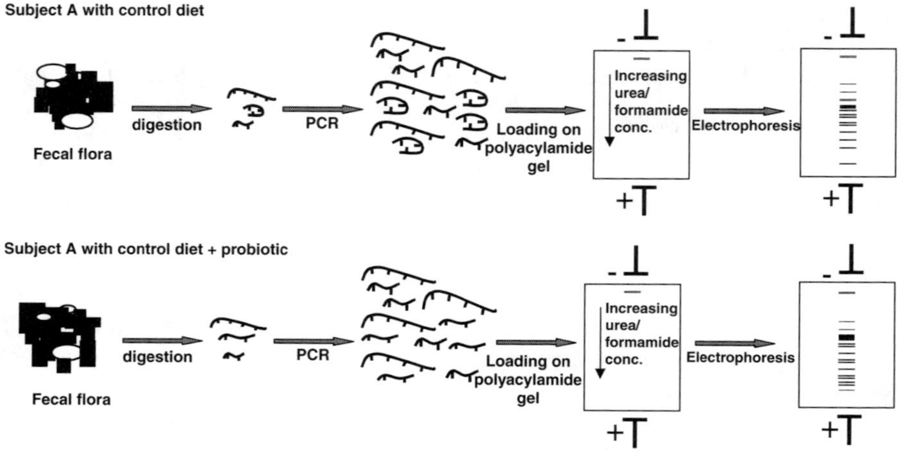

FIGURE 15.3 The DGGE method.

gradient of urea and formamide, which partially denatures (melts) the migrating 16S rRNA fragments and alters their structure and rate of migration. Bacteria with different 16S rRNA produce different 16S rRNA fragments, which in turn produce different migration patterns during electrophoresis. A characteristic pattern of 16S rRNA sequences from the original samples is obtained.[19] If identification of bacterial species is necessary, bands in the gel produced by DGGE can be cut out and the DNA fragments used to identify the bacterial species.

Tannock's group has used DGGE together with species-specific PCR primers to study fecal samples from mice, pigs, and humans.[20,21] They have shown that the DGGE method is able to identify bacteria down to the species level, and that in several cases, bacteria identified using DGGE were absent from traditionally plated samples. Changes in the DGGE pattern were used to identify the contribution of dietary bacteria to the total fecal population and to follow the time course of an administered probiotic bacterium (*Lactobacillus rhamnosus* DR20).

With the advent of techniques capable of identifying and enumerating bacteria at the species level, it is becoming apparent that a great variation exists among the microbial ecosystems of individuals, which means that in the future, probiotic products may have to be tailored to specific individual needs.

15.2.2 Bacterium-to-Bacterium Communication

The population of bacteria that inhabits the gastrointestinal tract is large and diverse, and over time a symbiosis is established that results in a relatively stable population.[22,23] Endogenous bacteria occupy specific niches, and bacteria that are ingested in probiotic foods often are unable to displace these resident bacteria. Bacteria have a limited number of options to use in order to repel other microorganisms that may be competing for niches along the GI tract. It has recently been shown that bacteria that are part of a large population of similar bacteria are able to act in unison against a potential danger posed by foreign bacteria. Bacteria communicate with each other, and they do not need to be in physical contact to do so.[24] Some bacteria use this communication system as a means of estimating the number of like-bacteria close by and then signaling the production of protective bacteriocins. Wirth et al.[24] have used the term bacterial pheromones to describe these chemical compounds that are produced within the bacteria, diffuse freely across the cell membrane, and accumulate in the surrounding medium. If the pheromone accumulates above a critical concentration extracellularly, an intracellular response regulator is activated. This system of communication signals bacteria to produce antibiotics and virulence factors.

Acylated homoserine lactone (AHL) was one of the first signal molecules to be identified and is perhaps the best understood.[25] The enzyme that makes AHL (Lux I-type protein) and the protein that detects it and responds by activating specific genes have been identified.[26–28] In one study, it has been shown that a bacterium that resides on wheat has an AHL system that it uses to determine when to make antibiotic that helps it suppress its microbial competitors, as well as protect its host.[28] This bacterium-to-bacterium communication mechanism has been termed quorum sensing, auto-induction, or population density-responsive gene regulation. It is believed that this attribute enables related bacteria to be competitive in a mixed bacterial population. It is also possible that the bacteria are exchanging more than just population density information.

It is evident that as these forms of communication between bacteria become better understood, it will be possible to choose probiotic bacteria that will be better able to compete with endogenous intestinal bacteria. At the present time, a major weakness of probiotic products is that the bacteria consumed are not able to displace endogenous bacteria. As a result, after the feeding regime is stopped, the food bacteria cannot be detected in the GI tract of the host. Bacteria that can sense the presence of pathogenic bacteria and are capable of releasing bacteriocins could also be used in place of broad-spectrum antibiotics to kill pathogenic bacteria that cause disease and infection.

15.2.3 Bacteria-to-Intestinal Cell Communication

The diverse bacterial population that inhabits the GI tract has the opportunity to interact with food bacteria, setting up cell-to-cell communication. However, even more important is the fact that the cells that line the GI tract come in contact with the contents of the digesta including both beneficial (probiotic) bacteria and enteric pathogens. Recent reports indicate that a complex host cell–bacterium communication occurs between the intestinal cells and GI tract bacteria. This bacterium–epithelial cell crosstalk is best understood in the case of pathogenic bacteria. It appears that a variety of mechanisms are used by the foreign bacteria to talk to cells.[29] This first step is followed by processes that allow the bacteria to adhere to and invade mucosal cells.

Researchers have shown that by using a group of mice reared in a germ-free environment and then exposing these mice to either intact mice microbiota or a selected group of organisms, the metabolism of the cells lining the intestine can be altered. In normal mice, the upper portion of the crypts and their associated villi in the mouse ileum produce fucosylated intestinal glycoconjugates. Bry et al.[30] have shown that germ-free mice lose the ability to produce these fucosylated glycoconjugates, but that it can be restored by inoculating the germ-free mice with gut flora from conventionally reared mice or with *Bacteroides thetaiotaomicron* — a bacterium that is normally found in the intestines of mice and humans. The *B. thetaiotaomicron* cells were only effective in restoring the fucosylation when their concentrations reached $>10^7$ colony forming units (CFU)/ml ileal contents. It is believed that the bacteria communicate with the mammalian cells without direct binding and that soluble mediators are involved in the crosstalk mechanism.

Lu and Walker[29] speculate that some of the beneficial effects that have been attributed to probiotic bacteria, such as disease prevention, immune system stimulation, and strengthening of the epithelial barrier, are the result of biochemical communication between the probiotic bacteria and enterocytes. If this is true, and the mechanism of communication can become understood, probiotic products could be designed that are capable of altering targeted metabolic pathways in intestinal cells.

15.3 THE ROLE OF INTESTINAL BACTERIA IN HUMAN HEALTH IN THE FUTURE

Many of the beneficial functions of probiotic bacteria have not yet been defined. However, researchers have come to identify some bacteria as "good" and some as detrimental to the host (Figure 15.1). It is the goal of all probiotic and prebiotic research to produce the most favorable balance of GI microflora and thereby improve the metabolism, digestion, health, and disease resistance of the host.

15.3.1 New Probiotic Products

At the present time, three bacterial genera dominate in terms of interest, research into their possible beneficial effects, and number of products already on the market. These bacteria are *Bifidobacterium* (including *B. bifidum, B. breve, B. infantis, B. longum,* and *B. adolescentis*), *Enterococcus* (including *E. faecium* and *E. faecalis*), and *Lactobacillus* (including *Lb. acidophilus, Lb. paracasei, Lb. rhamnosus,* and *Lb. reuteri*).[31] Tables 15.2 and 15.3 are lists of probiotic products currently on the market that contain lactobacilli and bifidobacteria. Although lactic acid producing bacteria are dominant today, in the future, other characteristics may become important. For example, some propionic acid producing bacteria produce vitamin B_{12} — a characteristic that would be important in the design of foods for vegetarians, who have few sources of vitamin B_{12} in their diets. Also, very little attention has been paid to the potential benefits of incorporating yeasts and molds into probiotic products.

Many of the probiotic bacteria of interest are of human origin, and therefore there is a tendency to believe that these bacteria are better suited to be included in human food, if the ultimate target is the human intestine. However, at this time, little scientific evidence exists to support this assumption. Very often, potential probiotic species do not grow well in milk, and given the fact that a large majority of commercially available probiotic products today are milk based, this presents problems in terms of maintaining viability until the product is consumed. In the future, other food matrices besides milk will be used as the basal ingredient in probiotic foods.

TABLE 15.2
Some Fermented Milk Products Containing *Lactobacillus acidophilus*[79]

Product	Microorganism(s)
Acidophilus milk	*Lb. acidophilus*
Acidophilus paste	*Lb. acidophilus*
Acidophilus buttermilk	*Lb. acidophilus,* mesophilic starter
Acidophilus natural buttermilk	*Lb. acidophilus,* mesophilic starter
Acidophilus yogurt	*Lb. acidophilus, Streptococcus thermophilus, Lb. delbrueckii* ssp. *bulgaricus*
Biogurt	*Lb. acidophilus, S. thermophilus*
Acidophilus–yeast milk	*Lb. acidophilus,* lactose–fermenting yeasts
Acidophilin	*Lb. acidophilus, Lactococcus lactis,* kefir culture
LC1	*Lb. acidophilus* La 1
ABS ferment	*Lb. acidophilus, Bifidobacterium, Lb. casei*
LA-7 plus	*Lb. acidophilus, Bifidobacterium bifidum*
Vita	*Lb acidophilus* LA-H3, *B. bifidum* LB-H1, *Lb. casei* LC-H2

Source: Kalantzopoulos, G., in *Encyclopedia of Food Microbiology,* Batt, C.A. and Patel, P.D., Eds., Academic Press, New York, 1999, pp. 1373–1379. With permission.

TABLE 15.3
Some Fermented Milk Products Containing Bifidobacteria[79]

Product	Microorganism(s)
Bifidus milk	*Bifidobacterium bifidum* or *B. longum*
Bifigurt	*B. bifidum, Streptococcus thermophilus*
Biogarde	*B. bifidum, Lb. acidophilus, S. thermophilus*
Biokys	*B. bifidum, Lb. acidophilus, Pediococcus acidilactici*
Special yogurt	*B. bifidum (B. longum), Lb. acidophilus, S. thermophilus, Lb. delbrueckii* ssp. *bulgaricus*
Cultura	*B. bifidum, Lb. acidophilus*
Mi-Mi	*B. bifidum, B. breve, Lb. acidophilus*
Progurt	*Lactococcus lactis* biovar. *diacetilactis, Lactococcus lactis* ssp. *cremoris, Lb. acidophilus* and/or *B. bifidum*

Source: Kalantzopoulos, G., in *Encyclopedia of Food Microbiology,* Batt, C.A. and Patel, P.D., Eds., Academic Press, New York, 1999, pp. 1373–1379. With permission.

15.3.2 Products for Specific Consumers

15.3.2.1 Infants

Studies of the bacterial profile of individuals have shown that as the diet changes from a milk-based one during infancy to a more varied one as we age, the population of gut bacteria changes. The newborn is inoculated by maternal bacteria, and this leads to colonization by large numbers of facultative anaerobes such as *Esherichia coli* and enterococci.[32] This in turn produces a highly reduced environment that favors the growth of strict anaerobes. At this stage of life, diet plays an important role in determining the microbiological population of the GI tract. Both breast-fed and formula-fed infants have a microbiota dominated by bifidobacteria. However formula-fed infants have a characteristic colon microflora that contains higher counts of other bacteria including bacteriodes, eubacteria, enterobacteria, peptococci, and streptococci.[33,34] The long-term health implications of these differences are not known at this time. However, much of the interest in bifidobacteria as possible probiotic bacteria stems from these observations.

The regulations governing the composition of milk replacers and infant food are very strict in most jurisdictions.[35] At the present time, regulations related to the microbiological composition of infant foods are limited to measures to ensure that infant formulas are not contaminated by pathogenic bacteria.[36] Human milk and the breast-fed infant are used as standards, and it is apparent that current milk replacers do not contain any live bacteria or any prebiotic material specifically added to encourage the establishment of a pattern of bacteria in the formula-fed infant that is closer to that of the breast-fed infant. It is conceivable that in the future, milk replacers and infant formulae will contain live bacteria that, when ingested by the infant, may produce a microbial population more like that of the breast-fed baby.

The intestinal microbial population once established is resistant to change. Therefore, targeting infants with specific beneficial bacteria may prove to be the best opportunity to change the endogenous bacterial population. As time passes, the endogenous microbiota become more established, and bacteria ingested in probiotic products have been shown to have a very low capacity to displace endogenous bacteria, thrive, and grow.[37] The factors that allow the members of the endogenous microbiota to establish and maintain their regional habitats are largely unknown at this time.

15.3.2.2 Aging Population

As humans age, their intestinal bacterial population changes. For example, it is generally agreed that *Bifidobacterium infantis* and *B. breve* predominate in the infant, but *B. longum* and *B. adolescentis* are most common in the adult, and *B. adolescentis* is most common in the elderly. These changes in the intestinal population are in part due to diet, but other factors including environment, disease state, and use of antibiotics and other drugs can also be important.[38] Although there is no accepted "ideal" pattern of GI bacteria, it is believed that it would be beneficial in the long term to increase or maintain high levels of bifidobacteria and lactobacilli. It is believed that with proper intervention, digestive system–related problems such as diarrhea, constipation, and gastroenteritis could be alleviated. In addition, as the metabolism and function of these bacteria become better understood, it is anticipated that other benefits will be realized, including an improved resistance to GI tract infections, improved immune response, prevention of cancer, production of vitamins, and increased calcium uptake.

At the present time, a multinational, multidisciplinary study is being carried out in Europe to characterize the "typical" gut microflora of an adult and an aged population.[39] This basic information is intended to be used as a baseline with which the effects of future interventions can be compared. Only by gathering such extensive data will it be possible to establish any relationships between the makeup of the gut microflora and health and disease status. It is believed that with the use of probiotic products (those that contain live bacteria), prebiotic products (those that contain specific nutrients, in particular complex carbohydrates), and synbiotic products (those that contain both probiotics and prebiotics), new food products can be developed that maintain a healthy gastrointestinal microflora even as life expectancy increases.

In the future, products will be specifically designed for particular age groups, those on special diets such as vegetarians, those with specific metabolic requirements, and even different probiotic products for men and women.

15.4 ASTRONAUTS

The preoccupation of nutritionists today is focused on the various sectors of the population here on Earth. Our experience with extraterrestrial travel has only a 41-year history,[40] but since the initial flights, astronauts have been under special scrutiny, and already several nutrition/metabolism related problems have been noted as a result of weightlessness, nutrient intake changes, and stress.[41–46]

The nutrition/metabolism problems that affect the performance of astronauts include gastrointestinal disorders, bone calcium loss, radiation damage, and immune response changes.[41,42,47–50] As space flights become longer in duration, health/metabolism problems will become even more important and may in fact put limits on how long we can live away from earth.[42,45,51] Some of the properties of fermented foods noted in this book may present partial solutions to these space travel-related problems.

15.4.1 Intestinal Problems (Diarrhea/Constipation)

Intestinal disorders of any sort could be very debilitating during space travel. Even a short bout of diarrhea could present many practical problems in the confines of a spacesuit or spaceship. Preflight precautions are taken in the preparation and packaging of food to avoid food-related infections, but it is inevitable that in spite of the high health status of the astronauts, diarrhea may strike space travelers. Several studies have been reported in which probiotic products containing particular strains of bifidobacteria, streptococcus, or lactobacillus have been successfully used to prevent diarrhea and to reduce the severity and duration of a variety of diarrheas.[52–55] Hove et al.[54] list 12 studies in which the consumption of fermented milk products or particular bacterial strains prevented or reduced the severity of diarrhea in infants and adults. The inclusion of probiotic products with a specific bacterial composition may be one way to reduce problems of diarrhea in space.

A general impression exists that constipation plagues astronauts, although this is not well documented.[56] The cause of this disorder is not clear, but it may be a combination of an inadequate water intake, confined living space, limited facilities for personal hygiene, and lack of gravity. In the case of the latter, the effects of zero gravity have been the subject of much speculation. Normally, it is assumed that peristalsis is responsible for facilitating the passage of food down the digestive tract. However, it is not clear whether this activity is assisted by or is in some way dependent on gravity. Changes in the perceived taste, texture, and bulk of meals may also contribute to digestive problems. In any case, it is common practice for astronauts to use mild laxatives during their missions to ensure proper bowel function.[56]

Probiotic products have a long history of effects at the intestinal level in terms of regulating bowel function.[57] Both stool frequency and stool softness were observed to increase when chronically constipated patients were given a strain of *Escherichia coli* daily.[58] Geriatric patients were found to have an increased defecation rate and decreased use of laxatives after consuming as little as 200 ml of acidophilus milk daily.[59] Relief of constipation and stimulation of intestinal motility are listed as beneficial effects of probiotic products.[60,61] Whether the positive effects of probiotics would be great enough to overcome the other constraints imposed during space travel will have to investigated.

The role of probiotics as often stated is to establish a balanced intestinal microbial population, and so it is clear that including probiotic products in the diet of astronauts may be one way to ensure proper intestinal function and reduce the possibility of diarrhea and constipation during space flights.

15.4.2 Calcium Metabolism

It has been well documented that during space flight, calcium metabolism is altered; calcium is lost from bones, serum levels of calcium rise, calcium excretion is increased, and some astronauts go into negative calcium balance.[47,48] This is a particular problem for the weight-bearing bones. Up to 0.4% of the total body calcium is lost per month, and concerns have been expressed that this rate may not decline as the time in space increases.[48] It appears that after return to Earth, calcium uptake by bones occurs, and blood levels and excretion rates return to normal values, but it is not clear at what point bone loss during space travel would be too great to replenish.

Dairy products in general are considered good sources of dietary calcium, and fermented milk drinks have been shown to be a good way to increase calcium intake.[62] Whether calcium absorption is improved after eating fermented foods is not clear. Early results using rats show (implied) improved calcium absorption.[63] Feeding yogurt has been shown to increase serum calcium levels,[64] but others have found no effect.[65] However, as Recker et al. have pointed out, the availability of calcium in fermented milk products may depend on the bacterial culture used to prepare the product.[65]

It is generally assumed that fermented foods can bring about a lowering of the pH in the GI tract, which should improve calcium absorption, and therefore including fermented milk products in the diet of astronauts may help improve calcium balance during long space voyages.

15.4.3 Radiation Damage

The effects of exposure to larger quantities of high energy radiation during space travel were anticipated as spacecraft ventured beyond the earth's atmosphere and geomagnetic field, which protect against high-energy charged particles. Based on dosimeter readings taken inside and outside spacecraft, it has been possible to calculate the amount of radiation astronauts are exposed to, and to calculate their probability of contracting radiation-induced cancer and radiation-induced cataracts.[49,66] This aspect of space travel could have a major impact on astronauts' health even when they return to earth, and ways of reducing the effects of this exposure therefore need to be found.

The antimutagenic properties of yogurt fractions, fractions of milk fermented by *Lactobacillus helveticus* L89, and a mixture of *Bifidobacterium longum, Lb. gasseri,* and *E. coli* have been demonstrated using *in vitro* tests.[67–69] In these experiments, the ability to prevent damage by known chemical and food-derived carcinogens and mutagens was tested. Both the bacteria themselves and fermentation products, which presumably contain active ingredients formed during fermentation, have been shown to be protective.

In animal models, both yogurt and kefir have been shown to suppress the growth of implanted tumors,[70–73] indicating that either the bacteria in these products or some product of the fermentation of milk is capable of slowing the progression of tumors.

Exposure to radiation produces several forms of acute injury. Exposure to 10,000 rad produces rapid death as a result of damage to the central nervous system. At lower doses (approximately 1000 rad), changes occur to the gastrointestinal bacterial population that also result in death, but over a period of days to weeks. Dong et al.[74] were able to show that feeding mice *Lactobacillus* GG before exposure to 1400 rad of radiation changed the intestinal flora and prolonged the animals' life.

It would appear that the consumption of fermented milk products may be a way to reduce the progression of cancer/tumors that result from exposure to ionization during space travel and may also be a way to restore the balance of intestinal microflora in astronauts who have been exposed to large doses of radiation. Whether fermented foods can also prevent the initial intra- and extracellular damage produced by radiation is not clear at this time.

15.4.4 Immune Function

Access to medical treatment during any voyage in space is extremely limited, and so it is of utmost importance that astronauts be in good health. To date, this has been achieved by selecting young, healthy individuals as candidates for space travel and monitoring their immunological responses before, during, and after space travel. Both *in vivo* and *in vitro* experiments have shown that cellular immune response is depressed during spaceflight.[75] As space voyages get longer, it is evident that every means possible must be used to keep astronauts healthy and disease free; maintaining or enhancing immune function will be important.

The impact of diet on the immune system is becoming clearer, and there is a growing consensus that the consumption of probiotics can improve immune status.[76]

The host immune system appears to be enhanced by activating macrophage function and increasing the activity of natural killer cells and T cells. Again, it is not clear whether the bacteria themselves or some metabolite(s) produced by the bacteria are stimulating immune function. Even if the exact mechanism is unknown, it may be prudent to include probiotic products in the diets of astronauts as a way to enhance their immune function as protection against disease and infection.

15.5 CONCLUSIONS

As our knowledge of the role microorganisms play in human nutrition, immune function, and disease resistance increases, so will the number of fermented products on the market. In some products, it will be necessary to ensure that the microorganisms consumed are alive, but in other products, metabolites and/or fermentation products may be the active ingredients. Fermented foods will be available for specific niche consumers.

Fermented foods have been part of the human diet for centuries, and it may well be that they will become important in the diets of future space travellers.

REFERENCES

1. Farnworth, E.R., Unique problems in designing and testing probiotic foods, *FoodInfo Online Features*, International Food Information Service, 2002, http://www.foodsciencecentral.com/library.html#ifis/3803.
2. Gibson, G.R., Nutritional modulation of the gut flora using probiotics, in *Probiotics and Health: The Intestinal Microflora.* Montreal International Symposium Proceedings, Roy, D., Ed., Edisem, Saint-Hyacinthe, Quebec, 2001, pp. 36–38.
3. Tannock, G.W., Exploring the intestinal ecosystem using molecular methods, in *Probiotics and Health: The Intestinal Microflora.* Montreal International Symposium Proceedings, Roy, D., Ed., Edisem, Saint-Hyacinthe, Quebec, 2001, pp. 14–35.
4. Welling, G.W., Elfferich, P., Raangs, G.C., Wildboer-Veloo, A.C.M., Jansen, G.J., and Degener, J.E., 16S RNA-targeted oligosaccharide probes for monitoring of intestinal tract bacteria, *Scand. J. Gastroenterol.,* 32 Suppl. 222, 17–19, 1997.
5. Suau, A., Rochet, V., Sghir, A., Gramet, G., Brewaeys, S., Sutren, M., Rigottier-Gois, L., and Doré, J., *Fusobacterium prausnitzii* and related species represent a dominant group within the human fecal flora, *Syst. Appl. Microbiol.,* 24, 139–145, 2001.
6. Woese, C.R., Bacterial evolution, *Microbiol. Rev.,* 51, 221–271, 1987.
7. Giovannoni, S.J., DeLong, E.F., Olsen, G.J., and Pace, N.R., Phylogenetic group-specific oligodeoxynucleotide probes for identification of single microbial cells, *J. Bacteriol.,* 170, 720–726, 1988.
8. Schwiertz, A., le Blay, G., and Blaut, M., Quantification of different *Eubacterium* spp. in human fecal samples with species-specific 16S rRNA-targeted oligonucleotide probes, *Appl. Environ. Micrbiol.,* 66, 375–382, 2000.
9. Franks, A.H., Harmsen, H.J.M., Raangs, G.C., Jansen, G.J., Schut, F., and Welling, G.W., Variations of bacterial populations in human feces measured by fluorescent *in situ* hybridization with group-specific 16S rRNA-targeted oligonucleotide probes, *Appl. Environ. Microbiol.,* 64, 3336–3345, 1998.

10. Langendijk, P.S., Schut, F., Jansen, G.J., Raangs, G.C., Kamphuis, G.R., Wilkinson, M.H.F., and Welling, G.W., Quantitative fluorescence *in situ* hybridization of *Bifidobacterium* spp. with genus-specific 16S rRNA-targeted probes and its application in fecal samples, *Appl. Environ. Microbiol.,* 61, 3069–3075, 1995.
11. Ames, J.M., Wynne, A., Hofmann, A., Plos, S., and Gibson, G.R., The effect of a model melanoidin mixture on faecal bacterial populations *in vitro, Br. J. Nutr.,* 82, 489–495, 1999.
12. Sghir, A., Gramet, G., Suau, A., Rochet, V., Pochart, P., and Dore, J., Quantification of bacterial groups within fecal flora by oligonucleotide probe hybridization, *Appl. Environ. Microbiol.,* 66, 2263–2266, 2000.
13. Apostolou, E., Pelto, L., Kirjavainen, V., Isolauri, E., Salminen, S.J., and Gibson, G.R., Differences in the gut bacterial flora of healthy and milk-sensitive adults, as measured by fluorescence *in situ* hybridization, *FEMS Immun. Med. Microbiol.,* 30, 217–221, 2001.
14. Welling, G.W., Elfferich, P., Raangs, G.C., Wildeboer-Veloo, A.C.M., Jansen, G.J., and Degener, J.E., 16S ribosomal RNA–targeted oligonucleotide probes for monitoring of intestinal bacteria, *Scand. J. Gastroenterol.,* 32 Suppl 222, 17–19, 1997.
15. Harmsen, H.J.M., Gibson, G.R., Elfferich, P., Raangs, G.C., Wildeboer-Veloo, A.C.M., Argaiz, A., Roberfroid, M.B., and Welling, G.W., Comparison of viable cell counts and fluorescence *in situ* hybridization using specific rRNA-based probes for the quantification of human fecal bacteria, *FEMS Microbiol. Lett.,* 183, 125–129, 1999.
16. Kruse, H-P., Kleesen, B., and Blaut, M., Effects of inulin on faecal bifidobacteria in human subjects, *Br. J. Nutr.,* 82, 375–382, 1999.
17. Tuohy, K.M., Kolida, S., Lustenberger, A.M., and Gibson, G.R., The prebiotic effects of biscuits containing partially hydrolysed guar gum and fructo-oligosacchaarides: a human volunteer study, *Br. J. Nutr.,* 86, 341–348, 2001.
18. Muyzer, G., De Waal, E.C., and Uitterlinden, A.G., Profiling of complex microbial populations by denaturing gradient gel electrophoresis analysis of polymerase chain reaction-amplified genes coding for 16S rRNA, *Appl. Environ. Microbiol.,* 59, 695–700, 1993.
19. Zoetendal, E., Akkermans, A.D., and De Vos, W.M., Temperature gradient gel electrophoresis analysis of 16S rRNA from human fecal samples reveals stable and host-specific communities of active bacteria, *Appl. Environ. Microbiol.,* 64, 3854–3859, 1998.
20. Walter, J., Hertel, C., Tannock, G.W., Lis, C.M., Munro, K., and Hammes, W.P., Detection of *Lactobacillus, Pediococcus, Leuconostoc,* and *Weissella* species in human feces by using group-specific PCR primers and denaturing gradient gel electrophoresis, *Appl. Environ. Microbiol.,* 67, 2578–2585, 2001.
21. Walter, J., Tannock, G.W., Tilsala-Timisjarvi, A., Rodtong, S., Loach, D.M., Munro, K., and Alatossava, T., Detection and identification of gastrointestinal *Lactobacillus* species by using denaturing gradient gel electrophoresis and species-specific PCR primers, *Appl. Environ. Microbiol.,* 66, 297–303, 2000.
22. Savage, D.C., Gastrointestinal microflora in mammalian nutrition, *Annu. Rev. Nutr.,* 6, 155–178, 1986.
23. Edwards, C., Interactions between nutrition and the intestinal microflora, *Proc. Nutr. Soc.,* 52, 375–382, 1993.
24. Wirth, R., Muscholl, A., and Wanner, G., The role of pheromones in bacterial interactions, *Trends Microbiol.,* 4, 96–103, 1996.

25. Eberhard, A., Burlingame, A.L., Eberhard, C., Kenyon, G.L., Nealson, K.H., and Oppenheimer, N.J., Structual identification of autoinducer of *Photobacterium fischeri* luciferase, *Biochemistry,* 20, 2444–2449, 1981.
26. Pierson, L.S., III, Keppenne, V.D., and Wood, D.W., Phenazine antibiotic biosynthesis in *Pseudomonas aureofaciens* is regulated by PhzR in response to cell density, *J. Bacteriol.,* 176, 3966–3974, 1994.
27. Fuqua, C., Winans, S.C., and Greenberg, E.P., Census and consensus in bacterial ecosystems: the LuxR–LuxI family of quorum sensing transcriptional regulators, *Annu. Rev. Microbiol.,* 50, 727–751, 1996.
28. Wood, D.W., Gong, F., Daykin, M.M., Williams, P., and Pierson, L.S. III, *N*-acyl-homoserine lactone-mediated regulation of phenazine gene expression by *Pseudomonas aureofaciens* 30–84 in the wheat rhibosphere, *J. Bacteriol.,* 179, 7663–7670, 1997.
29. Lu, L. and Walker, W.A., Pathologic and physiologic interactions of bacteria with the gastrointestinal epithelium, *Am. J. Clin. Nutr.,* 73 (suppl), 1124S–1130S, 2001.
30. Bry, L., Falk, P.G., Midtvedt, T., and Gordon, J.I., A model of host–microbial interactions in an open mammalian ecosystem, *Science,* 273, 1380–1383, 1996.
31. Fermented Foods and Healthy Digestive Functions, *Nutrition and Health Collection,* Danone Vitapole Research, 2001, pp. 47.
32. Yoshiota, M., Fujita, K., Sakata, H., Murono, J., and Iseki, K., Development of the normal intestinal flora and its clinical significance in infants and children, *Bifidobact. Microflora,* 10, 11–17, 1991.
33. Benno, Y., Sawada, K., and Mitsuoka, T., The intestinal microflora of infants: composition of fecal flora in breast-fed and bottle-fed infants, *Microbiol. Immunol.,* 28, 975–986, 1984.
34. Kawase, K., Suzuki, T., Kiyosawa, I., Okonogi, S., Kawashima, T., and Kuboyama, M., Effects of composition of infant formulas on the intestinal microflora of infants, *Bifidobact. Microflora,* 2, 25–31, 1983.
35. Infant Formula Act, Title 21 United States of America Code of Federal Regulations. Chapter 9, Subchapter IV, Section 350a, 1980, http://www4.law.cornell.edu/uscode/21/350a.html.
36. U.S. Food and Drug Administration, Food Composition, Standards, Labeling and Economics, Infant Formula Program — Import and Domestic, 1998, http://vm.cfsan.fda.gov/ ~comm/cp21006.html.
37. Rambaud, J.-C., Bouhnik, Y., Marteau, P., and Pochart, P., Manipulation of the gut microflora, *Proc. Nutr. Soc.,* 52, 357–366, 1993.
38. Mitsuoka, T., Intestinal flora and aging, *Nutr. Rev.,* 50, 438–446, 1992.
39. http://www.crownalife.be/html/framset.html.
40. U.S. National Aeronautics and Space Administration, Information Summaries: Astronaut Fact Book (NP-1997–03 008JSC), 1997.
41. Criswell-Hudak, B.S., Immune response during space flight, *Exp. Gerontol.,* 26, 289–296, 1991.
42. Seddon, M.R., Fettman, M.J., and Phillips, R.W., Practical and clinical concerns during spaceflight, *Am. J. Clin. Nutr.,* 60, 825S–830S, 1994.
43. Lane, H.W., Smith, S.M., Rice, B.L., and Bourland, C.T., Nutrition in space: lessons from the past applied to the future, *Am. J. Clin. Nutr.,* 60, 801S–805S, 1994.
44. Lane, H.W., Gretebeck, R.J., and Smith, S.M., Nutrition, endocrinology, and body composition during space flight, *Nutr. Res.,* 11, 1923–1934, 1998.
45. Sekiguchi, C. Issues of health care under weightlessness, *Acta Physiol. Scand.,* Suppl. 616, 89–97, 1994.

46. Lane, H.W. and Schoeller, D.A., Overview: history of nutrition and spaceflight, in *Nutrition in Space and Weightlessness Models*, Lane, H.W. and Schoeller, D.A., Eds., CRC Press, Boca Raton, FL, 1999, pp. 1–18.
47. Smith, M.C., Rambaut, P.C., and Vogel, J.M., Bone mineral measurement (experiment M078) in *Biomedical Results from Skylab (NASA SP-377),* Johnston, R.S. and Dietlein, L.F., Eds., National Aeronautics and Space Administration, Washington, 1977, pp. 183–190.
48. Whedon, G.D., Lutwak, L., Rambaut, P.C., Whittle, M.W., Smith, M.C., Reid, J., Leach, C., Stadler, C.R., and Sanford, D.D., Mineral and nitrogen metabolic studies, experiment M071, in *Biomedical Results from Skylab (NASA SP-377),* Johnston, R.S. and Dietlein, L.F., Eds., National Aeronautics and Space Administration, Washington, 1977, pp. 164–174.
49. Cucinotta, F.A., Manuel, F.K., Jones, J., Iszard, G., Murrey, J., Djojonegro, B., and Wear, M., Space radiation and cataracts in astronauts, *Radiat. Res.,* 156, 460–466, 2001.
50. Nachtwey, D.S. and Yang, T.C., Radiological health risks for exploratory class missions in space, *Acta Astronaut.,* 23, 227–231, 1991.
51. Grundy, D. and Scratcherd, T., Food for thought — nutritional problems in space, *J. Br. Interplanet. Soc.,* 42, 374–377, 1989.
52. Saavedra, J.M., Bauman, N.A., Oung, I., Perman, J.A., and Yolken, R.H., Feeding of *Bifidobacterium bifidum* and *Streptococcus thermophilus* to infants in hospital for prevention of diarrhoea and shedding rotovirus, *Lancet,* 344, 1046–1049, 1994.
53. Siitonen, S., Vapaatalo, H., Salminen, S., Gordin, A., Saxelin, M., Wikberg, R., and Kirkkola, A.L., Effect of *Lactobacillus* GG yoghurt in prevention of antibiotic associated diarrhoea, *Ann. Med.,* 22, 57–59, 1990.
54. Hove, H., Norgaard, H., and Mortensen, P.B., Review. Lactic acid bacteria and the human intestinal tract, *Eur. J. Clin. Nutr.,* 53, 339–350, 1999.
55. Gorbach, S.L., Lactic acid bacteria and human health, *Ann. Med.,* 22, 37–41, 1990.
56. Lane, H.W. and Schoeller, D.A., Eds., *Nutrition in Spaceflight and Weightless Models,* CRC Press, Boca Raton, FL, 1999.
57. Fuller, R., A Review. Probiotics in man and animals, *J. Appl. Bacteriol.,* 66, 365–378, 1989.
58. Sonnenborn, U. and Schulze, J., Actual clinical studies on the efficacy of a probiotic drug, *Microecol. Ther.,* 26, 211–220, 1997.
59. Alm, L., Humble, D., Ryd-Kjellén, E., and Setterberg, G., The effect of acidophilus milk in the treatment of constipation in hospitalized geriatric patients, *XV Symp. Swed. Nutr. Found.,* 1983, pp. 131–138.
60. Huis In't Veld, J.H.J., Bosschaert, M.A.R., and Short, C., Health aspects of probiotics, *Food Sci. Technol. Today,* 12, 46–50, 1998.
61. O'Sullivan, M.G., Thorton, G., O'Sullivan, G.C., and Collins, J.K., Probiotic bacteria: myth or reality?, *Trends Food Sci. Technol.,* 3, 309–313, 1992.
62. Lidbeck, A., Allinger, G.U., Orrhage, K.M., Ottova, L., Brismar, B., Gustafsson, J.-Å., Rafter, J.J., and Norn, C.E., Impact of *Lactobacillus acidophilus* supplements on the faecal microflora and soluble faecal bile acids in colon cancer patients, *Microb. Ecol. Health Dis.,* 4, 81–88, 1991.
63. McDonough, F.E., Wells, P., Wong, N.P., Hitchins, A.D., and Bodwell, C.E., Role of vitamins and minerals in growth stimulation of rats fed yogurt, *Fed. Proc.,* 42, 556, 1983.
64. Halpern, G.M., Vruwink, K.G., Van de Water, J., Keen, C.L., and Gershwin, M.E., Influence of long-term yoghurt consumption in young adults, *Int. J. Immunother.,* 7, 205–210, 1991.

65. Recker, R.R., Bammi, A., Barger-Lux, J., and Heaney, R.P., Calcium absorbability from milk products, an imitation milk, and calcium carbonate, *Am. J. Clin. Nutr.,* 47, 93–95, 1988.
66. Thompson, I., EVA dosimetry in manned spacecraft, *Mutat. Res.,* 430, 203–209, 1999.
67. Bakalinsky, A.T., Nadathur, S.R., Carney, J.R., and Gould, S.J., Antimutagenicity of yogurt, *Mutat. Res.,* 350, 199–200, 1996.
68. Renner, H.W. and Munzner, R., The possible role of probiotics as dietary antimutagens, *Mutat. Res.,* 262, 239–245, 1991.
69. Matar, C., Nadathur, S.S., Bakalinsky, A.T., and Goulet, J., Antimutagenic effects of milk fermented by *Lactobacillus helveticus* L89 and a protease-deficient derivative, *J. Dairy Sci.,* 80, 1965–1970,
70. Shahani, K.M., Friend, B.A., and Bailey, P.J., Antitumor activity of fermented colostrums and milk, *J. Food Prot.,* 46, 385–386, 1983.
71. Friend, B.A., Farmer, R.E., and Shahani, K.M., Effect of feeding and intraperitoneal implantation of yogurt culture cells on Ehrlich ascites tumor, *Milchwissenschaft*, 37, 708–710, 1982.
72. Tsuru, S., Shinomiya, N., Taniguchi, M., Shimazaki, H., Tanigawa, K., and Nomoto, K., Inhibition of tumor growth by dairy products, *J. Clin. Lab. Immunol.,* 25, 177–183, 1988.
73. Kubo, M., Odani, T., Nakamura, S., Tokumaru, S., Matsuda, H., Pharamacological study on kefir — a fermented milk product in Caucasus. I. On tumour activity, *Yakugaku Zasshi,* 112, 489–495, 1992.
74. Dong, M-Y., Chang, T-W., and Gorbach, S.L., Effects of feeding *Lactobacillus* GG on lethal irradiation in mice, *Diagn. Microbiol. Infect. Dis.,* 7, 1–7, 1987.
75. Criswell-Hudak, B.S., Immune response during space flight, *Exp. Gerontol.,* 26, 289–296, 1991.
76. Ouwehand, A.C. and Salminen, S.J., The health effect of cultured milk products with viable and non-viable bacteria, *Int. Dairy J.,* 8, 749–758, 1998.
77. Manz, W., Amann, R., Ludwig, W., Vancanneyt, M., and Schleifer, K.-H., Application of a suite of 16S rRNA–specific oligonucleotide probes designed to investigate bacteria of the phylum cytophaga-flavobacter-bacteroides in the natural environment, *Microbiology*, 142, 1097–1106, 1996.
78. Harmsen, H.J.M., Wildeboer-Veloo, A.C.M., Grijpstra, J., Knol, J., Degener, J.E., and Welling, G.W., Development of 16S rRNA–based probes for the *Coriobacterium* Group and the Atopobium Cluster and their application for enumeration for *Coriobacteriaceae* in human feces from volunteers of different age groups, *Appl. Environ. Microbiol.,* 6, 4523–4527, 2000.
79. Kalantzopoulos, G., Detection and enumeration of probiotic cultures, in *Encyclopedia of Food Microbiology,* Batt, C.A. and Patel, P.D., Eds., Academic Press, New York, 1999, pp. 1373–1379.
80. Gibson, G.R. and Roberfroid, M.B., Dietary modulation of the human colonic microbiota: introducing the concept of probiotics, *J. Nutr.,* 125, 1401–1412, 1995.

Index

C

D

E

F

K

L

R

S

T

U

V

W

Y

Z